UNDERSTANDING BASIC ELECTRONICS

FIRST EDITION

By: Larry D. Wolfgang, WR1B

PRODUCTION STAFF

Deborah Strzeszkowski, Graphic design, Typesetting and Layout

David Pingree, N1NAS, Technical illustrations

Steffie Nelson, KA1IFB, Proofreading

Michelle Bloom, WB1ENT, Production Supervisor

ARTWORK

Deborah Strzeszkowski, Unit title illustrations

Sue Fagan, Cover, Hammlett illustrations

Scott Johnson, Cartoons

Laurie Ingersoll, Line art

Printed in USA

Library of Congress Catalog Number: 92-074912

ISBN: 0-87259-398-3

First Edition
Fourth Printing, 1998

CONTENTS

Foreword

Electronics. The word conjures images of engineers and technicians huddled around their latest creation. Hot soldering irons and wire bits are scattered around. Cabinets full of resistors, transistors and other parts are nearby. Calculator in hand, an experimenter punches a few buttons and quickly selects a new component to try in the circuit.

But wait! You don't have to be an engineer or a math whiz to enjoy the thrill of experimenting with electronics circuits.

This book will teach you basic electronics principles. The only math skills you need are the ability to add, subtract, multiply and divide. The book is written in a light, easy-to-understand style that nontechnical readers will enjoy. Cartoons and drawings illustrate the electronics concepts to aid your understanding.

The text uses a modular approach to teaching. We normally present topics in two-page modules, and no module has more than four pages. This makes it convenient to study the book in small sections. It also makes it easy for you to skip modules that cover material with which you are already familiar.

The ARRL's *Understanding Basic Electronics* is *not* a study guide for an Amateur Radio license exam. Do you have an Amateur Radio license, but want to gain a more complete understanding of basic electronics principles? Then this book is for you. Maybe you just want to learn about basic electronics without studying the FCC rules and other operating practices. This book is also for you.

The ARRL has an excellent series of study materials to help you earn an Amateur Radio license. *Now You're Talking! Discover the World of Ham Radio* contains all the information you need to pass the Novice license written exam or the Technician license exam. (There is no Morse code exam required for the Technician license. The Novice license has a 5-WPM Morse-code exam requirement.) *The ARRL General Class License Manual, The ARRL Advanced Class License Manual* and *The ARRL Extra Class License Manual* will help you prepare for these license exams. If you already have a Novice license, *The ARRL Technician Class License Manual* provides all the material you need to upgrade to a Technician license.

David Sumner, K1ZZ
Executive Vice President

Newington, CT
September, 1992

Preface

The ARRL's *Understanding Basic Electronics* is an introduction to the exciting world of electronics. Readers with little or no previous electronics knowledge will be able to understand the text. Radio amateurs who want to build on their foundation of electronics knowledge will find this book helpful.

Writing this book has been no small task. It seems I've been working on it forever! Writing a book is *very* different from writing a short article or editing someone else's manuscript to produce a book.

ARRL Headquarters staff members were discussing the need for an introduction to electronics book when I began working here. Slowly a plan for such a book began to form.

Many people played a role in the project planning and development. It is difficult, if not impossible, to credit them all. Paul Rinaldo, W4RI, Charles Hutchinson, K8CH and Mark Wilson, AA2Z played major roles in helping develop the book plan. Without their assistance and encouragement, this book would not exist.

Deborah Strzeszkowski planned the book's graphic design. She also created the artwork that introduces each new Unit. Debby set the type and laid out the pages. No other publication in ARRL's library looks like *Understanding Basic Electronics*, thanks to Debby's talent. David Pingree drew all the schematic diagrams and technical illustrations on the ARRL's SUN CAD system. Jim Massara, N2EST suggested the cartoon pig, Hammlett, inspired by the nickname for Amateur Radio, ham radio. Sue Fagan drew Hammlett for the book. Two contributing artists shared their talents with us to help illustrate the book. Scott Johnson drew the book's many cartoon illustrations. Laurie Ingersoll drew the illustrations involving human figures and other line art. To all these artists I say a heartfelt "Thank You!"

Finally, to all who read this book, thank you. You are the real judges of whether the book meets its objectives. If it helps you learn some basic electronics, and leads you to further experimentation and enjoyment of ham radio, all the work has been worthwhile. Please share your suggestions and comments. There is a Feedback Form at the back of the book for this purpose.

Larry Wolfgang, WR1B
Senior Assistant Technical Editor, Books

Introduction

Modern technology touches every aspect of our lives. We must understand the basic concepts of technology to function effectively in our world. You can increase your knowledge and gain great enjoyment by using technology with a hobby. You explore new interests and learn about the world around you when you pursue a hobby.

Electronics is a fascinating part of modern technology. Wherever you look, there are electronic gadgets and gizmos. From automobiles to Post-Office ZIP code readers, nearly every machine uses some type of electronic control. You may not have to know anything about electronics to use most of the devices. You'll understand a lot more about the world around you if you know some basic electronics principles, however.

There is no better way to understand how something works than to build it yourself. Many Amateur Radio operators enjoy building all kinds of electronics devices. From gadgets for their radio stations to complete transmitters and receivers, many "hams" take great pleasure in saying, "I built it myself." You, too, can use your electronics knowledge to build useful projects of all kinds.

This book will help you learn the important basic principles of electronics. You will become more confident of your ability to build projects and understand how they work. By understanding these basic principles you can read and understand more advanced texts. You will be familiar with the technical terms and jargon associated with electronics and ham radio.

This book's main purpose is to teach you fundamental electronics principles. Many Amateur Radio examples throughout the text show you how to apply these principles. You will learn to work with simple direct-current circuits and simple alternating-current circuits.

Organization

This book is organized into four main sections, or *units*. These are:

Some Needed Math Skills

Elementary DC Electronics

Elementary AC Electronics

A Few More Building Blocks

Within each Unit, related topics combine to form *chapters*. For example, one chapter covers the metric system of measurement. Another chapter is devoted to Ohm's Law. There are chapters about capacitors and bipolar transistors. One or more *modules* comprise each chapter.

These modules form the heart of the book. Each module presents one important concept relating to the chapter. Most modules cover two pages, although a few extend to four pages. Text explanations are in the simplest possible language. Important words and electronics terms are printed in *italic type* the first time they are used. Simple definitions appear with these important terms. The Appendix also contains a glossary of electronics and Amateur Radio terms. Refer to this glossary whenever you come across an unfamiliar term.

We recommend you start at the beginning of the book, and work your way through the entire text. We took great care to organize the topics logically. Each section builds on the knowledge you gained from the last one. You will have a solid electronics foundation when you finish the book.

There are lots of examples showing step-by-step solutions to problems. Perform those calculations as you read the text. Be sure you understand each step and agree with the answers. The text also provides additional practice problems.

You may find that you are already familiar with certain topics. In that case, you might decide to skip over that section, going on to something with which you are less familiar. Each module forms a stand-alone piece of text. You can jump around, selecting those modules you are most interested in. If you find some unfamiliar terms or concepts, though, look those words up in the index and review the appropriate modules.

For example, if you have a strong math background, you may choose to skip Unit 1 entirely. Are you already familiar with the metric system of measurements? Have you solved lots of equations and worked with trigonometry functions? There is no need to spend time reviewing those areas, then.

Study any modules that cover topics with which you aren't as familiar, though. Perhaps you decide to skip most of the math unit, but you aren't as confident about your ability to work with logarithms. Be sure to review that chapter, then.

Good luck with your studies. We hope you enjoy this book, and have fun studying it.

Some Needed Math Skills

Learning electronics requires you to solve some simple arithmetic problems. Even if you don't want to design electronics circuits, you must still know how to do some basic calculations. Math skills will help you understand how a simple circuit operates. These calculations are not difficult, though. You just have to know some basic rules of arithmetic, such as how to add, subtract, multiply and divide. This unit explains how to handle just about any math you'll run into as you study basic electronics.

Numbers help us understand how a change made to one part of an electronics circuit affects the other parts. Lord Kelvin, a famous English scientist, understood how important numbers and calculations are. Lord Kelvin once said that until you can measure something, and express it in numbers, you have only the beginning of understanding. Kelvin may not have been talking about electronics, but his statement is true of modern electronics, just the same.

Many people feel a little frightened by mathematics. There are many tools to help you learn about math, however. Accept the challenge of learning. A scientific calculator is one tool that will be valuable as we study modern electronics. (The term *scientific* here means that the calculator includes keys for some important mathematical calculations.)

You'll need a few basic tools to perform simple experiments and gain "hands-on" experience with electronics circuits. A scientific calculator is also important for learning electronics. With a scientific calculator, you can solve even complicated problems with ease. A scientific calculator helps you think about what the numbers are telling you. Mathematics will increase your electronics knowledge and understanding.

Until you can measure something, and express it in numbers, you have only the beginning of understanding.
—*Lord Kelvin*

In This Unit You Will Learn:

- What numbers are.
- Tricks to help you use very large and very small numbers.
- The meaning of equality and how to solve equations.
- About the metric system of measurement.
- How to work with the mathematics of triangles, trigonometry.
- What logarithms are and why they are important to electronics.

CHAPTER 1 Learning to Work With Numbers

Numbers give us a way to describe things that happen around us. We use numbers every day. We use numbers to express time, to represent distances and to describe how heavy an object is. For instance, suppose it takes you 30 minutes to get to work or school and another 30 minutes to come home. Add the two times to find how much travel time you spend each day. (In this example, you spend 60 minutes or one hour traveling each day.) As another example, suppose the living room in your house is 10 feet by 12 feet. Multiply the dimensions to find that you need 120 square feet of carpet to cover the floor.

Numbers also represent various conditions about an electronics circuit. With some simple calculations those numbers help us predict other things about the circuit. These calculations are no more difficult than finding your daily travel time or how much carpet to buy for your living room. You will find it easier to understand the numbers and calculations used in electronics if you're familiar with the common terms used to describe the process. That's easily taken care of with a few examples, though.

Numbers are tools that help you learn electronics. To use these tools, be sure you understand what scientists and engineers mean when they talk about a number. When you say the word number, most people think of the group of "counting numbers" (1, 2, 3, 4 and so on). Mathematicians call these numbers *integers*.

What Are Numbers?

There are an unlimited number of integers available with which to count. If you think of the largest number imaginable, you can always add 1 to get a larger number. For example, think of the number nine hundred ninety nine million nine hundred ninety nine thousand nine hundred ninety nine. (That's 999,999,999.) If you add one to this number, you get one billion (1,000,000,000).

No matter how large a number you think of, you can always add 1 to it. Scientists have this idea in mind when they talk about *infinity*. So there are an infinite number of integers! This is a little bit like going outside on a clear, dark night and trying to count the stars. Just when you think you've counted them all, someone comes along with a telescope. They show you some stars that you couldn't see

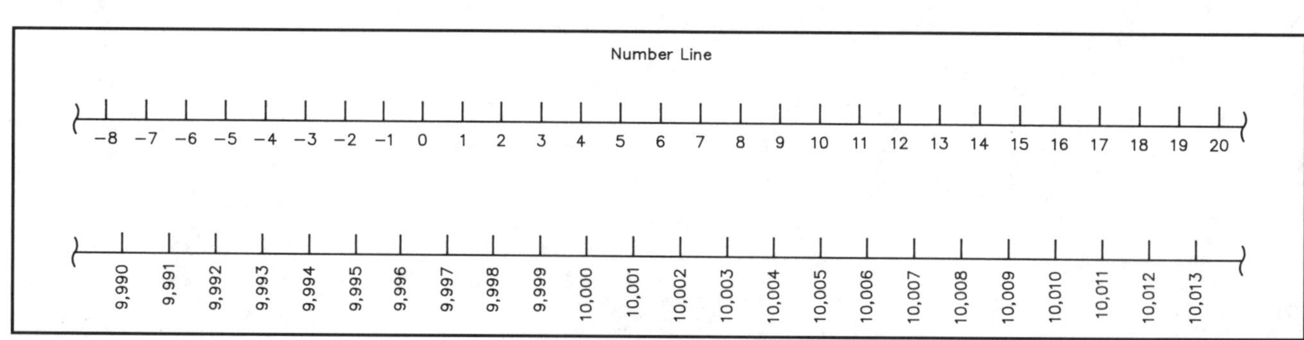

Figure 1—You can draw a simple number line like the one shown here.

with just your eyes. Then you notice that some of the stars you counted already are really two or more stars when viewed through the telescope. Even if there were some way to keep track of which stars you've counted, it's really quite impossible to count them all!

One convenient way to picture numbers is to draw a line across the page. Then pick a spot and make a small mark across that line. Label that mark as zero on your line. Now at some small distance to the right of zero (like 1/4 inch) make a mark and label that as 1. Continue across your line, making each mark the same distance from the last one. Label each new mark with the next integer, as shown in Figure 1. We call this a *number line.* Of course, you will run out of paper long before you run out of counting numbers, but you get the idea.

You probably have some space to the left of zero. You can mark off numbers here, too, but include a minus sign in front of them. Numbers less than zero are negative numbers. (Sometimes having less than nothing can be a problem. If your bank thinks you have less than nothing in your checking account, for example, they may be unhappy with you.)

There are as many negative integers as there are positive ones, so the numbers extend to infinity on both sides of zero. We often represent infinity by the symbol ∞. Using this symbol, we can write the mathematical expression for all integers:

$$-\infty < I < +\infty$$

If I stands for any integer number, then this expression tells us that the integers extend from $-\infty$ to $+\infty$.

Real Numbers

What happens if you go to a spot on your number line that is between two integers? Well, there is no integer to express this, but it is pretty easy to see that such a spot could exist. If you picked a spot halfway between 5 and 6, then you could call this spot $5^1/_2$. See Figure 1. Halfway between 1,250 and 1,251 is a spot you can call $1{,}250^1/_2$. Divide the space between these integers in half. Now there are two more spaces that you can divide in half. Each of these pieces represents a quarter of the distance between two integers ($^1/_4$).

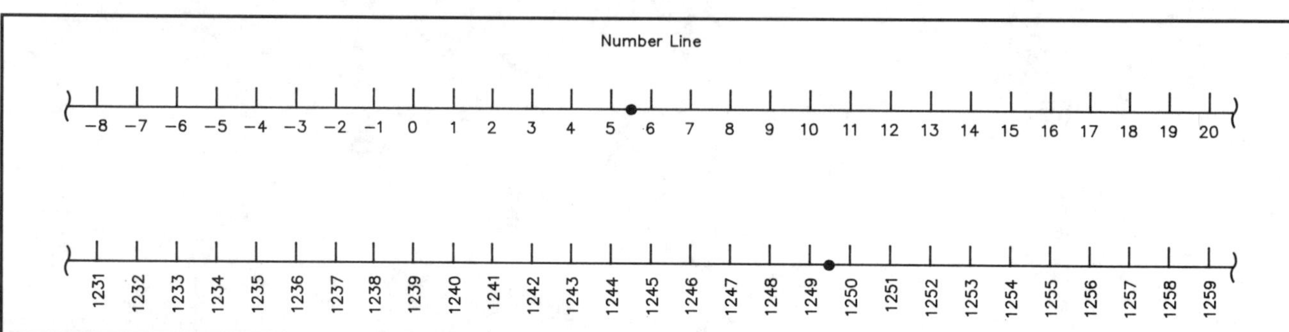

Figure 1—It's not hard to realize that some places on a number line fall between integers. Locate the point that is half way between 5 and 6, for example. We call this point $5^1/_2$.

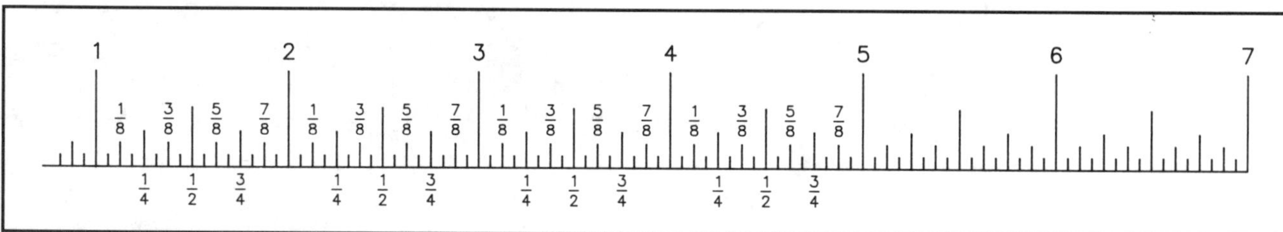

Figure 2—By expanding the scale on your number line, you can include more values between the integers. When mathematicians think of a number line, they imagine an infinite number of points between two integer values.

If you think your number line is beginning to resemble a ruler, you're right! This is exactly the way we divide inches into smaller and smaller pieces on a ruler. Divide each quarter of the integer space in half. Then eight pieces make up the entire space. In that case, each piece is one-eighth ($^1/_8$) of the space between integers.

By now you might begin to wonder how many times we can divide the space between two numbers. Well, there really is no limit! (Yes, that means we can divide the space an infinite number of times!) In practice, when you divide your number line, you soon come to a point where you can't divide the space any smaller. If your pencil line is wider than the space you're trying to divide, you have definitely reached the limit. But if you draw another line, leaving more space between integers, you can divide the space into more pieces. (Using a sharper pencil to draw a finer line also will help.) Figure 2 illustrates this idea.

Keep in mind that we are using the number line as a way to picture what a number is. No matter how small a piece of the number line you pick, you can always divide it into two more pieces!

It's often a little more convenient to write these fractions as decimal numbers. For example, $^1/_2$ is the same as 0.5, $^1/_4$ is 0.25 and so on. Using this technique, $1250^1/_2$ becomes 1250.5. In fact, by using decimal fractions, it's possible to write numbers that you couldn't express as a simple fraction using integers. For example, 3.141592654... (the dots mean that this fraction continues, and is not exact) is a number that is important for work involving circles. There is no simple fraction that can express this point on the number line.

Our number line, including all the imaginable fractions between each integer, now represents what mathematicians call the *real numbers.* We can write a mathematical expression for real numbers just like we did for integers. If R represents a real number (any point on your number line), then:

$$-\infty < R < +\infty$$

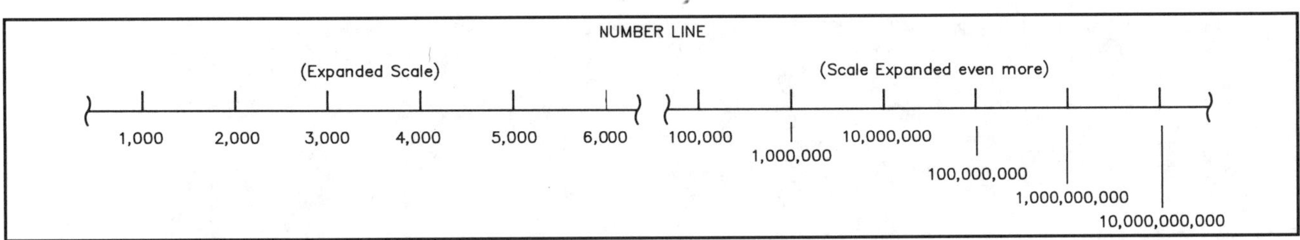

Figure 3—As the numbers get larger and larger, expanded scales can show portions of a number line that are far from zero. Similar scales can show negative numbers, to the left of zero.

So the real numbers include every possible value.

You already understand that numbers can be very large. As we add more places to express large quantities, working with the numbers can become confusing. Imagine trying to perform even simple addition with numbers like 4,567,835,251,000,000,000 or 85,432,164,749,000,000,000. Figure 3 shows how we might expand portions of our number line to show larger and larger values.

At the other end of the scale, close to zero, numbers get smaller and smaller. Decimal fractions with lots of zeros to the right of the decimal point, but before the nonzero digits, are common in electronics. For instance, you might find that you have to work with a number like 0.0000000000537 or 0.000000000491. The more zeros there are right after the decimal point, the smaller the number is (or the closer to zero it is). These numbers can be quite confusing to work with, and it's very easy to get mixed up with all those zeros. Figure 4 expands the number line between 0 and 1, to show the numbers getting smaller and smaller as you move toward zero.

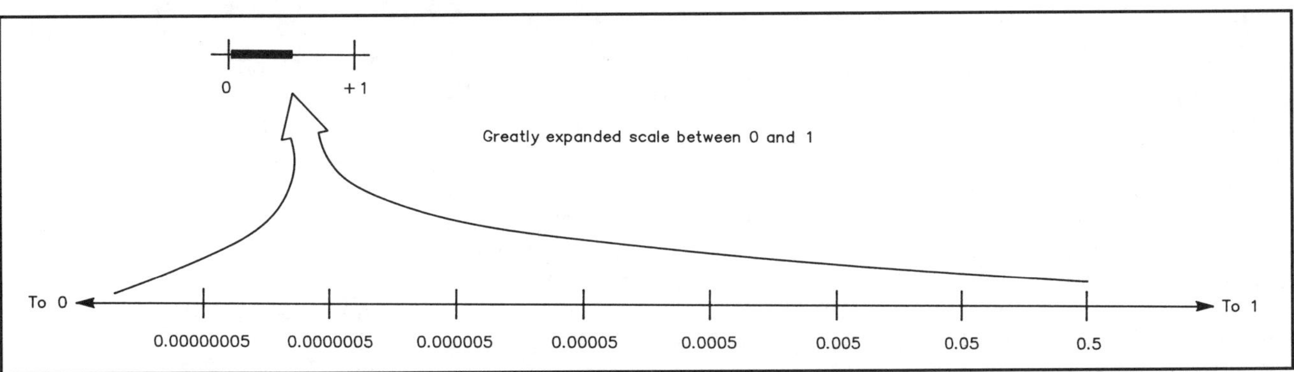

Figure 4—This portion of the number line is between 0 and 1. The more zeros there are between the decimal point and the first nonzero digit, the smaller the number is, and the closer it is to 0. We can expand the scale between any two integers in a similar way.

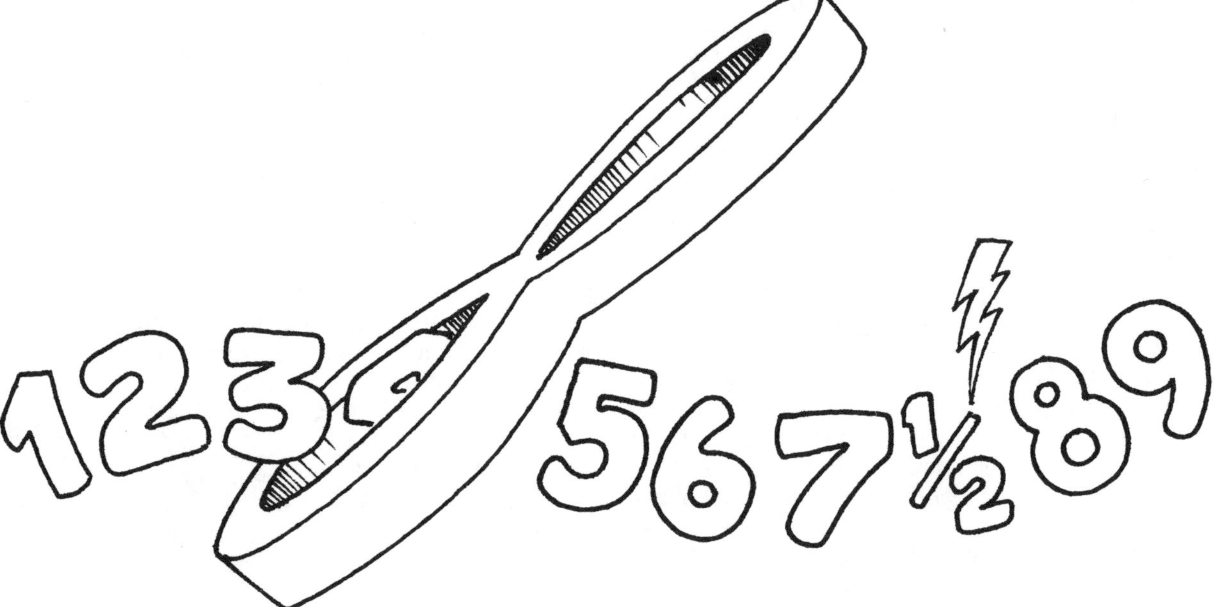

CHAPTER 2 Working with Large and Small Numbers

Exponential and Scientific Notation

Modern electronics often uses numbers that are either quite large or very close to zero. At either extreme, it is difficult to write the numbers because of all the zeros. Even the most careful person will occasionally drop a zero, or add an extra one when calculating with such numbers. For most of us, calculations like this are a real nightmare!

Suppose you want to multiply 250,000 times 500,000. You can see how easily you could skip a zero or get the numbers out of line when you are doing the multiplication. See Figure 1. With this many zeros in the answer (125,000,000,000) it is just too easy to make a mistake—even with a few commas thrown in to help you count the zeros.

```
       250000
   ×   500000
   ----------
       000000
      000000
     000000
    000000
   000000
  1250000
 ------------
 125000000000
```

Figure 1—This work may look extremely sloppy compared to the way you would do the problem—If you are just a little careless, though, you can become hopelessly lost.

There's an easier way to keep track of all the zeros, though. Every time you multiply a number by 10, you just move the decimal point one place to the right and add another zero. So you can represent a number by writing its digits up to the last nonzero digit. Then multiply that part by ten for every zero you omitted. In our example, 250,000 becomes 25 × 10 × 10 × 10 × 10. (You may not think this is much better right now, but be patient for a minute.)

If we multiply a number by itself several times, we say that the number is "raised to a power." If we raise 10 to the second power we say the number is 10 *squared*, or write it as 10^2. The 2 is an exponent of 10. Remember that this just means 10 × 10, or 100.

Going back to our example, we can write the first number in our multiplication problem (250000) as 25 × 10 × 10 × 10 × 10. Using the "power of 10" idea, you also can write this as 25×10^4 because we have 10 multiplied by itself four times. Remember that you just move the decimal point one place to the right (add a zero) every time you multiply by 10. Then we can "expand" 25×10^4 back out to the normal form by writing 25 followed by four zeros (250,000). This means that the exponent, or power of 10, always tells us how many places to move the decimal point. The second number in our example (500,000) becomes 5×10^5. When we write a number with a power of 10, we express it in *exponential notation.* As another example, we could write the number 670,310,000 as 67031×10^4.

$$250,000 = 25 \times 10^4$$

Figure 2—We can easily write large numbers in exponential form.

For numbers less than one, getting closer and closer to zero, we have a similar situation. In this case the number has many zeros right after the decimal point, such as 0.0000045. If it seems to you that dividing by 10 has the same effect as moving the decimal point to the left one place, you're right! And if you're also wondering, "Gee, does that mean I can write

$$5 \times 10^5 = 500,000$$

Figure 3—We can easily expand numbers written in exponential or scientific notation back to normal form.

0.0000045 as 0.45 ÷ 10 ÷ 10 ÷ 10 ÷ 10 ÷ 10 or as 45 ÷ 10 ÷ 10 ÷ 10 ÷ 10 ÷ 10 ÷ 10 ÷ 10?" the answer is Yes! You can write this number either way.

If we write the nonzero part of the number as 45, then we must divide by 10 seven times to get the original number. So we can combine the divide-by-ten factors as a power of 10, or 10^7. Then we can write $45 \div 10^7$. (Notice that we also could write this as $0.45 \div 10^5$ or $4.5 \div 10^6$.) If you prefer, you can write the number as a common fraction:

$$\frac{45}{10^7}$$

There is a nifty little trick that we often use when working with exponents. If the power of 10 is in the denominator (the bottom part) of a fraction, we can move it into the numerator (the top part) just by changing the sign of the exponent. So instead of writing $45 \div 10^7$ to show that we must divide by 10 seven times, we can write it as 45×10^{-7}. Now the −7 tells us to move the decimal point seven places to the left to write the number in its expanded form. If we move the decimal point two places to the left, so it is just before the 4, then it still has five places to go, and we have 0.45×10^{-5}.

$$\mathbf{0.0000045 = 4.5 \times 10^{-6} = 45 \times 10^{-7}}$$

Figure 4—When we write small numbers (close to zero) in exponential or scientific notation the power of 10 takes a negative value.

Think of the *number line* when you are working with numbers in exponential notation. If the power of 10 is a positive number, the decimal point moves to the right (like positive numbers). If the power of 10 is a negative number, the decimal point moves to the left (like negative numbers). See Figure 5.

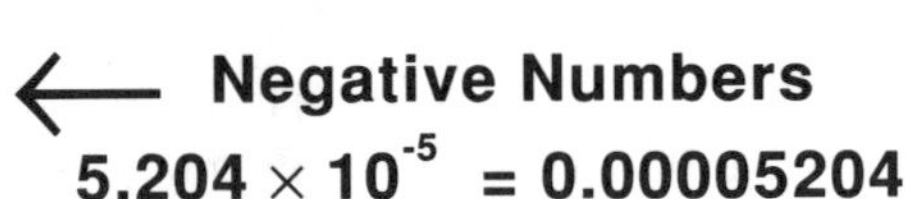

← Negative Numbers $5.204 \times 10^{-5} = 0.00005204$	Positive Numbers → $4.025 \times 10^{7} = 40,250,000$
Move the decimal point to the left when expanding negative powers of 10.	**Move the decimal point to the right when expanding positive powers of 10.**

Figure 5—Here is a simple key to remembering how to move the decimal point when expanding numbers out of exponential notation.

There is a special form of exponential notation, called *scientific notation.* Numbers expressed in scientific notation always have a single nonzero digit before the decimal point. The remaining nonzero digits come after the decimal point, followed by the power of 10.

Let's look at a few examples:

We can write 40,250,000,000 as 4025×10^7 or any number of other forms that include a power of 10. In each case we would say the number is in exponential notation. One possible way to write the number is as 4.025×10^{10}. We say the number is in scientific notation when it has this form.

Some Rules for Working with Exponential Notation

Let's become familiar with a few rules about working with numbers in exponential notation. You will enter numbers into your scientific calculator using exponential notation to solve electronics problems. The calculator will keep track of the powers of 10 for you as it calculates the answer. Knowing the rules for powers of 10, however, will make it easier for you to understand the calculations. Let's go over some of these basic rules before working a few problems with your calculator.

The first rule is about multiplication. To multiply numbers in exponential (or scientific) notation, first multiply the "plain number" part of the numbers. Next, add the exponents for the powers of 10. Put these two parts together, and you have the answer! Let's look at an example. To multiply 250,000 times 500,000, first write those numbers in exponential notation:

$250{,}000 = 25 \times 10^4$

$500{,}000 = 5 \times 10^5$

Next, set up the multiplication problem:

$$\begin{array}{r} 25 \times 10^4 \\ \times \quad 5 \times 10^5 \\ \hline \end{array}$$

Multiply the "plain number" parts to get 125 and add the powers of 10 to get 10^9. Combine these two parts for the answer:

125×10^9

You can expand this answer by moving the decimal point 9 places to the right:

125,000,000,000

Figure 1 shows this work. If you want to try multiplying these numbers without using exponential notation, go ahead. It's probably worthwhile for you to prove that the answer is the same whichever way you do it. Be extra careful with all those zeros, though. If the two answers do not agree, go back and check your work, because you probably made a mistake somewhere!

What happens if we have to multiply 500,000 times 0.0000045? Well, let's write the numbers in exponential form and set up the problem. See Figure 2.

$500{,}000 = 5 \times 10^5$

$0.0000045 = 45 \times 10^{-7}$

$$\begin{array}{r} 5 \times 10^5 \\ \times 45 \times 10^{-7} \\ \hline \end{array}$$

Multiply the numbers and add the exponents:

$5 \times 45 = 225$ and $10^{5 + (-7)} = 10^{-2}$

How did we get 10^{-2}? Don't forget that one of the exponents was –7, so when we add 5 and –7, we have:

$5 + (-7) = -2$

Our second important rule has to do with division. To divide numbers written in exponential or scientific form, first divide the "plain number" parts. Then subtract the denominator power from the numerator power. (The denominator is the bottom part of a fraction and the numerator is the top part.) Figure 3A shows a division problem, with the numbers converted to exponential and scientific notation. Part B shows how you can move the power of 10 from the denominator into the numerator. Then you can add the powers of 10. Part C shows the final answer.

You may have learned that you should "cancel" the same number of zeros from the numerator and denominator before dividing. That's following the rule for division with powers of 10! Suppose there are six zeros in the numerator and seven in the denominator. You can get rid of six of them right away. See Figure 4A. Part B shows how to solve this problem using exponential notation.

As another example, let's divide 250,000 by 500,000. We can write this as:

$$\frac{25 \times 10^4}{5 \times 10^5} = 5 \times 10^{-1} = 0.5$$

The 10^{-1} comes about because we have 4 – 5 for the exponents. See Figure 5.

It's very convenient to write numbers in exponential notation when we have to multiply or divide very large or very small numbers. When you must add or subtract numbers with a lot of zeros, you must remember a few different rules,

$$250{,}000 = 25 \times 10^4$$
$$500{,}000 = 5 \times 10^5$$

(A)

$$\begin{array}{r} 25 \times 10^4 \\ \times \quad 5 \times 10^5 \\ \hline 125 \times 10^{4+5} = 125 \times 10^9 \end{array}$$

(B)

$$125 \times 10^9 = 125{,}000{,}000{,}000$$

(C)

Figure 1—To multiply numbers written in exponential notation, first multiply the number parts, then add the exponents for the powers of 10.

$$500{,}000 = 5 \times 10^5$$
$$0.0000045 = 45 \times 10^{-7}$$

(A)

$$\begin{array}{r} 45 \times 10^{-7} \\ \times \quad 5 \times 10^5 \\ \hline 225 \times 10^{-7+5} = 225 \times 10^{-2} \end{array}$$

(B)

$$225 \times 10^{-2} = 2.25$$

(C)

Figure 2—When you must multiply a number that has a negative exponent, be sure to add the exponent as a *negative* value.

$$\frac{764{,}000{,}000}{382{,}000} = \frac{764 \times 10^6}{382 \times 10^3} = \frac{7.64 \times 10^8}{3.82 \times 10^5}$$

(A)

$$\frac{7.64 \times 10^8 \times 10^{-5}}{3.82} = 2 \times 10^{8 + (-5)}$$

(B)

$$2 \times 10^3 = 2000$$

(C)

Figure 3—To divide numbers written in exponential notation, first divide the number parts. Then move the exponent (power of 10) from the denominator of the fraction to the numerator by changing its sign (B). Finally, add the exponents in the numerator (C).

$$\frac{75{,}\not{0}\not{0}\not{0}{,}\not{0}\not{0}\not{0}}{50{,}\not{0}\not{0}\not{0}{,}\not{0}\not{0}\not{0}} = \frac{75}{50} = 1.5$$

(A)

$$\frac{75 \times 10^6}{5 \times 10^7} = \frac{75 \times 10^{6 + (-7)}}{5} = 15 \times 10^{-1}$$

$$= 1.5$$

(B)

Figure 4—You can divide numbers by canceling the same number of zeros in the numerator and denominator (A). You also can write the numbers in exponential form first (B).

$$\frac{250{,}000}{500{,}000} = \frac{25 \times 10^4}{5 \times 10^5}$$

(A)

$$5 \times 10^{4 + (5)} = 5 \times 10^{-1} = 0.5$$

(B)

Figure 5—If the denominator power of 10 is larger than the numerator power, the answer will have a negative power of 10. (This means the answer will be less than 1.) This is reasonable, because you know that you always get a value less than 1 when you divide by a larger number. For example, $^1/_2$, $^2/_4$ and $^3/_6$ all give answers of 0.5, which is less than 1.

however. To add or subtract, be certain to express all the numbers with the same power of 10. Do the arithmetic with the "plain number" part as you normally would. The power of 10 for your answer is the same power the numbers are expressed in. (When you add or subtract numbers, always be careful that the decimal point lines up in all the numbers. That is what you are doing by expressing all numbers in the same power of 10.) For example, if you have to add 2,000 and 508,000, you might write the problem as:

$$\begin{array}{r} 2 \times 10^3 \\ +\ 508 \times 10^3 \\ \hline 510 \times 10^3 \end{array}$$

You also can write this answer as 51×10^4, 5.1×10^5 or even as 510,000.

Figure 6 shows several other examples of addition and subtraction problems with numbers written in exponential notation. Study those examples, and be sure you understand the work.

This discussion gives you a good background in working with exponential notation. You can understand calculations with very large and very small numbers. Many quantities that we deal with every day in electronics involve numbers that we usually write in exponential form. It will be very helpful to work with numbers written in this way.

$$12{,}300{,}000{,}000 + 45{,}000{,}000$$

$$\begin{array}{r} 12300 \times 10^6 \\ +\ 45 \times 10^6 \end{array}$$

(A)

$$\begin{array}{r} 7.5 \times 10^4 \\ 126 \times 10^3 \\ 415 \times 10^5 \\ +\ 25 \times 10^2 \end{array} \qquad \begin{array}{r} 75 \times 10^3 \\ 126 \times 10^3 \\ 41500 \times 10^3 \\ +\ 2.5 \times 10^3 \\ \hline 41703.5 \times 10^3 \end{array}$$

(B)

$$7{,}930{,}000{,}000 - 7{,}137{,}000{,}000$$

$$\begin{array}{r} 7930 \times 10^6 \\ -\ 7137 \times 10^6 \\ \hline 793 \times 10^6 \end{array}$$

(C)

$$0.000000753 - 0.000000008$$

$$\begin{array}{r} 7.53 \times 10^{-7} \\ -\ 0.08 \times 10^{-7} \\ \hline 7.45 \times 10^{-7} \end{array}$$

(D)

Figure 6—Examples of addition and subtraction with numbers written in exponential notation. Notice that we must write all the numbers in a problem with the same power of 10, to align the decimal points.

Buying a Calculator

A scientific calculator is a valuable tool to help you solve electronics problems. You may already have a scientific calculator. If not, which one should you buy? Take a look at the wide range of calculators available at your local department store or electronics shop. You'll find many brands, and each company usually makes several different models. So how will you decide which one to buy? Well, we aren't going to name a specific brand and model, but we will suggest some features to consider. Look around in the stores in your area to see what specific models and brands are available. You're sure to find several suitable calculators in a price range you can afford.

Uh-Oh. Is this beginning to sound expensive? No, don't despair. It's not going to take a life's savings to buy a good calculator. There are several excellent calculators around for less than $20. Some even have solar cells, so they run on light energy, and never need batteries! Even the youngest readers could find some odd jobs around their neighborhood to earn the money for such an important tool!

If you already have a credit-card-sized pocket calculator, it probably will not be suitable for solving electronics problems. Most credit-card-sized calculators only include keys for addition, subtraction, multiplication and division. Some do include a key to calculate percentages (%) and have a one-number memory. A scientific calculator also includes keys to find some important mathematical functions. (A mathematical function is just a way of relating one number to another one. Don't worry about that for now. Just remember that the calculator will make it very easy to solve problems that involve these relationships.)

Be sure any calculator you buy includes keys labeled **log, sin, cos, tan,** $\sqrt{\ }$, $\mathbf{x^2}$, $\mathbf{10^x}$ and $\mathbf{1/_x}$. Normally, a calculator that has any of these keys will have all of them. Figure 1 shows typical arrangements for the function keys on some scientific calculators.

It's also important that your calculator can express numbers in exponential notation, or with powers of 10. (This is just a way to express very large or very small numbers without writing a lot of zeros.) You can tell if a calculator has this feature by looking for a key labeled **EE**, **EXP**, **EEX** or some similar notation. Ask the salesclerk to show you how to calculate 3,500,000 times 1,000,000. If the display shows 3.5 with a 12 in the right corner of the display, the calculator has automatically gone into exponential notation. That's what you want to see!

The exponential notation feature will make it easier when you are working electronics problems. You will find that quantities like radio frequencies in millions of hertz or capacitance values in millionths of farads are quite common. Key in the numbers carefully. The calculator keeps

Figure 1—The location and labeling of keys varies from one calculator brand to another. The calculator keypads shown here are representative of what you're looking for.

track of the exponent or power of 10 for the answer.

Most manufacturers use the same key for two or even three functions. This helps keep the number of calculator keys to a minimum. In this case you will see a key labeled **2nd, INV** or something similar. Besides the label printed on each key, there will be another label printed above (or below) some of the keys. The size and number of keys on a calculator are the main things that limit how small it will be.

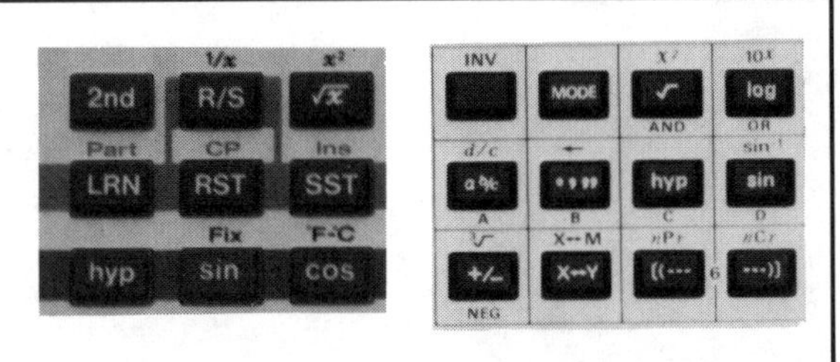

Figure 2—Many scientific calculators use a "second" or "inverse" key to allow some keys to serve more than one purpose.

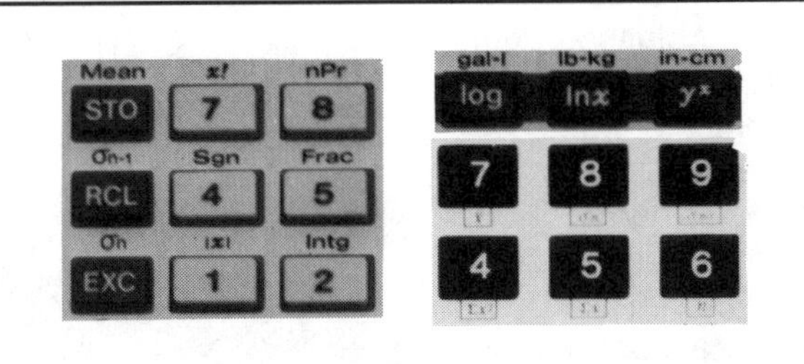

Figure 3—Some calculators have keys to solve statistics problems or to convert between values in the US Customary and Metric measuring systems.

Most scientific calculators also will include other features. For example, some calculators are "programmable." This means that you can teach the calculator to remember a certain set of steps in a long calculation. Suppose you have a problem with the following steps. "Multiply the number by 5, then add 20 to the result. Divide this by 10 and add another number that you will enter into the calculator. Take the square root of this answer, multiply it by 2 and then find the logarithm of this number." If you had to do that calculation more than once or twice, it would be a great advantage for your calculator to remember the steps! On the other hand, you will not run into calculations this complicated while you are learning basic electronics.

Some scientific calculators have keys to solve statistics problems, some will convert between values in the US Customary and Metric systems of measure and some will even allow you to work with numbers in the binary, octal and hexadecimal number systems. (This is most useful for people who work with computers.)

Should you buy a calculator with any of these extra features? Well, you won't really need them for the electronics covered in this book. The extra features usually add to the cost of the calculator, and that may be your biggest concern. You also should consider what other uses you might have for your calculator. If you may be able to use it at work or for school, consider what features would be helpful there. If you are a student, a calculator may be helpful when you take some math or science courses. In that case, you may want to spend the extra money for certain features.

In general, you should purchase the calculator with the most features for the price you can afford. There may be some buttons with labels that you don't understand. When you learn what those keys mean later, you'll probably be glad to have them! Of course another consideration is that the more buttons on your calculator, the more you will have to remember to operate it. That may be one good argument for buying a simple scientific calculator with only the basic math functions included.

There is one other item worth mentioning about scientific calculators. Most of the ones you will find in department stores use standard mathematical notation for performing the calculations. This means that if you want to multiply two numbers, you enter the values just as you would write them in an equation: 2×6=. There are some calculators that use a system called reverse Polish notation (RPN). With this type of calculator, you enter each number (there is a button labeled **ENTER**) and then press the button for the operation: **2 (ENTER) 6 (ENTER)** ×. There are some advantages to this type of notation, but we would not recommend such a calculator for an electronics beginner.

But I Already Own a Computer!

You may be saying, "Why should I rush out and buy a calculator? I have a home computer sitting on my desk!" (Maybe you stashed your computer in the attic until you find a use for it.) After all, one of the reasons you bought the computer was to help you with math. Besides, didn't that salesclerk tell you the BASIC computer language would handle all kinds of difficult calculations? Well, let's clear up that confusion right away. Yes, indeed, you can use your home computer as a scientific calculator for your work with electronics math.

The BASIC language that comes with most home computers includes the scientific functions that you should look for on a calculator. BASIC also will work with numbers expressed in exponential notation. So if you have a computer available, it can serve very well as your calculator.

There are programs for some computers (such as IBM-PC compatible computers) that operate just like a scientific calculator. In this section we will limit our discussion to BASIC. If you are using a calculator program, practice with the program and become familiar with it.

The first step in using your computer as a calculator, of course, is to turn it on and get it ready to run BASIC. If you are using a calculator program, you will have to start that program on your computer. This procedure varies from one machine to the next, so check the owner's manual that came with your computer. With some computers, such as a Commodore 64 or 128, BASIC is ready as soon as you turn them on. You may even have one of the older Sinclair or Timex computers, a Texas Instruments TI-99 or a Commodore VIC 20. These computers all offer BASIC language programming capability. Any of them will suit your needs for electronics calculations.

If your computer normally uses floppy disks to store programs, you may have to retrieve some information from the disk when you first turn it on. The Apple and IBM-PC computers (and others like them) even have more than one version of BASIC available for you to use. Again, you will have to look in the owner's manual that came with your computer to learn the specifics for your machine.

Your computer will execute BASIC commands immediately if you type them without using line numbers. By using this feature, the computer will work just like a calculator. If you want to multiply 250,000 times 500,000, you can just type:

PRINT 250000 * 500000

and then hit the RETURN key. The asterisk (*) is the BASIC

Figure 1—Many personal computers are ready to run BASIC programs as soon as you turn them on.

```
PRINT 250000 * 500000
1.25E + 11
```

Figure 2—After you type the command to print the result of a simple multiplication, your computer display should look something like the one shown here.

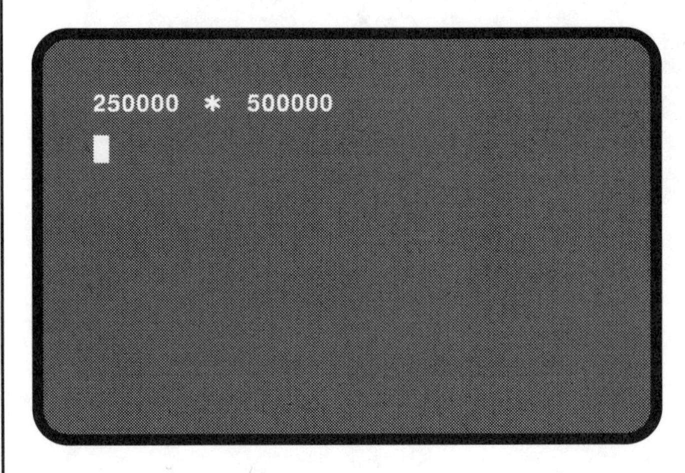

Figure 3—Does your computer display looks like this after typing the numbers and hitting the RETURN key? If so, it's because you forgot to tell the computer to display the answer!

multiplication symbol. Your computer display should look something like the one shown in Figure 2. (If you forget to type PRINT, the computer will still multiply the numbers. It won't display the answer on your monitor, however!)

You've done it! That E means the computer has displayed the number in exponential form, and the +11 means there are 11 zeros after the other numbers.

There are a few stumbling blocks, though. You may run into a few problems with some scientific functions if you jump in and start keyboarding numbers. These problems are pretty easy to solve, however. We'll show you how to handle the problems as we discuss the various functions in more detail.

If you're familiar with BASIC programming, you can use the scientific functions with ease. If not, look through the BASIC manual that came with your computer to learn more about the scientific functions. As you gain more experience with the computer and with BASIC, you'll discover many ways to use your computer for electronics calculations. In fact, you can do much more with a computer than with a calculator!

A scientific calculator does have some major advantages over a computer, though. The calculator is very easy to pick up and carry around with you. It runs on battery power (or light energy with solar cells). Also, the keys on a calculator normally give you a pretty good idea how to use them. You don't have to remember the exact spelling and format for each command, as you do with a computer.

Sooner or later, you'll find that it's to your advantage to own a scientific calculator as well as the computer!

Using Exponential Notation on Calculators and Computers

Learning to express numbers in exponential notation on your scientific calculator will be rather easy. When you bought the calculator, you made sure it had a key labeled **EE**, **EXP** or **EEX**. Hitting this key while entering a number alerts the calculator that the next one or two digits will be a power of 10.

Figure 1 gives you some simple problems with numbers that you should write in scientific notation. There are also some problems with numbers written in exponential notation for you to expand to normal notation. Some calculators have a feature that allows you to change

Write the following numbers in scientific notation:

a) 9,460,000,000,000

b) 186,000

c) 0.00001578

d) 300,000,000

e) 0.000000000001113

Expand the following numbers to normal notation, without a power of 10.

f) 147.1×10^{6}

g) 83.14×10^{2}

h) 6.67×10^{-11}

i) 6.022×10^{23}

j) 1.3806×10^{-23}

Figure 1—Here are some numbers for you to practice converting between scientific or exponential and expanded notations. Check your answers with the ones given in Figure 3.

Solve the following problems.

a) $12.5 \times 10^{-3} \times 6.8 \times 10^{3} =$

b) $625 \times 10^{-3} \times 5.6 \times 10^{3} =$

c) $6.28 \times 3.75 \times 10^{6} \times 2.48 \times 10^{-4} =$

d) $22.0 \times 10^{6} \times 0.045 \times 10^{-6} =$

e) $1.38 \times 10^{-23} \times 290 \times 5 \times 10^{2} =$

f) $\dfrac{1}{6.28 \times 21.15 \times 10^{6} \times 5.96 \times 10^{3}} =$

g) $\dfrac{7.125 \times 10^{6}}{1.2 \times 10^{2}} =$

h) $\dfrac{10.0 \times 10^{2} \times 2.00 \times 10^{3}}{50 + 2.00 \times 10^{2}} =$

i) $\dfrac{15 \times 10^{2} + 3.4 \times 10^{3}}{7.0 \times 10^{1}} =$

j) $\dfrac{6.67 \times 10^{-11} \times 6.02 \times 10^{24} \times 7.34 \times 10^{22}}{(3.925 \times 10^{8})^{2}}$

Figure 2—Work these practice problems with your scientific calculator or computer. Figure 4 gives the answers for them.

numbers between exponential and normal notations. Check your owner's manual to see if your calculator has this feature. If it does, follow the manual for instructions about working these problems. Otherwise, use the problems for practice without your calculator. Figure 3 lists the answers to the problems.

The problems in Figure 2 will give you some practice solving simple problems involving numbers expressed in exponential notation. These problems will help you become familiar with the basic operation of your calculator. After you solve all the problems, you'll know how to enter numbers expressed in exponential notation almost without thinking about it. Figure 4 lists the answers to these problems.

There may be some variation in the answers your calculator gives. The number of digits displayed by the calculator and the way the calculator rounds off numbers that are too large to fit the display affect the result. So your answers may not agree exactly with the ones given in Figure 4. Check your work if the last digit on your display differs by more than one or two from the same digit in the listed answer, however.

You should be aware of two important limitations to using a calculator. The first limitation is that the calculator can only work with numbers that have a power of 10 up to 99. If you ever have a calculation that includes a number like 5.24×10^{125}, your calculator won't be able to help. (But don't worry, you will never find such huge numbers in an electronics calculation!) You won't be able to work a problem with an answer that is larger than 10^{99}, either. You aren't going to find problems like this, either, so this limitation isn't serious.

If you try to enter more than two digits for the power of 10, most calculators will only use the last two digits entered. Get your calculator, and try entering this sequence of numbers:

2.56416 **EE** 346521

Chances are, the display shows:

2.56416^{21}

What if you strike a wrong key when entering a power of 10? You can correct the mistake by entering the *correct* numbers. This can be a lot quicker than clearing the calculator and starting over. In any case, you must know how your calculator will respond to such entries.

The second limitation is that your calculator will not work with more digits than its display can show. If you try to enter more digits, the calculator probably just ignores the extras. This is a problem because you might not notice that the calculator display doesn't include all the digits you entered. If you continue with the calculation, you are certain to get the wrong answer! The calculator will show an answer, but that answer will not be correct. Let's try an example. Enter the following sequence of numbers on your calculator:

123456789101112

a) 9.46×10^{12}

b) 1.86×10^{5}

c) 1.578×10^{-5}

d) 3×10^{8}

e) 1.113×10^{-12}

f) 147,100,000

g) 8314

h) 0.0000000000667

i) 602,200,000,000,000,000,000,000

j) 0.000000000000000000000013806

Figure 3—Compare your work on the exercises from Figure 1 with the answers shown here.

Now look at the display. It probably only shows the first eight or ten digits:
12345678
or
1234567891
Now enter the rest of the example problem:
$\times$ 3 =
The correct answer to this problem is:
370,370,367,303,336
What does your calculator show for the answer? If it can display ten digits, it is probably:
3703703673
and if it can only display eight digits, it probably shows:
37037034

Neither of these answers is even close to the correct one. With the ten-digit calculator, the answer is 10,000 times too small, and with the eight-digit one it is 1,000,000 times too small.

How can you solve the problem on your calculator, then? Write the number in scientific notation, and enter that in your calculator! In scientific notation, 123456789101112 becomes 1.23456789101112 X 10^{14}. Enter this number in your calculator, and then multiply by 3. Your calculator should show

3.703703673 ^{14}

or

3.7037034 ^{14}

The calculator has rounded the number off, but the power of 10 is correct. Either answer is reasonably close to the exact answer. This example points out why you must know how to write numbers in scientific or exponential notation.

Perhaps you decided to use your personal computer instead of buying a scientific calculator. Then you are probably wondering how you can do the problems presented in Figures 1 and 2. You already know that you have to turn the computer on and get it ready to run BASIC. Then you simply type in the problem you want to solve, and watch the answer pop up on the screen. (Remember that important PRINT at the beginning of the statement. Otherwise, you'll be waiting a long time for the answer to show up!)

If you are at least somewhat familiar with BASIC programming, you might have fun writing a short program to do the problems in Figure 1. It's not worth a lot of effort, though, because it's so easy to change between normal and exponential notations in your head.

To enter a number in exponential notation for a calculation on your computer, you type the digits of the number, an E, and the power of 10:

1.23456789101112E14

You also can use a plus sign between the E and the power of 10, but it isn't required. If the power of 10 is a negative value, you must use a minus sign, however:

1.2345E–14

In the calculator section, above, we mentioned two limitations to using a calculator. Similar limitations apply to using computers for the calculations. Your computer manual tells you how many digits the machine can use in calculations. In general, computers use between 8 and 12 digits. Some computers allow the use of "double precision," which means they will use 16 or more digits when performing calculations. Double-precision operation varies between different types of computers. You will have to check your owner's manual for the exact information about your machine. Type:

```
PRINT 123456789101112 * 3
```

and look at the result displayed on the screen. It may show the correct answer listed earlier, or it may show an answer expressed in scientific notation with fewer digits displayed.

Most computers can't work with numbers that have a power of ten greater than +38 or smaller than –39. Notice that this is more restrictive than with calculators, which can use powers between +99 and –99. This limitation is because of the way computers represent numbers. You won't find many practical problems that call for numbers outside this range, though.

Use your computer to solve the problems presented in Figure 2. Check your answers with those given in Figure 4.

a) 85

b) 35×10^{2}

c) 5.8404×10^{3}

d) 0.99

e) 2.001×10^{-18}

f) $\dfrac{1}{7.9161912 \times 10^{11}} = 1.263233763 \times 10^{-12}$

(The reciprocal, or 1/x key on your calculator is handy for problems like this.)

g) 5.9375×10^{4}

h) $\dfrac{2.00 \times 10^{6}}{2.50 \times 10^{2}} = 8000$

i) $\dfrac{4900}{7.0 \times 10^{1}} = 70$

j) $\dfrac{2.94725956 \times 10^{37}}{1.5405625 \times 10^{17}} = 1.913106128 \times 10^{20}$

(Notice that the denominator term is squared, or multiplied by itself.)

Figure 4—The answers you find to the problems in Figure 2 may be slightly different from the ones given here. Your calculator may display a different number of digits than the one used to get these answers. That will cause it to round off the numbers differently, giving a different result.

CHAPTER 3 Tricks for Manipulating Equations

The Meaning of Equality

We use numbers to express a quantity of some material or as a measurement of some type. The second step is to write an *equation* relating one quantity to another. The mathematical equations in this book are simple to work with, if you follow a few basic rules. What is an equation, anyway? Webster's Dictionary defines an *equation* as a "formal statement of the equality or equivalence of mathematical or logical expressions."[1] So now you are probably wondering what *equality* means. On the same page of the dictionary we discover that equality means "the quality or state of being equal."

We're getting closer to a definition that we can understand. How does Webster's Dictionary define *equal*? It says *equal* means something that is "of the same measure, quantity, amount or number" as something else. Equal also means identical, especially when we are talking about mathematical notations. We normally use a special symbol (=) to mean *equal* in mathematics. Then we write an *equation* by placing the two quantities on opposite sides of the equal sign.

If you're not familiar with mathematical equations you're probably wondering what value there is to writing an equal sign between identical terms. It is true that in its simplest form, an equation might look like Equation 1.

$$5 = 5 \qquad \text{(Equation 1)}$$

This isn't usually what we have in mind, however.

One interesting thing about a mathematical equation is that the two terms don't have to be *literally* identical. They have to be *mathematically* identical. The terms probably will *look* different but still be equal. We could write an equation like the one shown here.

$$1 + 3 = 2 + 2 \qquad \text{(Equation 2)}$$

These terms aren't literally identical, but adding the numbers on each side of the equal sign gives the same result, 4. This is an important concept for you to understand. The terms in an equation will seldom be literally identical, yet they will be mathematically the same.

Another important concept about equations is that usually there will be some unknown quantity as part of the equation. This means that you must understand what equality means so you can calculate the unknown. For example, we could rewrite Equation 2 as:

$$1 + 3 = 2 + X \qquad \text{(Equation 3)}$$

where

X is a symbol used to represent an unknown quantity

First you can add 1 and 3 to get 4. Next, you know that 2 plus the unknown must add up to this same value. Then you can quickly calculate the unknown value, which in this case must be 2.

The solution to an equation may not always be quite as obvious as this one was. Often, the numbers you are working with will include several digits. The equations you are working with may have decimal fractions, and some will even have powers of ten, especially with electronics problems. Even with these complications, you can solve most of the problems you'll run into if you follow a few simple rules.

Probably the most important rule for working with equations, and the one most often ignored, is to be neat! Keep your work organized in an orderly fashion, take your time, and write neatly so anyone could easily read it. It's just too easy to mistake one number for another or to drop part

[1]Merriam-Webster, Inc., *Webster's Ninth New Collegiate Dictionary* (Springfield, MA, 1984), p 420.

REMEMBER, BE NEAT!

of an equation as you work. Be neat and you'll avoid these and other careless mistakes.

Rule: Be Neat!

The order in which you perform calculations can be important. For example:

$3 \times 6 + 2 = 20$ (Equation 4)

clearly shows the answer we want, but if we write that as:

$2 + 3 \times 6$ (Equation 5)

you could incorrectly get 30 as an answer. So you must understand the correct order for calculations.

We often use parentheses to group the terms of an equation. The parentheses tell you to perform certain calculations before others. In the example shown in Equation 5, we could use parentheses to write:

$2 + (3 \times 6)$ (Equation 6)

Now you will know that you should multiply 3 by 6 first, and then add 2 to that result. This will give you the correct answer to the problem, 20. If we really intended to add 2 and 3 first, we would use parentheses to group these numbers together.

$(2 + 3) \times 6 = 30$ (Equation 7)

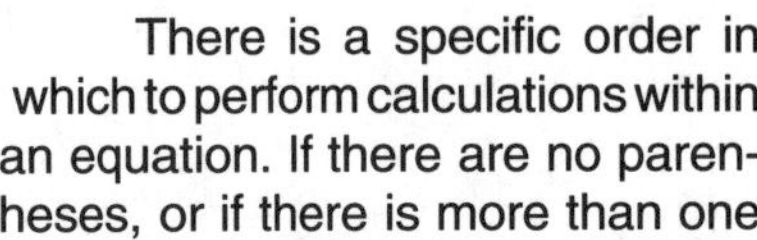

There is a specific order in which to perform calculations within an equation. If there are no parentheses, or if there is more than one operation inside parentheses, perform the calculations in the following order. First calculate any powers or roots of numbers required by the equation. Next, do the multiplications and divisions, moving from left to right through the expression. Finally, you should do any additions or subtractions. You can follow any order for the addition and subtraction.

Rule: Do calculations in the following order.
1) Powers and roots in any order
2) Multiplications and divisions from left to right
3) Additions and subtractions, in any order.

Sometimes you may find an equation that has parentheses inside of parentheses, or that uses brackets ([ ]) and braces ({ }). These symbols indicate several levels of grouping. In a case like this, start with the pair of parentheses inside all the others. Begin your calculations there and work your way out, one pair at a time. Be sure to follow the other rules about the order of calculation.

$5 + \{2 + 3 - [10 - 2 \times (3 + 1)]\} = x$ (Equation 8)

where

x is the unknown answer to this problem

$5 + \{2 + 3 - [10 - 2 \times 4]\} = x$

$5 + \{2 + 3 - [10 - 8]\} = x$

$5 + \{2 + 3 - 2\} = x$

$5 + \{3\} = 8$

Follow these simple rules, and always do the calculations in the proper order. With practice you'll soon find that you hardly think about the rules as you work through a problem.

Let's Not Call it Algebra

When it comes time to *solve* an equation for some unknown quantity, you'll have to follow a few basic rules. There are a few simple procedures that you should follow in working with equations. It doesn't hurt to know a couple of little tricks, either.

What does it mean to *solve* an equation? When you solve an equation, you normally want to have the unknown quantity by itself on one side of the equal sign. The other equation terms (the known values used to calculate the unknown) are on the opposite side of the equal sign.

The branch of mathematics that involves solving equations has its own name: *algebra*. That word may remind some readers of complicated procedures and difficult computations. The equations we'll be using in basic electronics are quite simple, however, and you'll be able to solve them easily. In this section we'll explain a few rules that you should follow when working with electronics equations. Also, we'll discuss the most common techniques that you'll use to solve these simple equations. You'll even learn a few tricks to make your task a bit easier.

The first rule, of course, is to be neat! Start with an organized work area, and have the materials you'll need within easy reach. Write carefully, being especially careful to form your numbers and letters properly. Don't crowd your work, and if you make a mistake be sure to erase completely. Do only one or two steps at a time, carefully copying the equation on a new line with each step. Later you can do more of the work mentally, and you won't have to write as many steps. Until you've gained more experience, though, you should follow this step-by-step technique carefully.

The first rule is to be neat!

Always keep your final goal in mind. You want the unknown term on one side of the equal sign and all other terms on the other side. Each step of simplifying the equation should help you work toward this goal. In general, you can add, subtract, multiply or divide the equation by just about any quantity that helps you meet that goal. (*Never* divide by zero. Also, it doesn't help to multiply by zero! Multiplying by 1, or adding or subtracting zero don't help much either.) Any operation you do to one side of the equation you also must do to the other side. Perform these operations on one side of the equation as a group rather than to individual terms. A couple of examples should help explain this process.

Any operation you do to one side of the equation you also must do to the other side.

Let's use the letter I to represent the current in an electronics circuit. Then suppose we have an equation that relates the current with some other circuit conditions. We want to solve the equation for the circuit current.

$$2I + 5 - 3I = 10 - 16I \qquad \text{(Equation 1)}$$

The first step here is to add 16I to both sides of the equation.

$$(2I + 5 - 3I) + 16I = (10 - 16I) + 16I$$

Since we are adding all these terms, we don't really need the parentheses. We can combine similar terms, simplifying the equation a bit.

$$15I + 5 = 10$$

The next step is to subtract 5 from both sides of the equation.

$$15I + 5 - 5 = 10 - 5$$

Then combine like terms again.

$$15I = 5$$

We have one more step to go before we have the solution to our equation. Divide both sides of the equation by 15, and we'll have the unknown by itself on the left side of the equation!

$$\frac{15I}{15} = \frac{5}{15}$$

I = 0.333

Here we have the solution to our equation. Since we measure electrical current in units called amperes, the current in our circuit must be 0.333 amperes.

What tricks could we use to save some time and a few steps in this solution? Well, you probably got a few ideas from the examples shown so far. Adding a term to both sides of the equation cancels the term on one side and moves it to the other. (Adding a negative value is the same as subtracting. So you can add or subtract a term to both sides of the equation.) On the opposite side of the equal sign, however, positive terms become negative and negative terms become positive. One time-saving trick, then, is to *transpose* such terms across the equal sign.

One time-saving trick is to* transpose *terms across the equal sign.

Suppose we have the equation:

$I + 25 = 30$ (Equation 2)

Instead of writing a step that shows 25 being subtracted from both sides of the equation, we can simply transpose the term, and write:

$I = 30 - 25 = 5$

Another trick involves a technique known as *cross multiplication.* This technique provides a quick way of multiplying or dividing both sides of the equation by some term.

Cross multiplication provides a quick way to multiply or divide both sides of an equation by some term.

Suppose you have an equation like the one shown here.

$$\frac{44x}{10} = \frac{80}{9} \qquad \text{(Equation 3)}$$

where

x is some unknown quantity

We could multiply both sides by 9 and then multiply both sides by 10. Instead, let's cross multiply. Move the denominator of both fractions diagonally across the equal sign. (The *denominator* of a fraction is the bottom part.)

$$\frac{44x}{10} = \frac{80}{9}$$

$44x \times 9 = 80 \times 10$

$396x = 800$

Now, we can cross multiply again by moving the 396 diagonally across the equal sign. (Leave the unknown, x, where it is.)

The 396 becomes the denominator of a fraction on the right side of the equal sign.

$$x = \frac{800}{396}$$

$x = 2.02$

Each type of problem is a little different from other types, so it's impossible to give you an exact procedure to follow. It's also possible to solve a problem by following several slightly different procedures. You'll have to study each problem to find a way to solve it. Follow the guidelines described in the examples of this section. With practice you'll soon gain the confidence to tackle problems that are even more difficult.

Simple Equation Circles

Equations show the relationships between various quantities. You can use an equation to calculate one quantity if you know the values of the other terms in the equation. You must know how to *solve* an equation to do this calculation. Equations sometimes look more complicated than they are because we use letter symbols to represent the various quantities. You'll learn to recognize what these letters represent in the equation, though. Then you must decide which are the known quantities and which are the unknowns. This process isn't difficult. Just keep in mind what the letters represent, and make a list of the quantities by letter symbol and their known values.

Here's an example of an equation that relates several quantities. Suppose you've been riding in a car for 4 hours. During that time the driver has kept an average speed of 40 miles per hour. Would you like to find out how far you have traveled? There is a simple relationship between speed, time and distance, but that probably won't surprise you too much. You may have even figured out that we might write an equation to show that relationship.

$$d = s \times t \qquad \text{(Equation 1)}$$

where:
d = distance = ?
s = speed = 40 mi/hr
t = time = 4 hours

This calculation is pretty simple because the equation has the unknown, d, by itself on one side of the equal sign.
$d = 40 \text{ mi/hr} \times 4 \text{ hours} = 160 \text{ miles}$

One possible point of confusion when you work with equations concerns the × symbol for multiplication. We often omit this symbol in an equation. You could easily mistake a hand-written × symbol for a letter X. We also often use a letter X to represent an unknown quantity in mathematics. Sometimes we use a raised dot (•) instead of the × to show multiplication, but often we don't use any symbol at all. When you see two letters written next to each other with no operation symbol between them, you'll know you should multiply them. Keep this in mind when you are working with electronics equations. Equation 1 might be written as:

$$d = s \cdot t \qquad \text{(Equation 2)}$$

or more often as:

$$d = st \qquad \text{(Equation 3)}$$

Many equations involve only multiplication or division of terms to calculate some unknown quantity. You can *cross multiply* to solve most of these equations for any of the terms. This is true for most of the basic electronics equations you will use. An *equation circle* gives you a visual reminder of how to do the cross multiplication. A few examples should help you understand what *equation circles* are and how to use them.

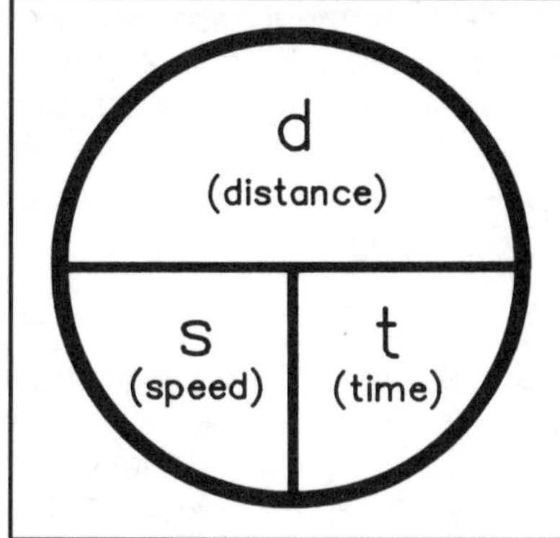

Figure 1—This is a simple equation circle. This circle relates distance speed and time from Equation 1. To use the equation circle, cover the letter of the quantity you want to find. The remaining letters show the multiplication or division needed to find your answer.

Figure 1 is an equation circle for the distance, speed and time equation, Equation 1. How do you use an equation circle? First identify the term you want to calculate. Cover that term with your finger, a piece of paper or another object. The remaining letters show you how to solve the equation.

What was your average speed on a bicycle ride, if it took you $2^1/_2$ hours to ride 20 miles? To solve this problem, we want to solve for the speed, so cover the s on Figure 1. Notice that you now have a fraction, with distance (d) divided by time (t). From this picture, we can write an equation solved for speed:

$$s = d / t \qquad \text{(Equation 4)}$$

$s = 20 \text{ miles} / 2.5 \text{ hours}$
$s = 8 \text{ miles} / \text{hour}$

See how easy it was to use an equation circle to solve that problem?

You can make an equation circle for other equations, too, under certain conditions. First, the basic equation should have one quantity by itself on one side of the equals sign. The terms on the second side of the equal sign must all be multiplied. There can't be any terms added or subtracted in the equation to make an equation circle. You'll find many equations like this as you study basic electronics, though, so equation circles are very useful.

Let's use some examples of electronics equations. Chances are you won't know all the quantities used in these equations. You'll learn more about them later in the book. For now, our main interest is in learning how to solve some basic equations.

We'll start with an equation known as *Ohm's Law*. Ohm's Law relates voltage (E), current (I) and resistance

(R) in an electronics circuit. Ohm's Law is normally written as:

$E = IR$ (Equation 5)

This is fine if you know what the current and resistance are (I and R), and want to calculate the voltage. What if you want to know either I or R, though? You'll have to solve the equation for another quantity.

Figure 2 shows an equation circle for Ohm's Law. Identify the term you want to calculate and cover it with your finger, a piece of paper or another object. The remaining letters show you how to solve the equation.

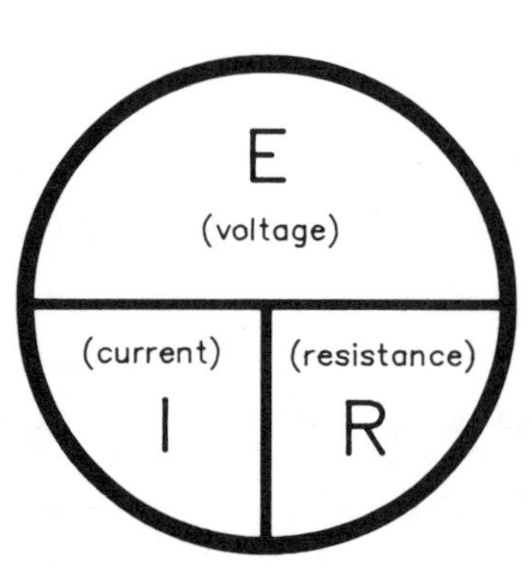

Figure 2—An equation circle will help you solve Ohm's Law problems. The text describes how to use such an equation circle to solve a problem. You can make equation circles for other equations that have a similar format.

Suppose a circuit has a voltage (E) of 12 volts applied to it. When you measure the current (I), you find 0.5 amps flowing through it. You want to calculate the circuit resistance (R). (Remember, you may not be familiar with these electronics terms, but our purpose here is to teach you how to solve some equations.) Notice that when you cover the R in the equation circle of Figure 2 you're left with a fraction. The E is the numerator and I is the denominator. From this we can write:

$R = E / I$ (Equation 6)

Then we can calculate the circuit resistance by substituting the numbers from our problem into the equation.

R = 12 volts / 0.5 amps = 24 ohms

As another example, let's choose the equation that relates the wavelength and frequency of a radio wave. (You shouldn't worry if you don't understand what these terms mean. You'll learn more about them later in the book.) The equation is:

$c = f\lambda$ (Equation 7)

where:

c = speed of the radio wave
f = frequency of the wave
λ = wavelength

The speed of a radio wave through space is the same as the speed of light. You can look up the speed of light in a reference book if you need to. As a close approximation, we'll use the value of 3.00×10^8 metres/second for the speed of a radio wave. Now suppose we want to know the wavelength of a radio wave that has a frequency of 3,700,000 Hz. (That's the same as 3.70×10^6 Hz.) We can make an equation circle for Equation 7. In the top half of a circle write a c for the speed of light. Then divide the bottom half into two parts, and write an f on one side and a λ on the other. Your completed equation circle should look like the one shown in Figure 3.

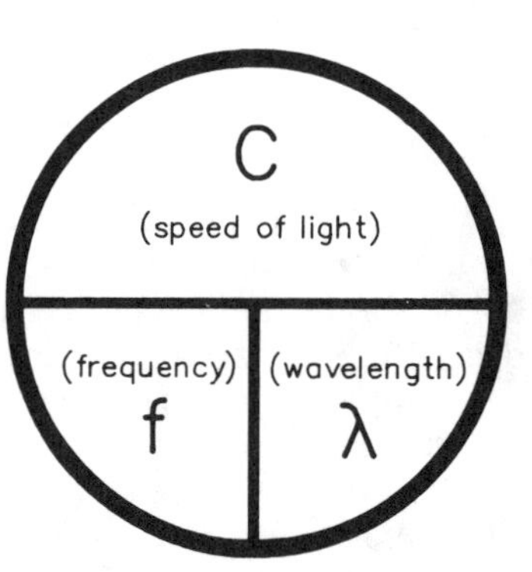

Figure 3—This is another example of an equation circle. The text also describes how to use this equation circle to solve a problem. You can make equation circles for other equations that have a similar format.

To solve the equation for the wavelength, cover the λ in the equation circle. Now write λ =, and complete the equation from the remaining portion of your equation circle.

$\lambda = c / f$ (Equation 8)

Then perform the calculation using the known values.

$\lambda = 3.00 \times 10^8$ m/s / 3.70×10^6 Hz
λ = 81.1 metres

Equation circles also serve as a memory device to help you remember an equation. You can form a mental picture of the letter symbols arranged inside an equation circle. This often makes the equation much easier to remember.

Solving Equations by Substitution

We often write equations with letter symbols to represent measured quantities. Sometimes information seems to be missing, and that makes it difficult for you to solve the problem. Often this just means you must perform an extra calculation to find the missing data, though.

When you have a problem to solve, start by listing all of the information you know from the problem. Then identify what you are trying to calculate. Write your list using the letter symbols that usually represent the quantities. That makes it easier to choose an equation to solve your problem. After you have identified the known and unknown quantities, you're ready to find an equation. You may not know an equation to solve your problem directly. In that case, try to find other equations that relate the information you have to the unknown quantity. (We'll work through an example to show how to identify related equations.)

You might proceed in one of two ways after you identify the equations you need. You can calculate separate pieces of information needed to solve your problem and then use those numbers to calculate your result. While this method may seem like the most direct approach, it has one big snag. Every time you write a number and perform a calculation, you run the risk of making an error. If you don't write the numbers clearly, it can be very easy to mistake one number for another. When you go to the next step you must be very careful to copy the numbers correctly. Another problem can be adding an extra zero or dropping a zero out of a number. This is especially easy when you work with numbers that include more than three or four zeros.

A better way to tackle these problems is to solve the literal equations. What does that mean? Well, a *literal equation* is one written with letter symbols rather than numbers. So if you are going to solve a literal equation, you work with the letter symbols. Put the numbers into the equation only when you have solved it for the unknown letter. Sometimes you'll find a second equation that provides a piece of information missing in the main equation. In that case, substitute the letter symbols that represent the information you need. Let's work through an example to see how to do this.

We'll use an electronics problem as our example, but you may not be familiar with the equations and terms we use. In this section, our goal is to teach you the mathematical methods needed to solve similar problems as you study electronics.

In a certain electronics circuit you measure the voltage across a 50-ohm resistor. The voltage is 125 volts. What is the power in the resistor? Our first step should be to make a list of the known and unknown quantities, as shown in Figure 1. Equation 1 is the most common equation for calculating power in a circuit.

$$P = I\,E \qquad \text{(Equation 1)}$$

where:

P = power in the circuit, measured in watts
I = current through the circuit, measured in amperes (or amps)
E = voltage applied to the circuit, measured in volts

You can tell right away that we have run into trouble here. The list of information shown in Figure 1 doesn't include a current, so we can't use Equation 1 directly. You also should notice there was a resistance given in our problem, but Equation 1 doesn't use that quantity. Your next question should be "How can I use voltage and resistance to find current?" Someone familiar with electronics would immediately answer, "With Ohm's Law!" Equation 2 shows Ohm's Law.

Voltage = E = 125 volts

Resistance = R = 50 ohms

Power = unknown

P = I E

Figure 1—The first step in solving any problem is to make a list of the problem information. Include the letter symbol used to represent each quantity, and the value of any known quantity.

$$E = I\,R \qquad \text{(Equation 2)}$$

where:

E = voltage across the circuit
I = current through the circuit
R = circuit resistance

We want to use this equation to find current, so we have to solve the equation for I. Figure 2 shows an equation circle for Ohm's Law. From that figure we can write Equation 3.

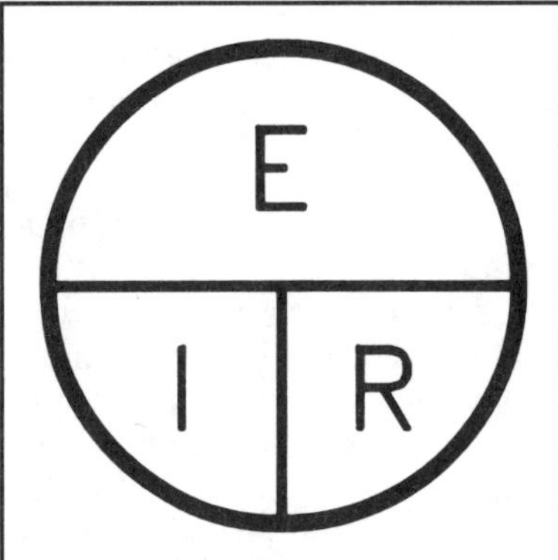

Figure 2—This Figure shows an equation circle for Ohm's Law. The module just before this one explains equation circles and how to use them.

$$I = \frac{E}{R} \qquad \text{(Equation 3)}$$

Now we can use this *literal equation* to replace the I in Equation 1 with another expression. We'll combine Equations 1 and 3 in the next step.

$$P = I\,E \qquad \text{(Equation 1)}$$

$$P = \left(\frac{E}{R}\right) E \qquad \text{(Equation 4)}$$

$$P = \frac{E^2}{R} \qquad \text{(Equation 5)}$$

Here we have another equation to calculate power, and it only uses quantities that we know from our problem. Now we can write the equation with our numbers, and calculate the power.

$$P = \frac{(125\text{ V})^2}{50\text{ Ohms}}$$

$$P = \frac{15625\text{ V}^2}{50\text{ Ohms}} = 312.5\text{ watts}$$

The technique shown here involves substituting an expression of known quantities for an unknown quantity in another equation. You can use this technique to solve some rather difficult problems after you have practiced it and gained some experience with it.

CHAPTER 4
The Metric System of Measure

System International (SI) Units

When we measure something, we find a number to express the size or quantity of the measured item. We also use numbers to represent the results of simple (or complex) calculations. In addition to these numbers, there is always a unit, or expression to describe what the numbers mean. In electronics, those units are part of a system of measurements known as the metric system.

Most people have heard of the metric system, but it is not the system most often used in the United States. The unfamiliar terms of the metric system frighten some people. You won't be one of them, though, after you read this section!

Metric system units make up an internationally recognized measuring system used by most scientists throughout the world. We call this the International System of Units, abbreviated SI (for the French, Système International d'Unités). The International Bureau of Weights and Measures oversees the use of, and recommends changes to, this system. Most countries of the world use this system as their standard measuring system. Scientists use the metric system to make measurements and conduct experiments.

The metric system is easy to understand after you learn the meaning of a few basic terms. Before we go into the details of the metric system, however, let's take a brief look back through history. We'll see how measuring systems changed through the years.

From earliest recorded history, people found it necessary to measure quantities of various materials. Each culture developed its own set of measuring units.

One popular system used parts of the human body as the measuring tools. When these people wanted to know how long something was, they always had a set of measuring instruments with them. The length of a foot or a step measured distances. The distance between fingertips on their outstretched arms measured length. People also found convenient ways to measure other items. For example, they measured the quantity of fruits and vegetables by using baskets or other containers.

Imagine how much difference it makes when two people make the same measurement with a system like this. Let's suppose a seven-foot basketball star and a ten-

Figure 1—You always carry a variety of measuring tools with you, and can use them to estimate dimensions.

Figure 2—Farmers often measure quantities of fruits and vegetables in various-sized baskets.

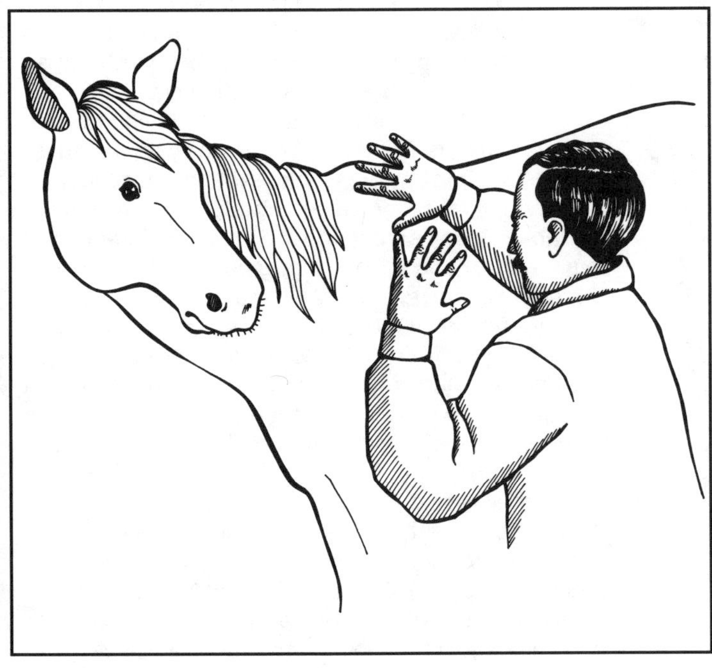

year-old child measured the length of a basketball court. The seven-footer would say it was quite a few feet shorter than would the ten-year old! Or if you went to the market to buy some potatoes, you would want to find the farmer with the biggest baskets.

As it became important to make more accurate measurements, people had to adopt s*tandard measuring units.* They used the same units they were familiar with, but selected standard sizes. So there was a s*tandard inch*, a *standard foot*, a *standard quart*, and so on. These standards had to be physical objects so people could compare them to other measuring instruments to ensure accuracy. A manufacturer of foot rulers would compare them to the *standard foot*, for example, to be certain that their rulers were accurate.

In the United States, our National

Institute of Standards and Technology (NIST) keeps a set of standard weights and measures. We can compare these *standards* to the international standards, kept at the International Bureau of Weights and Measures in Sèvres, France.

The US Customary System (sometimes called the English System) evolved from the original units related to commonly available measuring tools. Because of this, there is no simple relationship between the units. There are 12 inches in a foot, 16 yards in a rod, 5280 feet in a mile, and so on. In addition, we divide most of these units into smaller units by repeatedly dividing the unit in half. So we end up with fractional parts of a unit, measured in halves, quarters, eighths, sixteenths and smaller fractions. Calculations involving accurately measured quantities are difficult. This system is not well suited to scientific experiments.

Figure 3—Foot size or step length varies a lot depending on the individual's size. Imagine the confusion that results when we use different sized measuring tools to measure the same quantity.

As early as 1790, scientists were developing a measuring system based on multiples of 10, which they called the metric system. To measure a quantity, you start with a basic unit, dividing that unit into ten equal pieces to make smaller measurements. You can even divide each of these ten pieces into ten more pieces. Continue this process until you have a unit that is convenient for the size of the measurement you are making. Likewise, you can multiply the basic unit by ten, a hundred, a thousand or even more, to get larger, convenient units.

So how many different units will you have to learn to be familiar with the metric system of measure? Well, in any measuring system you need units for *length*, *mass* (or weight) and *time*. We also must be able to measure *temperature* (or differences in temperature) and *electric charge*. We call these units *fundamental units*. You can't measure fundamental units in terms of some other unit, so you must define them to start with. You can measure all other quantities in terms of these units. (For example, once you've defined a length unit and a time unit you can measure speed. Speed is distance traveled divided by the time it takes.)

That makes five *fundamental units* so far. Actually, there are two other quantities that require fundamental units to measure them. They are *luminous intensity* (the brightness of a light) and *molecular quantity* (the number of atoms present in an object). Since we won't be measuring either of these quantities in our study of basic electronics, we won't worry about those units now.

What are the metric units for these fundamental quantities? Well, the basic unit of length is the *metre*. (Most people in the US spell this *meter*, but in the rest of the world it is spelled metre.) The basic unit of mass is the *gram*. The basic time unit is the *second*, the same as in the US system. We often use the *Celsius* temperature scale to relate "normal" everyday temperatures, although the *Kelvin* scale is the official metric scale. Finally, the *coulomb* is a measure of the quantity of electrical charge.

Figure 4 shows some common items that will help you understand these fundamental units. The ruler (shown actual size) will help you estimate how long a metre is. The section from the edge of the ruler to the number 10 is 1/10th of a metre long. (You'll learn what those numbers represent later.) So if you mark that section off along the edge of your desk ten times, you'll have the length of a metre. A common paper clip (also shown actual size in Figure 4) has a mass of about 1 gram. On the Celsius temperature scale, water freezes at 0°C and boils at 100°C. The Kelvin temperature scale is very similar to the Celsius scale. The size of the unit is the same on both scales, but water freezes at 273.15 kelvins (K) and boils at 373.15 K. (Notice there is no degree symbol used with the Kelvin scale, and we do not refer to degrees kelvin.) A coulomb of electric charge is equivalent to 6.24×10^{18} electrons.

Now you have at least heard of the fundamental units used in the metric system. There is no reason for metric system units to frighten you. From these units we can define other quantities that we may want to measure. You're ready to learn how to multiply or divide the common units to obtain a conveniently sized unit for any measurement.

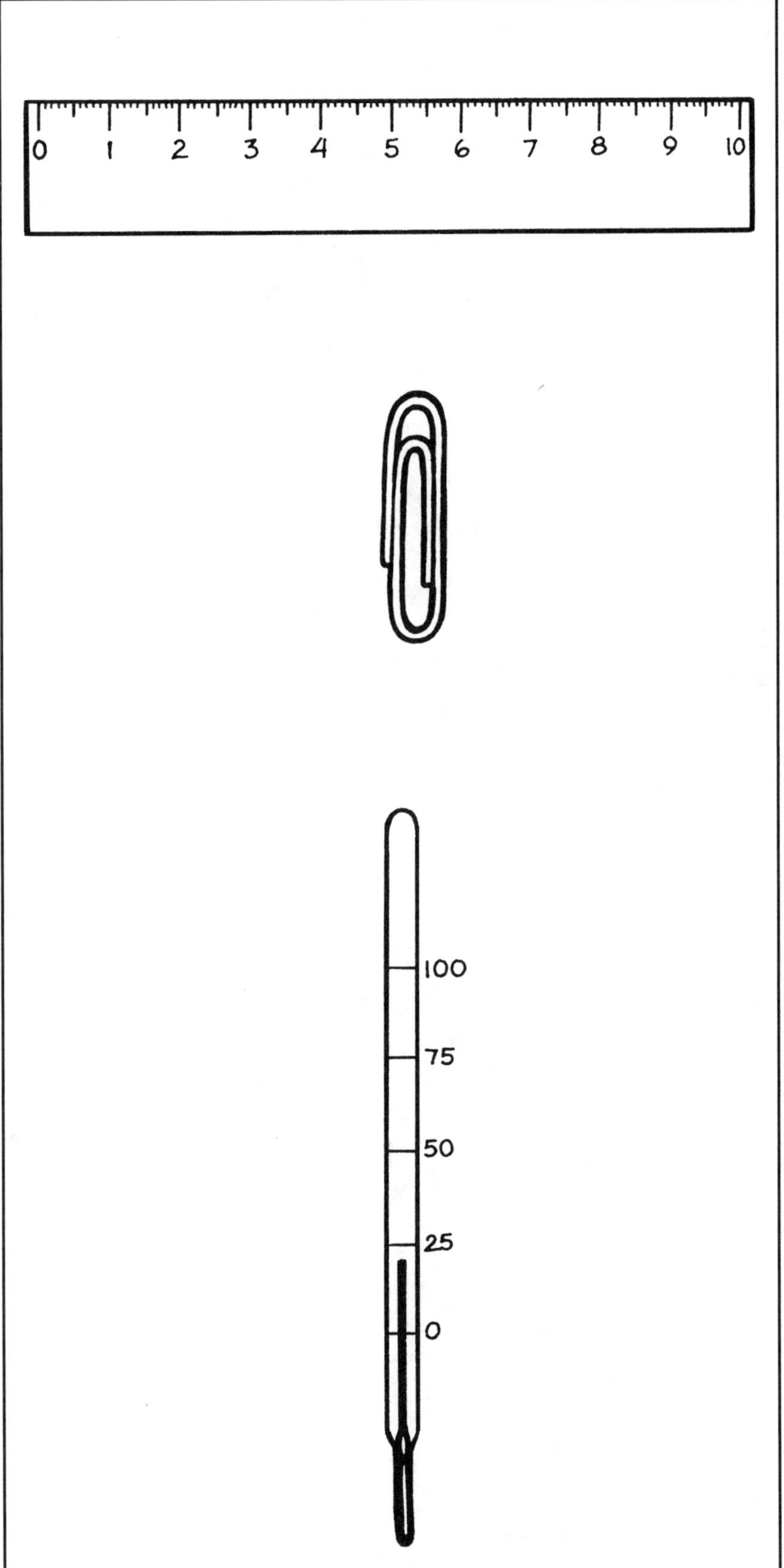

Figure 4—Each numbered mark on the ruler represents 0.01 metre, so the distance from the edge of the ruler to the number 10 represents 0.1 metre. A standard-sized paper clip has a mass of about 1 gram. On a Celsius thermometer, water freezes at 0°C and boils at 100°C. Room temperature (68°F) is 20°C. The picture shows the actual size of the ruler and paper clip. Most thermometers are longer than the one shown here.

Common Metric Prefixes

When you are working in the metric system there are only a few basic units that you must be familiar with. You can simply multiply or divide the basic unit by factors of 10 if you need a larger or smaller measuring unit. These factors of 10, or multiples of 10, have special names in the metric system. We use these names as prefixes, attaching them to the beginning of the names of basic units.

Here are some examples to make this easier to understand. First, we'll measure the length of a farmer's property line. Remember, the basic length unit is the metre. This is a large property, so it's going to be difficult to measure the line using a metre stick! (A metre stick is a ruler that is one metre long.) A tape measure that is 10 metres long might be more convenient. The prefix that means 10 times the basic unit is *deca*, so we could say our tape is one *decametre* long. (Notice that we simply added the prefix deca to the basic unit, the metre.) You may see this prefix spelled deka.

In this example, a longer length unit would be convenient. We could use one that is equal to 10 decametres or 100 metres. The prefix meaning 100 is *hecto*, so a hundred metres is a hectometre. *Kilo* means a thousand, so a kilometre is 1000 metres long.

Now let's suppose we want to measure something smaller, like the width of a door. This time we'll need to divide the metre into at least ten pieces. The prefix that means one tenth of a unit is *deci*. A decimetre, then, is 0.1 metre. We can divide the unit into 100 equal pieces. The prefix *centi* describes each of those 100 pieces. A centimetre equals 0.01 metre and 10 centimetres equal 1 decimetre. The numbered marks on the ruler of Figure 1 show centimetre divisions. Notice that each of the centimetre divisions also has ten sections marked off. It takes 1000 of these smaller pieces to make one metre (each mark is 0.001 metre). We call these smaller divisions millimetres, so the prefix that means one one-thousandth is *milli*.

Let's take a moment to review what you have learned so far in this section. You're already familiar with the metric prefixes used most. Kilo means thousand, hecto means hundred and deca means ten. Deci is one tenth, centi is one

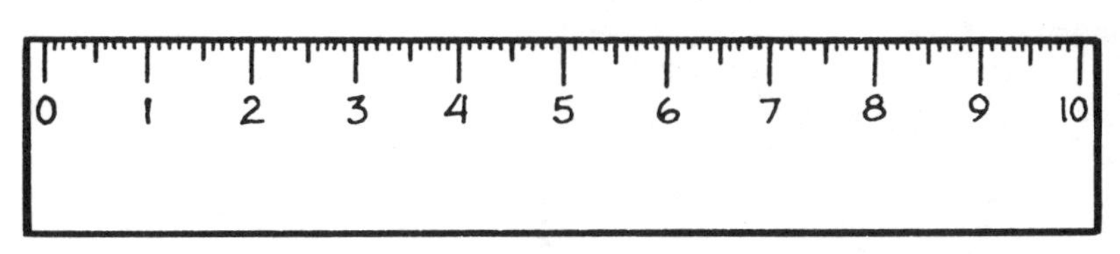

Figure 1—The metric ruler shown in this picture is life sized. Each numbered mark represents one centimetre. Notice that each centimetre is divided into ten parts. Each of those smaller parts is one millimetre. This ruler is 10 centimetres, or one decimeter, long.

one-hundredth and milli is one one-thousandth. We can use these same prefixes with any unit in the metric system. The gram is the basic unit of mass. A mass of one kilogram equals 1000 grams. Likewise, a millisecond is equal to 0.001 second. Your car's speedometer may have markings calibrated in kilometres per hour. The speedometer tells how fast the car is traveling, in thousands of metres per hour.

When you see an unfamiliar metric unit, look first for the prefix, and be sure you understand what that means. Then you can check the basic unit, to learn what it represents. If you see a container in the supermarket with a label that claims it holds 2000 millilitres, what does that mean? It sure sounds like a lot of something, doesn't it? Well, first notice the prefix milli. Since milli means 0.001, if we multiply this by 2000, we get 2:

$$2000 \times 0.001 = 2 \qquad \text{(Equation 1)}$$

From this you can see that the "2000 milli" part is equal to 2 of whatever the units represent.

Now what does the unit litre mean? Perhaps you already know that it is a volume unit, used mostly for liquid measure. How big is it? Well, a litre is just another name for a cubic decimetre. One way to picture this is to imagine a cube that is one decimetre on a side. (That's 0.1 metre or 10 centimetres on a side. The ruler of Figure 1 is one decimetre long.) A 2000 millilitre container holds the same amount as a 2 litre bottle.

You may have noticed the syringe a doctor uses to give someone medicine with a needle. The doctor often measures medicine in cubic centimetres (abbreviated cc). A cubic centimetre is the same as a millilitre. Since there are 1000 millilitres in a litre, there are also 1000 ccs in a litre.

Most of the time we abbreviate the metric prefixes and units when we write them. Each prefix has a common abbreviation. You should be familiar with the prefixes, so you can recognize them easily. With one exception, we use the first letter of a prefix as an abbreviation. We use m to abbreviate milli, c for centi, d for deci, h for hecto and k for kilo. The exception is deca, which we abbreviate da. These abbreviations combine with the unit abbreviations to specify a measurement. For example, cm means centimetre, km means kilometre and dag means decagram.

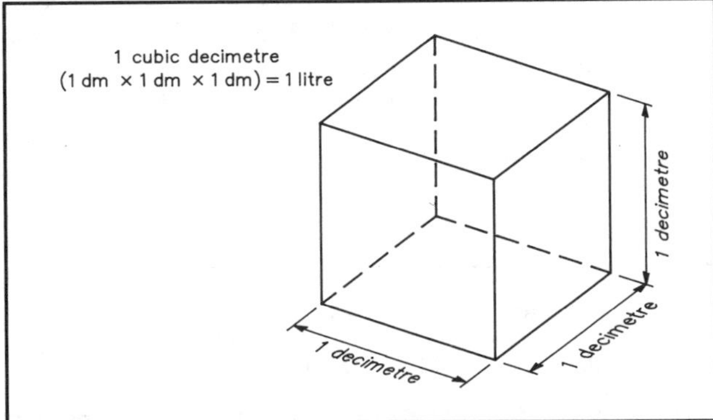

Figure 2—A cubic decimetre is the same as one litre. A cube that is one decimetre on each side forms a one litre container. One decimetre equals 10 centimetres, so the cube is also 10 centimetres on each side.

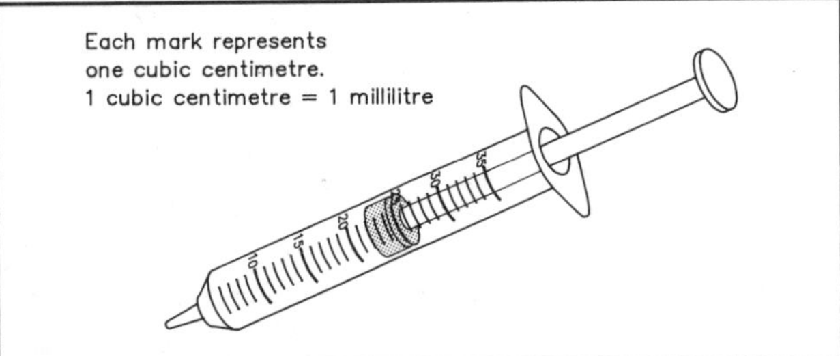

Figure 3—A syringe used by doctors to inject medicine with a hypodermic needle has marks every cubic centimetre to measure the medicine. One cubic centimetre is equal to one millilitre.

You should be familiar with the technique of writing very large and very small numbers in exponential notation. If you need a refresher, review that module. When we write numbers in this way, we include a *power of ten*. This is a way of showing how many places to move the decimal point to write the number in *expanded* (normal) notation. For example, we could write 405,000 as 405×10^3. The power of ten here is 3. It tells us that the decimal point has to move three places to the right after the 5 in 405. (That's three more zeros in this case.) Multiplying a number by a factor of 10^3 is the same as multiplying it by 1000. From this you can see that a factor of 10^3 is the same as the metric prefix kilo. Hecto is the same as a factor of 10^2 and deca is the same as 10^1. Deci is equal to 10^{-1}, centi is equal to 10^{-2} and milli is 10^{-3}. (Negative powers of ten mean the number was divided by that factor of ten. They also tell you to move the decimal point to the left if you write the number in expanded form.)

Quite often in electronics we will be working with either very large or very small quantities. Many times we would like to have units that are more than a thousand times larger than the basic unit. There also are times when we would like units that are much smaller than one one-thousandth of the basic unit. There are other metric prefixes, defined in multiples of a thousand, for units larger and smaller than we have discussed so far. Table 1 shows the complete list of metric prefixes, including the standard abbreviation and the power of ten for each.

Table 1
International System of Units (SI)-Metric Prefixes

Prefix	*Symbol*			*Multiplication Factor*
exa	E	10^{18}	=	1,000,000,000,000,000,000
peta	P	10^{15}	=	1,000,000,000,000,000
tera	T	10^{12}	=	1,000,000,000,000
giga	G	10^{9}	=	1,000,000,000
mega	M	10^{6}	=	1,000,000
kilo	k	10^{3}	=	1,000
hecto	h	10^{2}	=	100
deca	da	10^{1}	=	10
(unit)		10^{0}	=	1
deci	d	10^{-1}	=	0.1
centi	c	10^{-2}	=	0.01
milli	m	10^{-3}	=	0.001
micro	μ	10^{-6}	=	0.000001
nano	n	10^{-9}	=	0.000000001
pico	p	10^{-12}	=	0.000000000001
femto	f	10^{-15}	=	0.000000000000001
atto	a	10^{-18}	=	0.000000000000000001

Changing Metric Prefixes

With the metric system, the relative size of a measurement unit differs by powers of ten from other similar units. After you have learned the metric prefixes, it is quite easy to express a measured quantity using any of these prefixes. If we measure one distance as 12.5 kilometres and another as 100 decametres, how can we add these two numbers? Well, first we must change one of them, so we can specify both in the same units. We can change 12.5 km to 1250 dam, or we could change 100 dam to 1 km. (We'll give more details about making these conversions a little later.) Now we can add the numbers:

12.5 km	or	1250 dam
+ 1 km		+ 100 dam
13.5 km		1350 dam

These answers are equal to each other even though they seem to have different units. We could even change both measurements to some different unit, and then add. For example, we could choose to express both numbers in metres before adding them.

Right now, you may not be sure how to change from one metric prefix to another. From this example, however, you can see that the non-zero digits remain the same; only the decimal point moves. All metric prefixes are equivalent to a power of ten. Always change from one prefix to another simply by moving the decimal point.

There are two simple ways to make changes in metric system prefixes. The first one that we will cover involves using the power of ten factor for each metric prefix. Let's suppose you have several measurements all specified with different prefixes. You can convert them all to the basic unit simply by replacing the prefix with its power of ten. For example, here is a set of measurements that must all be specified with the same prefix for some further calculations:

0.05 megagrams, 125 kilograms, 25 grams and 45 decagrams. We can write these numbers with the power of ten factor instead of the metric prefix:

0.05×10^6 grams, 125×10^3 grams, 25 grams and 45×10^1 grams.

There is one snag that you still have to deal with after replacing the prefixes with their powers of ten. Before you can add or subtract the numbers, or do some other calculations, they also must have the same power of 10. That means you'll have to move the decimal points around a bit. For example, to go from mega (10^6) to kilo (10^3), there is a difference of 10^3. This is like multiplying by 1000, or moving the decimal point 3 places to the right. So 0.05×10^6 is the same as 50×10^3. (The decimal point moved 3 places to the right.)

Let's suppose you want to change from kilo (10^3) to mega (10^6). There is a difference of 10^{-3} between these two, which is the same as dividing by 1000. We could change 125×10^3 to 0.125×10^6 by moving the decimal point 3 places to the left. Table 1 (see page 4-6) summarizes the metric prefixes, their abbreviations and the powers of ten associated with each prefix.

Let's try changing the numbers in our example to several different powers of ten. First we'll change them all to the basic unit (10^0).

0.05 megagrams	=	0.05×10^6 grams	=	50000 grams
125 kilograms	=	125×10^3 grams	=	125000 grams
25 grams	=	25×10^0 grams	=	25 grams
5 decagrams	=	45×10^1 grams	=	450 grams

As another example, we'll move the decimal points to change all the original prefixes to kilo (10^3).

0.05 megagrams	=	0.05×10^6 grams		
	=	50×10^3 grams	=	50 kg
125 kilograms	=	125×10^3 grams	=	125 kg
25 grams	=	25×10^0 grams		
	=	0.025×10^3 grams	=	0.025 kg
45 decagrams	=	45×10^1 grams		
	=	0.45×10^3 grams	=	0.45 kg

By now you should begin to understand how to change prefixes in the metric system. First write the numbers with the proper powers of ten, and then move the decimal point. With a little more practice you'll be changing prefixes with ease.

The other method you can use to convert between metric prefixes involves a little trick. See Figure 1. Learn to write this chart on a piece of paper when you are going to make a conversion. Always start with the large prefixes on the left and go toward the right with the smaller ones. Sometimes you can make an abbreviated list, using only the

10^{18}		10^{15}		10^{12}		10^{9}		10^{6}		10^{3}	10^{2}	10^{1}	10^{0}	10^{-1}	10^{-2}	10^{-3}		10^{-6}		10^{-9}		10^{-12}		10^{-15}		10^{-18}
E	••	P	••	T	••	G	••	M	••	k	h	da	U	d	c	m	••	μ	••	n	••	p	••	f	••	a

Figure 1—This chart shows the symbols for all metric prefixes, with the power of ten that each represents. Write the abbreviations in decreasing order from left to right. The dots between certain prefixes indicate there are two decimal places between those prefixes. You must be sure to count a decimal place for each of these dots when converting from one prefix to another. When you change from a larger to a smaller prefix, you are moving to the right on the chart. The decimal point in the number you are changing also moves to the right. Likewise, when you change from a smaller to a larger prefix, you are moving to the left, and the decimal point also moves to the left.

10^3	10^2	10^1	10^0	10^{-1}	10^{-2}	10^{-3}
k	h	da	U	d	c	m

Figure 2—Sometimes you can use a smaller chart, listing only some of the prefixes. This is the case if you are working with numbers from a thousand to one one-thousandth of the basic unit.

units from kilo to milli, as shown in Figure 2. If you need the units larger than kilo or smaller than milli, be sure to include the dots as shown in Figure 1. (They mark the extra decimal places between the larger and smaller prefixes, which go in steps of 1000 instead of every 10.) Once you learn to write the chart correctly, it will be very easy to change prefixes.

Let's change 12.5 kilometres to metres, just for practice. Since we are starting with kilometres (kilo), start with the k on the chart. Now count each symbol to the right, until you come to the basic unit (U). Did you count three places? Well that's how many places you must move the decimal point to change from kilometres to metres. Which way do you move the decimal point? Which way did you move on the chart? To the right. Move the decimal point in the same direction. Now you can write the answer: 12.5 km = 12500 m!

Suppose you add the volume of a group of small containers, and find that they have a combined volume of 5,436,000 millilitres. How many litres is that? First, we will write the list of metric prefixes. Since we won't need those smaller than milli or larger than kilo, let's write the abbreviated list. See Figure 3. We won't write the powers of ten this time. To change from milli to the unit, we count 3 decimal places toward the left. This tells us to move the decimal point in our number three places to the left.

5,436,000 mL = 5,436 L

What other prefix might we conveniently use to express this volume? Probably the most obvious one would be kilo, although we could select any prefix that we wanted to use. Let's change this to a volume expressed in kilolitres. First count the places from the unit (U) to kilo (k) on our chart. We'll find that we must move the decimal point three more places to the left.

5,436 L = 5.436 kL

Let's try another example, for more practice changing metric prefixes by moving the decimal point in a number. For this example, we'll use a term that you will run into quite often in your study of electronics, *hertz*. Hertz (abbreviated Hz) is a unit that refers to the frequency of a radio or television wave. What if someone told you to tune your radio receiver to 75,635,000,000 Hz? You probably won't find any radio receiver with a dial marking like this! To make the number more practical, we'll write the frequency with a prefix that's more likely to appear on a receiver dial.

Figure 3—For this example, we will use the smaller chart. You'll have to move the decimal point three places to the left to change from the prefix milli to the basic unit. To change from the basic unit to kilo, you'll have to move the decimal point three more places to the left. That's a total of six places to change from milli to kilo.

Our first step is to select a new prefix to express the number. We can write the number with only one or two digits to the left of the decimal point. It looks like we'll need the entire prefix chart for this one, so write it down as described earlier. See Figure 4.

The next job is to count how many places you can move the decimal point. The number you end up with should have one or two digits to the left of the decimal point. Remember that metric prefixes larger than kilo represent multiples of 1000, or 10^3. Did you count ten places to move the decimal point in our example, 75,635,000,000 Hz? That would leave us with 7.5635×10^{10} Hz. There is no metric prefix for 10^{10}, so that won't be a convenient way to write this number. Since the larger metric prefixes represent multiples of 10^3, it would be convenient to move the decimal point nine places instead of ten. In that case we would write the number as 75.635×10^9 Hz.

Now we can go back to the chart of Figure 4 and count nine places to the left. (This is the same number of places and the same direction as we moved the decimal point.) The new spot on the chart indicates our new metric prefix. Figure 4 shows that the new prefix is giga, abbreviated G. Replacing the power of ten with this prefix, we can write our frequency as 75.635 GHz.

Follow the procedures described in this section and you will soon be changing metric prefixes with ease. Just remember to write the prefix chart and count off the number of places to move the decimal point.

Figure 4—For this example, we'll need the complete prefix chart. We must move the decimal point in a number nine places to the left, starting with the basic unit. We end up at the symbol G, which stands for the prefix giga. In this example, we are changing from the basic unit to a prefix of giga. 75,635,000,000 Hz becomes 75.635 GHz.

E • • P • • T • • G • • M • • k h da U d c m • • μ • • n • • p • • f • • a

CHAPTER 5 Basic Trigonometry

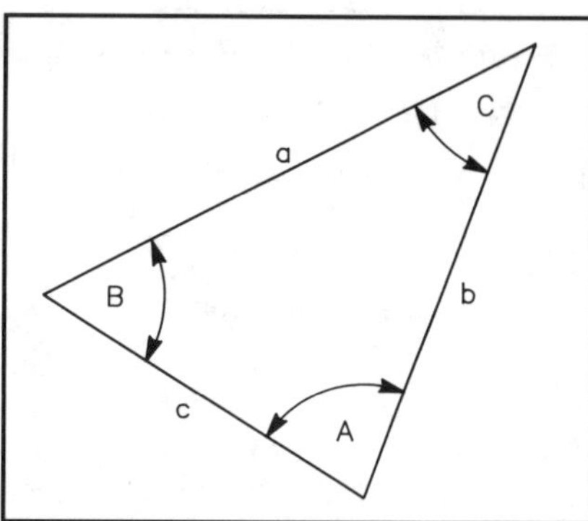

Figure 2—Label the sides of a triangle with small letters a, b and c. Then label the angles with capital letters A, B and C. An angle and the side across from the open end of that angle have the same letter.

Right Triangles

Draw two straight lines end to end, but at some angle so they do not form a single straight line. (See Figure 1A.) Now draw a third line between the free ends of the first two. These three lines form a *triangle*. (See Figure 1B.) When two lines touch each other they form an angle, so the three sides of a triangle form three angles. *Tri* means three. Triangle, then, means three angles. The number of degrees in the three angles of a triangle add up to 180 degrees. (There are 360 degrees around a complete circle.)

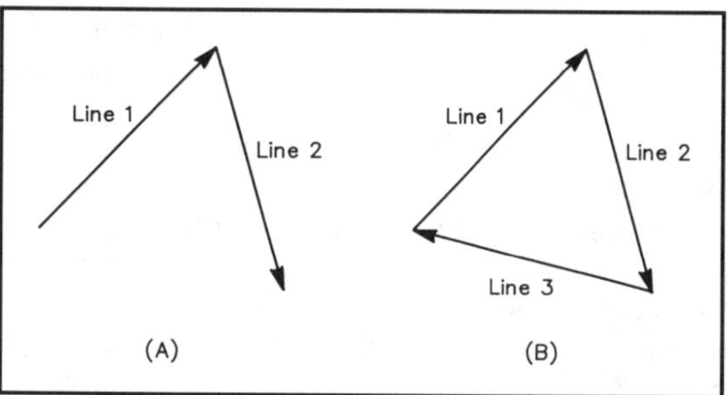

Figure 1—Three straight lines drawn end to end, so the last line comes back to the starting point, form a *triangle*.

We often label the sides of a triangle with lower case letters, a, b and c. Then we label the angles with upper case letters, as shown in Figure 2. Notice that the corresponding letters appear across from each other on the drawing. Angle A is across from side a so we usually say that side a is opposite angle A. (We also could say that angle A is opposite side a.) Likewise, B is opposite b and C is opposite c. It makes no difference which side of the triangle is labeled a, which is b or which is c. You can name the sides in any order you like, and can even choose other letters. Most of the time we label the longest side of the triangle as side c, however. Always be sure to use the same capital letter to represent the angle opposite any side.

If two lines form a 90° angle, we say they are *perpendicular* to each other. Perpendicular lines form a *right angle*. (See Figure 3A.) A right angle is a 90° angle. If the lines form an angle that is less than 90°, we have an *acute angle*. Figure 3B shows several acute angles. An angle that is greater than 90° is an *obtuse angle*. Figure 3C shows some lines that form obtuse angles.

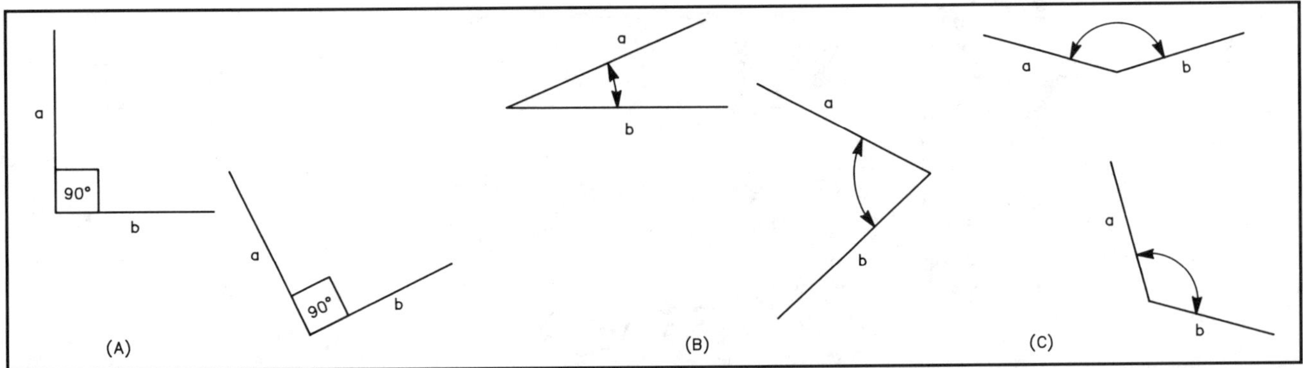

Figure 3—Whenever two straight lines meet, they form an angle. Part A shows *right angles*, which form 90 degree angles. Notice that we have indicated the right angles by marking a small square in the corner of the angle. The angles shown at B are *acute angles* because they are less than 90°. Part C shows two *obtuse angles*, or angles that are greater than 90°. Acute and obtuse angles have a small arc drawn inside the angle. The arc usually has arrow heads on both ends.

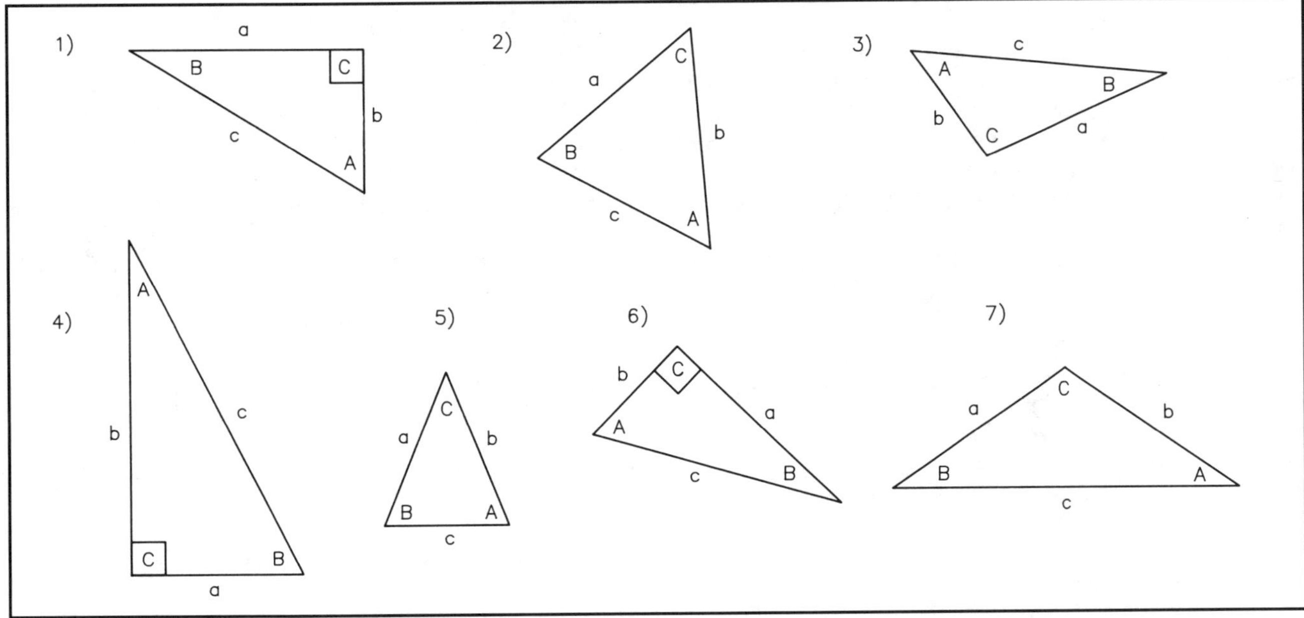

Figure 4—Triangles come in many shapes and sizes. This drawing shows a mixture of *right triangles*, *acute triangles*, and *obtuse triangles*. Can you identify each triangle by its type?

A curved line drawn inside the angle identifies acute and obtuse angles. This curved line has an arrow head on both ends. (Notice that we could measure another angle where the lines meet. For all the examples shown, that angle would be greater than 180°.) A small square drawn in the corner of right angles indicates that they are different from other angles. It might be difficult to tell the difference between a right angle and another angle between about 85 to 95 degrees otherwise. So this small square in the corner is helpful to positively identify a right angle.

Table 1
Types of Triangles Shown in Figure 4

Triangle	*Type*
1)	Right
2)	Acute
3)	Obtuse
4)	Right
5)	Acute
6)	Right
7)	Obtuse

We can form triangles that have many different shapes, as shown in Figure 4. We often describe a triangle by the type of angles it contains. For example, if a triangle contains only acute angles, we call it an *acute triangle*. If one of the angles is obtuse, however, then it is an *obtuse triangle*. By now you have probably figured out that if a triangle includes a right angle, then we call it a *right triangle*. Can you identify which of the triangles shown in Figure 4 are acute, which are obtuse and which are right triangles? Check your answers with the ones given in Table 1.

"Trig" Functions Defined

Many electronics problems lead to diagrams of right triangles for their solutions. This is why you will have to know some of the basics of working with right triangles. The branch of mathematics that deals with right triangles has a special name; *trigonometry*. Trigonometry gives us a way to calculate the lengths of the sides and the size of the angles in a right triangle.

Trigonometry involves a series of *functions* that relate the sides and angles of a right triangle to each other. A *function* is a relationship that, when applied to a number, gives another number in return. Let's make up a function, and call it SAMPLE. Table 1 shows the result of applying our function SAMPLE to the digits 1 to 10, and 100. Can you figure out how to calculate the results of SAMPLE for other vaules from this table? Did you notice that the result is always the input number multiplied by itself? SAMPLE gives us the square of the input number. You can find the result of applying SAMPLE to any value. Just multiply the input value by itself. We might write:

SAMPLE (2) = 4
SAMPLE (6) = 36
SAMPLE (20) = 400

Table 1
SAMPLE Function

Input Value (N)	*SAMPLE(N)*
1	1
2	4
3	9
4	16
5	25
6	36
7	49
8	64
9	81
10	100
100	10000

You should recognize that we can take the results of applying the function SAMPLE, and get the original number back. When we take the function value and go back to the original number, we are using the inverse of the function. When we take the inverse of a function, we might write that as:

Inverse SAMPLE (400) = 20
Inverse SAMPLE (900) = 30

The inverse of a function is often written with an exponent of –1 next to the function name:

SAMPLE^{-1} (100) = 10

Don't be confused by the word *function*. It is simply a way to describe a certain relationship between numbers. You can find the values for a certain function by looking the number up in a table. Most scientific calculators and computers can find the value that important functions associate with a given number. (They can find the inverse functions, too!)

Now, what are those trigonometric functions that we mentioned earlier? Well, we have the sine, cosine, tangent, cotangent, secant and cosecant. These are usually abbreviated sin, cos, tan, ctn, sec and csc. With a little practice you'll become quite familiar with the first three of these functions. Then you should be able to solve just about any electronics problem that involves a right triangle. (You'll solve most electronics problems without using ctn, sec or csc.)

Look at Figure 1. The side of the triangle that is opposite the right angle (the side labeled c) is the *hypotenuse* of the triangle. We know angle C is 90° because we are working with a right triangle. Next look at angle A. Side a is the *side opposite* the angle and side b is the *side adjacent* to the angle. If we look at angle B, side b becomes the *side opposite* and side a becomes the *side adjacent*. These terms are important. You must learn these definitions so you can understand the functions relating the sides and angles.

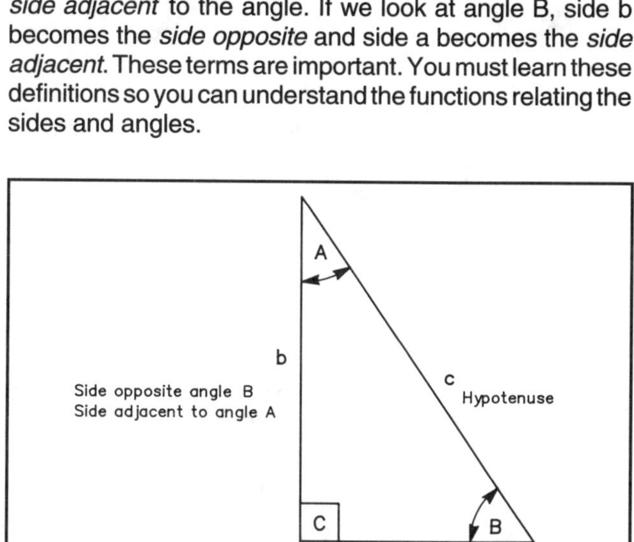

Figure 1—This diagram shows the parts of a right triangle. Angle C is the right angle and side c is the *hypotenuse* of the right triangle. Side a is the side *opposite* angle A and *adjacent* to angle B. Similarly, side b is adjacent to angle A and opposite angle B.

Now we can define the three trigonometric functions that we will use to solve electronics problems.

$$\text{sine} = \sin = \frac{\text{side opposite}}{\text{hypotenuse}} \qquad \text{(Equation 1)}$$

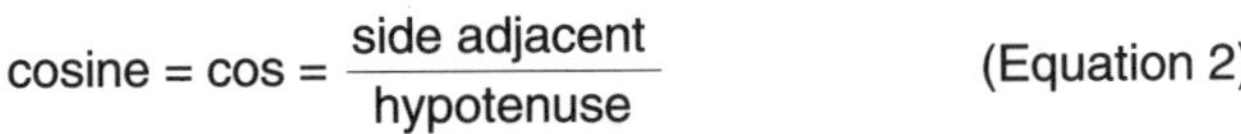

$$\text{cosine} = \cos = \frac{\text{side adjacent}}{\text{hypotenuse}} \qquad \text{(Equation 2)}$$

$$\text{tangent} = \tan = \frac{\text{side opposite}}{\text{side adjacent}} \qquad \text{(Equation 3)}$$

Each trigonometric function associates a particular number with any angle in a right triangle. Some reference books include tables of values for these trigonometry functions. You can use such "trig tables" to find a particular value. You won't need to use such tables, though, because your calculator will tell you what the values are. If you know the lengths of the triangle sides, the divisions indicated by Equations 1, 2 and 3 will give you these same numbers.

Table 2 lists values of these three trig functions for some angles. Don't try to memorize the values listed there.

Table 2
Values of Selected Trigonometry Functions

Angle (X)	*Sin(X)*	*Cos(X)*	*Tan(X)*
0°	0	1	0
10°	0.174	0.985	0.176
20°	0.342	0.940	0.364
30°	0.500	0.866	0.577
40°	0.643	0.766	0.839
45°	0.707	0.707	1
50°	0.766	0.643	1.192
60°	0.866	0.500	1.732
70°	0.940	0.342	2.747
80°	0.985	0.174	5.671
90°	1	0	Undefined at 90°

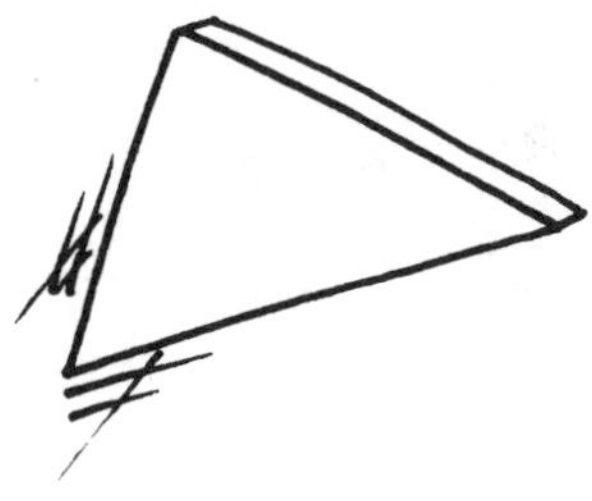

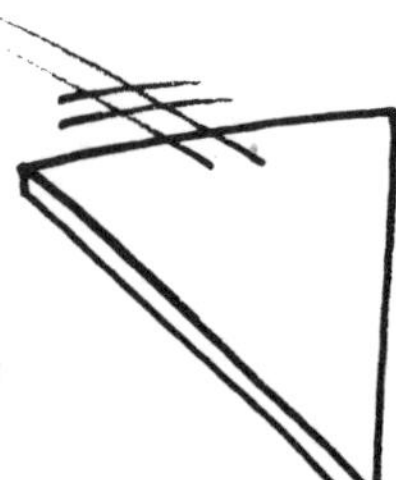

You can always find the values when you need to use them. Table 2 will help you see how the values of the functions change for various angles. It also will help you gain a better understanding of what a function is.

There is one more equation that proves to be very powerful for solving right triangle problems. This equation is the *Pythagorean Theorem*, named after a Greek mathematician, Pythagoras.

$$c^2 = a^2 + b^2 \qquad \text{(Equation 4)}$$

When you square a number (or raise it to the second power) you multiply the number by itself.

To calculate the length of any side we must solve the equation for that letter symbol. Solving the equation means getting that symbol on one side of the equal sign by itself. All the other symbols must be on the opposite side of the sign. Then we must take the square root of both sides of the equation.

For example, to calculate the hypotenuse (side c) we simply take the square root of Equation 4:

$$c = \sqrt{a^2 + b^2} \qquad \text{(Equation 5)}$$

Taking the square root (symbolized by the $\sqrt{\ }$ sign) is the inverse of the square function. You are finding the number that would give this value if you multiplied it by itself. Don't worry about how to find square roots. Your calculator has a button to do this for you. Just enter the value whose square root you want to know, and hit the $\sqrt{\ }$ key.

To find side a, first subtract b^2 from both sides of Equation 4:

$$c^2 - b^2 = a^2 + b^2 - b^2$$

Notice that $b^2 - b^2$ on the right hand side of the equal sign will eliminate b^2 there, leaving a^2 by itself. Next, take the square root of both sides, to solve the equation for side a:

$$\sqrt{c^2 - b^2} = a \qquad \text{(Equation 6)}$$

You can find a similar equation to calculate the length of side b. Follow the same procedure, but subtract the a^2 term from both sides of the equation instead of the b^2 term.

Are you ready to learn how these trigonometry functions can help us calculate the sides and angles of a right triangle? Turn the page and look at the next section. There are several examples there to show you how to calculate the sides and angles of any right triangle.

Using Trig Functions

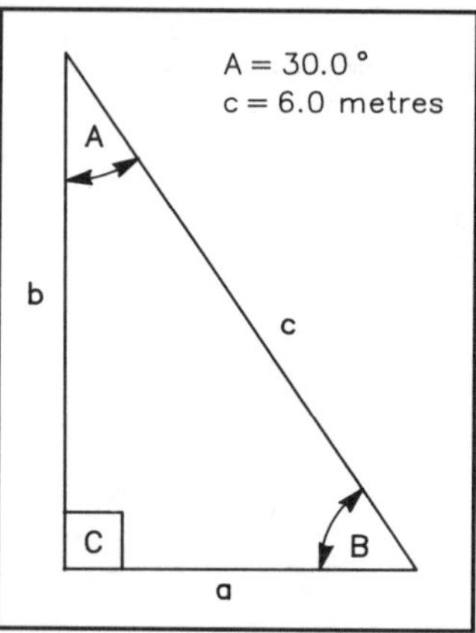

Figure 1—Here is a right triangle that has one angle of 90° and another one of 30°. We also know that the hypotenuse is 6.0 metres long. The examples in this section use trigonometry to calculate the other two sides and the third angle of this triangle.

How can we use trigonometry functions to calculate the parts of a triangle? Well, let's look at a couple of examples. Suppose that angle A of the triangle shown in Figure 1 is 30° and that the hypotenuse is 6 metres long. Since we know angle A and the hypotenuse of the triangle, the sine function will help us calculate the length of side a. From the definition of sine:

$$\sin(A) = \frac{\text{side opposite}}{\text{hypotenuse}} \qquad \text{(Equation 1)}$$

$$\sin(30°) = \frac{a}{6\text{ m}}$$

Now, let's move the length of the hypotenuse to the other side of the equal sign, so only the unknown remains on the right side:

$\sin(30°) \times 6\text{ m} = a$

What do we do with the sin(30°) part? This is where you turn to your trusty scientific calculator. Enter 30 on your calculator, then press the button labeled SIN. Does the display show 0.5 now? It should! Now press the x (multiplication sign) key on your calculator, then enter the 6. When you press the = key your display should show 3. Side a is 3 metres long.

How do we find the length of side b? Well, there are several ways to get that answer. One way is to use angle A and the cosine function:

$$\cos(A) = \frac{\text{side adjacent}}{\text{hypotenuse}} \qquad \text{(Equation 2)}$$

$$\cos(30°) = \frac{b}{6\text{ m}}$$

Next, solve for b the same way we solved for a in the previous example.

$\cos(30°) \times 6\text{ m} = b$

If you enter 30 into your calculator and then press the COS key, the display should show 0.866. (It probably shows more digits than this, but you could round the value off to three digits.) Next hit the x key, enter 6 and press the = button. The answer should be 5.196 metres. (That rounds off to 5.20 metres.)

Well, now we know all three sides of our triangle, but we only know two of the angles. What about the third angle? Again, there are several ways that we could go about calculating that angle. Perhaps the easiest method is to remember that all three angles total 180°. One angle is a right angle (90°) and the other is 30°. We can calculate the third angle from this information.

First, let's add the two angles we know:

$$\begin{array}{r} 90° \\ +\ 30° \\ \hline 120° \end{array}$$

Next, subtract this total from 180° to find the third angle.

$$\begin{array}{r} 180° \\ -120° \\ \hline 60° \end{array}$$

Now we know all three angles and all three sides of our right triangle.

What other ways could we have chosen to solve this problem? If we had found the third angle first, we could have used the sine and cosine of 60° instead of 30°. We also could use the tangent function to find the last side of the triangle.

$$\sin(B) = \frac{\text{side opposite}}{\text{hypotenuse}}$$

$$\sin(60°) = \frac{b}{6\text{ m}}$$

$\sin(60°) \times 6\text{ m} = b$

$0.866 \times 6\text{ m} = 5.196\text{ m}$

(This is the same value we found for side b using the cosine function and the 30° angle.)

Now let's use the tangent function to solve for the last side:

$$\tan(A) = \frac{\text{side opposite}}{\text{side adjacent}} \qquad \text{(Equation 3)}$$

$$\tan(30°) = \frac{a}{b}$$

$$\tan(30°) = \frac{a}{5.196\text{ m}}$$

$\tan(30°) \times 5.196\text{ m} = a$

$0.577 \times 5.196\ \text{m} = 3.00\ \text{m}$

(The value of 3.00 comes from rounding off the calculator's answer—2.998—to 3 digits.)

If we know the two sides of a right triangle, we can find the hypotenuse by using the Pythagorean Theorem. Suppose we know that the two sides of a right triangle are 3.00 m and 5.196 m. What is the hypotenuse?

$$c = \sqrt{a^2 + b^2} \qquad \text{(Equation 4)}$$

$$c = \sqrt{(3.00\ \text{m})^2 + (5.196\ \text{m})^2}$$

$$c = \sqrt{9.00\ \text{m}^2 + 27.0\ \text{m}^2}$$

$$c = \sqrt{36.0\ \text{m}^2} = 6.00\ \text{m}$$

Notice that this answer agrees with the value given for the hypotenuse in the original problem.

Use the ideas presented in this section to help you solve for the parts of a right triangle. Your calculator will do the math and find the values for the trig functions. You will run into many problems that involve solving right triangles as you learn more about electronics.

Rectangular Coordinate Systems

Numbers help us specify measured or calculated quantities. In working with those numbers, it is often helpful to draw a picture or graph to show the relationships between the numbers. A *coordinate system* is a type of graph scale, which provides a way for us to draw such a picture.

A *coordinate system* uses a set of numbers to represent the location of a point on a surface. Plotting a group of *coordinates* or numbers on the scale can help you recognize patterns and relationships that you might otherwise miss. These patterns can be especially helpful in solving some kinds of problems.

One coordinate system that is often helpful in solving electronics problems is the *rectangular coordinate system.* Figure 1 will help you understand how this system got its name. You can specify a point on such a graph by giving the distance from each *axis* or base line. We often refer to the horizontal axis as the X axis and the vertical axis as the Y axis.

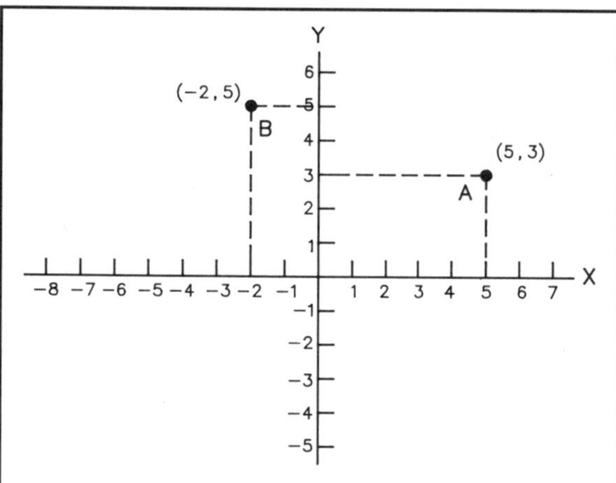

Figure 1—A set of horizontal and vertical lines, marked off with a scale, forms a rectangular coordinate system. We usually label the horizontal line, or axis, as the X axis. Then we call the vertical axis the Y axis. By drawing lines from a point on the graph at right angles to the X and Y axes, we can form a rectangle. This is how the rectangular coordinate system gets its name.

Suppose you want to specify a point in the rectangular coordinate system, such as point A shown in Figure 1. First, give the distance along the horizontal axis, then the distance along the vertical axis. The coordinates for point A, then, are (5, 3). This means that the point is 5 units along the horizontal axis and 3 units along the vertical axis.

Notice that point B is on the negative side of the horizontal, or X axis. The X coordinate for point B will have a minus sign in front of it. We can express the location of this point as (–2, 5).

The scale on the graph of a rectangular coordinate system can have just about any value that is practical. The horizontal and vertical scales do not have to be identical. You can choose the actual values for the scale to help solve a particular problem. Figure 2 shows an example of a coordinate system with different scale factors.

A rectangular coordinate system extends as far as you can imagine in all directions. You can't possibly show the entire system on a piece of paper. We normally show only the portion of a graph that includes the numbers needed to represent a solution to a problem. Figure 3 shows a portion of a coordinate system that is far from the center, or *origin* as we call it.

Most of the time we will be working with a two-dimensional coordinate system. That means there are only two numbers to specify any point on the graph. It also means that the graph is a flat surface. We could add a third dimension to our rectangular coordinate system. All that we need is a third line or axis that forms a 90° angle with

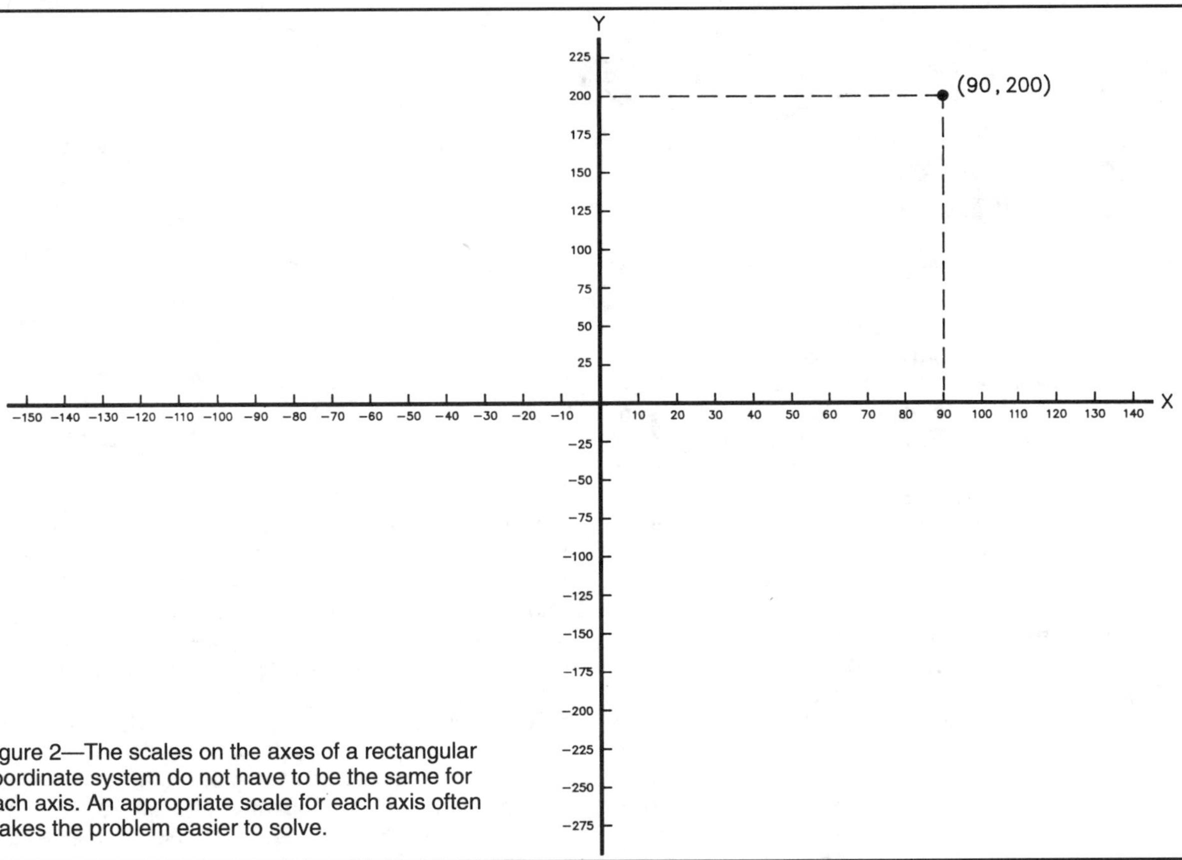

Figure 2—The scales on the axes of a rectangular coordinate system do not have to be the same for each axis. An appropriate scale for each axis often makes the problem easier to solve.

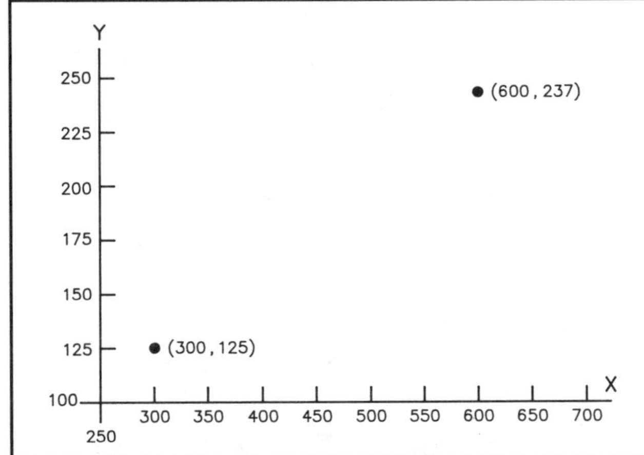

Figure 3—Normally, we only show a small portion of the complete coordinate system. Depending on the problem, you may need to show a section of the system that is far from the center, or origin.

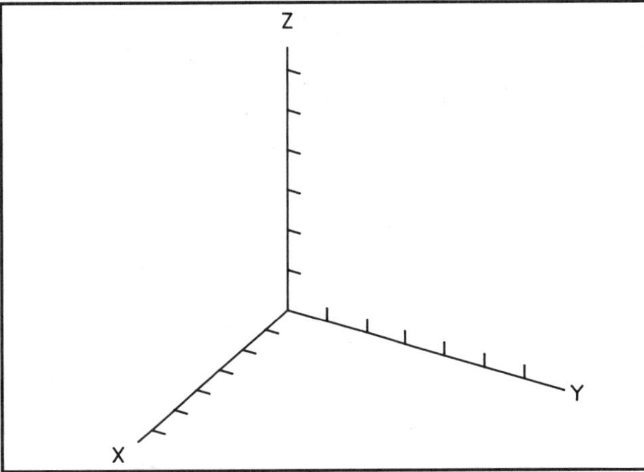

Figure 4—If you draw three axes, each at right angles to the others, you will have a three-dimensional coordinate system. Such a system can be useful for representing a spot in a room or a location in space.

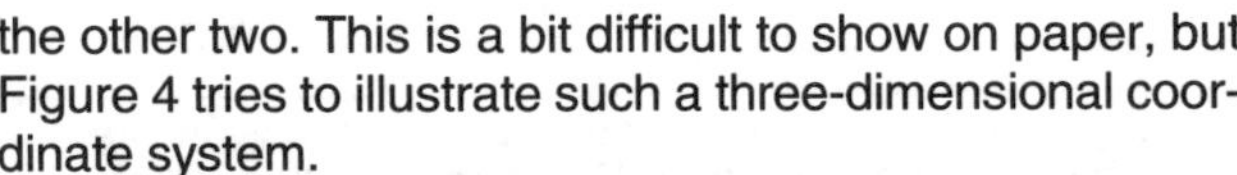

the other two. This is a bit difficult to show on paper, but Figure 4 tries to illustrate such a three-dimensional coordinate system.

Another way to picture a system like this is to look at one corner in a room. Where the walls touch the floor, there are two lines that form a 90° angle. These are like the ones we might draw on a piece of paper for the X and Y axes of a coordinate system. The line formed where the walls meet is like the third axis at a 90° angle to the other two. You could use this system to specify any spot in your room. Just measure how far that spot is along both walls and how high it is off the floor.

Polar Coordinate Systems

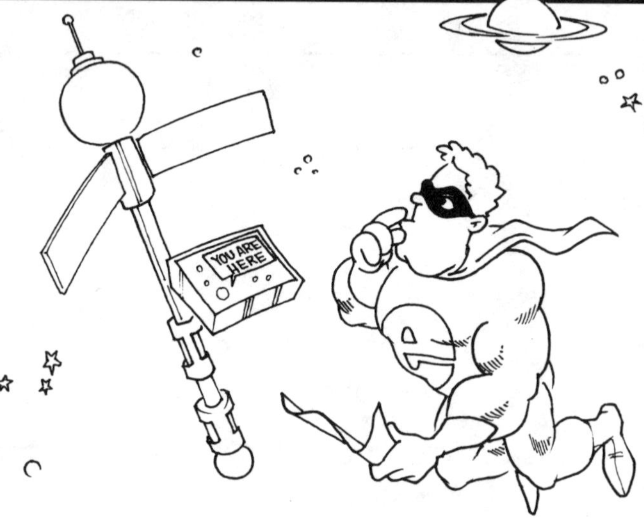

We should consider one of the limitations of using a rectangular coordinate system. The numbers that specify a point on the graph represent the distance from the origin to that point. We always measure this distance along the X coordinate and then along the Y coordinate. We all know that the shortest distance between two points is a straight line. So a rectangular coordinate system does not tell us the shortest distance from the origin to the point. That is one disadvantage of using the rectangular coordinate system.

There is a coordinate system that does give the shortest distance between the origin and a point on the scale. The *polar coordinate system* uses a series of circles drawn from a common center to measure distance. See Figure 1. The radius of each circle represents a distance from the origin. So the distance to point A in Figure 1 is 5, indicated by the circle it falls on.

By now you've probably realized that you can't just give the distance from the origin to specify the location of a point. After all, the point could be any place along that circle. We also must specify an angle, measured from a reference line. As Figure 1 shows, this reference line usually extends horizontally to the right from the origin. We normally measure the angle counterclockwise from that line. We often include crossed horizontal and vertical lines with the polar coordinate scale. These lines provide reference points for the 0°, 90°, 180°, 270° and 360° angles. Point A in Figure 1 is at an angle of 45°, so we specify its coordinates as (5, /45°).

Take a look at point B in Figure 1. You'll notice that this point is on the circle with a radius of 3. You also should notice that this point is more than half way around the circle from the reference line. It's at an angle of 210°. You would give the coordinates for point B as (3, /210°).

The radius scale on the graph of a polar coordinate system can have just about any value that is practical. The actual values that you choose should be convenient for solving the problem you are working on. Of course, the angles will always vary from 0° to 360°, counterclockwise around the circle.

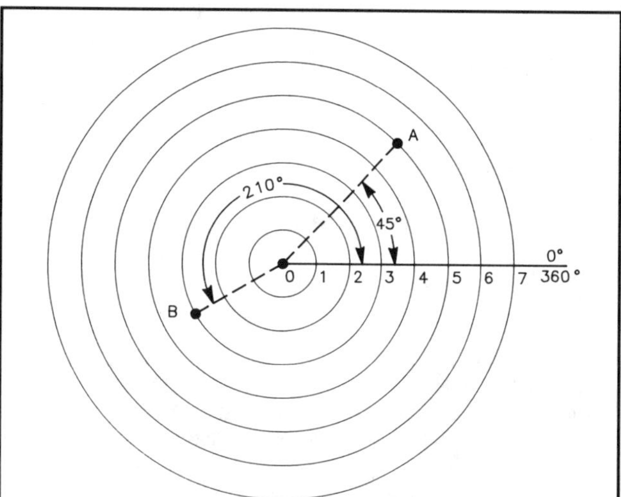

Figure 1—This drawing shows a *polar coordinate system.* A coordinate system like this helps us specify the location of a point as measured from the origin or center point. One coordinate specifies the distance measured along a straight line from the origin to the point. We also must specify an angle, measured counterclockwise from a reference line that extends horizontally to the right of the center point.

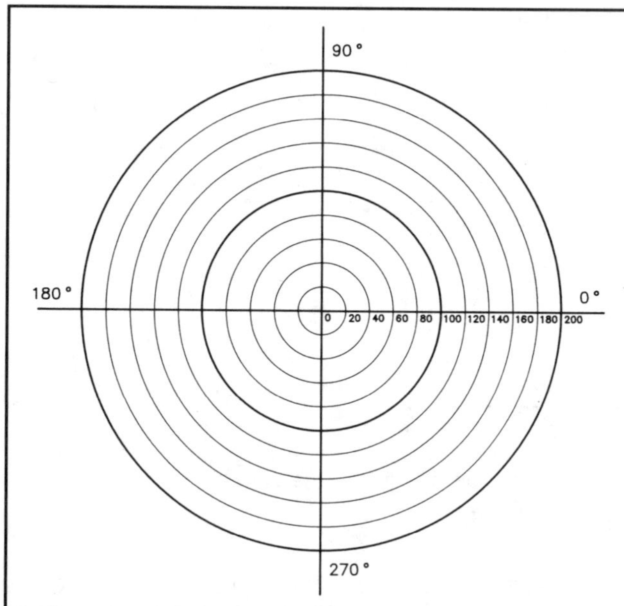

Figure 2—You can choose a scale that is appropriate for the problem you are trying to solve. There are always 360° around the complete circle, no matter what radius scale you choose, however. So the angle measurement won't change from one graph to another.

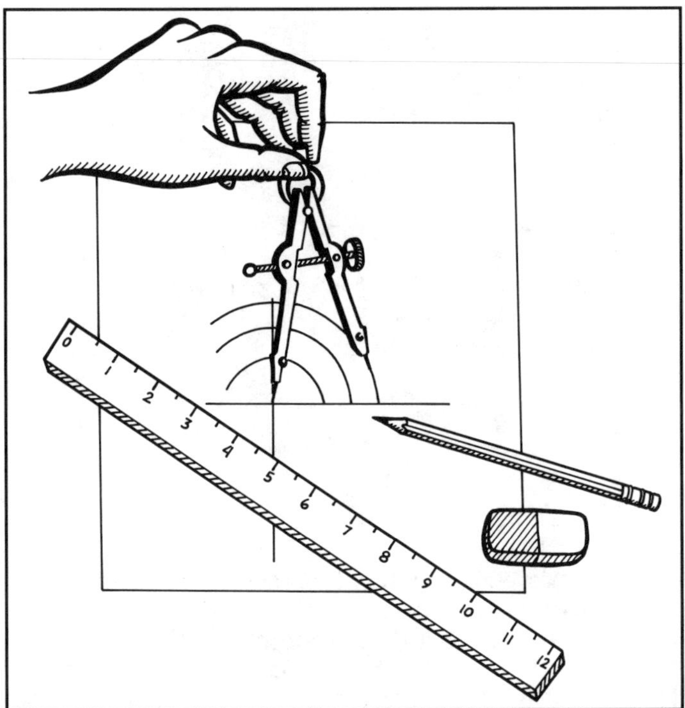

Figure 3—A drawing compass can be most helpful for drawing an *arc*, or part of a circle.

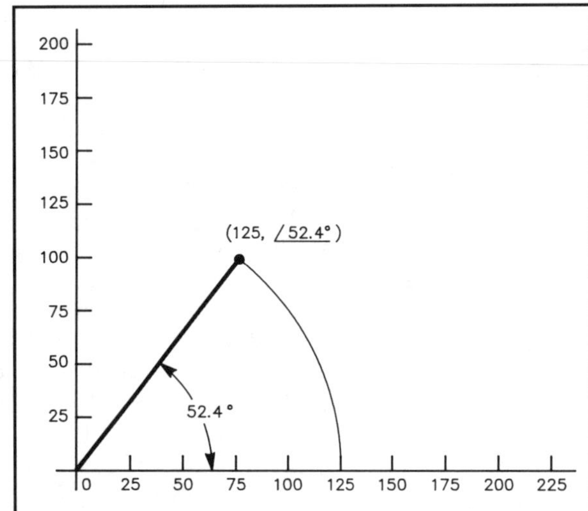

Figure 4—You don't always have to show the complete circle around a polar coordinate system. When making the sketch to solve a problem, you often need only half or one quarter of the entire circle. Similarly, you often don't need to draw in the circles. First, mark a scale along the crossed lines that go through the center of the graph. Then you can use that scale to measure the distance to any point on the graph.

You can always add more circles to your polar coordinate graph, making the system larger. The scale you choose limits the portion of the coordinate system that you can show on a certain piece of paper. Figure 2 shows another possible choice of values for a polar coordinate scale.

You can buy graph paper marked with a polar coordinate system. You don't need graph paper for most electronics problems that you will have to solve, however. Quite often a simple sketch is all you will need. In fact, most of the time you won't even have to draw complete circles on the graph. You can simply mark a scale along the crossed lines through the center of the circle. Then you can use those marks to find which radius line your point falls on. A drawing compass can be helpful for marking the circles or measuring the distance to a point.

It's not always convenient to show a full circle around the center point. Sometimes you'll know that all values for a problem are likely to stay within a certain half or quarter of the circle. That's the only part of a polar coordinate system you'll have to show to solve the problem then. Figure 4 shows a polar coordinate graph with the scale indicated along the crossed lines through the center of the graph.

The polar coordinate system is a two-dimensional, or flat-surface system. We can add a second direction of rotation to make a three-dimensional system. This enables us to describe the position of objects around us. Such a *spherical coordinate system* is especially useful when it comes to describing our universe. With a spherical coordinate system, we can give the positions of stars and planets in relation to the earth. Figure 5 shows a spherical coordinate system. You'll have to use a bit of imagination when you look at this drawing, though. It tries to show a three-dimensional view on a two-dimensional surface, and that can get confusing. Don't worry too much if you don't understand how to specify a location with a system like this, though. You won't run into many electronics problems that require spherical coordinates for a solution!

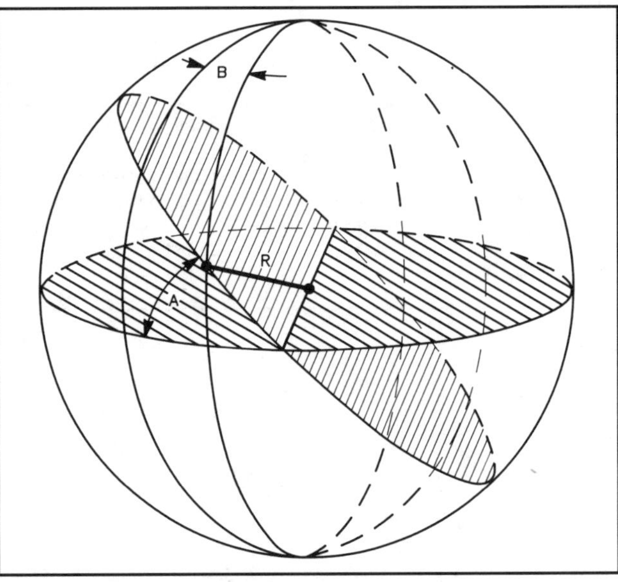

Figure 5—To form a three-dimensional coordinate system, start with a polar coordinate system. Then add another direction of rotation that comes out of the page. This new system is the *spherical coordinate system.* Such a system allows you to specify locations in our three-dimensional world. This is especially helpful for describing locations on or above the earth, including other stars and planets in our universe.

Converting Between Coordinate Systems

Have you been studying this chapter from the beginning? If so, you're probably wondering what rectangular and polar coordinate systems have to do with trigonometry. Well, read on, and we'll answer that question now!

You can solve many electronics problems rather easily by making a graph or drawing. The lines of the graph represent the electrical quantities involved with a particular problem. For example, Figure 1 shows a simple circuit diagram. There is a voltage source, or signal generator, a resistor, and a coil or inductor. Someone measured the voltages across the resistor and inductor for us, and wrote them on the diagram. The voltage across the resistor, E_R, is 26.0 volts and the voltage across the inductor, E_L, is 15.0 volts. Our problem is to find the generator voltage.

Now look at Figure 2. The line along the X axis represents the voltage across the resistor. The line along the Y axis represents the voltage across the inductor. A line drawn from the origin to point P will represent the voltage applied to the circuit. (You'll learn more about problems like this and how to solve them in a later section of this book.)

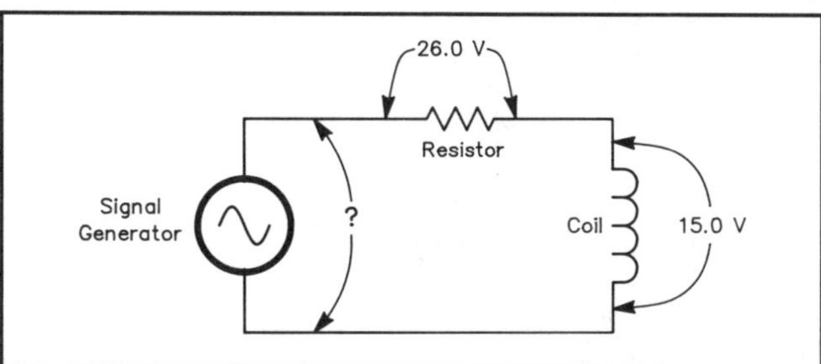

Figure 1—Here is a simple electronics circuit. It consists of a signal generator or voltage source, a coil or inductor and a resistor. The diagram shows the voltages measured across the coil and resistor in this circuit. Our task is to calculate the voltage produced by the generator.

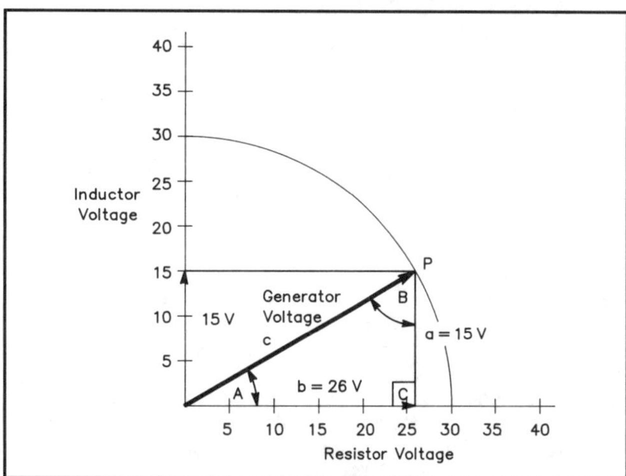

Figure 2—This graph represents the relationship between the voltages found in the circuit of Figure 1. Side a represents the voltage across the coil and side b represents the voltage across the resistor. Side c is the hypotenuse of a right triangle, and represents the voltage supplied by the generator. Angle A is also of interest in problems like this. You'll learn more about why this angle is important in later sections of this book.

Figure 2 is a graph that represents the relationship between the voltages found in the circuit of Figure 1. You've probably recognized that we have a right triangle here. The hypotenuse of that triangle represents the voltage applied to the circuit. By using the rectangular coordinate system, we have drawn a graph of our problem. If you are careful enough when you draw such a graph, you can measure the length of side c along the scale. We also need to know angle A on Figure 2. Once again, we could measure this angle on the drawing, if it were carefully drawn on graph paper. We would need a special tool, called a protractor, to make this measurement. Figure 3 shows how you could measure an angle with a protractor. Graphical solutions to problems like these are not difficult as long as you measure and draw the diagram carefully. You can find an answer that is a bit more accurate, though, by doing a simple trigonometry calculation.

In this problem you were given the lengths of two sides of a right triangle and asked to find the hypotenuse. Perhaps the most direct way to calculate the hypotenuse is by using the Pythagorean Theorem. Do you remember that equation?

$$c = \sqrt{a^2 + b^2} \qquad \text{(Equation 1)}$$

$$c = \sqrt{(15.0\ \text{V})^2 + (26.0\ \text{V})^2}$$

$$c = \sqrt{225\ \text{V}^2 + 676\ \text{V}^2}$$

$$c = \sqrt{901\ \text{V}^2} = 30.0\ \text{V}$$

Next, we need to calculate angle A. We can calculate this angle by using one of our trigonometric functions. Often, you will have to solve a problem for several quantities. It's always a good idea to use the original information for as many parts of the solution as possible. That way, if

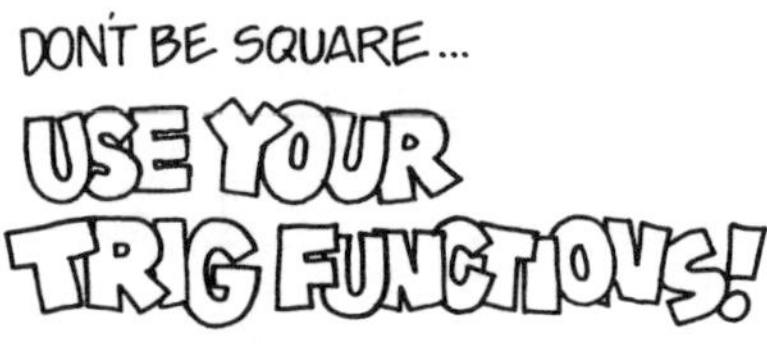

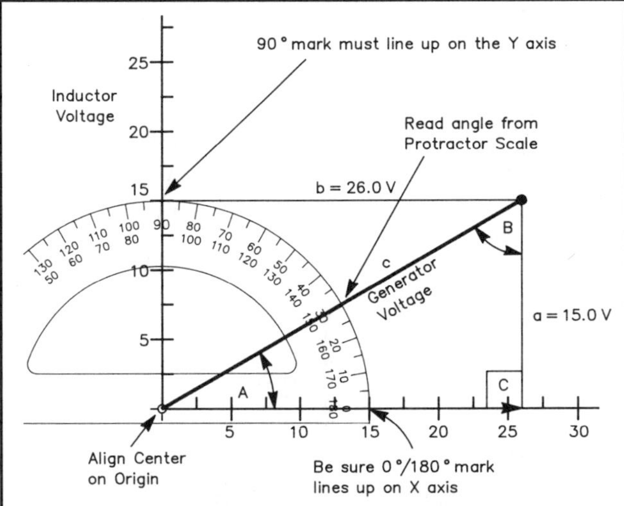

Figure 3—You can use a protractor to measure the angles of a triangle on a carefully measured and accurately drawn graph or diagram. Such a graphical solution to a problem may provide a check on your mathematical calculations. Carefully align one edge of the protractor with one side of the angle. Be sure the center mark on the protractor lines up exactly over the point where the two sides of the angle meet. Then the line from the second side of the angle will cross the degree markings along the round side of the protractor. You simply read off the mark that matches the line to measure the angle.

you make a mistake in one calculation, that error won't carry through the rest of your solution. The tangent function seems to fit this requirement because it relates one angle to the two sides of a right triangle. That's just the information given at the start of our problem.

$$\tan (A) = \frac{\text{side opposite}}{\text{side adjacent}} \qquad \text{(Equation 2)}$$

$$\tan (A) = \frac{15.0\ \text{V}}{26.0\ \text{V}} = 0.577$$

Now we have to find the inverse of the tangent. When you take the inverse of this function, you are really looking for the angle that has 0.577 for its tangent. Look for the key labeled INV, ARC or 2ND. The owner's manual for your calculator should tell you which keys to press to find the inverse trig functions. When we write the inverse tangent in an equation, we use an exponent of –1 with the tangent function. The –1 means that we are taking the inverse of the function.

$\tan^{-1} (\tan (A)) = \tan^{-1} (0.577)$

$A = 30.0°$

It's always a good idea to check your work. We can use the sine function to double check the hypotenuse, which represents the voltage applied to our circuit.

$$\sin (A) = \frac{\text{side opposite}}{\text{hypotenuse}} \qquad \text{(Equation 3)}$$

$$\sin (30.0°) = \frac{15.0\ \text{V}}{c}$$

Solve this equation for c by moving the c and sin(30.0°) terms diagonally across the equal sign.

$$c = \frac{15.0\ \text{V}}{\sin (30.0°)}$$

$$c = \frac{15.0\ \text{V}}{0.500} = 30.0\ \text{V}$$

Since this agrees with the answer we calculated by using the Pythagorean Theorem, we can be pretty sure it is correct. We could represent the voltage applied to our circuit with the coordinates expressed in the rectangular coordinate system: (26.0, 15.0 volts). Most of the time, however, we prefer to express the voltage in its polar-coordinate form: (30.0 volts, $\angle 30.0°$).

There is a very interesting thing that you should notice about this problem. We started by using a rectangular coordinate system to represent two voltages. We prefer to express the answer to the problem in polar-coordinate form, however. We used the trigonometry functions to convert from the rectangular coordinate system to the polar coordinate system. That conversion was all we needed to solve our problem. You'll find other electronics problems with information you can represent in polar coordinate form. Those will often require you to convert to rectangular coordinates to express the answer. You can solve many electronics problems in this simple, direct manner.

Solving Trigonometry Problems on Your Calculator

You learned about trigonometry functions earlier in this Chapter. Your scientific calculator helped you solve a few "trig" problems. Chances are, you could use more practice with your calculator, though. In that case, here is just what you need.

The first step is to be sure you can find all the necessary keys on your calculator. Look for keys labeled SIN, COS and TAN. These will give the values for the sine, cosine and tangent functions at any angle.

You also will need to find the keys that give the inverse trig functions. Many calculators have an INV or ARC key. Hitting this key just before the SIN, COS or TAN key gives the angle associated with that function value.

We often write the inverse trig functions as *arcsine*, *arccosine* and *arctangent*. Sometimes we also use an exponent of –1 with the function name to write the inverse function. Equation 1 is an example of how we use this exponent.

$$\mathrm{SIN}^{-1}(0.707) = \mathrm{ARCSIN}(0.707) = X = 45^\circ \quad \text{(Equation 1)}$$

Some calculators use a button labeled 2ND to call functions printed above the keys. In that case, look for ARCSIN, ARCCOS and ARCTAN labels above the SIN, COS and TAN keys.

Figure 1 lists 10 angles. Use your calculator to find the values of the sine, cosine and tangent functions for those angles. Figure 4 shows the answers to the Figure 1 problems. Look at Figure 4 only after you find all the answers on your calculator. Then check your work.

Figure 2 lists 10 trigonometry functions. For each value of sine, cosine or tangent, find the angle that corresponds to that function. Figure 5 lists the answers to these problems.

Figure 3 shows 4 triangles. There is some information about each triangle, but not all three sides and angles. Use your knowledge of trigonometry and your calculator to find the missing information about each triangle. Compare your answers with the ones given in Figure 6. Each solution shown in Figure 6 represents one of many possible ways to find the values for that triangle. You may find the quantities in a different order. You may use different equations to find the answers. All solutions should lead to the same results, however.

	Angle	SINE	COSINE	TANGENT
1)	10°	_____	_____	_____
2)	25°	_____	_____	_____
3)	30°	_____	_____	_____
4)	40°	_____	_____	_____
5)	45°	_____	_____	_____
6)	50°	_____	_____	_____
7)	60°	_____	_____	_____
8)	65°	_____	_____	_____
9)	70°	_____	_____	_____
10)	80°	_____	_____	_____

Figure 1—Use your scientific calculator to find the value of the sine, cosine and tangent function for each angle.

1) Find the angle whose sine is 0.2588. **arcsine(0.2588)** = ___
2) Find the angle whose cosine is 0.2588. **arccosine(0.2588)**=
3) Find the angle whose tangent is 4.7046.**arctangent(4.7046)** =
4) What angle has a cosine of 0.7880? **cos^{-1}(0.7880)** = _____
5) What angle has a sine of 0.8910? **sin^{-1}(0.8910)** = _______
6) Find the angle whose cosine is 0.01745. **arccos(0.01745)** =
7) What angle has a tangent of 3.4874? **tan^{-1}(3.4874)** = _____
8) What angle has a sine of 1.000? **sin^{-1}(1.0000)** = ________
9) Find the angle whose tangent is 1.0000. **tan^{-1}(1.0000)** = __
10) What angle has a cosine of 1.0000? **cos^{-1}(1.0000)** = _____

Figure 2—Your calculator will help you find the angle associated with each of these trigonometry functions. You must find the inverse functions, or arc functions to answer these questions.

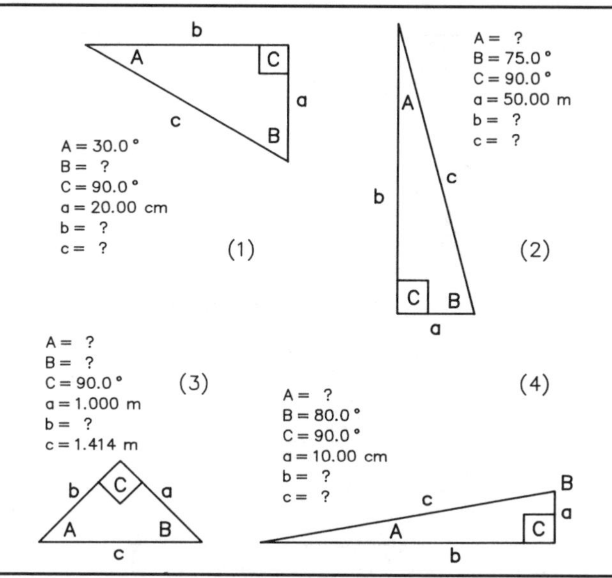

Figure 3—Find the missing information for each of these triangles. Use your knowledge of trigonometry to find this information. Your scientific calculator will help you find the trigonometry-function values.

	Angle	SINE	COSINE	TANGENT
1)	10°	0.1736	0.9848	0.1763
2)	25°	0.4226	0.9063	0.4663
3)	30°	0.5000	0.8660	0.5774
4)	40°	0.6428	0.7660	0.8391
5)	45°	0.7071	0.7071	1.0000
6)	50°	0.7660	0.6428	1.1918
7)	60°	0.8660	0.5000	1.7321
8)	65°	0.9063	0.4226	2.1445
9)	70°	0.9397	0.3420	2.7475
10)	80°	0.9848	0.1736	5.6713

Figure 4—This table lists the sine, cosine and tangent functions for each of the angles given in Figure 1. Your calculator probably displayed more digits than shown here. Your answers should agree with these when you round them off to 4 digits.

1) arcsine(0.2588) = 15°
2) arccosine(0.2588) = 75°
3) arctangent(4.7046) = 78°
4) $\cos^{-1}(0.7880) = 38°$
5) $\sin^{-1}(0.8910) = 63°$
6) arccos(0.01745) = 89°
7) $\tan^{-1}(3.4874) = 74°$
8) $\sin^{-1}(1.0000) = 90°$
9) $\tan^{-1}(1.0000) = 45°$
10) $\cos^{-1}(1.0000) = 0°$

Figure 5—This Figure shows the angles associated with each of the trigonometry functions listed in Figure 2. Your calculator may display more digits than shown here. Your answers should agree with these when you round them off, however.

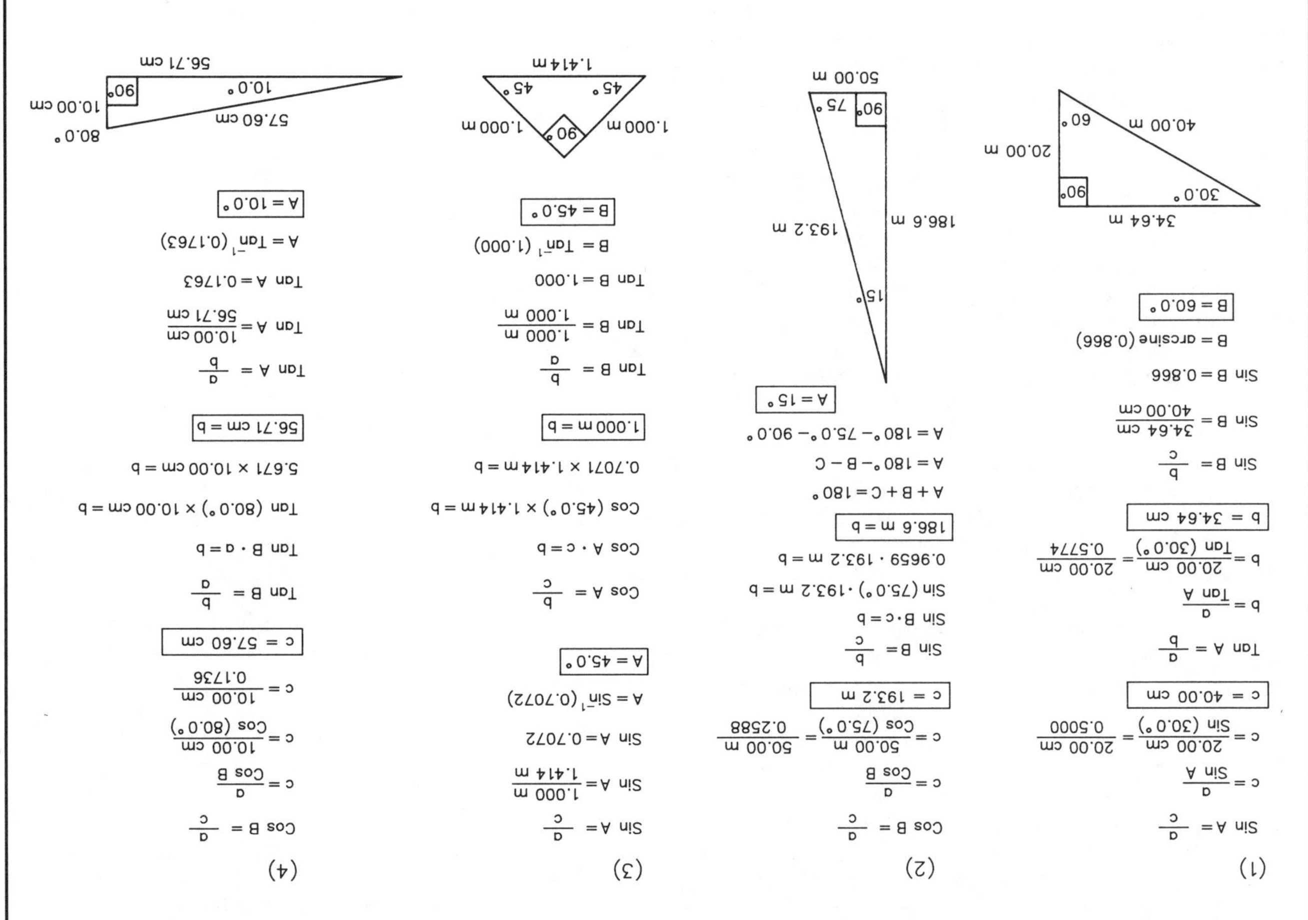

Figure 6—The triangles from Figure 3 are shown here with all the missing information filled in. The solutions shown represent one of many possible ways to solve each problem. You may use different trigonometry functions to find the triangle's sides and angles. You might even find the information in a different order. The answers should be the same, however.

Trig Problems and Your Computer

Earlier in this chapter you learned about trigonometry functions and how to use them. There were even some practice problems for you to solve using your scientific calculator. What if you don't have a calculator, though? Perhaps you have a personal computer, and want to use it to solve the problems. Even if you have a calculator, you may still want to know how to solve trig problems with your computer.

Almost any personal computer can solve trigonometry problems. After all, a computer is a much more powerful machine than a calculator. You may have to learn a few additional skills, however. With the help of this book, and with your computer manuals as a guide, you will soon have your computer solving trigonometry problems.

There are many personal computers available. Each computer brand is slightly different from the others. We can't give you specific information about using your computer in this book. That's the computer manual's job. We can give you general guidelines that you can use with nearly any computer.

Several methods can help you solve these problems. Probably the most common one is to write programs in the BASIC computer language. BASIC is an acronym for Beginner's All-purpose Symbolic Instruction Code. Each computer's BASIC is a little different. Some computers have several versions of BASIC available.

Most personal computers have one or more *spreadsheet* programs available. A spreadsheet program creates a large table in your computer and displays part of it on your screen. Each section of the spreadsheet table is a *cell* that allows you to enter text or equations for calculations. Spreadsheets have trigonometry functions and other functions built in, so you just type the proper command as part of an equation.

Some computers have calculator programs available. These may be individual programs, or they may come as part of a package with another program. Calculator programs are most common with the IBM PC and compatible computers.

If you have a calculator program for your computer, try it out. Is it a scientific calculator? Some are only simple math calculators. Some are financial or statistical calculators that include many special features for those intended purposes.

A scientific-calculator program is probably the simplest way to use your computer to solve trigonometry problems. Try solving the problems in the last section with your computer calculator. It should work similarly to a small calculator. You enter the numbers and press the keys associated with the various functions shown on the screen.

With a spreadsheet program, you type an equation into one

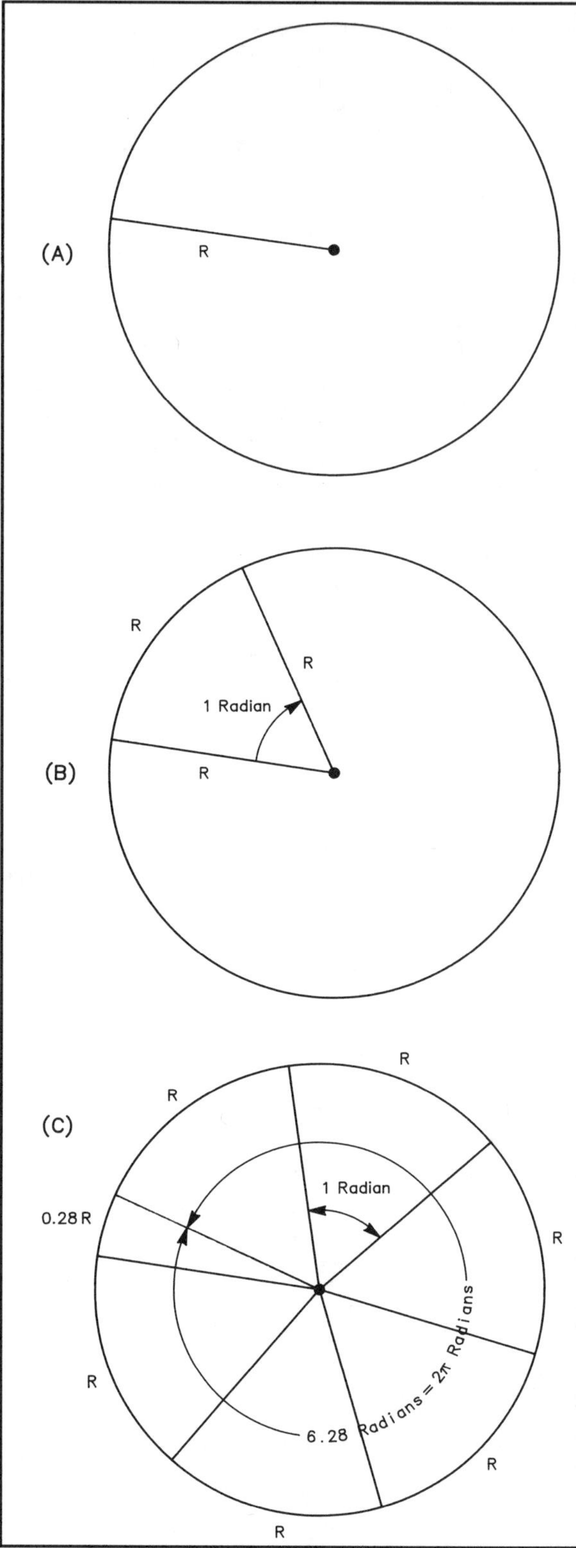

Figure 1—Part A shows a circle of radius R. The pie-shaped piece marked out on the circle at Part B defines an angle of 1 radian. You can continue marking off the length of the radius along the circumference as shown in Part C. The radius will reach around the circle 6 times. You will need an extra 0.28 times the radius to fill in the last section.

or more cells. When you hit the ENTER key, the computer displays the results of your equation in the cell.

Most spreadsheets use the @ character to identify a mathematical function. Typing @SIN, @COS or @TAN calls those trigonometry functions. You call the inverse functions by typing @ASIN, @ACOS or @ATAN. These functions also need an *argument.* The *argument* of a function is that part on which the function acts. It is the angle for a trigonometry function or the function value for the inverse functions.

For example, you might type @ASIN(0.500) to find what angle has a sine of 0.500. Your computer probably won't print 30.0 degrees as the answer to this problem. Chances are, it will display 0.5236 (at least if you round the answer off to four places). That's because computers don't normally think in degrees. They use another angle measurement, called *radians.* You can easily teach your computer to convert its answer into degrees, though. Trig functions in the BASIC language also work with angles measured in radians. Continue reading to learn about radians. You will use the same methods for your spreadsheet program or for a BASIC program.

You can use the BASIC computer language two ways. You can type a single BASIC instruction, and have the computer execute that command immediately. You also can type a series of commands to create a *program.* Then you can save the program for later use. A program executes each step in sequence, performing the command and then going on to the next step.

Figure 1 is a circle with a radius of R. Measure the length of the radius along the outside of the circle. Now draw another radius, forming a wedge-shaped piece of the circle. Figure 1B shows this segment. The angle marked off this way is one *radian.*

What do you think will happen if we mark off sections like this all the way around the circle? Figure 1C shows the result. The radius will fit around the circle 6 full times. There will be a small section left over, which will take a little more than one quarter of the radius to fill.

The circumference of a circle is 2π times the radius. Pi (π) is not an exact number. It is approximately 3.14, so 2π is about 6.28. Therefore, we can mark off the radius around a circle 6.28 times. Another way to say this is that there are 2π radians in a circle, the same as there are 360 degrees in a circle. How many degrees make up one radian? Divide 360 by 2π to find out.

$$360° / 2\pi \text{ radians} = 360° / 6.28 \text{ radians} \qquad \text{(Equation 1)}$$
$$1 \text{ radian} = 57.3°$$

How many radians are there in one degree? Divide 2π by 360°.

$$2\pi \text{ radians} / 360° = 6.28 \text{ radians} / 360° \qquad \text{(Equation 2)}$$
$$1° = 0.01744 \text{ radians}$$

If you have an angle in degrees and want to change it to radians, multiply by 0.01744 radians per degree. If you have an angle in radians and want to convert it to degrees, multiply by 57.3 degrees per radian.

Type each equation in your spreadsheet and then type the ENTER key. Your results should be similar to the answers given on the right side of the equal sign.

@ASIN(0.500) * 57.3 = 30

@SIN(30 * 0.01744) = 0.500

Since π is not an exact number, all the values given here are rounded off. Your computer can use more digits in its calculations. If your spreadsheet or BASIC has a function called PI, you can use that value. Write equations to calculate the number of radians per degree and the number of degrees per radian. Use those equations in your spreadsheet or BASIC program. Equations 1 and 2 will serve as examples to write your computer equations.

BASIC has built-in functions to find the sine, cosine and tangent of an angle. Convert the angle to radians using Equation 1.

Figure 2 is a BASIC program that prompts you for an angle, and then prints the values for sine, cosine and tangent. It converts your input angle in degrees to an angle in radians. The program is very simple. You can add many enhancements to improve the display and usefulness of the program. It will run on most versions of BASIC without modification.

Figure 3 is a BASIC program that asks if you want to find the arcsine, arccosine or arctangent of a value. Then it prompts for the value and finds the appropriate inverse function. Finally, it converts the angle from radians to degrees and displays the answer. This program derives the arcsine and arccosine functions from the arctangent (ATN) function built into BASIC. Most versions of BASIC do not have the arcsine or arccosine functions built in.

Use one of the methods described in this section to answer the trigonometry practice problems given in the previous section. Your answers should be close to those given there. There may be some variation, however, because of the way computers round numbers off.

```
10 CLS
20 PI = 3.141592654
30 PRINT "ENTER AN ANGLE IN DEGREES."
40 INPUT AN
50 AR = AN * (2 * PI / 360)
60 PRINT "SINE = "; SIN(AR)
70 PRINT "COSINE = "; COS(AR)
80 PRINT "TANGENT = "; TAN(AR)
```

Figure 2—This BASIC program prompts you to enter an angle. It converts that angle from degrees to radians and calculates the sine, cosine and tangent for that angle.

```
10 CLS
20 PI = 3.141592654
30 PRINT "ENTER ARCS FOR ARCSINE, AC FOR ARCCOSINE
   OR AT FOR ARCTANGENT"
40 INPUT TR$
50 PRINT "ENTER A VALUE FOR THIS FUNCTION"
60 INPUT X
70 IF TR$ = "ARCS" THEN PRINT "ARCSINE("; X; ") = "; (2 *
   ATN(X / (1 + SQR(1 - X * X)))) * (360 / (2 * PI))
80 IF TR$ = "AC" THEN PRINT "ARCCOSINE("; X; ") = ";
   (-ATN(X / SQR(-X * X + 1)) + 1.57079) * (360 / (2 * PI))
90 IF TR$ = "AT" THEN PRINT "ARCTANGENT("; X; ") = ";
   ATN(X) * (360 / (2 * PI))
```

Figure 3—This BASIC program asks whether you want to find an arcsine (ARCS), arccosine (ACS) or arctangent (ATN). Then it prompts you to enter the function value and finds the appropriate inverse function. Finally, the program converts the angle from radians to degrees and displays the results.

CHAPTER 6 Logarithms for Electronics

Compressing the Number Scale

You will use a wide range of numbers when you make electronics measurements. You will use numbers smaller than 1×10^{-12} and larger than 1×10^{12}. This represents a range of more than a trillion between the smallest and largest numbers. Can you draw a number line that represents these values? Figure 1 shows pieces of such a number line. It really is just about impossible to draw a line with each *integer* (counting number) represented, though, isn't it?

We use units of microfarads and picofarads to measure the value of capacitors. We use millihenrys and microhenrys to measure the values of inductors. You'll see terms like kilohms and megohms when we talk about resistor values. We measure frequency in units like megahertz and gigahertz. (Don't worry, you'll learn what capacitors, inductors and resistors are later in this book. You also will learn what we mean by frequency.)

Do you recognize the metric prefixes in these terms? Mega (M), kilo (k), milli (m), micro (μ) and pico (p) should be familiar. Look at Table 1 on page 4-6 to refresh your memory about these prefixes and their values.

Let's try writing our number line another way. Figure 2 is a number line with only multiples of 10 shown. This looks less cluttered, and we can show a wide range of numbers on a single line. We still haven't shown all the integers. That may not be critical, as long as we can estimate values between those listed. After all, what's a few thousand one way or the other when you are talking about numbers in the trillions?

Take a careful look at Figure 2. You should notice several features about this number line. First, we've spaced all the multiples of 10 equally along the line. There are 8 small marks between each multiple of 10. Between 1 and 10 (10^0 and 10^1), these marks represent the values 2 through 9. Between 10 and 100 (10^1 and 10^2), they represent 20 through 90. The marks represent 20,000 to 90,000 between 10^4 and 10^5. For values less than one, the marks represent values of 0.02 to 0.09 between 10^{-2} and 10^{-1}. What numbers do the marks represent between 10^{-6} and 10^{-5}?

You also should notice that the marks between multiples of 10 get progressively closer as you go from 2 to 9. This is evidence of the number-scale compression produced by the Figure 2 line. What do we mean by *number-scale compression*? Values much larger than 1 are closer than the values near 1. It also puts the values much smaller than 1 farther apart. We move the same distance along the number line each time the number gets larger or smaller by a factor of 10. This results in a squashed or shortened number line, compared to the Figure 1 number line.

One measured quantity often varies over a wide range when another value changes a small amount. The output signal strength from a low-pass filter is a good example. Your Amateur Radio transmitter or transceiver probably

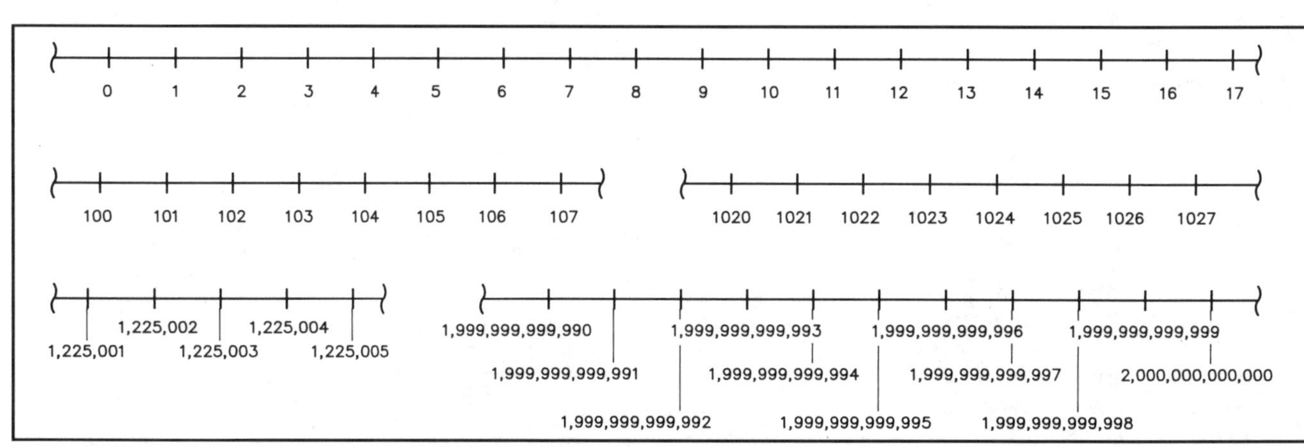

Figure 1—Here are some pieces of a number line that extends from 0 to 2 trillion. It is almost impossible to include all counting numbers in this range on a number line.

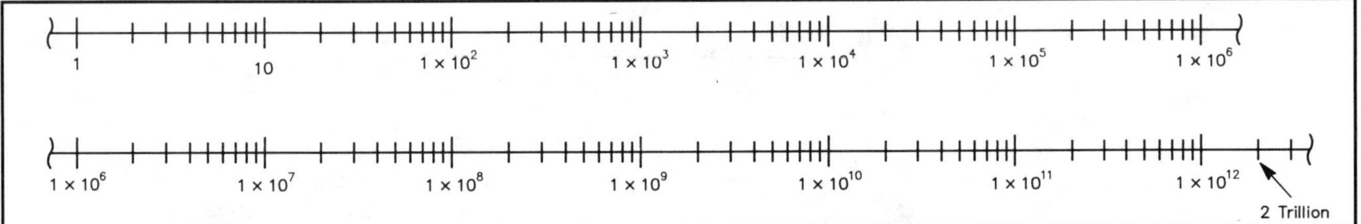

Figure 2—We can write our number line showing powers of 10 instead of all those integers. Marks between the powers of 10 help estimate values between the multiples shown.

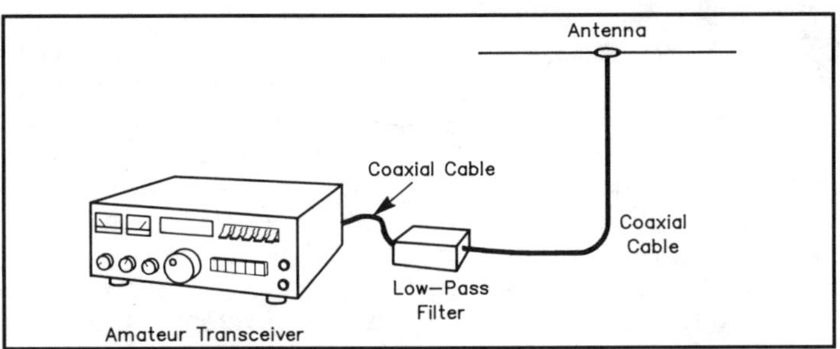

Figure 3—You can add an external low-pass filter to your Amateur-Radio transmitter output. Such a filter allows the desired amateur signals to go on to your antenna, while it blocks unwanted higher-frequency signals.

has a low-pass filter at its output. Such a filter allows the desired transmitted signals to go through. The filter blocks higher-frequency signals, preventing them from being transmitted. Many amateur transmitters and transceivers include a built-in low-pass filter. As a precaution, especially with older transmitters, many hams use an external filter. Figure 3 is a block diagram showing how you might connect an external filter to a transmitter.

One test for filter effectiveness is to feed a signal into the filter and measure the output signal. Then we change the input-signal frequency, but keep the same input power. We measure the output power, and plot it on a graph. This is what Figure 4 represents.

Figure 4 is a graph of the filter-output signal strength. The horizontal, or X, axis covers a small range of frequencies. We use a normal number scale for this axis. The vertical, or Y, axis covers a wide output-power range. We use a compressed scale, similar to the Figure 2 number line for this axis.

The frequency scale starts at 28 megahertz (MHz). Below this frequency a 100-watt input signal results in a 100-watt output signal. The filter has no effect at frequencies below 30 MHz. At 31.7 MHz the filter begins to reduce the output-signal strength. Just above 34 MHz, the filter reduces a 100-watt input signal to a 50-watt output. At 40 MHz the output signal is less than one watt for a 100-watt input signal. By the time we get to 49 MHz, the filter reduces a 100-watt input signal to about 0.0017 watts!

Can you imagine trying to draw the graph of Figure 4 using a linear number line for the vertical axis? The compressed scale suits this job well. You will find many other uses for a compressed number scale.

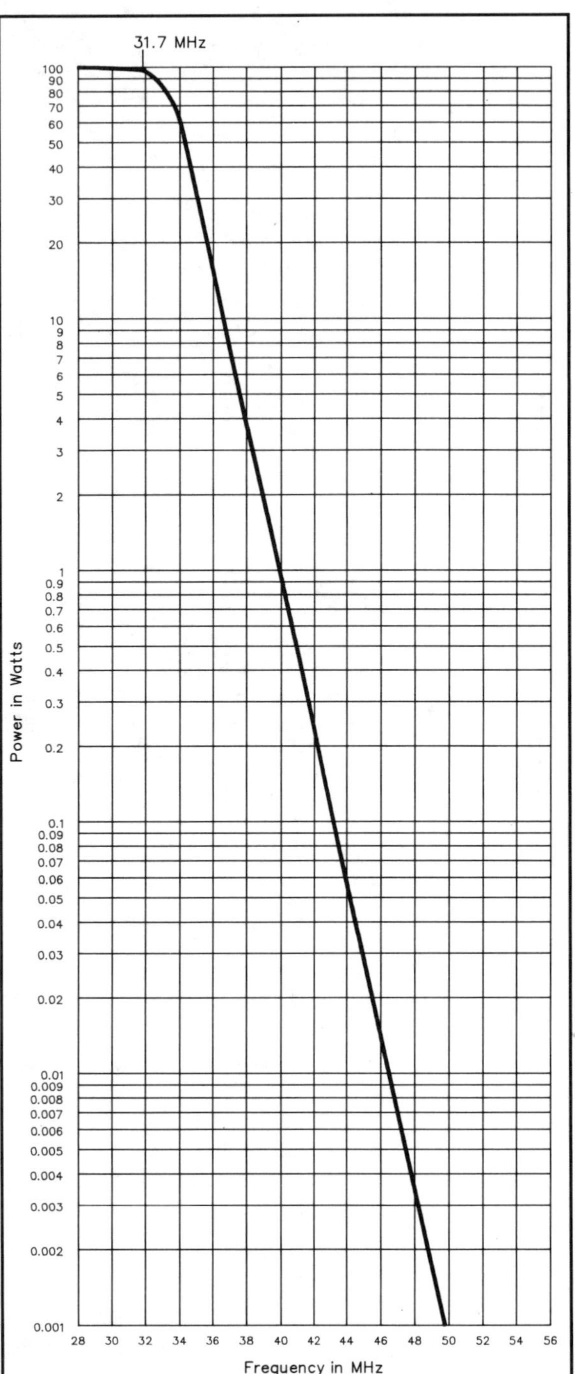

Figure 4—This graph represents the output from a low-pass filter. The vertical, or Y, axis uses a compressed scale similar to the one of Figure 2. The filter power output varies over a wide range when the frequency changes a small amount.

Defining the Logarithm

Powers of 10 help us work with large and small numbers. They help us compare numbers that cover a wide range of values. Powers of 10 also give us a way to compress the number scale. By now you probably have begun to realize that we use powers of 10 for many electronics applications.

We can write any number in *exponential* or *scientific notation*. This means we can write the number with any nonzero digits and a power of 10. The power of 10 tells how many places to move the decimal point to return to the number in *expanded* form. Expanded form shows the number without a power of 10.

For example, we write 45,300 as 453×10^2 or 4.53×10^4. We also can write this number with the decimal point in other positions, with the correct power of 10.

We also write numbers less than 1 with powers of 10.

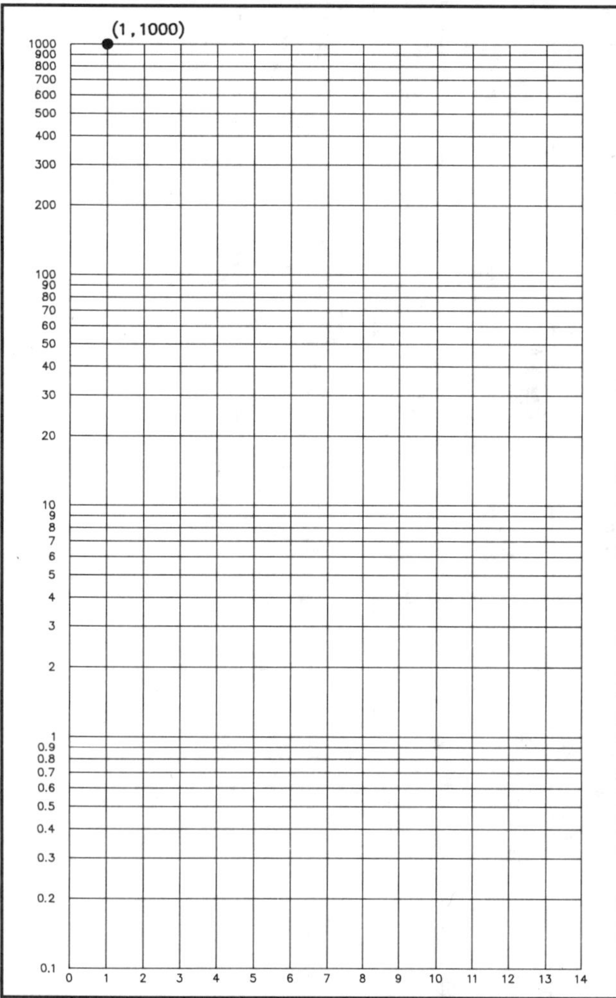

Figure 1—This graph uses a compressed number scale along the Y, or vertical, axis. Plot the values given in Table 1 on this graph, and connect the data points with a line.

Remember that 0.000625 is the same as 6.25×10^{-4}. We can write this number as 625×10^{-6}.

You learned how to write numbers using powers of 10, and practiced using exponential and scientific notation earlier in this book. If you are uncertain about writing numbers with a power of 10, review that section now. Work the practice problems again, too.

Let's look at writing some multiples of 10 in scientific notation. How would you write 10? In scientific notation, it is 1×10^1. Write 100 as 1×10^2. We can write 1000 as 1×10^3 and 1,000,000 as 1×10^6. Notice that we can drop the "1 X" from each of these expressions. Anytime you multiply a number by 1, it has no real effect, and you can drop the 1. This means we can write 10 as 10^1, 100 as 10^2, 1000 as 10^3 and 1,000,000 as 10^6. We express the number completely by using only a power of 10.

Can we express numbers less than 1 with only a power of 10? Sure. The *exponent*, or power, has a negative value, as you saw earlier. We will write 0.1 as 10^{-1}, 0.01 as 10^{-2} and 0.000001 as 10^{-6}.

Writing numbers as a power of 10 is really what we did to compress the number scale. The powers of 10 are spaced equally along the number scale, instead of spacing the integers evenly along the scale. The graph shown in Figure 1 uses a Y, or vertical, axis that is a compressed number scale.

Table 1
Data Points for Figure 1 Graph

Data Given as (X, Y) Values
(1, 1000)
(2, 100)
(3, 10)
(4, 1)
(5, 0.1)

Plot the data given in Table 1 on the graph of Figure 1. The data is given as a series of number pairs. The first number is the value along the X, or horizontal, axis. The second number in each pair is the value along the Y axis. We plotted the first point for you, and printed the X, Y values alongside the point. Now connect the data points with a line. Do all of your points fall on the same straight line?

Now we will try to plot this same data on the linear graph of Figure 2. Right away you'll notice that the vertical scale is too small. You can't plot data from near zero to 1000 on a scale that only goes to 10! First we'll have to compress the Y data values. You're probably thinking, "We can use scientific notation to do that!" We write the Y values as: $1000 = 10^3$, $100 = 10^2$, $10 = 10^1$, $1 = 10^0$ and $0.1 = 10^{-1}$. (We write 1 as 10^0 because you don't have to move the decimal point to expand the number. Remember there is a "1 X" in front of that power of 10.)

Are you ready for the tricky part? We will use only the *exponent*, or power, to graph our data on Figure 2! Table 2 shows the data values to plot on the graph. Again, we plotted the first value for you, and printed the X, Y values alongside that point. Plot the remaining values, and connect them with a line. Do the data points all follow the same straight line

again? Both graphs are really the same. The compressed number scale of Figure 1 saves us the trouble of finding the power of 10 to plot the data.

We can write whole-number multiples of 10 as only a power of 10. You might be curious to know if we can write other numbers only as a power of 10. The answer is yes, we can. This gives us a helpful mathematical tool for solving many problems.

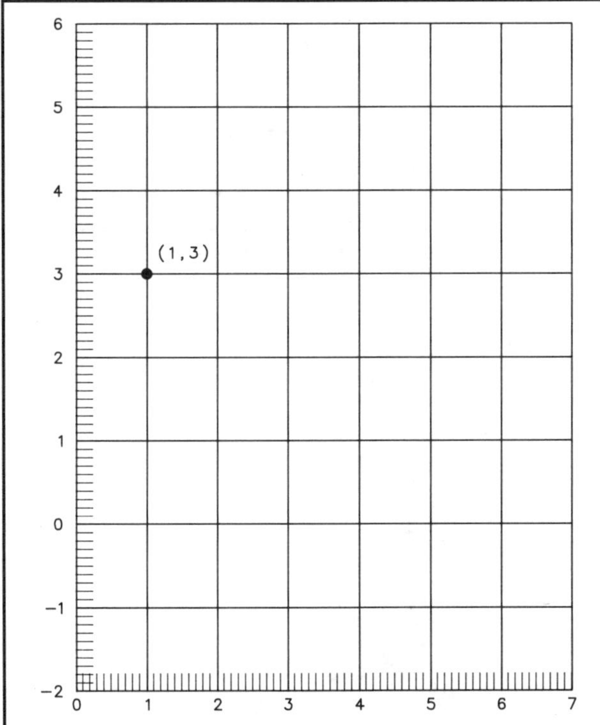

Figure 2—We can't plot the data of Table 1 directly on this graph because the Y axis isn't long enough. We'll have to compress the data first, using powers of 10. Table 2 shows this compressed data for you to plot on this graph. After you plot the data, connect the points with a line.

Look at the vertical axis of Figure 2. There is space between the integers along that axis, and you could estimate values like 1.5 or 3.25. These numbers represent larger values on our compressed scale. Values between 2 and 3 represent numbers between 100 and 1000. Values between 0 and 1 represent numbers between 1 and 10.

This process of letting one number represent another one is a mathematical *function.* You learned about trigonometry functions earlier. The *logarithm function* gives a number that is the power to which you must raise ten, to get the original number. For example, you must raise 10 to the third power (10^3) to get 1000. Therefore, the logarithm of 1000 is 3. What is the logarithm of 100? That's right, it's 2. The logarithm of 0.1 is −1 ($0.1 = 10^{-1}$) We plotted the logarithms of our data along the Y axis in Figure 2. The Y axis of Figure 1 is a logarithmic scale. Equation 1 gives a mathematical definition of the logarithm (*log*) function.

If $X = 10^Y$, then $\log(X) = Y$ (Equation 1)

where X is any number and Y is the power of 10 that gives X as the result.

Your scientific calculator will find logarithm values for you. This is especially helpful for numbers other than whole multiples of 10. Enter 20 on your calculator, and then hit the button labeled LOG. (On some calculators this may require you to type the FUNCtion or 2ND button first.) The display should show 1.30103. The log of 20 is 1.30103. Now enter 200 and hit the LOG button. Your display should show 2.30103 as the log of 200.

Table 2
Data Points for Figure 2 Graph

Data Given as (X, Y) Values
(1, 3)
(2, 2)
(3, 1)
(4, 0)
(5, −1)

The log function is only defined for positive values. So you cannot take the log of 0. If you try, your calculator will tell you there is an error in your calculation. You cannot take the log of a negative value, either. If you try, your calculator will display an error message.

You will learn more about logarithms and how to use them in the next few sections of this book.

Logarithms give us a way to compress the number scale. We also use logarithms (*logs*) in some types of electronics calculations. You'll learn more about using *logs* in later sections of this book.

A logarithm is a function. Functions follow a pattern to associate one number with another number. You are familiar with the sine, cosine and tangent functions. These are trigonometry functions. You also may be familiar with other mathematical functions.

You should remember that the trigonometry functions can work in reverse. Given a function value, you can find the inverse function. The inverse function is the angle associated with a function value.

Logarithms also can work in the reverse direction. When you have the log of a number you find the original number by taking the *antilog*. We sometimes write *antilog* as log^{-1}, similar to the way we use the –1 exponent with trig functions.

To define the logarithm function we write Equation 1.

If $X = 10^Y$, then $Y = \log(X)$ (Equation 1)

The opposite is also true.

If $Y = \log(X)$, then $X = 10^Y$ (Equation 2)

Equation 2 defines the antilog. It tells us that if we raise 10 to the power of the log value, we will get the original number again.

Let's try an example to see how this works. The logarithm of a certain number is 2. What is the number? We'll use Equation 2 to find the value of the unknown number.

$X = 10^Y$ (Equation 2)
$X = 10^2 = 100$

Notice that we didn't need a calculator or any other assistance to find this number. You can tell the log of a number like 100 or 1000 without any help. Converting back to the original number is just as easy.

Try another example. What is the antilog of –2? Equation 2 will help again.

$X = 10^Y$ (Equation 2)
$X = 10^{-2} = 0.01$

You can find the log of any whole multiple of 10 by writing the number in scientific notation. The power of 10 will be the log of your number. Finding the antilog of any whole number value is just as easy. Write the value as a power of 10 and then expand the scientific notation.

You'll need your calculator to find the antilog of a number that includes a fraction. What is the antilog of 2.5? You probably can't raise 10 to the 2.5 power in your head.

Look at your calculator keypad. Do you see a label like ANTILOG? If so, that's the key to use. Is there a key labeled INV? Then you probably can use that key with the LOG key to find antilogs. If you don't see either of these keys, you probably will find one labeled 10^X, Y^X or some-

What is the antilog of 2.5? Key 2.5 into your calculator. Then hit the ANTILOG key. Your calculator display should show 316.22777 or something very close to this. The antilog of 2.5 is 316.22777. Another way of saying this is "10 raised to the 2.5 power is 316.22777."

Figure 1—This example shows how to find an antilog with a calculator that has a key labeled ANTILOG.

Find the number whose log is 2.75. In this example we'll assume your calculator has a key labeled INV. Enter 2.75, then press INV followed by LOG. Your calculator display should show 562.34133. The antilog of 2.75 is 562.34133.

Figure 2—This procedure shows how to find antilogs with a calculator that includes an INV key.

The Exponential Function—Inverse Logarithms

thing similar. Figures 1 through 4 show how to use each of these key combinations.

Follow the Figure 1 example if your calculator has an ANTILOG key. If your calculator has an INV key, follow the Figure 2 example. Figure 3 shows you how to find an antilog if your calculator has a 10^x key. Figure 4 is an example that uses a y^x key to find antilogs. After you study the example that applies to your calculator, find the antilogs presented in the other examples.

Follow this procedure if your calculator has a 10^x key. What is the antilog of 3.25? By definition, the result of raising 10 to a power is an antilog. So enter 3.25 into your calculator and then press the 10^x key. Your calculator should display 1778.2794. If the 10^x label is printed above a key, such as the LOG key, you may have to press a key labeled INV or 2ND before you press 10^x. Check your calculator owner's manual if you aren't sure about the INV or 2ND keys.

Figure 3—These steps show how to find an antilog on a calculator with a 10^x key.

You will have to enter an extra number to use the y^x key on your calculator. Your calculator has two or more registers for numbers it will use in a calculation. Each register is just a storage area to hold the number. With most calculators, you enter the first number, then an operation key, and then a second number. The first register is known as the X register. Whenever you enter a number on the keypad, it goes into the X register. When you enter a second number, the first number moves to the Y register.

To tell your calculator that you want to raise 10 to a power, you will enter 10, then hit the y^x key. Next you must enter the power, or the number whose antilog you want to find. Let's use 1.398 as an example.

After you enter 10 and hit the y^x key, enter 1.398. To complete the operation, you must press the = key. Now your calculator should display 25.003454.

Figure 4—Some calculators have a y^x key. Here are the steps to find an antilog using such a calculator.

Sharpening Your Calculator's Logging Axe

Your calculator has several keys to help you work with logarithms and antilogs. It is a valuable addition to your toolbox. You must hone your operating skills to gain the most benefit from the calculator, though. With practice, you'll soon be splitting those log problems with ease.

Figure 1 lists 10 numbers. Use your calculator to find the log of each number. Compare your results with those shown in Figure 3. We've shown the results rounded to five places after the decimal point. Your answers should be similar after you round off the values.

Splitting logs is only one of your challenges. Sometimes you also have to put them back together, by finding the antilog of a number. Your calculator has just the glue for that task, too. It can be a tricky process, though, unless you practice the skill.

Find antilogs for the 10 numbers given in Figure 2. Compare your results with those given in Figure 4. You'll have to round off your calculator's answers again.

You will find logarithms several places in your study of electronics. The first place is likely to be for comparing power levels in a circuit. You'll learn about using logs to compare power levels later in this book. Logs provide some interesting results when we use them in calculations.

Let's try a few calculations. Find the log of 25 and round it off to five places after the decimal point. Did you get 1.39794? What is the log of 2? Round that value off to five decimal places, too. Is your result 0.30103? Good. Now let's add those values.

$1.39794 + 0.30103 = 1.69897$

What is the antilog of 1.69897? Your answer should round off to 50.

Adding logarithms is like multiplying numbers. What do you think would happen if we subtracted logs? Let's try an example, and find out.

What is the log of 5400? When you round the answer off to five places after the decimal you should get 3.73239. Let's subtract the log of 200 from this value. In the last example you found the log of 2. You could find the log of 200 without further assistance from your calculator, using that information. After all, 200 is just 2×10^2. Put a 2 in front of the log of 2, and you have 2.30103. (Prove that is the correct value for the log of 200 with your calculator if you aren't sure.)

$3.73239 - 2.30103 = 1.43136$

What is the antilog of 1.43136? Your answer should round off to 27. ($5400 \div 200 = 27$)

Find the log of 25. Does your calculator display show 1.39794? Now multiply that value by 2 and find the antilog.

$\log^{-1}(2.79588) = 625$

Here we have written the antilog as log with an exponent of –1. This is a common way to abbreviate "antilog."

In this example you squared a number by multiplying its log by 2. Multiplying a number's logarithm by 3 cubes the number. (That is the same as multiplying the number by

1) log(625) = ________
2) log(1250) = ________
3) log(3.1623) = ________
4) $\log(2.512 \times 10^6)$ = ________
5) $\log(4.00 \times 10^{-4})$ = ________
6) log(1024) = ________
7) $\log(6.25 \times 10^{18})$ = ________
8) log(75) = ________
9) $\log(6.67 \times 10^{-11})$ = ________
10) log(9.80665) = ________

Figure 1—Find the logarithm for each of the numbers listed here. Compare your answers with those shown in Figure 3.

1) antilog(8.38021) = ________
2) $\log^{-1}(-1.11351)$ = ________
3) $\log^{-1}(10.95521)$ = ________
4) antilog(1.78364) = ________
5) antilog(2.92620) = ________
6) $\log^{-1}(0.08988)$ = ________
7) $\log^{-1}(3.214)$ = ________
8) antilog(–4.71444) = ________
9) $\log^{-1}(2.4864)$ = ________
10) $\log^{-1}(1.76043)$ = ________

Figure 2—Use your calculator to find the antilog of each value given here. Figure 4 lists answers for you to check your results.

itself three times. $A^3 = A \times A \times A$.) You can raise a number to any power using logarithms. Simply multiply the log by the power to which you want to raise the number. The antilog of the result gives you the answer.

Let's try one more sample logarithm calculation. Find the log of 27 and divide your answer by 3.

log(27) = 1.4313638

1.4313638 ÷ 3 = 0.4771213

Now find the antilog of 0.4771213

$\log^{-1}(0.4771213) = 3.00$

In this last example, we found the cube root of 27. The cube root of a number is the value you multiply by itself three times to get the number. In our example, if you multiply 3 × 3 × 3 you get 27.

You can find the cube root of any number by dividing its log by 3. Dividing the log by 2 gives the square root. You can find any root of a number as easily. Divide the number's logarithm by the desired root and take the antilog. If you ever want to know the cube or fourth root of a number, logarithms will be the easiest way to find the answer.

1) log(625) = 2.79588
2) log(1250) = 3.09691
3) log(3.1623) = 0.50000
4) $\log(2.512 \times 10^6)$ = 6.40002
5) $\log(4.00 \times 10^{-4})$ = −3.39794
6) log(1024) = 3.01030
7) $\log(6.25 \times 10^{18})$ = 18.79588
8) log(75) = 1.87506
9) $\log(6.67 \times 10^{-11})$ = −10.17587
10) log(9.80665) = 0.99152

Figure 3—These logarithms of the numbers in Figure 1 should be similar to the answers you found. Your calculator may show more or fewer decimal places, but the rounded values should be the same.

1) antilog(8.38021) = 2.40×10^8
2) $\log^{-1}(-1.11351)$ = 7.70×10^{-2}
3) $\log^{-1}(10.95521)$ = 9.020×10^{10}
4) antilog(1.78364) = 60.763
5) antilog(2.92620) = 843.723
6) $\log^{-1}(0.08988)$ = 1.23
7) $\log^{-1}(3.214)$ = 1636.82
8) antilog(−4.71444) = 1.930×10^{-5}
9) $\log^{-1}(2.4864)$ = 306.48
10) $\log^{-1}(1.76043)$ = 57.60

Figure 4—Here are the antilogs of the numbers listed in Figure 2. Your answers should round off to these values.

Computers Can Find Logs Too

Do you have a personal computer? Perhaps you want to use it for electronics calculations. Can computers find logarithms? Sure they can. You can use the same methods available for trigonometry functions. You might have a program that works like a scientific calculator on the screen. Spreadsheet programs usually have a logarithm function built in. Nearly every personal computer uses some form of BASIC, so you can write a simple BASIC program to find logs and antilogs.

If you have a scientific calculator program for your computer, use it. Check the operator's manual for specific instructions. It probably works just like the calculators described earlier. You should have no difficulty with the practice problems in the last section.

When you learned to solve trigonometry problems with your computer there was an extra complication. Computers use angles measured in radians. You at least must know how to change from degrees to radians and from radians to degrees.

There also is a minor complication to using your computer for logarithm problems. In the earlier sections of this chapter we described *common logarithms.* Common logs aren't the only kind, however. Another type of logarithm is the *natural logarithm*, also called the *Naperian logarithm.* Computers use natural logarithms.

Equation 1 defines common logarithms.

If $X = 10^Y$, then $Y = \log(X)$ (Equation 1)

Natural, or Naperian logarithms use another number instead of 10. Equation 2 defines natural logarithms. We abbreviate natural log as *ln*. That helps distinguish between natural and common logs.

If $X = e^Y$, then $Y = \ln(X)$ (Equation 2)

$X = e^Y$ is a special function, called the *exponential function.* This special function appears in many calculations, including several electronics applications. The value of e is not an exact number. It is a repeating decimal. If you raise e to the power of 1 in your calculator or with your computer you will get an approximate value.

$e^1 = 2.718281828$ (Equation 3)

The decimal values continue to repeat the 1828 pattern for as many places as you care to write.

So how can we find a common log instead of the natural logarithm with a computer? Luckily, there is a simple solution. All you have to do is divide the natural logarithm of your number by the natural logarithm of 10. This converts the natural logarithm to a common log.

$$\log(X) = \frac{\ln(X)}{\ln(10)} \qquad \text{(Equation 4)}$$

We can simplify this equation by finding the natural logarithm of 10. Then we find the *reciprocal* by dividing that result into 1.

$$\log(X) = \frac{\ln(X)}{\ln(10)} = \frac{\ln(X)}{2.3025851}$$

$$\log(X) = 0.4342945 \ln(X) \qquad \text{(Equation 5)}$$

Some spreadsheet programs include the log function. Others may only include the natural logarithm function. Figure 1 shows a few cells of a spreadsheet. The labels in row 1 identify the functions in row 2. Type any value in cell A2. If your spreadsheet program has a log function, cell B2 calculates the log. Cell C2 finds the natural logarithm. Cell D2 divides that value by the natural logarithm of 10.

	A	B	C	D	E	F
1	Value	@LOG(A2)	@LN(A2)	C2/@LN(10)	10^(B2)	@EXP(C2)
2	2	0.301029	0.693147	0.301029995	2	2
3						
4						

Figure 1—This figure shows several cells of a sample spreadsheet. You can program equations into the spreadsheet to calculate logarithms and antilogarithms.

To find the antilog of a value, you must raise 10 to the power of that value. Most spreadsheets use the ^ (caret) to raise a number to some value. This is a SHIFT key combination on most computer keyboards. Cell E2 calculates the antilog of the value in cell B2. You can use this same spreadsheet to find antilogs by entering a value in cell B2 instead of A2. (Copy the row of cells to another row first, so you don't lose the equation in cell B2.) Cell F2 calculates the exponential function of the value in cell C2. The exponential function is the inverse of a natural logarithm.

You can use the exponential function to find the antilog of a common logarithm. First you must change the common log value to a natural log value. Can you guess how to do that? Yes, just divide the common log value by the common log of e.

$$\ln(X) = \frac{\log(X)}{\log(e)} = \frac{\log(X)}{\log(2.718281828)} \quad \text{(Equation 6)}$$

$$\ln(X) = \frac{\log(X)}{0.4342945} = 2.3025851 \log(X) \quad \text{(Equation 7)}$$

Use the example shown in Figure 1 to program this equation into your spreadsheet. Remember, you already know the value for log(X). You won't need to use the @LOG function. Make a call to another cell that holds the value whose antilog you want to find. Then use the exponential function as shown in cell F2 to calculate antilogs.

You also can use the BASIC computer language to calculate logarithms. Some versions of BASIC include the common log function. Most BASIC versions only include natural logarithms, however. BASIC allows you to raise a number to a power with the ^ (caret). Find antilogs with a command like:
PRINT 10^X
where X is the log value whose antilog you want.

The methods described for spreadsheets also apply to BASIC programs. Figure 2 lists a simple BASIC program that will find common logarithms. Your BASIC programmer's manual will guide you if you find any difficulties with this program. You can make many enhancements to the program. This one is simplified to ensure it will run in most BASIC versions without modifications.

Figure 3 lists a BASIC program to find antilogs. BASIC uses the ^ (caret) to indicate that you want to raise a number to some power. When we raise 10 to a power, we find the antilog of the value used as the power.

Use one (or more) of the methods described in this section to find the logs and antilogs listed in Figures 1 and 2 of the last section. Your answers should be similar to the answers listed there.

When you try to solve those problems you will find that some of the log problems involve numbers in scientific notation. Spreadsheets and BASIC use a capital E in the number to indicate a power of 10. You would enter 2.512×10^6 as 2.512E+6. Numbers with a negative power of 10 require a minus sign before the exponent. Enter 6.67×10^{-11} as 6.67E–11.

```
10 REM COMMON LOG PROGRAM
20 PRINT "ENTER A POSITIVE NUMBER, OR 0 TO END: ";
30 INPUT N
40 IF N = 0 THEN GOTO 100
50 A = LOG(N)
60 B = LOG(10)
70 L = A / B
80 PRINT "THE COMMON LOG OF "; N; " IS "; L
90 GOTO 20
100 END
```

Figure 2—This BASIC program will run in most BASIC versions. It finds common logs by converting the computer's natural logarithm value to a common log.

```
10 REM ANTILOG PROGRAM
20 PRINT "ENTER A VALUE TO FIND AN ANTILOG, OR 0 TO END ";
30 INPUT N
40 IF N = 0 THEN GOTO 80
50 A = 10 ^ (N)
60 PRINT "THE COMMON ANTILOG OF "; N; " IS "; A
70 GOTO 20
80 END

100 REM ROUTINE TO FIND ANTILOG USING EXPONENTIAL FUNCTION
150 A = EXP(2.3025851 * N)
```

Figure 3—This program finds antilogs by raising 10 to the log power. You also can find antilogs by changing the common log value to a natural logarithm value and applying the exponential function. You can replace statement 50 with the one shown as 150 to calculate the antilog this way.

Elementary DC Electronics

Your First Task:

Study the effects of steady, unchanging signals in an electronics circuit.

In This Unit You Will Learn:

- About voltage, the pressure that makes electricity move through a circuit.
- About current, a measure of the amount of electricity in a circuit.
- What an electric field is by comparing it with magnetism and by observing some effects of electric force acting through space.
- That we can control how easy it is for electricity to flow though a circuit by selecting conductors, insulators or resistors to make up the circuit.
- There are two kinds of electrical circuits—series circuits and parallel circuits.
- How to use Ohm's Law, an important electronics principle, to calculate circuit conditions.
- About two other important electronics principles, Kirchhoff's Laws.
- What electrical energy is.
- How to calculate power, the rate of using that energy.
- How to use decibels when comparing power in circuits.

You are ready to learn about electronics. You have the important math skills needed to calculate electronics-circuit conditions.

Most electronics circuits process a signal in some way. A signal can be as simple as the electrical pulse created by closing the contacts of a switch. Most signals are more complex, however. A radio receiver takes signals it picks up from an antenna and turns them into sound. A television takes signals it picks up and forms pictures on the screen. Radio and TV signals change constantly. The radio or TV receiver makes more changes to the signal as it converts it to sound or a picture. There are many electronics parts in such a circuit. When you look inside a radio you might think it will be impossible for you to understand how such a circuit works.

Each part has a specific job in the whole circuit. Each part behaves in a certain way when those signals come along. We analyze circuits by looking at each part's task. Then we study the effects it will have on the signal coming into it. In this way, we can understand even the most complicated circuits.

We won't begin our study of electronics by looking at how circuits affect changing signals. Just keeping track of what the signal is doing can be difficult. We'll begin by looking at what happens to a constant, unchanging signal. That way you can concentrate on learning the basic principles without trying to keep track of what the signal is doing.

Your first task, then, is to study the effects of steady, unchanging signals in an electronics circuit. Such signals tell us much about how electronics circuits operate. Once you know how to analyze a circuit with steady signals, it will be much easier to understand the effects of changing signals.

For an unchanging signal, the electricity flows only in one direction. Electricity that flows only in one direction is a *direct current* signal. Electricity that changes direction, flowing first in one direction and then in the other, is an *alternating current* signal. We often use the abbreviations *dc* and *ac* when we describe these two types of electricity. Sometimes the amount of electricity in a dc signal changes, but it still flows only in one direction. In this unit we will concentrate on dc signals that do not vary.

Voltage
Current
Fields
Flows
Circuits
Decibels
Laws
Elementary DC Electronics
Current
Direct... DC
Alternating... AC
Help ...
BLOOM BUILDING
73
LARRY'S ELECTRIC SUPPLY
ARRL HQ
ONE WAY

CHAPTER 7 Voltage—The Pressure of Electricity

It is dark outside, so you pick up a flashlight before stepping out the door. When you turn on the switch, a bright beam of light shines ahead of you. Have you ever stopped to think about the magic that goes on inside that flashlight? What happens during that moment when you push the switch? The battery inside the flashlight stores electricity. When you close the switch, you complete a path for that electricity to travel to the bulb, and it lights. The battery stores the electricity as chemical energy. When the chemicals react, they produce electricity. (You can learn more about how batteries work later.)

Figure 1—This photo shows a variety of common batteries. You must select the proper battery for each application. Otherwise you risk destroying part of the circuit, or at least not having the circuit work properly.

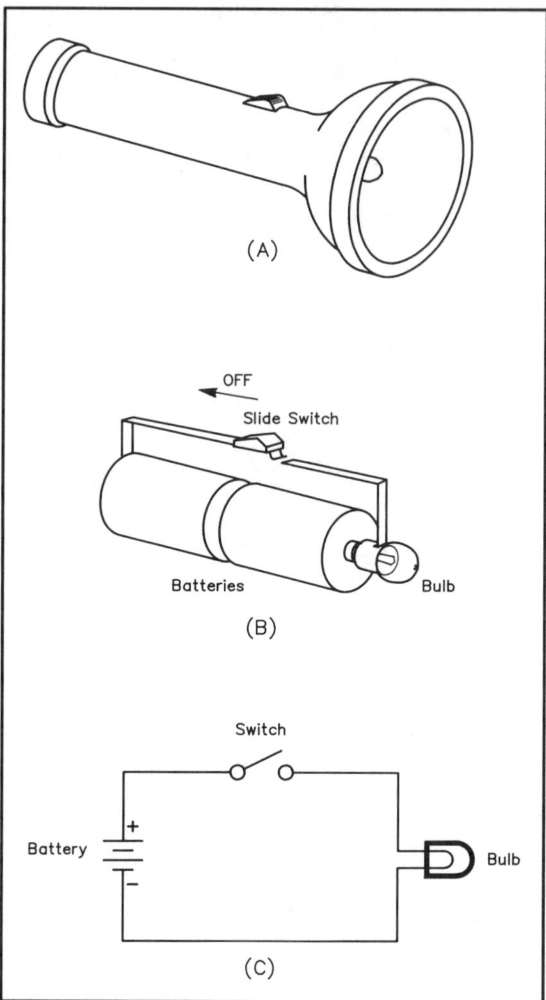

Figure 2—A flashlight is a good example of a simple electronic circuit. There are three main ingredients to any electronic circuit. You must have a voltage source, which in this case is the flashlight battery. You also must have some wires or other conductors to carry the electricity. Finally, there must be a load, or something for the electricity to do. In this example, the flashlight bulb serves as the load. Part A shows a flashlight, and B shows a view of the circuit inside the flashlight. Part C is a schematic diagram of the circuit. Schematic diagrams use special symbols to represent circuit components.

A battery moves electricity through the bulb. Something about the battery pushes the electricity, forcing it to light the flashlight. You also may have noticed that flashlights produce different amounts of light. A small light that uses a single AA-cell battery makes a pretty small beam. A flashlight that uses four or more D-cell batteries usually produces a much larger, brighter light beam. Eventually the batteries begin to die, however, and the light beam becomes smaller and weaker.

Batteries such as AA, AAA, C and D cells have a rating of 1.5 volts. Other batteries have different voltage ratings. For example, small "transistor batteries" have a

9-volt rating. Lantern batteries (the kind with a coiled spring on top) have a 6-volt rating. Car batteries have a 12-volt rating. Still others, like small watch batteries, have various voltage ratings. Figure 1 shows several battery types.

These voltage ratings tell us something about how much electricity that battery will push through a circuit. You must select the proper battery for a particular job, or the circuit won't receive the proper amount of electricity. If you tried to connect a car battery to your AA-cell flashlight, the bulb would burn out. A single AA cell wouldn't do much good trying to light your car's headlights, either.

Find some type of battery and take a close look at it now. You will see two "terminals" or places to connect wires. One terminal has a + and the other a –. This shows there are two sides to the voltage. One side is positive and the other is negative. It also indicates that two wires must connect the voltage to the circuit.

An electrical circuit forms a complete path for the electricity. It flows from the battery, through the circuit, and back to the battery. If the circuit is not complete, the electricity will not flow through it. This is why the switch on your flashlight turns the light off. It opens, or breaks, the circuit. Figure 2 shows the electrical circuit of a flashlight.

The *volt* measures the amount of pressure, or push in an electric circuit. This push is the force that moves electricity, so we sometimes refer to a voltage as an *electromotive force*, abbreviated *EMF*. Another name we often use for voltage is *electrical potential*, or just *potential*.

Figure 3 shows a container full of water. Some of the water is pouring out through a hole near the bottom of the container. What is forcing the water out of the container? The pressure of the water in the container is exerting a force to push some water through the hole. When the container is full, as shown in part A, the pressure is highest. The water will shoot out farthest from the bottom when the container is full. When the container is nearly empty, as shown in part B, the pressure is much lower. Now the water doesn't shoot out as far. The amount of water pressure depends on how high the water is above the hole in the container. The more water in the container, the greater the pressure pushing water out a hole in the bottom.

This example can help you better understand electrical pressure because it is similar to water pressure. The amount of EMF, or voltage, indicates how much electricity that voltage source can move in a circuit. An EMF of 100 volts is much stronger than one of 10 volts. A higher voltage can move more electricity than a lower one.

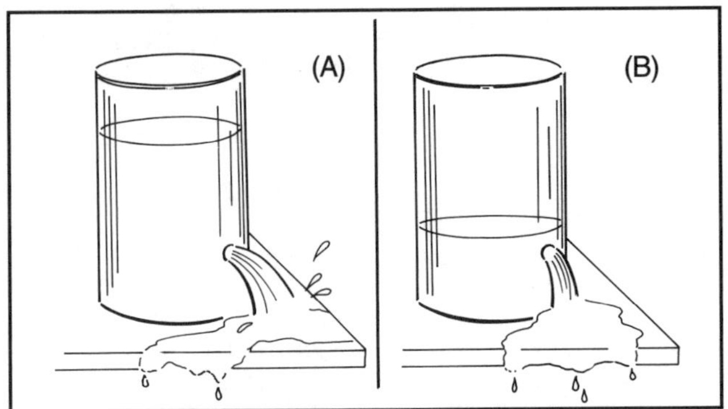

Figure 3—The height of water in a container determines how much pressure the water exerts on the container bottom. In Part A, the container is full, so there is a lot of pressure. The water shoots far out of the hole in the container. In Part B, the container is nearly empty, so there is only a little pressure. The water does not shoot out as far.

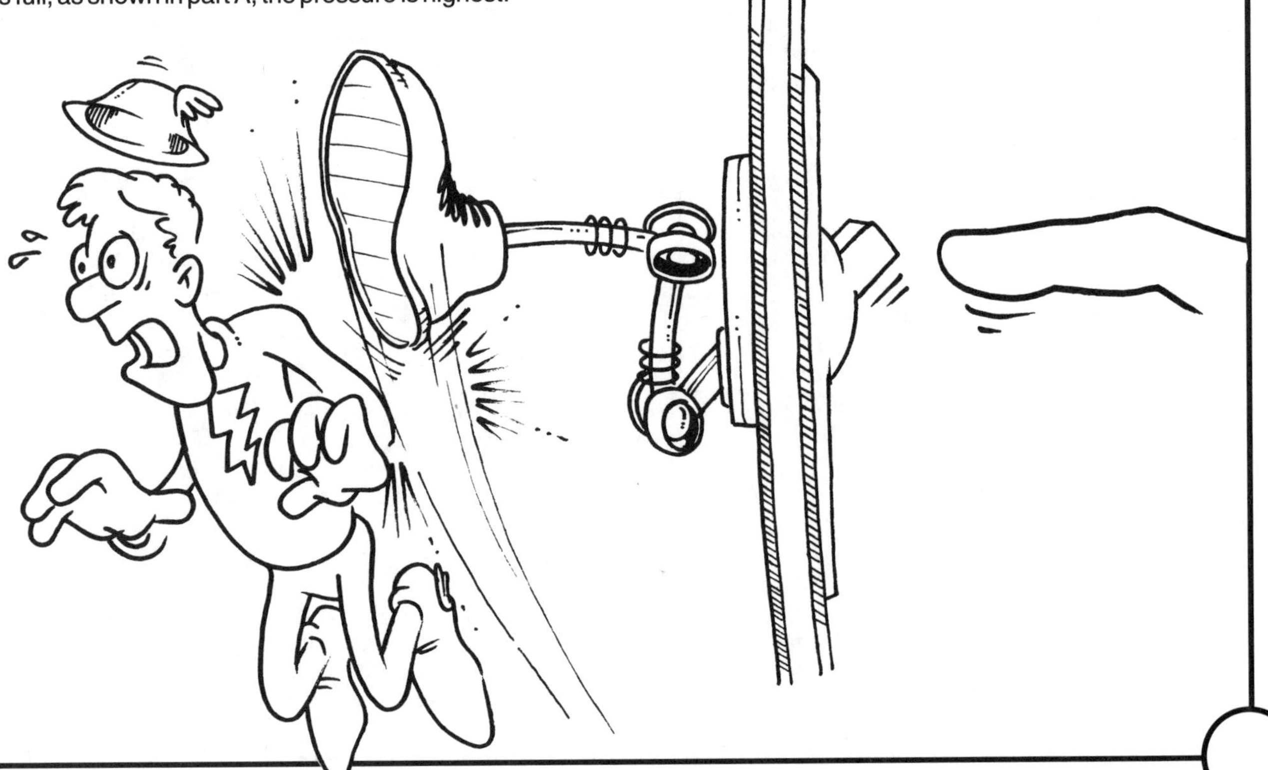

CHAPTER 8
Current—The Flow of Electricity

When you switch on an electronics circuit, something happens. A flashlight shines your way in the dark, a radio begins to play, or a computer comes to life. A voltage source supplies the force, or pressure, to move electricity through a circuit. We measure this force in *volts*. A higher voltage pushes more electricity through the circuit. How do we measure the amount of electricity, though?

A unit called the *ampere* measures the amount of electricity flowing in a circuit. We often abbreviate ampere as *amp*, or simply *A*. The amount of electricity in a circuit is called the circuit current. Besides requiring a specific voltage, electronics circuits also require a certain current to operate properly. For example, an Amateur Radio transceiver may require a circuit that supplies a 20-ampere current at 12 volts of electrical pressure. A 60-watt light bulb requires $^1/_2$ amp with 120 volts applied to it. Small portable radios often use 9-volt transistor batteries. Such a radio may only require a few milliamperes to operate.

Milliamperes? What's a *milliampere*? Remember that the units we use in electronics are part of the metric system. So *milli* is the metric prefix for one thousandth, or 10^{-3}. A milliampere (or milliamp) is a thousandth of an amp. Some circuits may even have a current as small as a few *microamps*. You must be familiar with all the metric prefixes. They show up in all kinds of electronics measurements. Table 1 serves as a reminder of the metric prefixes we commonly use in electronics measurements.

Table 1
Metric Prefixes Common to Electronics Circuits

Prefix	*Abbreviation*	*Value*
Giga	G	10^9
Mega	M	10^6
Kilo	k	10^3
Deca	da	10^1
(Unit)		10^0
Deci	d	10^{-1}
Centi	c	10^{-2}
Milli	m	10^{-3}
Micro	µ	10^{-6}
Pico	p	10^{-12}

Figure 1 is the *schematic diagram* of a simple electronics circuit. A schematic diagram uses special symbols to represent the parts of a circuit. The diagram shows how those parts connect to form a circuit. Do you remember that a voltage source has two connections? Which way does the electricity flow out of the battery? That is an important question because the answer determines which direction the current flows. Unfortunately, there are two answers to the question!

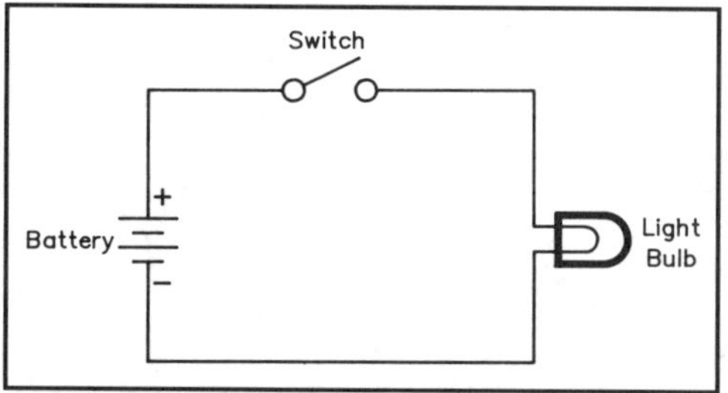

Figure 1—This drawing is a *schematic diagram* of a simple electronics circuit. This circuit represents a flashlight. We labeled each symbol on the diagram to help you identify the circuit parts.

Benjamin Franklin conducted some experiments with electricity. Franklin thought that the electricity came from the positive side of the battery. So he described an electric current flowing from the positive terminal, through the circuit, and back to the negative terminal. Figure 2 shows the current going from the positive battery connection, around the circuit and back to the negative terminal. Electrical engineers refer to this as *conventional current.*

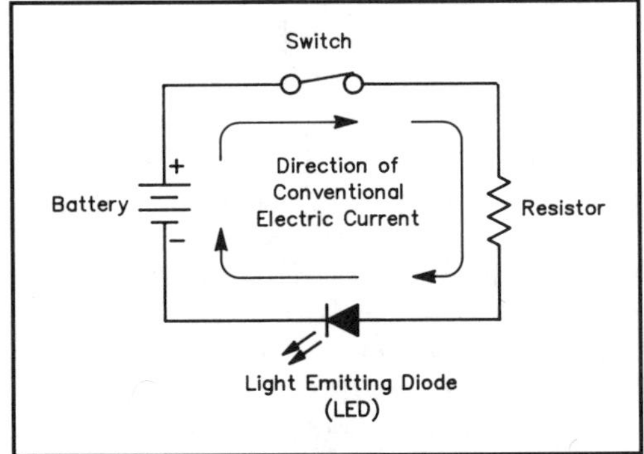

Figure 2—This drawing shows a circuit with the conventional current direction added. Conventional current moves from the positive battery terminal, through the circuit and back to the negative terminal.

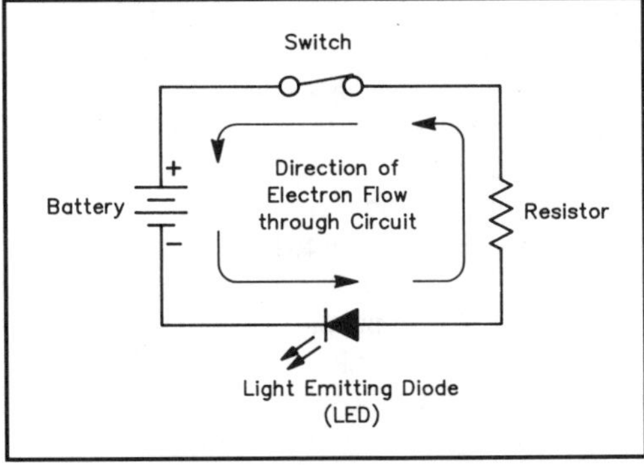

Figure 3—This diagram shows the direction of electron flow through the circuit of Figure 2.

Much of the mathematics of modern electronics follows this convention, or agreement among engineers. They say current goes from positive to negative.

Many years after Franklin's experiments, scientists learned that negatively charged particles called *electrons* actually make up the electrical current. (You'll learn more about the important role of these electrons in the next few chapters.) Today, we know that electrons flow from the negative terminal, through the circuit and back to the battery. Figure 3 shows the direction of this *electron flow* through our circuit.

Can't we just ignore the old *conventional current*? Well, often we can. You must be aware of conventional current, however, because engineers still use it for many calculations. We continue to use conventional current because it provides a convenient mathematical solution to many problems. Many schematic diagram symbols include some type of arrow. These arrows point the direction of conventional current through that part. For example, Figure 4 shows the schematic diagram symbol for a diode. Conventional current flows through the diode in the direction shown by the arrow.

In this book, we have been very careful to distinguish between conventional current and electron flow. Most references in *The ARRL Handbook for Radio Amateurs*, and in other ARRL publications, also carefully make this distinction. Not all publications are as careful, however. As you study other books to learn more about electronics, be sure you understand whether the book means electron flow or conventional current when it refers to current.

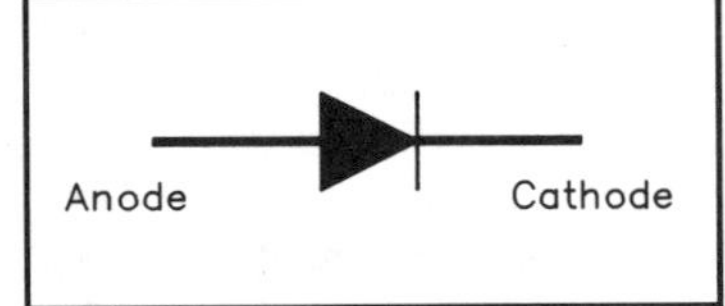

Figure 4—This drawing shows the schematic diagram symbol for a diode. The symbol's arrow shows the conventional current direction through the diode.

CHAPTER 9 Electricity and Magnetism

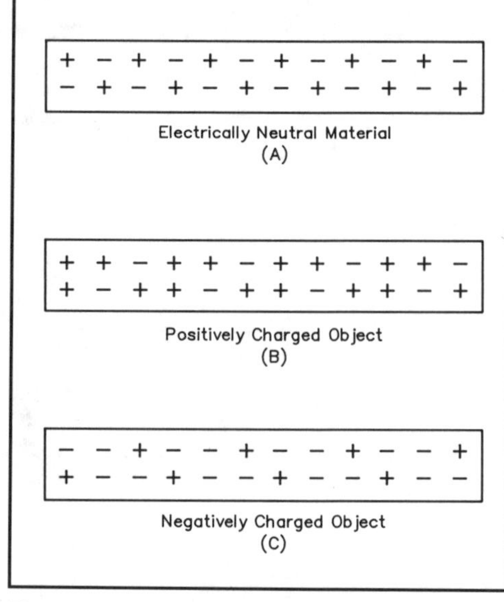

Figure 1—The normal condition for an object, like the metal bar shown, is to have the same number of positive and negative charges. The bar shown in part A is electrically neutral. If an object has more positively charged particles than negative ones, the object has a positive charge (Part B). If an object has more negatively charged particles than positive ones, the object has a negative charge (Part C).

Electric Charge

We usually think of electricity as a modern development. It wasn't until the late 1800s that people began to use electricity for the benefit of society. Actually, the discovery of electricity took place more than 2000 years ago! Ancient Greece is known for the philosophers, mathematicians and scientists who lived there. They wanted to discover more about the world around them. They found a kind of rock that seemed to have magical properties when they rubbed it with a piece of fur or cloth. This rock, which they called *elektron*, would attract small bits of dried leaves, wood shavings, straw and other materials. (Today we call this rock *amber*.) Our word *electricity* comes from the Greek name for this rock.

It wasn't until 1733 that anyone realized there might be two kinds of electric charge, or force. Charles DuFay noticed that a charged piece of glass attracted some electrically charged objects but repelled others. He concluded there are two types of electricity. Then in the mid 1700s, Benjamin Franklin conducted some further experiments with electricity. It was Franklin who gave the names *positive* and *negative* to the two kinds of electric charge.

In Ben Franklin's time, scientists thought that electricity was a kind of fluid found in all objects. If there were more of one type of fluid, the object would have the charge of that type, either positive or negative.

Today's scientists have identified very tiny particles that carry the electric charges. *Protons* carry the positive charges and *electrons* carry the negative charges. These particles are much too small for you to see with your eye. In fact they are even too small to see with a powerful microscope. Scientists have found ways to detect these particles by the effects they produce. Protons and electrons are present in every material known. In fact, they are some of the basic building blocks of all matter.

The amount of negative electric charge on an electron is the same as the amount of positive electric charge on a proton. So the relative numbers of electrons and protons in an object determine what kind of electric charge that object has. (If it has more electrons it has a negative charge and if it has more protons it has a positive charge. The normal condition is to be electrically neutral, or to have the same numbers of electrons and protons.) See Figure 1. A charged object will exert a force on other objects. The objects do not have to be in direct contact for this force to be exerted, since the force acts through space.

By now you are probably wondering how an object can become electrically charged, if its normal condition is to be neutral. Let's consider a couple of examples. The ancient Greeks discovered that they could charge a piece of amber, or *elektron* as they called it, by rubbing it with a piece of animal fur. The friction between the fur and the rock causes the rock to pull some extra electrons away from the fur. Thus, the rock becomes negatively charged. (Actually, it is the heat produced by the friction of rubbing the fur that frees some electrons from the fur.)

You're probably familiar with what happens when you shuffle your feet across a carpet and then touch something metal, like a doorknob! You also can conduct a simple experiment based on this principle. First, tear up a few small pieces of paper and place them on a table. Next, take a comb made from hard rubber or nylon, and run it quickly through your hair a few times. The friction that this rubbing produces will pull electrons away from your hair and onto

Figure 2—A negatively charged object will attract small bits of paper or other lightweight objects.

the comb. In this way, the comb gets a negative charge. Now bring the comb close to the bits of paper. Notice how the paper jumps onto the comb? The electric charge on the comb exerts a force on the paper, attracting it to the comb.

Would you like to try another example? Blow up a balloon and rub that against your hair. Then carefully stick the balloon to a wall or ceiling. Again, the friction of rubbing the balloon through your hair has given the balloon a negative electric charge. That charge exerts an electric force to attract the balloon to the wall.

All materials don't become negatively charged when you rub them. Rub a piece of glass, like a glass rod, with a piece of silk cloth. The heat produced by this friction will free some electrons in the glass. That will leave the glass with a positive electric charge. The glass rod will still attract small bits of paper and other lightweight items, however. An object with a positive charge exerts an electric force, just as a negatively charged object does.

It would be helpful to have a number that we could associate with the amount of electric charge on an object. A simple measurement would allow us to compare one charged object with another. In electronics we use the *coulomb* to measure electric charge. A charge of 1 coulomb is approximately equal to 6.24×10^{18} (that's 6,240,000,000,000,000,000) electrons. Of course it's impossible to count the number of extra electrons on a charged object. That's not how we measure a coulomb. Instead, we measure the effects of the charge. The effects are greater for an object with more charge.

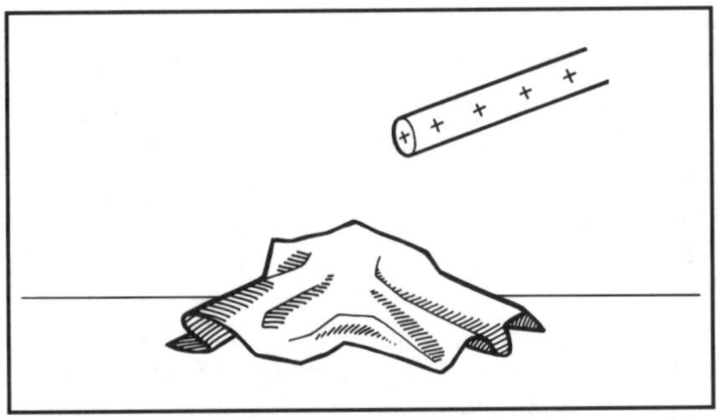

Figure 3—You can give a positive electric charge to a glass rod by rubbing it with a piece of silk cloth.

Rules of Attraction and Repulsion

In the last section you learned there are two types of electric charge, *positive* and *negative.* If an object has more of either type of charge than the other, it will exert an electric force on nearby objects. In this section you will learn how to determine whether an object has a positive or a negative electric charge. You also will learn some rules about the electric forces exerted by such charged objects.

Figure 1 shows an electroscope. This is a simple device that illustrates the effects of electric charge on an object. The metal rod and gold leaves are inside a glass jar, mainly to protect the delicate gold foil. When the two leaves of gold foil are electrically neutral, they hang straight down, as shown.

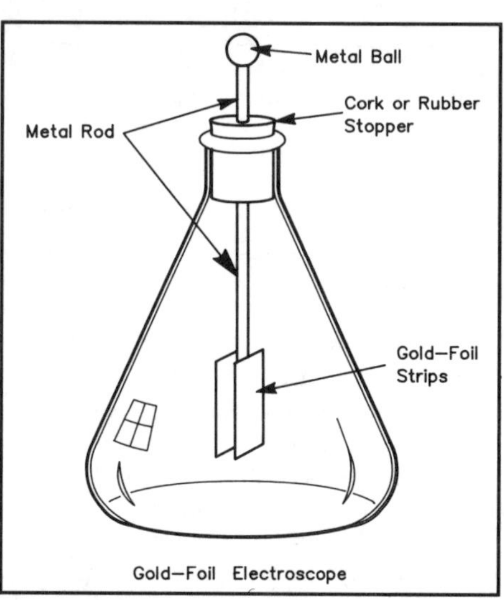

Figure 1—An electroscope illustrates the effects that electrically charged objects cause. The main purpose of the glass jar is to protect the delicate gold-foil leaves.

Now let's bring a negatively charged object close to the metal ball on top of the jar. (This might be a hard-rubber rod that we rubbed with fur.) Some electrons move away from the ball and onto the gold foil. When the two gold leaves have extra electrons, they push away from each other. This force causes the leaves to stand out to the side, as shown in Figure 2A.

Remove the hard-rubber rod without touching the electroscope. The electrons will return to their normal positions and the gold-foil leaves will go together. Figure 2B shows the electroscope in an uncharged, or neutral, position.

If we touch the rod to the top of the electroscope, some electrons will flow from the rod onto the metal ball. Now take the negative rod away from the electroscope. The extra electrons keep the gold-foil leaves negatively charged. The leaves remain apart until they become electrically neutral again. See Figure 3.

These examples illustrate an important rule for electric charges: *Like charged objects repel.* If the objects have only a small amount of charge, there will be a small force trying to push them apart. If the objects have a large electric charge, though, there will be a bigger force trying to push them apart. With an electroscope, you also can observe how the strength of the force varies. As you bring the charged rubber rod close to the electroscope, the gold-foil leaves begin to move apart. As you bring the rod closer to the electroscope, they move farther apart. If the rod has quite a large amount of charge, you can make the gold leaves stand almost straight out.

Let's go back to the negatively charged electroscope, as shown in Figure 3. Now suppose we take a glass rod and rub it with a piece of silk cloth. The silk cloth pulls electrons away from the glass, leaving it with a positive charge. When we bring the charged glass rod close to the electroscope, the leaves go down, as shown in Figure 4. Why is this? The positively charged glass rod attracts electrons from the gold leaves to the top of the electroscope. That removes the negative charge from the gold foil, allowing the leaves to hang straight down again.

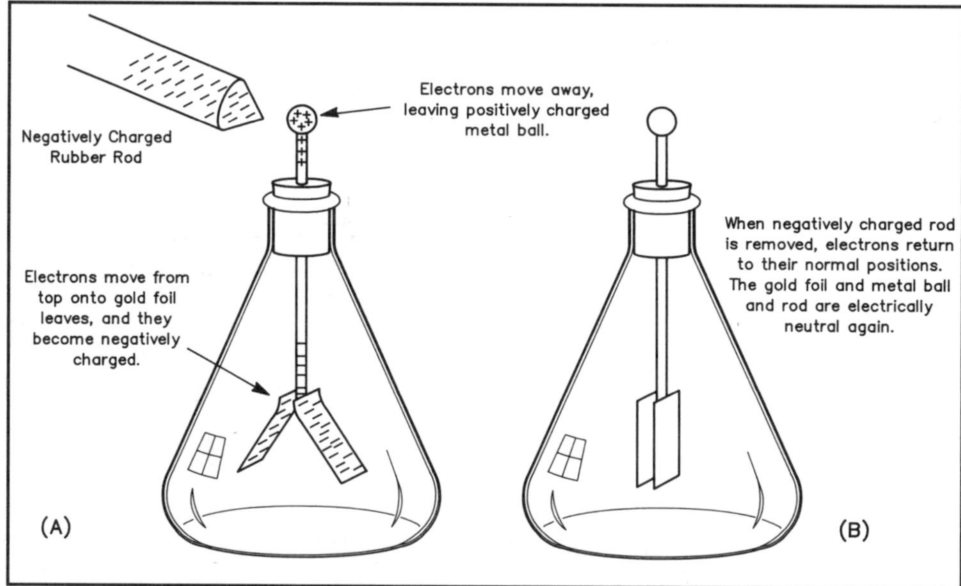

Figure 2—When we bring a negatively charged object near the top of the electroscope, electrons from the metal ball move onto the foil. With extra electrons on them, the gold leaves push away from each other. The leaves stand out to the sides, as shown in part A. When we take the negatively charged object away, the electrons move back to the top of the electroscope. Then the gold foil is neutral again and the leaves hang straight down, as shown in part B.

Figure 3—Touch the negatively charged rod to the ball on top of the electroscope. Some of the excess electrons will move onto the gold foil, as shown in part A. Then, when we remove the rod, the charge will remain on the electroscope. The gold leaves will stay apart, as shown in part B.

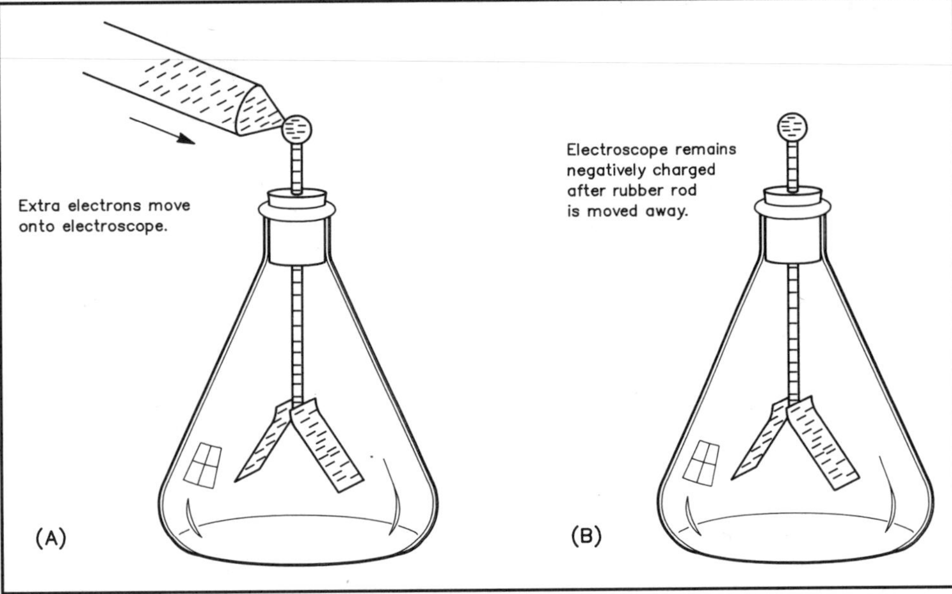

Now we'll touch the ball on top of the electroscope with the positively charged glass rod. Figure 5 shows how the gold leaves will spring out to the side again. Some electrons jumped from the electroscope to the glass rod, replacing a few of those pulled off by the silk cloth. The glass rod has less positive charge now than before we touched it to the electroscope. The gold foil has a positive charge, though, because it gave up some of its electrons.

From this discussion you should realize there is a force of attraction between positive and negative electric charges. This brings us to the second important rule for electric charges and the forces they exert: *Oppositely charged objects attract.*

It is very important that you understand and remember these two rules of electric forces. They will help you understand many important principles of electronics. Remember: *likes repel, opposites attract.*

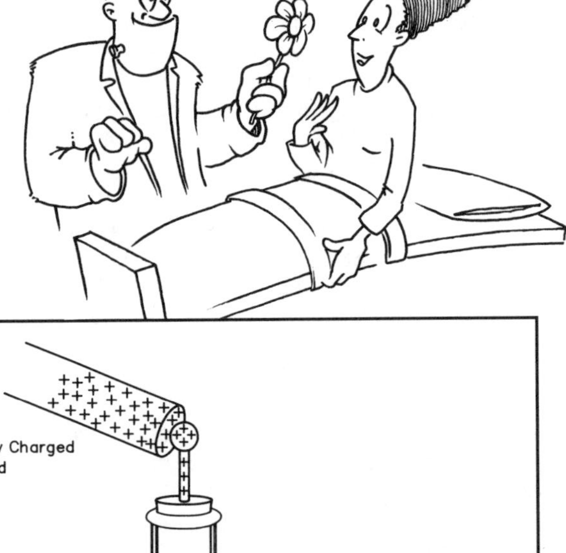

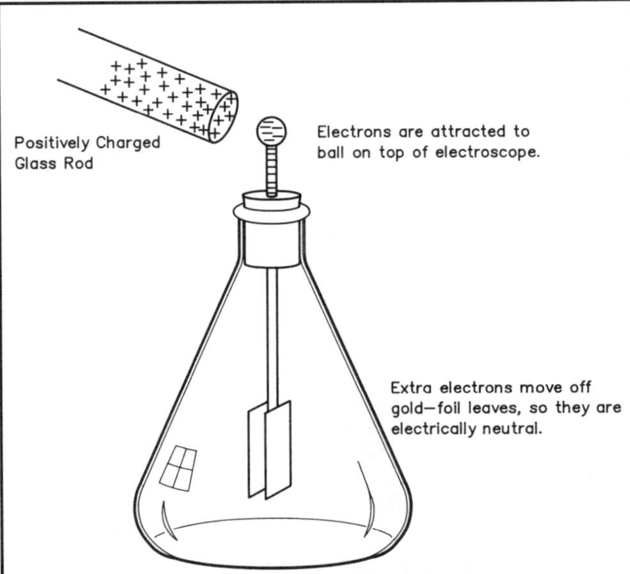

Figure 4—When we bring a positively charged glass rod close to the negatively charged electroscope, the gold leaves will go down. This is because the positively charged glass rod pulls electrons to the top of the electroscope, leaving the foil leaves neutral again.

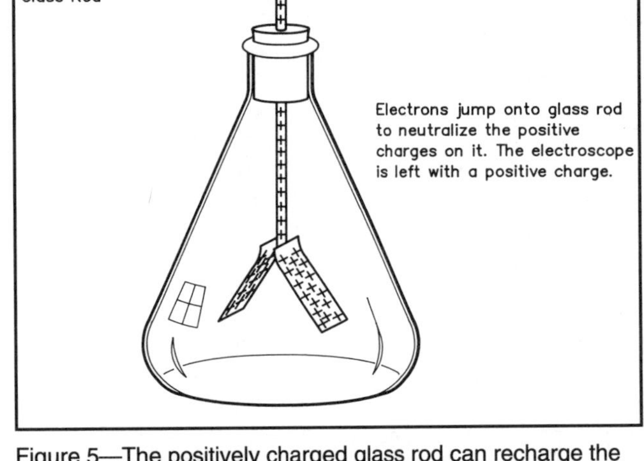

Figure 5—The positively charged glass rod can recharge the negative electroscope when it touches the ball on top. The gold leaves of the now positively charged electroscope again repel each other, and stand out to the side.

The Structure of Matter

We have been studying electric charges and electrically charged objects. You know there are two kinds of electric charge, and that electrons and protons are small particles that carry these charges. You are probably wondering how these electric charges move around on an object and how they move from one object to another. To understand the movement of these charges better, we'll have to look at the makeup of all matter.

Matter is a scientific term describing every kind of material. Matter can be a solid, a liquid or a gas.

Let's begin our brief study by looking at a cup of water. Normally, we think of water as a liquid. Keep in mind, though that we can turn water into a solid by freezing it. We also can turn water into a gas by boiling it. After any of these changes, the material is still water. (When we boil the water, it turns to steam and the steam leaves the cup, but it is still water.) See Figure 1.

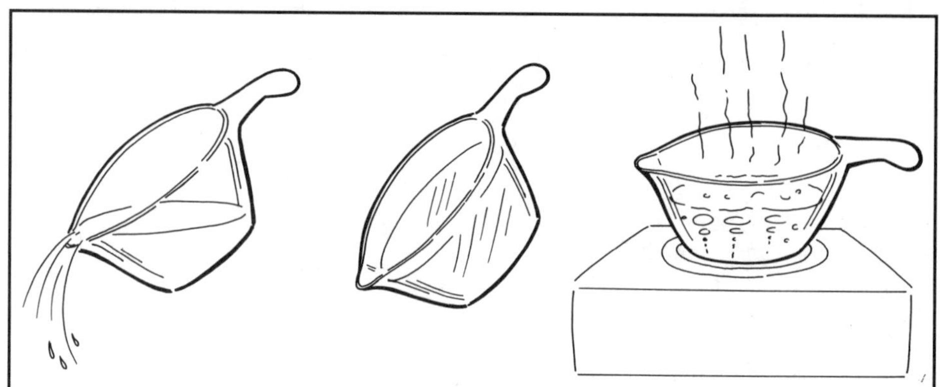

Figure 1—A cup of water shows the three states of matter. At room temperature, water will be in its liquid form. If the temperature is low enough, the water will turn to its solid form, ice. If we heat the water on a stove it will boil and turn into a gas, steam.

Imagine taking smaller and smaller amounts of water, until eventually we have a single *molecule* of water. A *molecule* is the smallest particle of any material that will still have the properties of that material.

We can't make a water molecule any smaller. We could split the molecule, however, and end up with an *atom* of oxygen and 2 atoms of hydrogen. Figure 2 illustrates a water molecule splitting into hydrogen and oxygen atoms. *Atoms* are the smallest particles of *elements*, which combine to form molecules of other materials. An *element* is the simplest form of matter. Scientists have found more than a hundred different elements. Each element has certain characteristics to help identify it. Copper, gold, nitrogen, silicon and uranium are all elements.

Scientists who study matter look for ways to break particles into smaller and smaller pieces. From our discussion so far, you can see they have been quite successful in finding smaller particles that make up everyday materials.

As you might imagine, scientists weren't content to stop when they found atoms. They wanted to know what atoms were made from, too. Individual atoms are so small they are impossible to see, even with a very powerful microscope. Through some experiments, however, scientists learned a lot about how atoms are put together.

Atoms are mostly empty space. There is a small solid center in an atom, and that center, or *nucleus* has a positive charge. Surrounding the nucleus is a rather large, (nearly) empty space that has a negative charge. See Figure 3. There are the same number of positively charged particles in the nucleus as there are negatively charged particles around it. The overall effect is that atoms are electrically neutral.

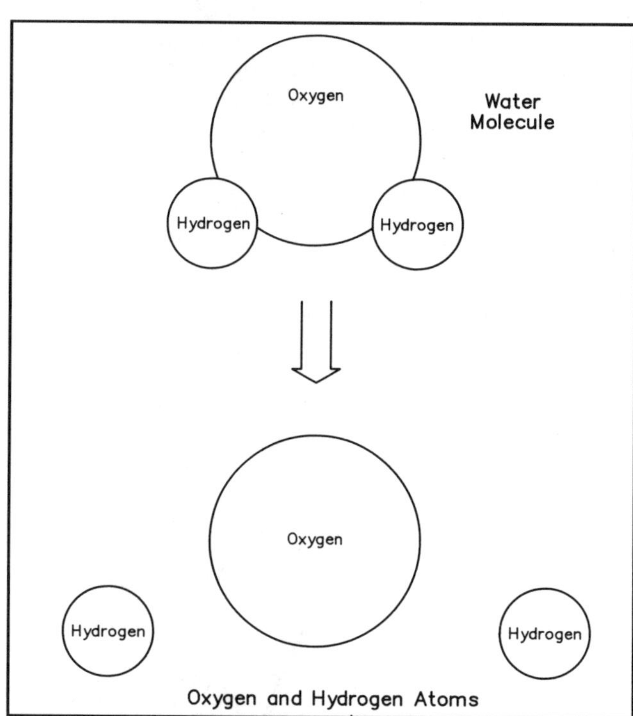

Figure 2—One oxygen atom and two hydrogen atoms make up each water molecule. We can divide water molecules into these individual atoms, but the resulting material will be oxygen and hydrogen rather than water.

The negatively charged particles surrounding the nucleus are *electrons*. These electrons move very fast. Scientists found two kinds of particles inside the nucleus. The positively charged particles in the nucleus are *protons*. You also might find *neutrons*, which are electrically neutral, in the nucleus. The protons and neutrons are almost 1840 times heavier

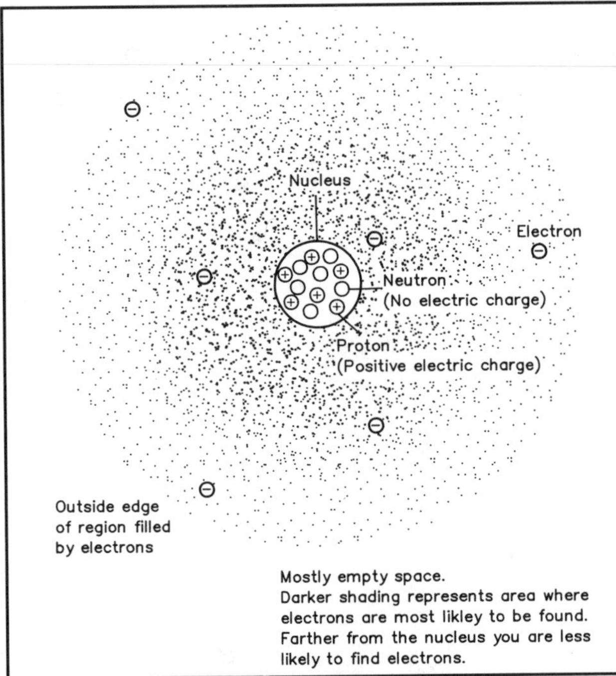

Figure 3—This drawing shows a simplified model of an atom. There is a small central core, called the *nucleus*. The nucleus contains positively charged protons and electrically neutral neutrons. Surrounding the nucleus is a larger region of (mostly) empty space. Negatively charged electrons occupy this space. The number of electrons surrounding the nucleus is the same as the number of protons in the nucleus. Atoms are electrically neutral.

than the electrons. That helps explain why the electrons can move so much more easily than protons or neutrons.

The number of protons is the same as the number of electrons in any atom. The number of neutrons varies. It is the number of protons (or electrons) that determines what element a certain atom is. For example, a hydrogen atom has one electron and one proton. A helium atom has two electrons and two protons while a carbon atom has six of each. Oxygen has eight protons in its nucleus and gold has 79.

Figure 4 shows a drawing of a simple model of an oxygen atom. The nucleus contains eight protons and eight neutrons. There are eight electrons flying around in the space surrounding the nucleus. Although most of the space around the nucleus is empty, the electrons effectively fill the space because of their motion. It might help to think of the space occupied by the electrons as a cloud surrounding the nucleus. There isn't a definite edge, or boundary, to the cloud. Sometimes an electron might fly out a bit farther from the nucleus and at other times it might be in much closer.

Maybe you could use some help understanding the relative size of an atom and its nucleus. Imagine a tiny grain of sand in the middle of a football field. Now let the grain of sand represent the nucleus of an atom. Then the electrons will fill a space the size of the football field, in all directions.

The model of an atom developed by scientists includes a definite, detailed electron structure. You don't need to know about those details right now, however. Just remember that

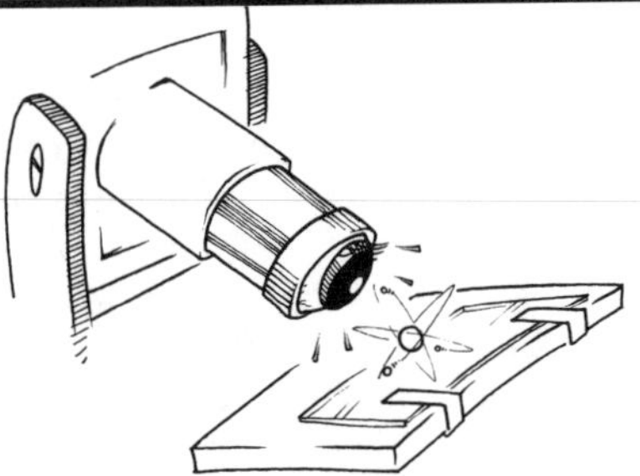

some elements have electron structures that are more stable than others. With some elements it is not very difficult for an electron to leave one atom and move on to another one. The atoms of some elements will gladly accept an extra electron or two. There are other elements, however, that simply won't allow any more in their electron clouds.

If an atom loses an electron, then it has one more positive charge than negative charges. When we rub a glass rod with a silk cloth, we pull electrons away from some of the atoms. These atoms now have a positive charge, and so the glass rod itself has a positive charge.

Similarly, if an atom gains an electron it has one more negative charge than positive charges. Some of the atoms in a hard rubber rod each pick up an extra electron when you rub the rod with fur. The atoms in the rod have a negative charge, and so the rubber rod itself also has a negative charge.

It is important to remember that electrons move from one atom to another. Normally, atoms have as many electrons as protons, so they are electrically neutral. Electrons give atoms a negative charge if there are too many of them. They also give atoms a positive charge if there aren't enough of them. The atoms of some elements, like copper, gold and silver allow their electrons to move from one atom to another rather easily. The atoms of some other materials hold on to their electrons very tightly. This prevents electrons from moving from atom to atom.

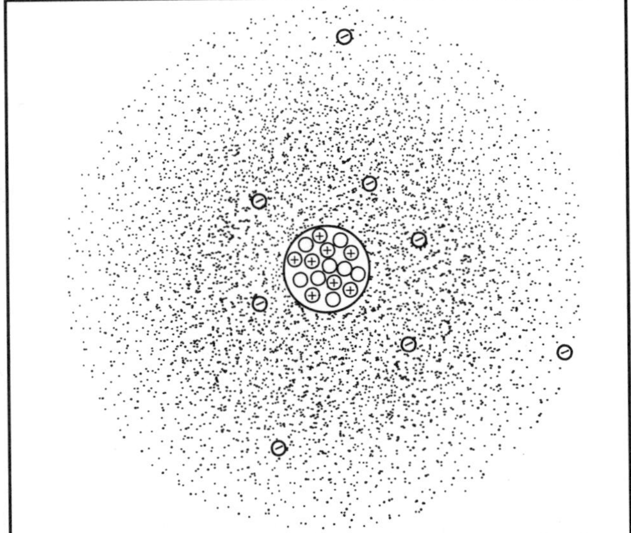

Figure 4—This drawing is a simplified sketch of an oxygen atom. There are eight protons in the nucleus and eight electrons in the *electron cloud* surrounding it. There are also eight neutrons in this particular oxygen atom. There could be anywhere from five to twelve neutrons in an oxygen atom!

Magnets and Magnetic Fields

Most readers are familiar with some of the basic properties of magnets. There are two ends to any magnet, and these ends represent "opposite" kinds of magnetism. These opposite kinds of magnetism are the north-seeking pole and the south-seeking pole of the magnet. (The north-seeking pole of a magnet will turn to point toward the Earth's north pole, if it is free to move. This is why we call that pole the north pole of the magnet.)

We know that two magnetic poles of the same kind repel each other. Two opposite magnetic poles attract each other. Either end of a magnet will attract a piece of iron that is not magnetized. Some metals besides iron can be magnetized or attracted by a magnet. Many metals are nonmagnetic. (Cobalt and nickel are slightly magnetic. Aluminum and copper are examples of metals that are not magnetic.)

There is one other important property of magnets that we should think about here. The forces of attraction and repulsion that magnets exert on each other and on other materials act through space. The objects do not have to be in contact for magnetic forces to act. This is different from many other natural forces with which you are probably

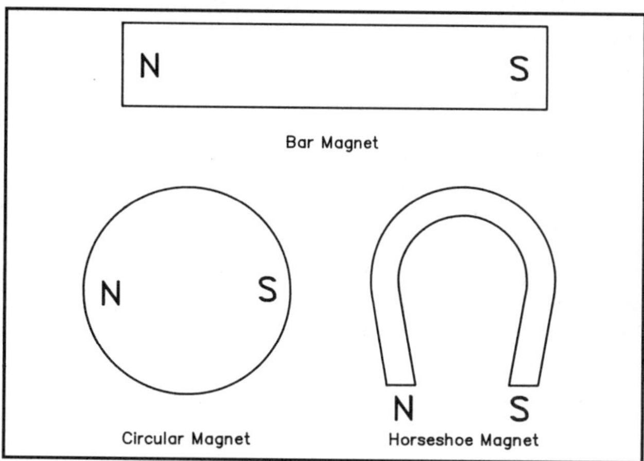

Figure 1—Every magnet has two ends, called *poles.* These poles represent "opposite" kinds of magnetism.

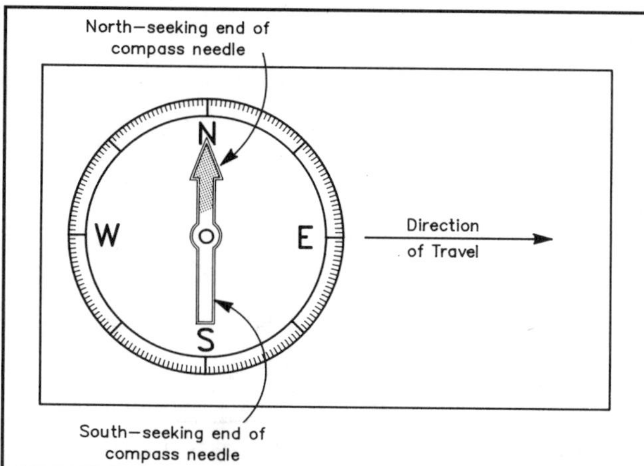

Figure 2—The poles of a magnet are "*north-seeking*" and "*south-seeking*," depending on which of the earth's magnetic poles they point towards.

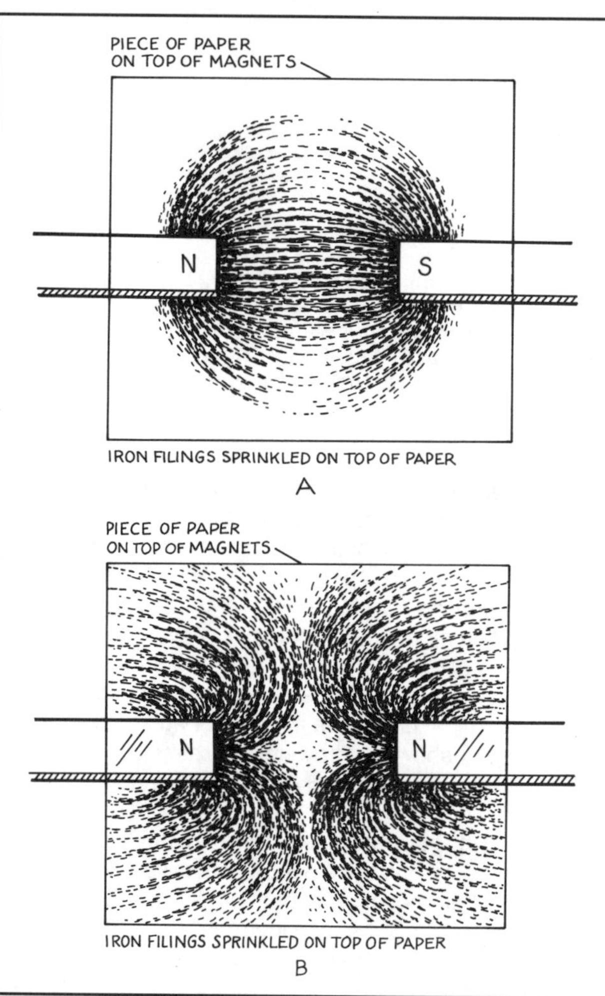

Figure 3—To illustrate the lines of force in a magnetic field, we can place a piece of paper on top of two magnets. Then if we sprinkle some iron filings on the paper, they will line up with the lines of force. If there are many lines between the two magnets, the magnetic field is strong. If there are only a few lines, the field is not as strong. The number of lines represents how strong the force of attraction or repulsion will be. If the lines extend from one magnet to the other, as at A, the magnets attract each other. If the lines leaving one magnet return to the same magnet, as at B, the magnets repel each other.

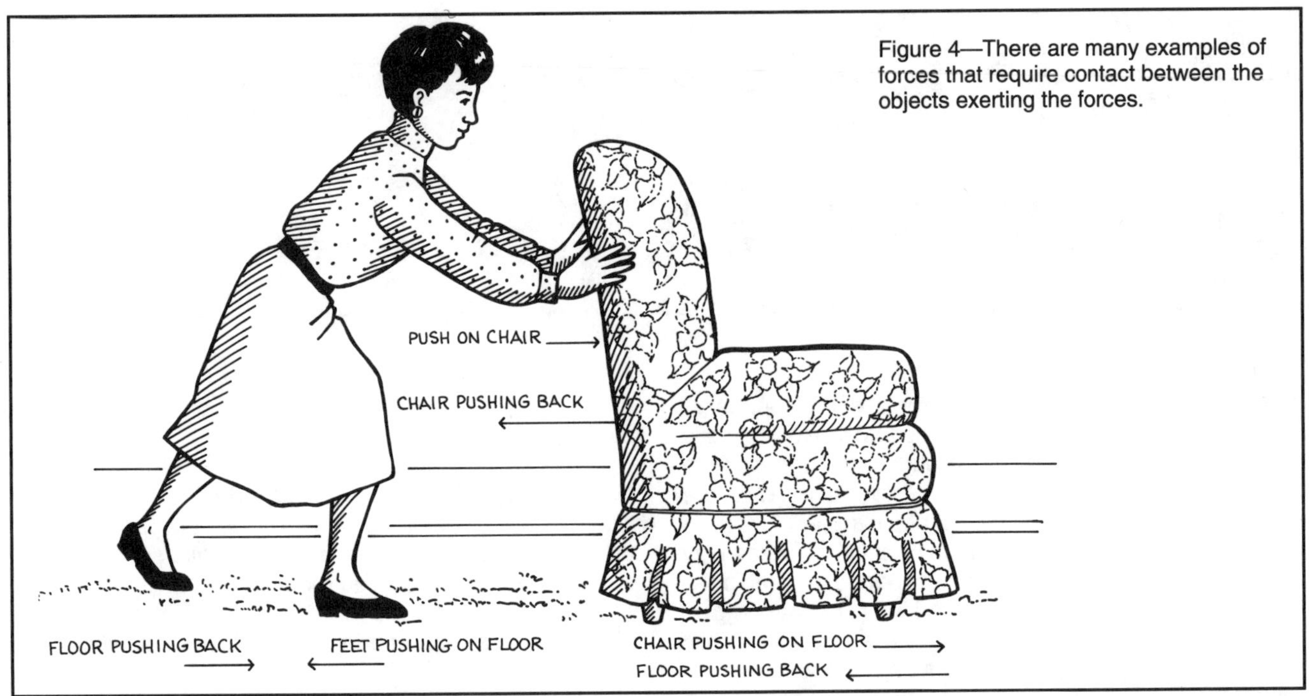

Figure 4—There are many examples of forces that require contact between the objects exerting the forces.

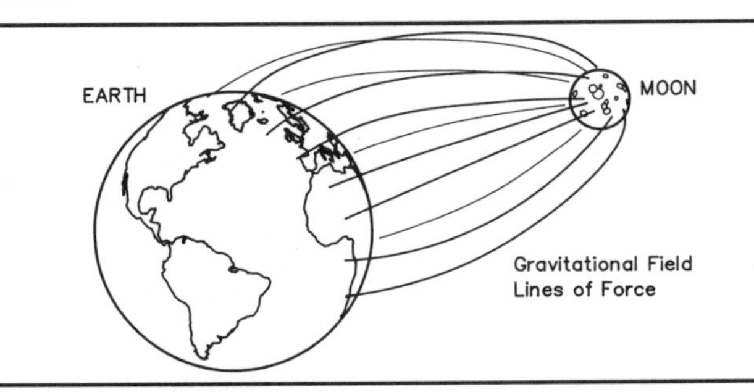

Figure 5—There is a gravitational field between the earth and the moon that holds the moon in orbit around the earth. The lines represent the lines of force for this gravitational field.

familiar. For example, consider what happens when you push or pull on a chair to slide it across the floor. There are several forces acting because of the objects in contact with each other. You must be touching the chair to move it. You can't look at the chair from across the room and make it move! There is also a friction force that tries to keep the chair from moving. It acts because the chair legs are in contact with the floor.

There are many other examples of forces that act only when objects are in physical contact. The force of exploding gasoline vapor pushing down on the piston in a car engine is one. The difference in air pressure that results in the "lift" that makes an airplane fly is another.

Gravity is a natural force that acts without contact. Pick up a small stone and hold it in the air. Can you feel the earth trying to pull the stone back to the ground? If you let go, the stone will fall. Gravity forces also can act over very large distances. The pull of gravity between the earth and the sun keeps the earth in orbit around the sun.

You might call the forces that act through space *invisible forces*. Scientists have a special name for these invisible forces. They call them *fields*. A field also refers to the region of space through which the force acts. When scientists describe a force of gravity, they call it a *gravitational field*. Likewise, they describe a magnetic force as a *magnetic field*.

Electric Fields Are Similar to Magnetic Fields

The two ends of any magnet represent "opposite" kinds of magnetism. The force that these magnetic poles exert on each other produces a *magnetic field*. The two kinds of electric charge, *positive* and *negative*, are similar to the two magnetic poles. A positive electric charge attracts a negative electric charge in much the same way that a magnetic north pole attracts a magnetic south pole. Two electric charges of the same kind push each other apart in much the same way as two like magnetic poles would. The electric charges do not have to be in contact with each other to produce these forces. The electric force acts through the space between the charges. As you probably guessed, we call this electric force an *electric field*.

Let's consider some of the properties of an electric field. Like the magnetic field, the electric field is invisible. Tear a few small bits of paper and place them on your desk or table. Then run a plastic comb quickly through your hair four or five times. Now bring the comb slowly toward the paper bits. What happens? Does the paper jump one or two centimetres off the table to meet the comb? Can you see anything reaching out from the comb to pick up the paper?

When the comb runs through your hair, friction pulls electrons off your hair and onto the comb. In this way the comb becomes negatively charged. You can collect more electrons on the comb by running it through your hair 15 or 20 times instead of five times. (There is a limit to how much negative charge you can collect on the comb, however. You probably won't collect a bigger charge by running the comb through your hair 50 or 100 times rather than 20 times!) Does the paper jump farther off the table when you approach it with the comb now? It should. When the comb has a larger negative charge, it exerts a larger force on the paper bits. The electric field between the comb and paper must be stronger when there are more electrons.

We can conduct the same experiment with a positively charged object. Rub a glass rod with a silk cloth. The silk pulls electrons off the glass in much the same way the comb pulls electrons off your hair. The glass rod will have a positive charge after you rub it.

We represent the electric field between charged objects by drawing a series of lines. This is similar to the

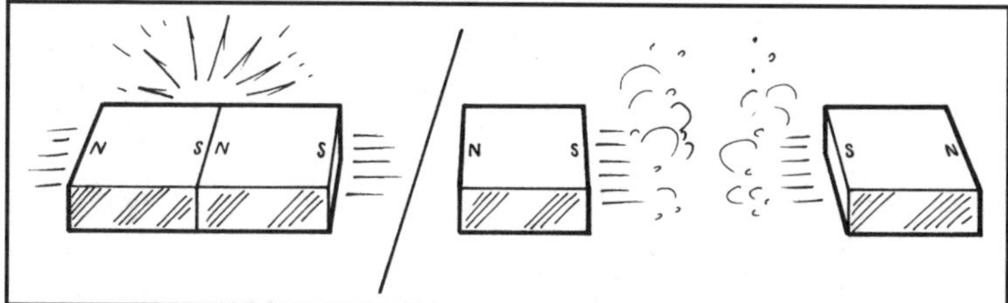

Figure 1—A magnetic field represents the invisible force between two magnets. When two opposite magnetic poles face each other, there is an attraction force between the magnets. If like poles face each other the force would try to push the magnets apart, or repel each other.

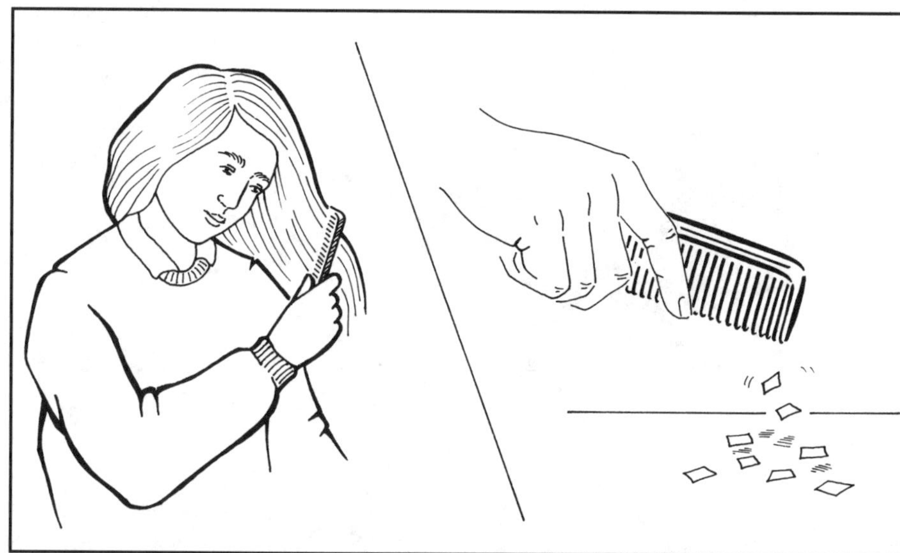
Figure 2—Try this simple experiment. Tear a few small bits of paper and put them on a table. Then run a comb through your hair a few times. Slowly bring the comb toward the paper bits and watch what happens.

way we draw a magnetic field. Figure 3 shows the electric field between a positively charged ball and a negatively charged ball. Notice that the arrows on the lines point from the positive charge to the negative charge. These arrows represent the direction of the electric field.

Figure 4 shows the electric field between two negatively charged balls. A similar picture would represent the electric field between two positively charged balls.

The number of lines indicates the strength of the electric field. A stronger field has more lines than does a weaker one. An electric field represents the strength of the force between the charged objects.

The force of an electric field can move electrical charges. The paper bits in our comb-and-paper experiment jumped up to the comb. The comb pushed electrons away and attracted the positive charges. We refer to this ability to move electric charges as an *electrical potential.* (The field has the *potential*—or possibility—of moving charges.) Voltage is another name for electric potential. Voltage is like a pressure that moves electrons through a wire.

Is an electric field the same as a voltage? No. They are related, but they are not the same. You don't have to know the exact details to understand basic electronics, so we'll stay away from the more difficult mathematics in this book. Just remember this important idea: Any time there is a voltage, or potential difference, between two charged objects there is also an electric field.

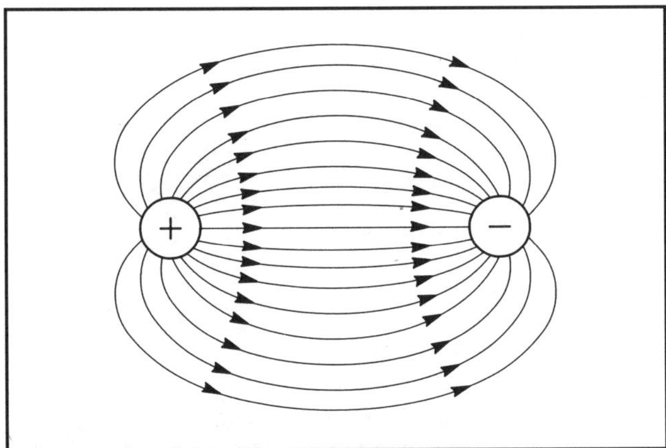

Figure 3—An electric field represents the invisible force between two electrically charged objects. This diagram shows the electric field between a positively charged ball and a negatively charged one. Notice that the electric-field lines go *from* the positive charge *to* the negative charge.

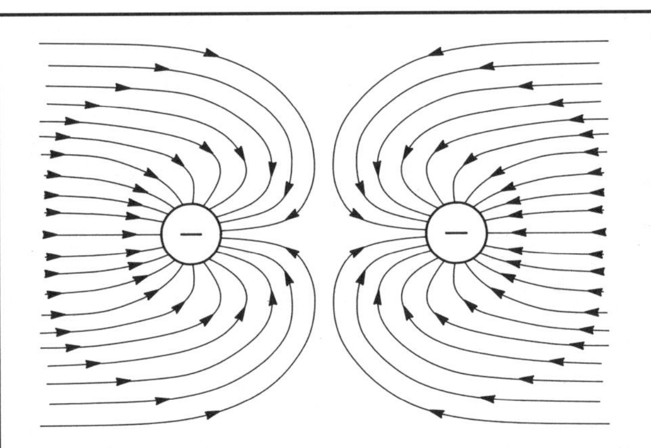

Figure 4—This drawing shows the electric field between two negatively charged balls. Notice that the electric-field lines point toward the negatively charged balls. The electric field between two positively charged balls looks similar, but the lines point away from the charged objects.

Electric Fields— An Experiment for You to Do

In the earlier sections of this chapter you learned about electric charges and electric fields. You can't see an electric charge, nor can you see an electric field. Yet both are real. You can observe some effects of charges and fields, however. It can be difficult to understand something you can't see. Hands-on observations of their effects may help you gain some insight into these concepts.

This experiment will take about 5 minutes to set up. Don't be surprised if you spend much more than that observing the results, however. The results are fascinating!

Begin with a strip of aluminum foil about 1 to 1½ inches wide. A standard roll of foil is 12-inches wide. Tear your foil strip into two 6-inch lengths. Roll these pieces into two small foil balls. Next, you'll need two pieces of sewing thread. You don't need an exact length; 10 to 20 inches each should be plenty. You'll also need a sewing needle, a pencil and a nylon or hard-rubber comb. Figure 1 shows all the pieces ready for assembly.

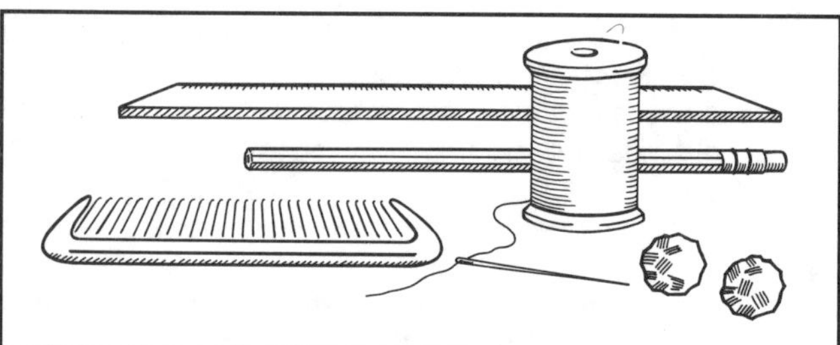

Figure 1—Here is the equipment you will need for a simple electric fields experiment. Make the aluminum-foil balls by tearing two pieces of foil, about 1 to 1½ inches wide and 6 inches long. The thread length doesn't have to be exact; two pieces 10 to 20 inches long will serve. You will attach the balls to the string and suspend them from the pencil. A pen or piece of wooden dowel also will work.

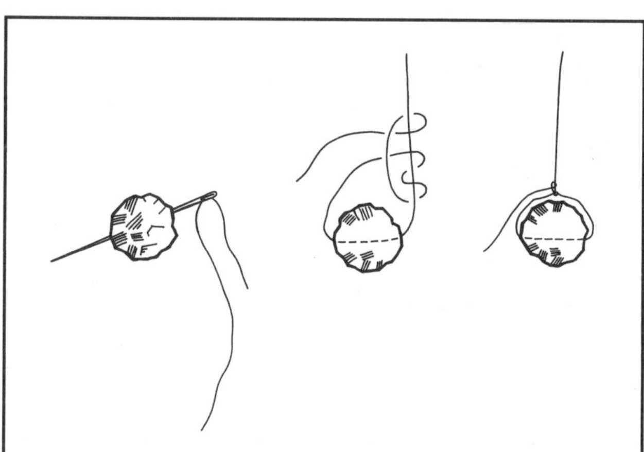

Figure 2—Attach one piece of thread to each aluminum-foil ball. Use a sewing needle to pass the thread through the foil. Then secure the thread by going around the ball and tying a knot.

Now for the hard part. Use a needle to pull the thread through the foil balls. Pull the end through, and leave a few inches to work with. Take the short end around the ball and tie it to the longer piece using a small knot. Tie the strings to a pencil or a small piece of wooden dowel. Figure 2 shows the detail of the string tied to one foil ball. Now tie the free string ends to the pencil, so both balls hang to the same level.

Finally, find some way to hang your experiment so the foil balls can swing freely. The balls should hang about ½ inch apart. Figure 3 shows one way, using a ruler under some books on a shelf. In this example, the pencil balances on the ruler. You may think of better ways to suspend the experiment.

Now you are ready to conduct your experiment. Run the comb through your hair a few times. This puts a negative electric charge on the comb. Bring the comb close to the foil balls and watch what happens.

At first, the balls swing toward the comb, because the negative charge attracts them. You can pull the balls around without touching them because of the effects of the

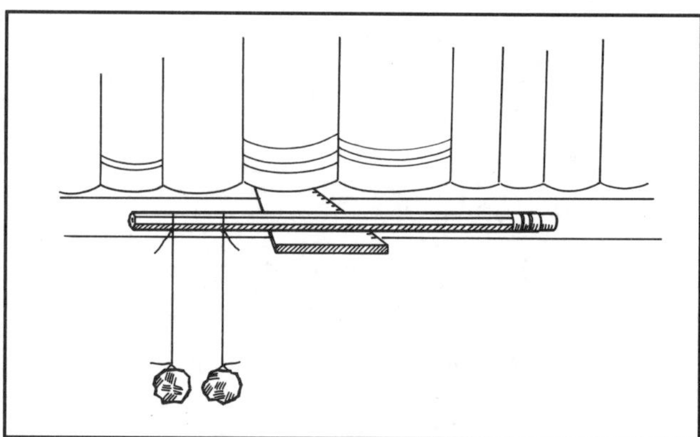

Figure 3—Tie the strings to the pencil so the aluminum-foil balls hang at the same height. Suspend the pencil so the foil balls are away from any objects that might interfere with their movement. Here, some books hold a ruler, and the pencil balances on the ruler. Keep the strings toward one end so both strings are on the same side of the ruler. The foil balls should be about ½ inch apart.

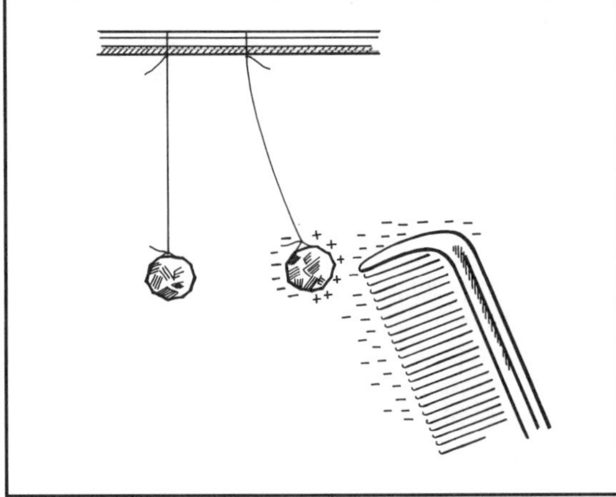

Figure 4—Run the comb quickly through your hair to give it a negative charge. Now bring the comb close to the foil balls. Notice how the comb attracts the balls? What happens if you run the comb through your hair longer or faster? Try running the comb through other family members' hair. You also can experiment with your dog's or cat's fur.

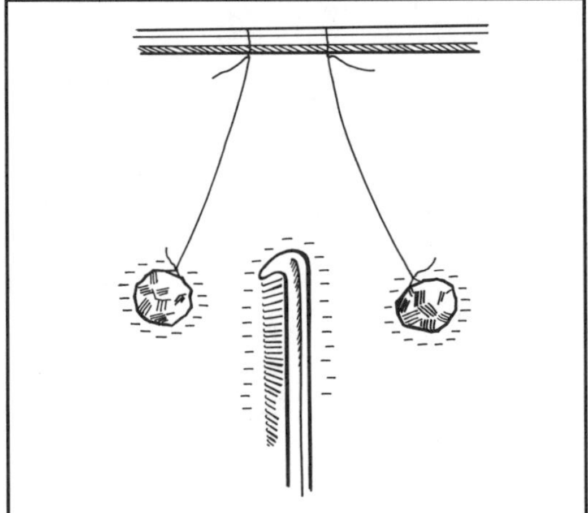

Figure 5—After you experiment with the comb attracting the foil balls, allow the balls to touch the comb. Notice how they immediately swing away from the comb? Experiment with pushing the foil balls around with the comb. Swing the balls out to the side and make them touch the comb again. Do the balls become more negatively charged? How can you tell? Repeat the process a few times to see how much stronger you can make the force between the foil balls and comb.

electric field surrounding the comb. Figure 4 shows that the negative charge on the comb pushes electrons away from the side of the ball it approaches. This leaves that side of the ball positively charged. That is why the comb attracts the ball.

Allow one ball to touch the comb. Notice how the ball jumps away immediately after touching the comb? When the ball touches the comb, it takes some electrons from the comb. Then the foil ball becomes negatively charged. Now the comb repels the foil ball, because they have like charges. See Figure 5.

Run the comb through your hair again to recharge it. You can push the foil balls around without touching them now, because of the electric fields. Of course if you try to push the balls too far to the side, the force of gravity will overcome the electric force. Then the balls will fall against the comb. Carefully bring the comb between the two balls and push one to the side until it touches the comb. Do the same with the other ball. You can add more charge to the balls by repeating this procedure a few times. (There is a limit to how much charge you can put on the foil balls, though.)

If you can find a piece of glass rod and some silk cloth, you can repeat the experiment with positive charges. Be careful with the glass rod, though.

CHAPTER 10 Conductors, Insulators and Resistors

Some Materials Permit Electrons to Flow More Easily Than Others—Conductors

There are some materials that allow their electrons to move from one atom to another rather easily. Other materials hold tightly to their electrons, and are not so willing to let go. The amount of electrical pressure, or voltage, it takes to move electrons through a material depends on how free its electrons are. If lots of electrons flow through a material with only a small voltage applied, we call that material a *conductor.* The main reason for using conductors in an electronics circuit is to provide a path for electrons to follow.

Most metals are good conductors. Copper is probably the most common conductor. Silver is one of the best conductors available, although it is expensive. Gold is also a good conductor, but not as good as copper. The main advantage of using gold is that it won't corrode like copper will. A light coating of silver or gold on switch contacts or connector surfaces will help maintain good contact. Aluminum is not as good a conductor as copper, but it is less expensive and lighter.

Iron conducts electricity, although it is not a very good conductor. (It takes only a small voltage to produce a certain current through a copper wire. It takes quite a bit more voltage to produce the same current in an iron wire.) Mercury, a metal that is a liquid at normal temperatures, will conduct electricity. While it is not one of the best conductors, mercury has some interesting uses because it is a liquid.

The *conductivity* of a material tells us how good a conductor that material is. We won't go into the details of how to measure conductivity here. Just remember that a material with a large conductivity is a better conductor than one with a small conductivity. In fact, we seldom use the actual conductivity of a material in electronics calculations. It's common practice to use a number that compares the conductivity of one material with the conductivity of copper. Any material with a relative conductivity greater than 1 is a better conductor than copper. Materials with a relative conductivity of less than 1 are not as good as copper.

Table 1 lists many of the metals commonly used as electric conductors. This table also shows the relative conductivities of the metals listed. The best conductors appear at the top of the list, with the poorer ones shown last.

Figure 1 shows a section of copper wire. To help you better understand how electrons move in a conductor, this figure shows a simplified picture of the copper atoms. It only shows one electron with each atom. That one electron is free to move from atom to atom. Suppose the electron

Table 1

Some Common Conductors of Electricity	*Relative Conductivity (as compared to copper)*
Silver	1.064
Copper	1.000
Gold	0.707
Aluminum	0.659
Zinc	0.288
Brass	0.243
Iron	0.178
Tin	0.151
Mercury	0.018

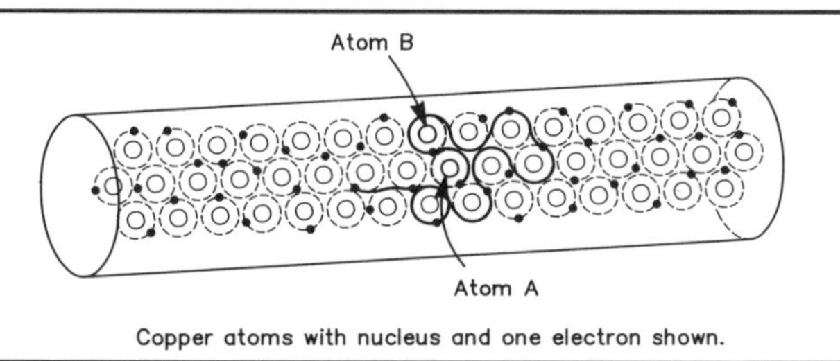

Figure 1—This drawing shows a simplified model of the atoms in a copper wire. It shows the nucleus of some atoms, along with one electron from each atom's electron cloud. The electrons shown can move from atom to atom with little difficulty. The shaded line represents the path followed by one electron as it moves around.

from atom A moves over to atom B. Atom A immediately attracts an electron from another atom to fill the space left by the moving electron. Electrons normally wander randomly from atom to atom, but don't move away from the general area of the atom they started with.

Now let's apply a voltage to the wire. The positive side of the voltage immediately attracts electrons. As the electrons move from atom to atom, they drift toward the positive end of the wire. Figure 2 shows this electric current in a wire. We use the word drift when describing the movement of individual electrons because each electron moves rather slowly through the wire. If you could watch an electron

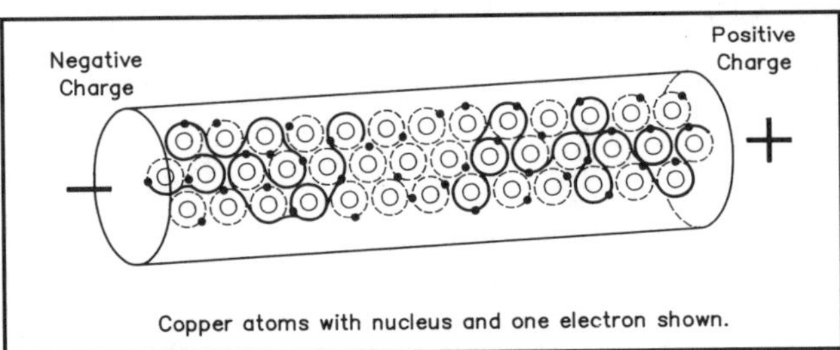

Figure 2—If we connect a piece of copper wire to a voltage source (like a battery) the electrons drift toward the positive terminal. Individual electrons may only move a few inches each second. Electrons are constantly moving from the negative side of the voltage source onto the wire. Other electrons are constantly moving from the wire to the positive voltage source.

move through a wire, you'd find that the electron only moves a few inches every second.

This sounds like a pretty slow speed for electrons to move through the wire. How can an electric light come on so fast when you turn the switch, if the electrons move so slowly? What you have to remember is that when you turn the switch, electrons are already in the wire at the light bulb. They are just waiting for the voltage to push them through the bulb. So as soon as we apply a voltage to the wire, electrons begin to move. Electrons in the center of the wire begin moving toward the positive end. Electrons at the negative end move farther into the wire and electrons from the negative voltage source move onto the wire. At the positive end of the wire, electrons immediately begin to move off the wire and into the positive voltage source. We don't measure an electric current by how *fast* individual electrons move through a wire. We measure current by how *many* electrons are moving.

Figure 3 shows a thin slice taken out of a copper wire that is conducting an electric current. By counting the number of electrons that move through this slice each second, we can measure the current. There is 1 ampere of current for every 6.24×10^{18} electrons moving through this slice in one second! (Yes that sure is a lot of electrons! Atoms are very small, though. There are many times that number of electrons in a thin slice of even a small wire.) Of course we don't actually have to count electrons to measure the current in a wire. We'll discuss more of the details about measuring current in a later section.

You might also realize there are many more atoms in a slice from a thicker wire than from a thinner one. Larger

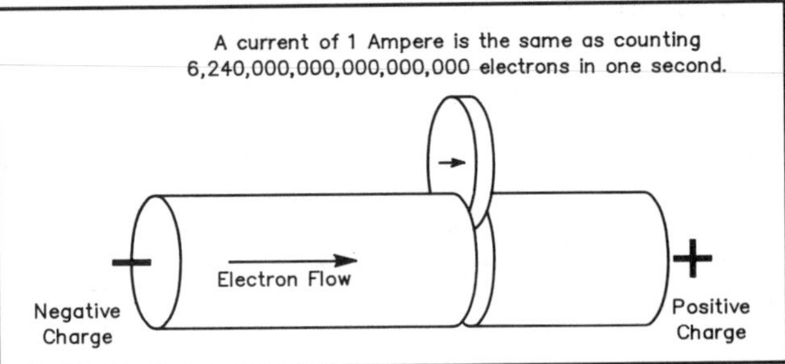

Figure 3—We've taken a thin slice out of a copper wire so we can watch the electrons more easily. Imagine being able to count the electrons as they moved through this section. If you could do that, it would give us a measure of the current in the wire. There is 1 ampere of current for every 6.24×10^{18} electrons that move through this section each second.

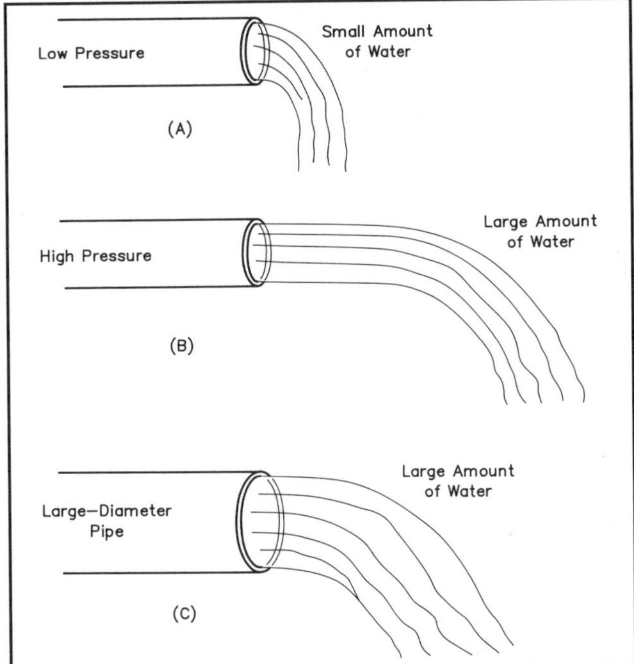

Figure 4—We can compare electric current flowing through a wire with the example of water flowing through a pipe. More water flows if there is more pressure, or if we use a larger diameter pipe.

currents can flow through the thicker wire because there are more electrons available to move.

It's interesting to note that the electrons don't follow a straight path through the wire. Instead, they move from atom to atom, possibly even moving back towards the negative side of the voltage sometimes. As we increase the voltage applied to the wire, however, the electrons move more directly toward the positive end.

We can compare the movement of electrons through a wire with water flowing through a pipe. If there is more pressure pushing the water, more will flow. A higher voltage pushes more electrons through a wire. If we use a larger diameter pipe, it is easier to force more water through the pipe. This is similar to using a thicker wire to carry more current.

It Takes a Lot of Force to Move Electrons in Some Materials—Insulators

Some materials hold tightly to their electrons. These materials don't allow their electrons to move when there is a small voltage across them. Anything that doesn't conduct electricity easily is an *insulator*. You should know, however, that it is possible to make an electric current flow through every material. If the voltage is large enough, even the best insulators will break down and allow their electrons to move.

Insulators are important in electronics because they help to keep the voltages and currents where you want them in the circuit. Insulators also help to protect us from the dangerous voltages used in some circuits.

Figure 1 shows a spark jumping through an insulator. Once the voltage causes a spark to jump, a current will often continue to flow even at much lower voltages. The spark actually punches a hole in the insulator, destroying it.

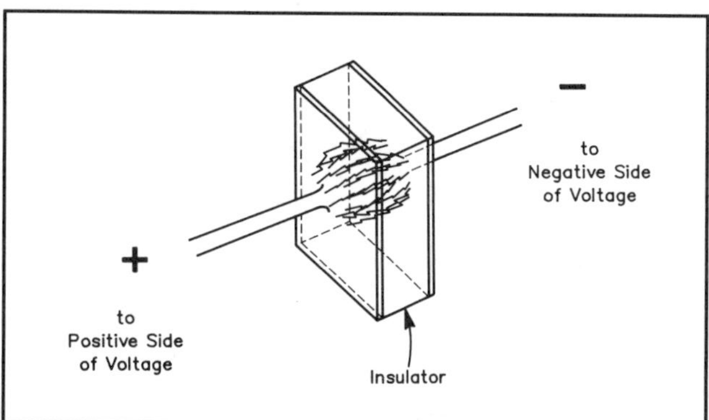

Figure 1—You can make a spark jump through any insulator by applying a large enough voltage. Such a spark will punch a hole in the material, often destroying the insulator.

Figure 2

Just as not all conductors are as good as others, all insulators aren't equally as good either. Some will break down and conduct at lower voltages than others. Any insulator will hold up against a higher voltage if we use a thicker layer of the material, however.

Let's discuss a couple of examples of insulation breakdown that you're probably already familiar with. Nearly everyone has had the experience of walking across a carpeted floor and receiving a shock while reaching for a metal doorknob. You'll notice that you don't receive the shock while you are half way across the room. You do receive it before you actually touch the doorknob, though. In this example, your body builds up an electric charge as you walk across the carpet. The air between your finger and the doorknob serves to insulate that charge from the metal conductor. As you move closer to the doorknob, you reduce the thickness of air insulating you from it. At some point the voltage is large enough to cause a spark to jump through that insulator.

You could conduct a small experiment by first shuffling your feet across a room with a thick pile carpet. Carefully approach a metal surface and point your finger as you reach toward the metal. Move gradually closer, until a spark jumps between your finger and the metal, giving you a small shock. How close did your finger get before the spark jumped? Try walking around the room to build up a bigger electric charge on your body. If the charge is larger, what do you know about the voltage between the metal doorknob and your body? Is the spark likely to jump a bigger or smaller distance this time?

Lightning gives us another example of an insulator breaking down when there is a large voltage across it. Lightning is an electric spark that jumps between clouds or from a cloud to earth. How does the voltage that produces lightning compare with the voltage that gives you a shock after walking across the carpet? Of course the voltage that produces lightning is **much** larger, because it moves electrons through a much thicker layer of air!

The example of water pressure in a pipe may help you understand insulator breakdown. The hollow opening in the pipe is like a conductor and the pipe itself is like the insulation on a wire. If we block one end of the pipe and increase the water pressure, eventually the pipe will burst. Water will shoot out through the break, instead of remaining inside the pipe. It won't take nearly as much pressure to burst a garden hose as it will to burst a heavy iron pipe. No matter how strong the pipe is, if the pressure is large enough, the pipe will burst.

Table 1 lists some materials that we often use as insulators in electronics circuits. The list is in decreasing order of ability to withstand high voltages without conducting. Rubber is one of the best insulating materials known. Mica is another excellent insulator. (Mica is a mineral that easily separates into thin layers and is almost clear, with slight color variations.) Paper can be a good insulator, especially if it is first soaked in some types of oil or wax. Porcelain and most ceramic materials are very good insulators. Bakelite is a brittle plastic material often used as an insulator in older electronics equipment. Glass and fiberglass are also common insulators. Air can serve as an insulator, although its voltage breakdown characteristics vary quite a bit depending on temperature and humidity. You can use wood as an insulator as long as it is dry. Modern electronics circuits also make use of various types of plastics as insulators.

Table 1
Common Insulators

Common Insulators
Rubber
Mica
Wax or Paraffin
Porcelain
Bakelite
Plastics
Glass
Fiberglass
Dry wood
Air

Remember that each material has a limitation on how high a voltage you can safely apply to it. Table 2 lists the approximate voltages that will cause a spark to jump through some of the common insulators. These numbers aren't exact, but they will give you some idea of how much voltage the insulators can withstand.

There is another item that you should consider about insulators. Dirt and moisture may serve to conduct electricity around an insulator. So if an insulator is dirty and there is quite a bit of moisture around it, you could have some problems. Notice that this is not a case of the insulator itself breaking down from the applied voltage. The insulator could probably withstand a much higher voltage, but the dirt provides a path for the electrons to flow around it. See Figure 4. Sometimes accumulated dirt can lead to problems with electronics circuits, then. It's important to keep things clean so the insulators can do their job.

Table 2

Insulator	*Voltage required to spark through 1 cm thickness*
Glass	1,000,000
Mica	500,000
Oiled paper	400,000
Paraffin	350,000
Air	30,000

Figure 3—Lightning is a natural source of electricity. A layer of air makes an insulator between clouds or a cloud and the earth. An electric charge builds up on the clouds and the earth, producing a voltage between them. When the voltage is large enough, a spark jumps through the layer of air separating them.

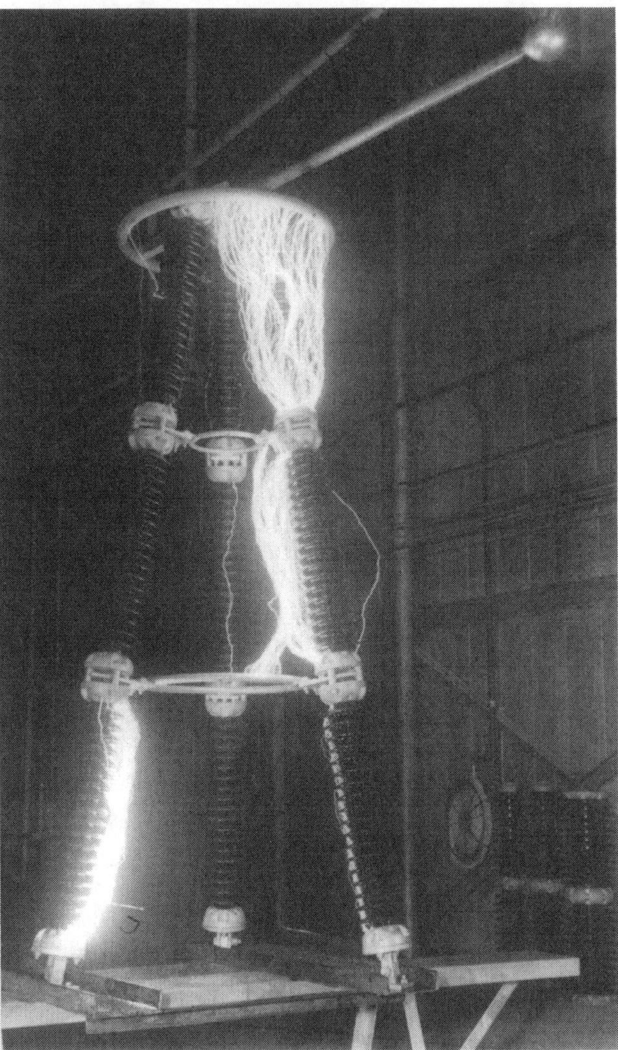

Figure 4—Sometimes the dirt and moisture on an insulator will conduct an electric spark, allowing the spark to jump around the insulator. *(Photo courtesy of Lapp Insulator Company, Le Roy, NY.)*

Controlled Opposition to Electron Flow—Resistors

All materials present some opposition to the flow of electrons. If that opposition is relatively small, the material is a conductor. Insulators are materials that have a large opposition to electron flow. Some materials have too much opposition to be good conductors, yet they allow too much current to be insulators. Such materials are useful in electronics circuits because of their opposition to electron flow. We call materials like this *resistors*. The materials resist, or oppose, the flow of electrons through them.

When we describe electronics circuits, it is helpful to make drawings of the various parts and their connections. We call drawings like this *schematic diagrams*. Special *schematic* symbols represent each part in a circuit. Figure 1 shows the schematic symbol for a resistor. The zig-zag line seems to create a traffic jam for electrons. It's like the electrons must slow down to wind their way through the resistor.

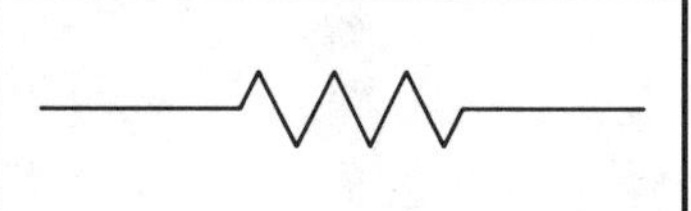

Figure 1—This symbol represents a resistor on schematic diagrams of electronics circuits.

Figure 3—All electronics circuits produce some heat. Most of the time we don't want this heat, but sometimes producing heat is the main purpose of a circuit.

You can control the amount of current in an electronics circuit by carefully selecting the amount of resistance in that circuit. So one important reason to use a resistor in a circuit is to limit the current in some part of the circuit. Carefully selected and properly placed, resistors will guide exactly the right amount of current to each part of the circuit. See Figure 2.

Current flowing through any resistance will produce heat. There will be more heat when there is a larger current. Most of the time this heat is a waste product that we must remove from the circuit. Occasionally, however, we will *want* to produce some heat with a circuit. See Figure 3.

Some metals have a resistance that is quite a bit greater than the resistance of copper, which is the most common conductor. By selecting a type of wire with higher resistance, you can control the amount of current in a circuit. Nichrome, a mixture of the metals nickel, chromium and iron, is one such material. There will be much less current through a nichrome wire than there would be through a copper wire of the same thickness.

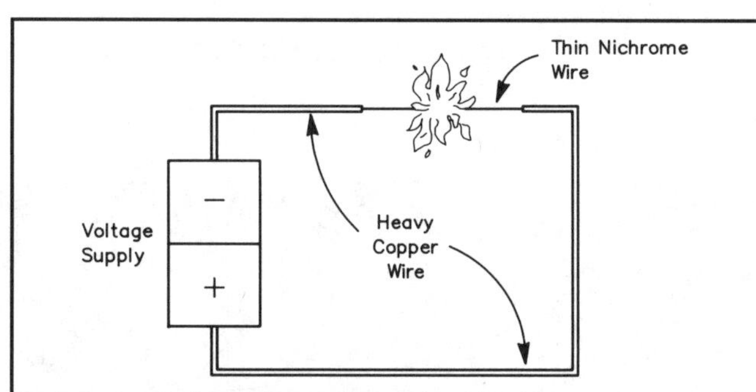

Figure 4—If the current through a resistor produces too much heat, the resistor may melt, or even start a fire.

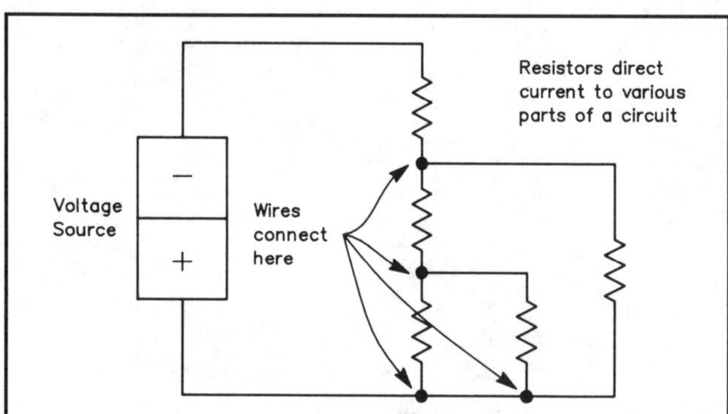

Figure 2—Resistors give us a way to control the amount of current going to different parts of an electronics circuit.

Another way to control the amount of current in a circuit is to select the proper diameter wire. It is much harder for electrons to flow through a thin wire than through a heavy one. So by carefully selecting the diameter of wire used in a circuit you can allow just the right amount of current. Remember though, if it's harder for the electrons to flow through the thinner wire, then the current will produce more heat. If there is too much heat the wire may start a fire, or it may melt and break. It's important to be sure that any resistor can get rid of the heat it produces without damaging itself, or anything nearby.

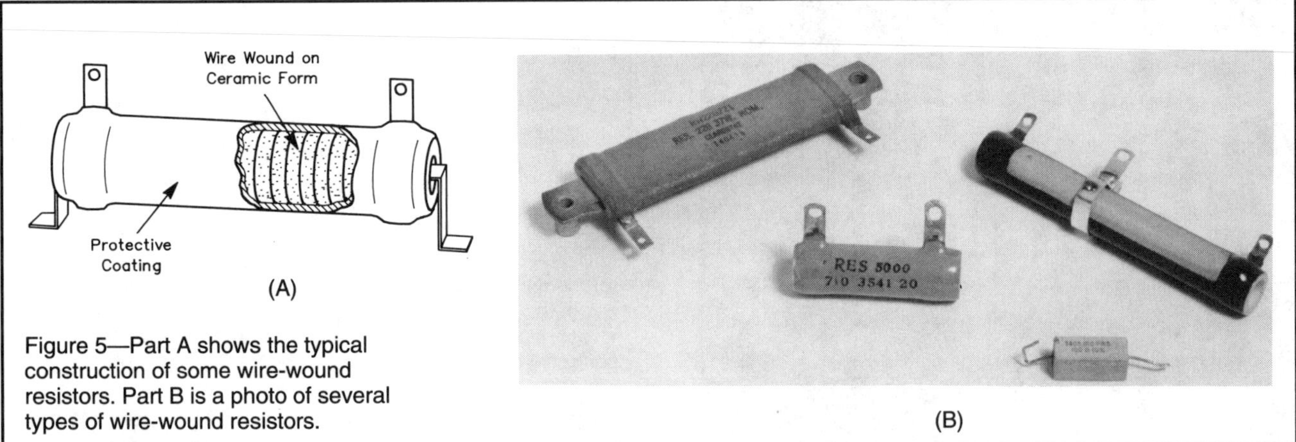

Figure 5—Part A shows the typical construction of some wire-wound resistors. Part B is a photo of several types of wire-wound resistors.

The length of wire that the electrons must flow through helps determine the amount of resistance for a particular piece of wire. You can wind a length of thin wire around a piece of wooden dowel or other form to make a resistor. The exact amount of resistance depends on the type of wire and on its length and diameter. Can you guess why we call resistors like this *wire-wound resistors*? Figure 5 shows several commercially manufactured wire-wound resistors.

Carbon, a black, powdery substance, has a resistance that is quite a bit greater than copper. Manufacturers mix carbon with clay and form it into thin cylinders, or pellets. Then they attach wire leads to the ends, forming resistors that we can conveniently connect into an electronics circuit. The clay serves to hold the powdery carbon together. By changing the amount of clay in the mixture, manufacturers can also control the resistance. As you might have guessed, we call resistors made in this way *carbon-composition resistors*. Figure 6 is a cutaway view of a carbon-composition resistor.

A method used more and more to make resistors is to start with a cylinder of ceramic material, which is an insulator. Next, the manufacturer places a thin film of carbon over the ceramic. Then they attach leads and coat the resistor with a layer of epoxy sealer to insulate and protect it. Resistors like this are *carbon-film resistors*. A similar construction method uses a thin film of metal on the ceramic base. Resistors made in this way are *metal-film* types. Figure 7 shows a cutaway view of film-resistor construction.

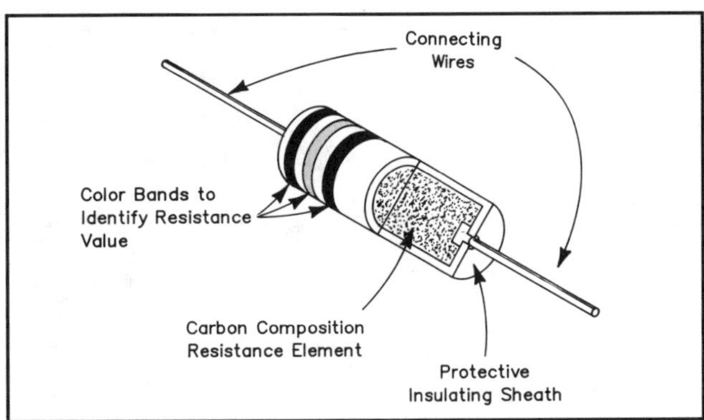

Figure 6—This cutaway view of a carbon-composition resistor shows the construction details of this type of resistor. The resistance material is a mixture of powdered carbon and clay.

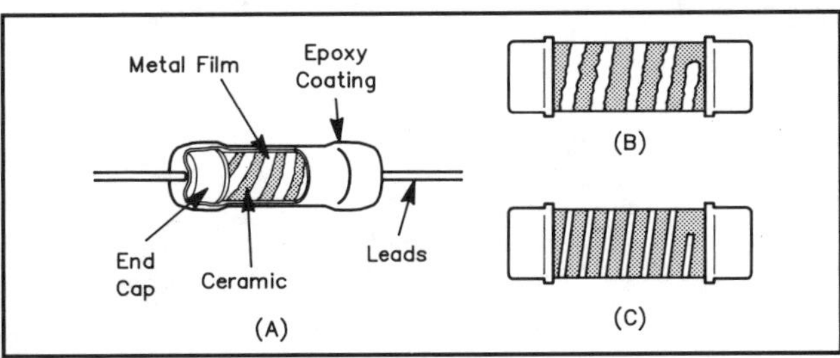

Figure 7—This is a cutaway view of a film resistor. Both metal-film and carbon-film resistors use the same basic construction techniques. The metal or carbon film is trimmed away in a spiral to control the resistance. Part B shows a resistor trimmed with a lathe. The resistor at Part C was trimmed with a laser.

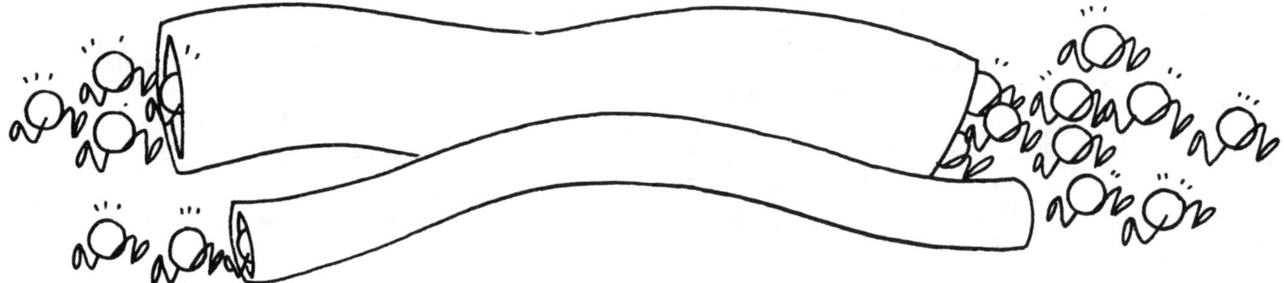

CHAPTER 11 Electrical Circuits

In the last chapter, we described the use of conductors, insulators and resistors in electronics circuits. Just what is an *electronics circuit*, anyway? One simple definition is that an electronics circuit is a path that electrons follow. An electronics circuit usually includes more than a wire from the negative to the positive side of the voltage source, however. Other parts help the circuit do something useful. Figure 1 is a drawing of a flashlight, which is a simple electronics circuit.

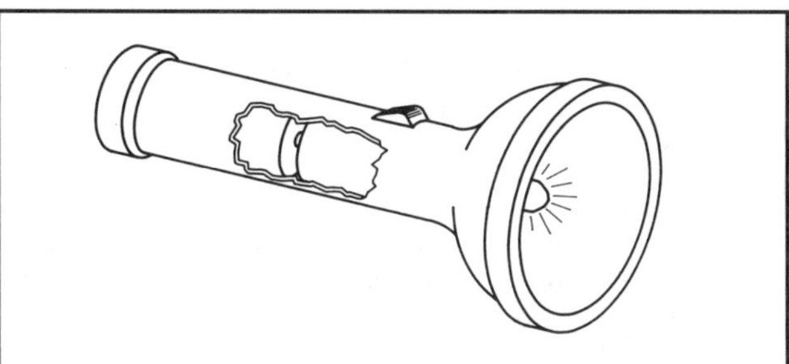

Figure 1—A flashlight is one example of a very simple electronics circuit.

Figure 2 shows this flashlight circuit drawn as a schematic diagram. Two D cells provide the voltage to make electrons flow in our circuit. The label on Figure 2 identifies the schematic symbol for any battery. Notice that the long and short bars on the battery symbol identify the positive and negative ends of the voltage. The schematic symbol shown for an ON/OFF switch is one of several symbols used to represent switches. As you become more familiar with electronics circuits, you will learn the symbols for other types of switches. The lamp symbol shown on the figure can represent just about any type of light bulb.

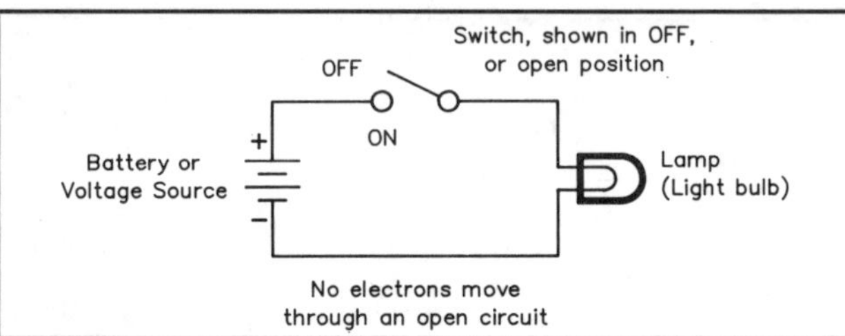

Figure 2—This drawing shows the schematic diagram representing the flashlight of Figure 1. The diagram labels explain the meaning of each special symbol. These symbols represent parts of the circuit. In this diagram the switch is open, so we have an open circuit, and the lamp is off.

In Figure 2, the switch is *open*, and the light bulb is off. With the switch in this position, we have an *open circuit*. No electrons can flow in an open circuit. Let's close the switch, as shown in Figure 3. Now there is a complete path for electrons to move from the negative side of the battery to the positive side. The light bulb glows brightly when electrons flow through it. With the switch in the closed position, we have a *closed circuit*, also sometimes called a *complete circuit*. Electrons can only flow in a closed circuit.

Now let's review what we need to make a useful electronics circuit. First we need a voltage source, the electrical pressure to move electrons through the circuit. Next we need some wire or other conductor for the electrons. Finally, there should be some purpose for the circuit. The light bulb provides the purpose for our circuit of Figures 2 and 3. We often refer to the part that gives a purpose to the circuit as the *load*. So we say the light bulb in our flashlight is the load for the circuit. Any useful electronics circuit has at least these three parts.

We often draw a resistor to represent the load in a circuit. Sometimes a resistor also represents the combined resistance of all the other parts of a circuit. This resistor can include the resistance of the wire that connects the various parts. Even batteries, often used as the voltage source for circuits, have some resistance to the flow of electrons. These resistances are not commercially manufactured

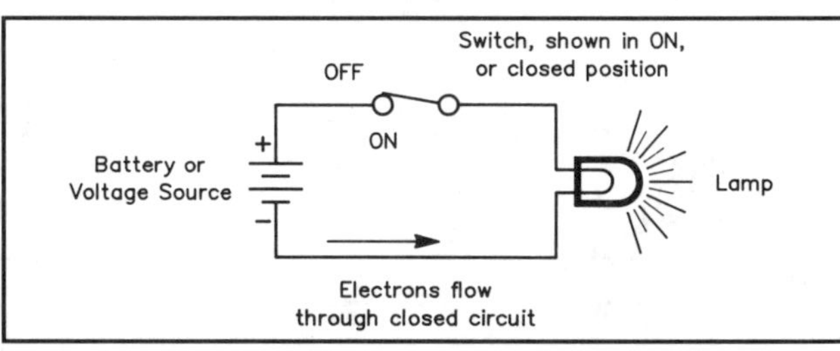

Figure 3—This schematic diagram of a flashlight shows the switch in a closed position. This time we have a closed, or complete circuit, and the lamp is on.

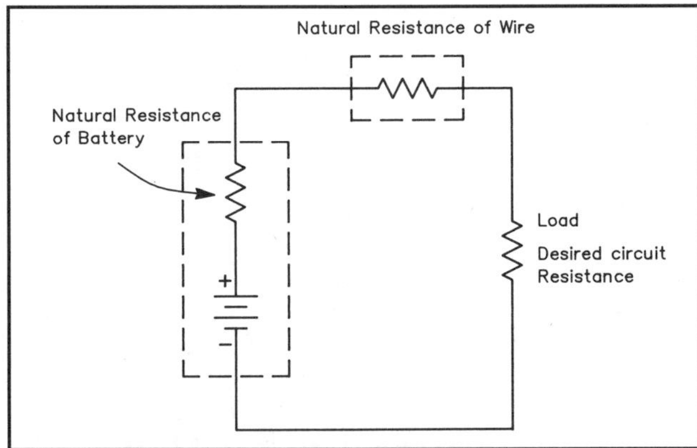

Figure 4—All parts of an electronics circuit have some resistance to current flowing through them. This includes batteries, wires and other parts, along with resistors. The diagram shown here includes resistors to represent the natural resistances of the circuit, as well as the resistance of the load.

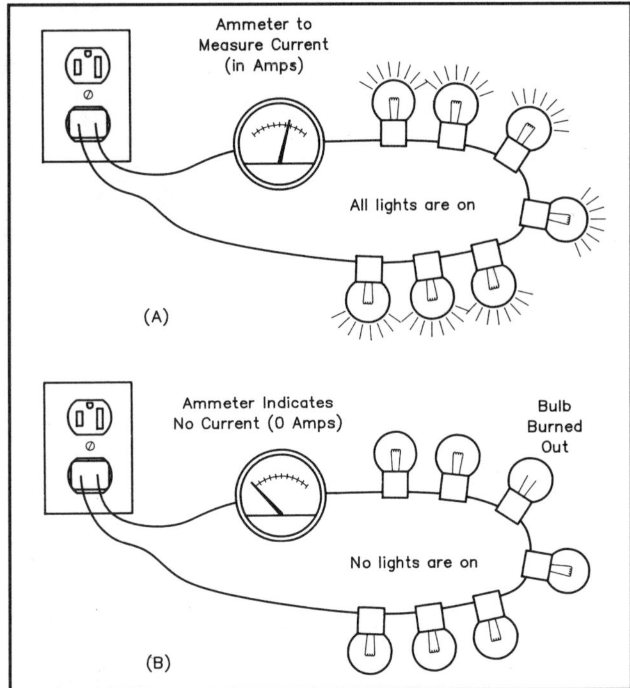

Figure 5—The bulbs for some Christmas-tree lights connect in series. When one bulb burns out, none of the bulbs will light.

resistors placed in the circuit. They represent the unavoidable natural resistance of the materials. Figure 4 shows a simple circuit that includes resistors to represent the load and the other circuit resistances.

All the circuits shown in this section have one other important feature. Any electrons that leave the negative battery terminal must flow through every part of the circuit to reach the positive battery terminal. There is no place in the circuit where the electrons have a choice of directions. This is a bit like getting on an interstate highway and finding there are no exits to get off the highway. You just have to keep going until you reach the end of your journey.

We have a special name for electronics circuits that provide no alternate paths for electrons to follow. A *series circuit* is one in which all electrons must follow the same path, going through every part of the circuit. The current in one part of a series circuit is the same as in every other part. If you measure the current at any place in a series circuit, you'll know the current everywhere in the circuit.

You can have more than one load in an electronics circuit. If you connect several loads so the same current must flow through each load, you have a series circuit. Christmas tree lights give us a common example of a series circuit with several loads. The light strings with small bulbs are often connected in series. If you are familiar with these lights, you also know there is one big disadvantage to such series circuits. If there is a break in the circuit at any point, it becomes an open circuit, and no electrons can flow. So if one bulb burns out, all the lights go out. It can be quite a problem to find the bad bulb in a case like this. Often you have to change one bulb at a time, until you change one and the lights all come on again! If more than one bulb burned out at the same time, you can have a real problem.

Parallel Circuits

Many electronics circuits include alternate paths for the current to follow. Figure 1 is an example of such a circuit. The electrons can follow one of two different paths to move from the negative to the positive side of the voltage supply. *Parallel circuits* are electronics circuits that have alternate paths, so all the electrons don't follow the same path. Each possible electron path forms a *branch* of the parallel circuit.

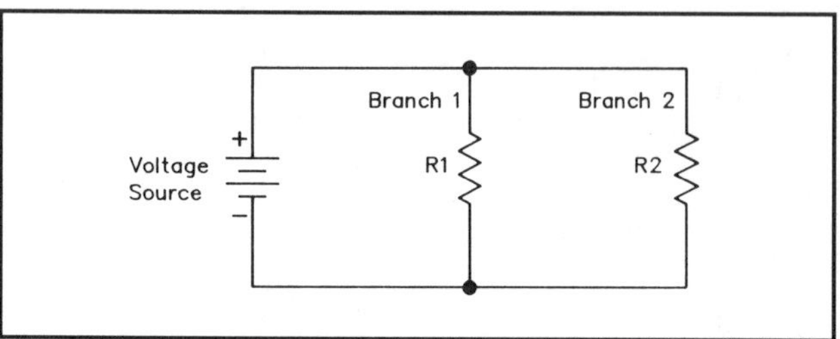

Figure 1—Here is a simple example of a *parallel circuit.* There are at least two paths that the electrons can follow to move from the negative to the positive battery terminal.

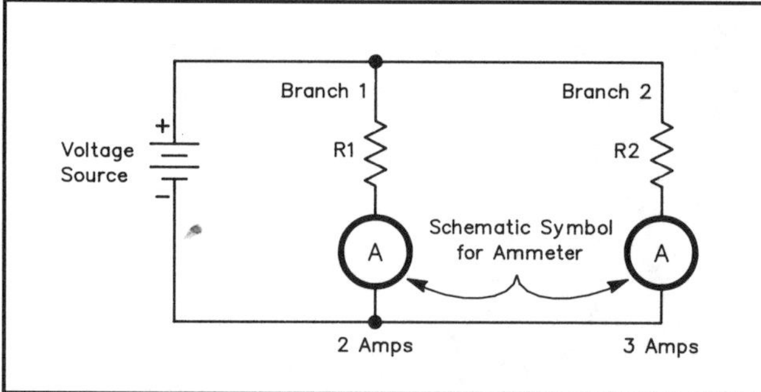

Figure 2—Ammeters measure current in a circuit. There is an ammeter in each branch of the circuit shown in this diagram. Add the current in each branch. This total current is equal to the current out of the battery. In this example, there is a current of 5 amps from the battery.

The current that flows through each individual path may be different. Each branch of the circuit shown in Figure 2 has an ammeter to measure the current in that branch. The current in branch 1 is 2 amperes while in branch 2 it is 3 amperes, in this example. In a parallel circuit, the branch currents add up to the total current from the voltage supply. At the negative battery terminal, we should measure a total of 5 amps for the circuit of Figure 2. (Remember, in a *series circuit* the current in every part of the circuit is the same as in every other part.)

It's often helpful to compare an electronics circuit and water flowing through a pipe. A garden hose is like a series circuit. The faucet that you connect the hose to is like a voltage source, providing pressure to push the water through the hose. The other end of the hose is like the circuit load, where the circuit does something useful. It doesn't matter if you are using the hose to water your lawn, wash a car or fill a swimming pool. All the water flowing into the hose must come out the other end. There is no other way for the water to get out of the hose (at least if you don't have a leaky hose)!

Similarly, we can compare the water pipes through your house to a parallel circuit. Some of the water coming into your house goes to the water heater. Some water goes to the kitchen sink, some to the bathtub and some to an outside faucet. There are many paths for the water to follow. There may be water running from several faucets at the same time, or there may only be one running. This is much the same as a parallel electronics circuit.

Look at the circuit shown in Figure 3. Here, we've connected several light bulbs in parallel. Notice that one of those bulbs has burned out, but the others remain lit. This is an important characteristic of parallel circuits. An open circuit in one part may not stop current from flowing in other parts of the circuit. A switch placed in one parallel branch can turn the current in that branch on or off. Having one branch turned off won't affect the current in other branches.

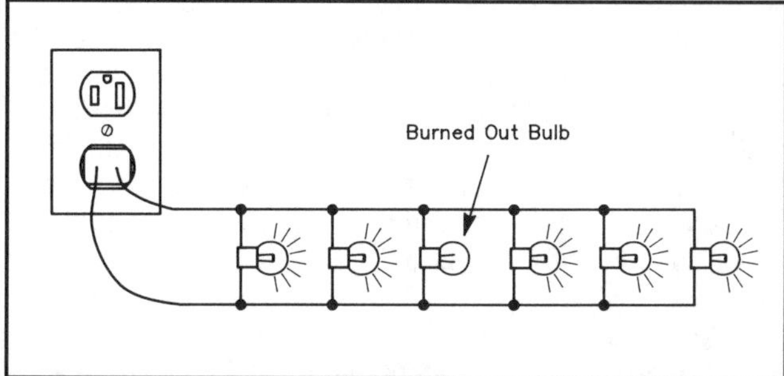

Figure 3—These light bulbs are in parallel. If one bulb burns out, current can still flow in other branches of the circuit.

Even with a parallel circuit, there are parts of the circuit through which all the current must flow. The battery and wires connecting it to the rest of the circuit, for example, must carry all the current. Figure 4 also includes a resistor that must carry all the current. We wire those parts of a circuit that must carry all the current in *series.* Those parts that carry only some of the current are in *parallel.*

You may sometimes see a reference to a *series/parallel circuit.* This simply means that the circuit has some parts wired in series and other parts wired in parallel. Figure 5 shows another example of a series/parallel circuit.

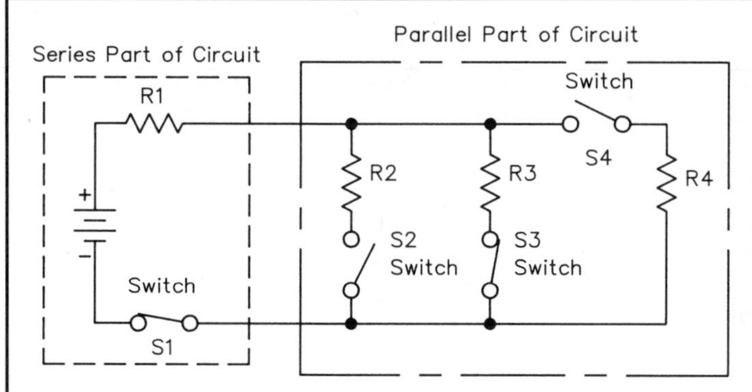

Figure 4—This is an example of a *series/parallel circuit*. There are some parts of the circuit through which all the current must flow. In other parts of the circuit, the current has a choice. Switches in the parallel branches can add or remove that branch from the circuit. Switch S1, in series with the entire circuit, can turn off all the current.

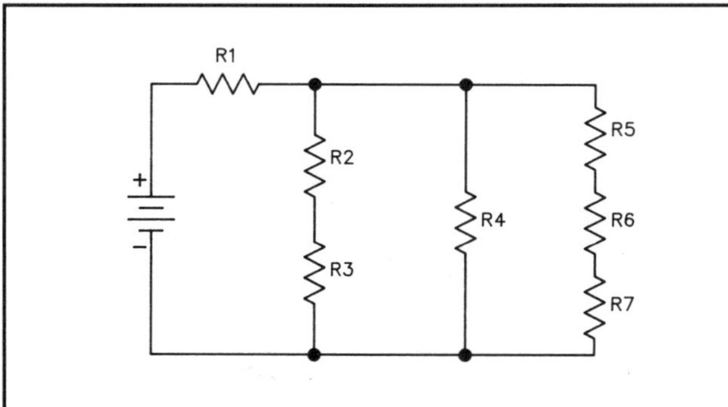

Figure 5—Parallel circuit branches can include parts wired in series.

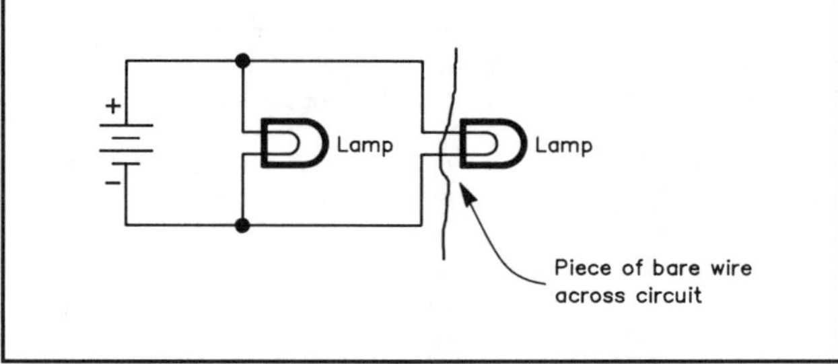

Figure 6—If a conductor provides a current path around the circuit load, there is a *short circuit*. A short circuit can damage other electronics parts in a circuit, and may produce enough heat to start a fire.

Notice that you can have some parts connected in series as a part of a parallel branch.

Some electronics circuits look very complicated. Just remember that all the parts connect either in series or in parallel with the parts around them. With this idea, you can always simplify a circuit to make it easier to understand.

There is one other special type of parallel circuit that you should know about. Suppose that a bare wire or other conductor falls across the circuit, as shown in Figure 6. The alternate electron path provided by this wire will have very little resistance and a large current will flow from the battery. We say the bare wire *shorts* the circuit, because it conducts the electrons around the desired path. This is a *short circuit.*

There are many possible causes for short circuits. Frayed insulation on an electrical power cord is one common cause. A situation like this is dangerous. First is the danger of someone touching the bare wires and receiving a shock or being electrocuted. When the bare wires touch, there will be sparks, and they may cause a fire. The large current flowing in a short circuit can cause the wires to become hot. That heat also can start a fire.

Many circuits include *fuses* to protect against short circuits. A fuse has a thin piece of wire that will get hot and melt if a large current goes through it. Fuses have a rating for how much current they can carry. If the current is larger than the rated amount, the fuse wire melts, and opens the circuit. (No current can flow in an open circuit.) After you correct the cause of the short circuit, you can replace the fuse with a new one. Figure 7 shows a circuit that has a fuse to protect it from short circuits.

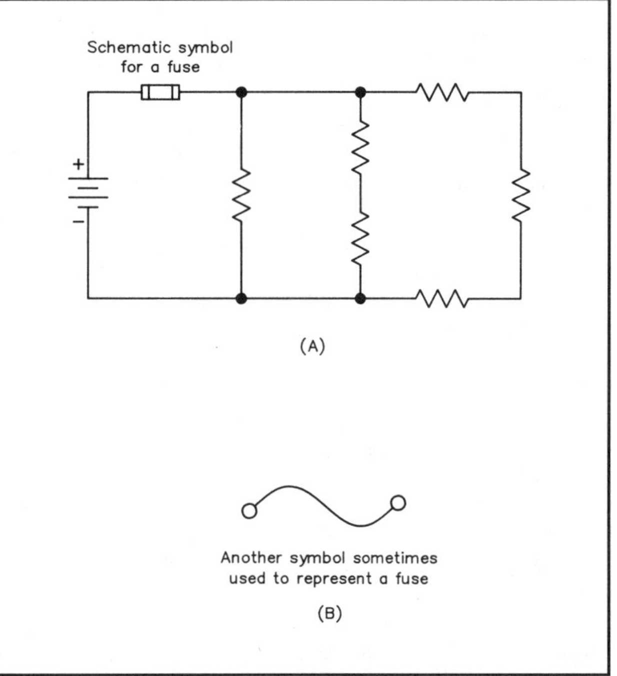

Figure 7—A *fuse* protects a circuit from accidental shorts. The wire in the fuse melts if there is more current than its rating allows. In this way, the fuse produces an *open circuit*, stopping all current. Part B shows another symbol sometimes used to represent a fuse on schematic diagrams.

CHAPTER 12 Ohm's Law

Voltage, Current and Resistance are Related

Let's review a few definitions before getting into the details of Ohm's Law. *Voltage* is the force or pressure that causes electrons to move. *Current* is the movement of electrons through a wire or other conductor. *Resistance* is the opposition to electron movement that any material has. In our earlier discussions, we mentioned that if there is more electrical pressure, more electrons will move in a circuit. We also mentioned that if a conductor presents more resistance to the flow of electrons, there will not be as much current. Obviously, there is some relationship between these three quantities.

In the late 1820's, Georg Simon Ohm, a German scientist, conducted some experiments with simple electric circuits. Ohm discovered a simple mathematical relationship between the voltage applied to a circuit, the resistance of the circuit conductor and the current through the circuit. Other scientists did not accept Ohm's work at first. By the 1840's they began to understand the importance of his discovery, however. Today, *Ohm's Law*, as we call it, is perhaps the most important key to understanding modern electronics.

Look at the simple electronics circuit shown in Figure 1. There is a voltage source and some resistance in the conductor between the positive and negative voltage terminals. A voltmeter in the circuit gives us a convenient way to measure the voltage. An ammeter indicates circuit current.

Let's assume that the resistance in the circuit is a constant value, but we can change the voltage over a range from 0 to 20 volts. We can read the circuit current from the ammeter as we change the voltage. Table 1 shows a series of measurements that we could make with such a circuit. Take a look at the numbers in that table. Do you recognize a pattern? For one thing, you should notice that as the voltage increased, the current also increased. There is also another pattern that may not be so obvious. Every time we double the voltage, the current also doubles. The graph of Figure 2 may help you recognize these patterns. There is a direct relationship between the voltage applied to the circuit and the current through the circuit. That direct

Table 1
Measurements from the circuit of Figure 1 as the voltage changes.

Voltage (V)	*Current (A)*
0	0
1	0.5
2	1.0
3	1.5
4	2.0
5	2.5
6	3.0
7	3.5
8	4.0
9	4.5
10	5.0
11	5.5
12	6.0
13	6.5
14	7.0
15	7.5
16	8.0
17	8.5
18	9.0
19	9.5
20	10.0

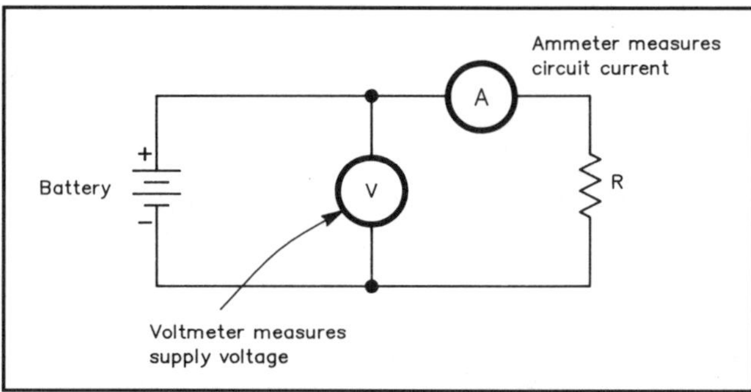

Figure 1—This simple electronics circuit will help us study the relationship between voltage, current and resistance. The voltmeter and ammeter provide a convenient way to measure conditions in the circuit. The text describes some measurements that we can make with such a circuit. Tables 1 and 2 list the results of these measurements for two values of circuit resistance.

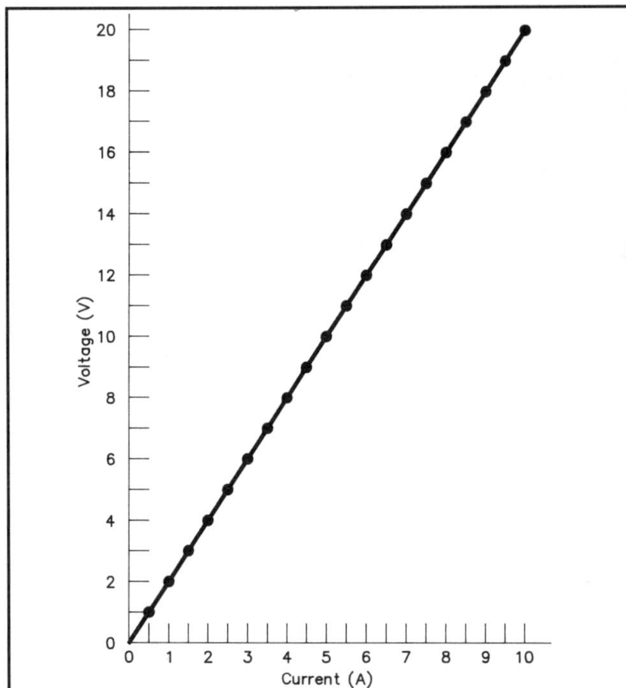

Figure 2—This graph presents the data from Table 1 to illustrate the straight line, *direct proportion* relationship between voltage and current. The values in Table 1 represent an ideal set of data. Actual measured values would vary somewhat from these ideal values.

results of our measurements, as shown in Figure 2. In mathematical terms, we would say that the voltage and current are *directly proportional* to each other.

When two quantities are directly proportional, you can write a simple mathematical equation. For any value of one quantity (let's use current for our example), you can find the value of the other quantity (voltage) by multiplying by some number. So to find the value of voltage for any value of current, we just have to know the correct number, or constant. We can write this direct proportion as an equation:

voltage = current × constant (Equation 1)

From the data shown in Table 1, you can tell that the correct constant for our example is 2. If you multiply any of the listed current values by 2 you will get the voltage value which produced that current.

Now let's suppose there is twice as much resistance in the circuit of Figure 1. Table 2 shows a new set of values for the current as the voltage varies from 0 to 20 volts. Again, you should notice that the current increases as the voltage increases. Also, again, the current doubles every time the voltage doubles. You might try plotting a graph of voltage and current similar to the graph of Figure 2. You will get a straight line graph, proving that voltage and current are still directly proportional. This time, however, you must multiply the current values by 4 to get the corresponding voltages.

We are making the same discovery here that Georg Simon Ohm made in his experiments! We define the amount of resistance in a circuit as the constant that you must multiply current by to find the voltage. So in our first example, the circuit had *2 ohms* of resistance and in the second example it had *4 ohms* of resistance. With the constant in Equation 1 defined as the circuit resistance, we can rewrite that equation as:

voltage = current × resistance (Equation 2)

We use the letter E to represent voltage (or electromotive force). The letter I represents current (from the French word for current, *intensité*) and R represents resistance. Using these letters instead of the words in Equation 2, we can write:

$E = I \times R$ (Equation 3)

Here we have Ohm's Law expressed in its most common form.

The equation for Ohm's Law defines a unit for measuring resistance. The name given to this unit, the *ohm*, is in honor of Georg Simon Ohm and his experiments. We often use the upper case Greek letter omega (Ω) to represent ohms. If a certain resistor has 200 ohms of resistance, we might write this as 200 Ω.

If we know any two of the three quantities in Equation 3, we can calculate the other one. For example, given current and resistance, we can calculate voltage. If we know the voltage and resistance we should be able to calculate the circuit current. If we know the circuit current and voltage, we can calculate the resistance of the circuit. In the next section, we will discuss the variations on the basic Ohm's Law equation, and ways to help you remember those variations.

Table 2

Measurements from the circuit of Figure 1 as the voltage changes.

Voltage (V)	*Current (A)*
0	0
1	0.25
2	0.50
3	0.75
4	1.00
5	1.25
6	1.50
7	1.75
8	2.00
9	2.25
10	2.50
11	2.75
12	3.00
13	3.25
14	3.50
15	3.75
16	4.00
17	4.25
18	4.50
19	4.75
20	5.00

Finding the Unknown— The Ohm's Law Circle

Ohm's Law is a tool you will use to study most electronics circuits. There are other important rules and conditions that you will study as you learn more about modern electronics. Most of those rules will require an understanding of Ohm's Law, however.

We usually write Ohm's Law as an equation.

$E = I \times R$ (Equation 1)

We often omit the multiplication sign in such equations, simply writing the letters of quantities to be multiplied next to each other:

$E = IR$ (Equation 2)

When you know any two of the three quantities in a circuit (voltage, current and resistance) you can calculate the third quantity.

Suppose the voltmeter in the circuit of Figure 1 reads 10 volts and the ammeter reads 0.5 ampere. How can we calculate the circuit resistance? With Ohm's Law, of course. It's not too difficult to *solve* Equation 2 for R, if you remember how to work with such equations. Some of you may have already written a new equation:

$R = E/I$ (Equation 3)

and calculated the circuit resistance.

$R = 10\ V / 0.5\ A = 20$ ohms

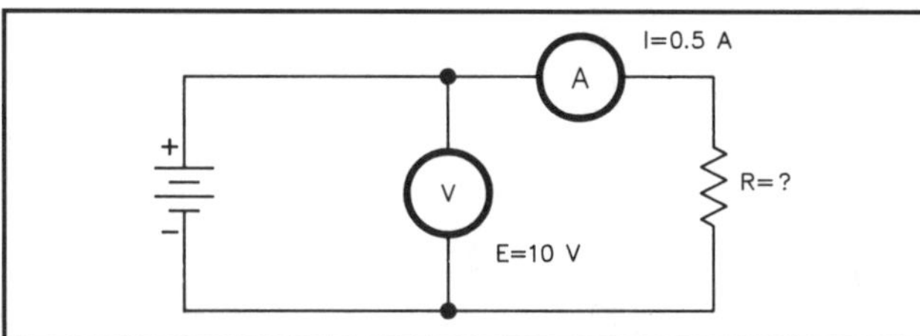

Figure 1—This circuit presents a simple Ohm's Law problem. You want to find the circuit resistance, and are given the voltage and current.

If you're not familiar with solving equations, you're probably wondering what sort of magic we used here! This is a good time to review the modules about solving equations. Figure 2 also shows a very helpful memory device for working with Ohm's Law. Let's take a closer look at how this *Ohm's Law Circle* works.

The basic operation of such a diagram is rather

Figure 2—The Ohm's Law Circle is a memory aid, or mnemonic to help you solve the Ohm's Law equation for either voltage, current or resistance.

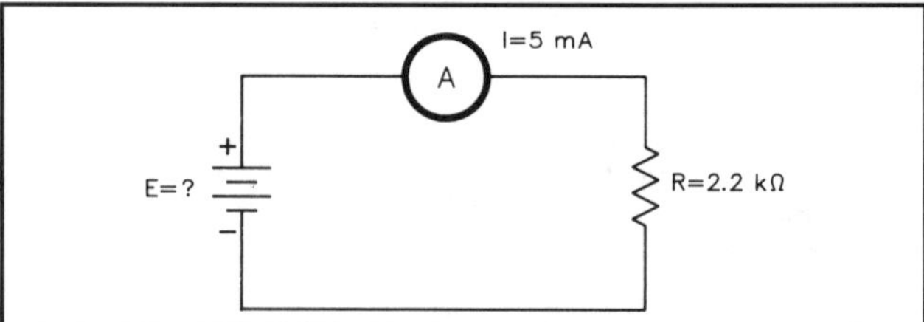

Figure 3—Here is another circuit that presents an Ohm's Law problem. This time you are given the current and resistance, and have to find the circuit voltage.

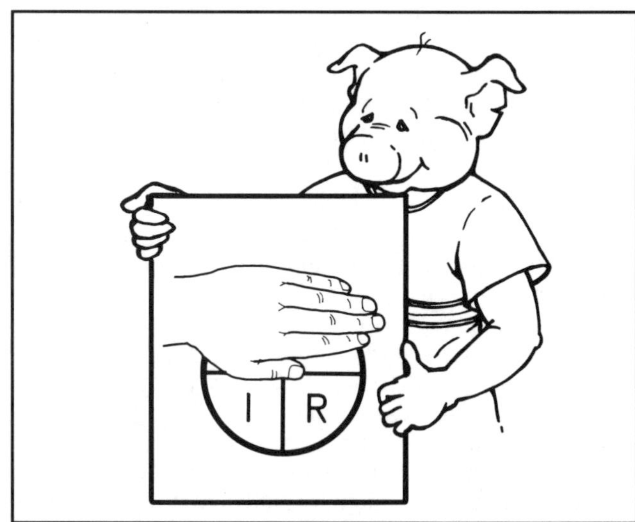

Figure 4—Hammlett shows us how to use the Ohm's Law Circle to solve a problem that asks us to calculate a circuit voltage. By covering the letter of the quantity we want to find, the remaining letters show us what calculation to perform. Here, we must multiply current (I) and resistance (R) to calculate voltage (E).

simple. First decide which piece of information you want to calculate and which two you already know. (You may have to examine the circuit or diagram to find this information.) Figure 3 shows another problem that we'll use as an example. We're trying to find the voltage, and the diagram gives us the circuit current and resistance. In this case, cover the E on the Ohm's Law Circle, as Hammlett is showing us in Figure 4. Notice that you can still see the I and R on the diagram. They are next to each other, indicating multiplication. Now it is time to write an equation. The letter you covered goes on the left side of an equals sign, and the other two letters go on the right side:

$E = IR$

Next substitute the values from the problem into the equation and do the arithmetic.

$E = 5\ mA \times 2.2\ k\Omega$
$E = 5 \times 10^{-3}\ A \times 2.2 \times 10^{3}\ \Omega = 11\ V$

(Do you remember how to write numbers using powers of ten instead of metric prefixes? If not, this is a good time to review "Working with Large and Small Numbers" in Unit 1.)

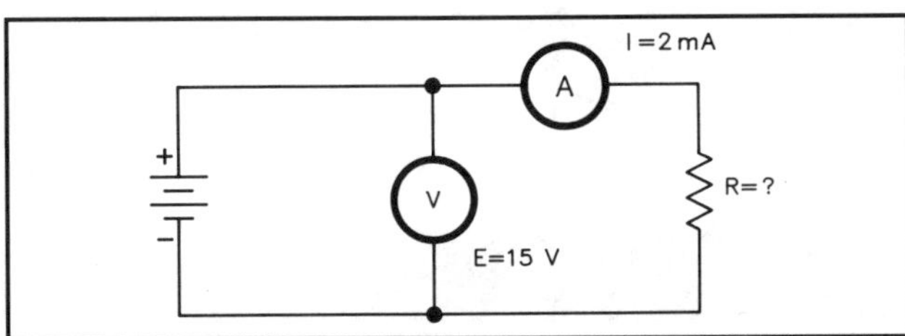

Figure 5—This diagram shows an electronics circuit in which we know the voltage and current, but not the resistance. Ohm's Law will help us calculate that resistance using the other information about the circuit.

Shall we try another example? Look at Figure 5. This problem tells us the circuit voltage and current, and asks us to find the resistance. We'll use the Ohm's Law Circle again. This time we have to cover the R (see Figure 6). The rest of the diagram shows E in the top of the circle and I on the bottom. This represents a division problem, and the equation is the same as Equation 3:

$R = E/I$
$R = 15\ V / 2\ mA = 15\ V / 2 \times 10^{-3}\ A$
$R = 7.5 \times 10^{3}\ \Omega = 7.5\ k\Omega$

Let's work one more example problem using Ohm's Law. Figure 7 presents a problem in which you know the circuit voltage and resistance, and you are to calculate the current. Figure 8 shows how to use the Ohm's Law Circle to find the equation:

$I = E/R$ (Equation 4)
$I = 12\ V / 270\ k\Omega = 12\ V / 270 \times 10^{3}\ \Omega$
$I = 4.44 \times 10^{-5}\ A = 0.0444 \times 10^{-3}\ A = 0.0444\ mA$
or
$I = 4.44 \times 10^{-5}\ A = 44.4 \times 10^{-6}\ A = 44.4\ \mu A$

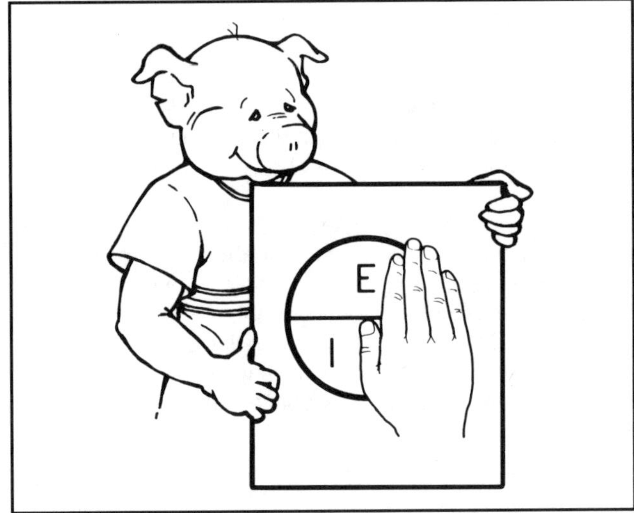

Figure 6—In this drawing Hammlett uses the Ohm's Law Circle to find the equation needed to solve the problem given in Figure 5.

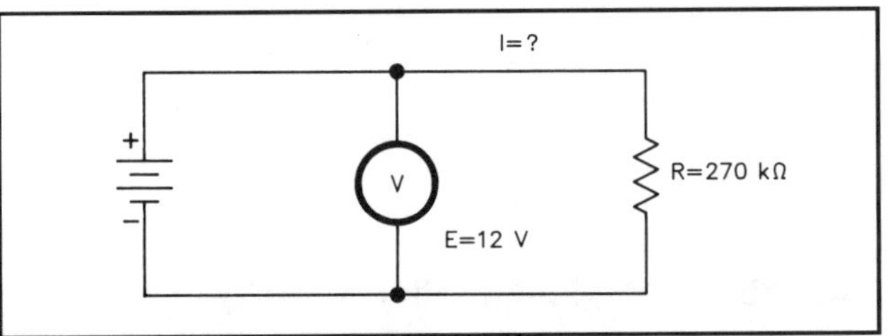

Figure 7—Here is another electronics circuit. In this one, we know the voltage and resistance, but not the current. Again, Ohm's Law will rescue us with an equation to calculate current.

Through these examples, you worked one of each type of Ohm's Law problem. You calculated a voltage in one circuit, a resistance in another and a current in the third example. Remember the Ohm's Law Circle, because it shows you the proper equation for each type of problem. After a while, you probably will remember the three ways of writing the equation. While you are learning about Ohm's Law, you won't have to worry about remembering these equations. Just picture this diagram each time you work a problem.

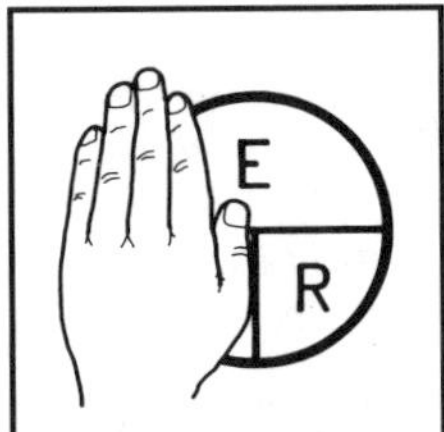

Figure 8—Find an equation that will solve the problem of Figure 7 by covering the I on the Ohm's Law Circle.

Some Practice Problems—And Their Solutions

You will become more confident in your ability to solve Ohm's Law problems as you get more practice solving them. It's very important that you improve your skill at solving Ohm's Law problems, and gain some confidence before we proceed much further. The practice problems on this page should help you meet those goals.

Each problem is presented using a schematic diagram, with circuit values labeled on the diagram. There is also a brief description of each problem. Figure 1 shows the Ohm's Law Circle for your convenience in solving these problems. After you have completed this exercise you can check your work with the solutions shown on the next page. (You'll have to turn the book upside down to read those solutions, though!)

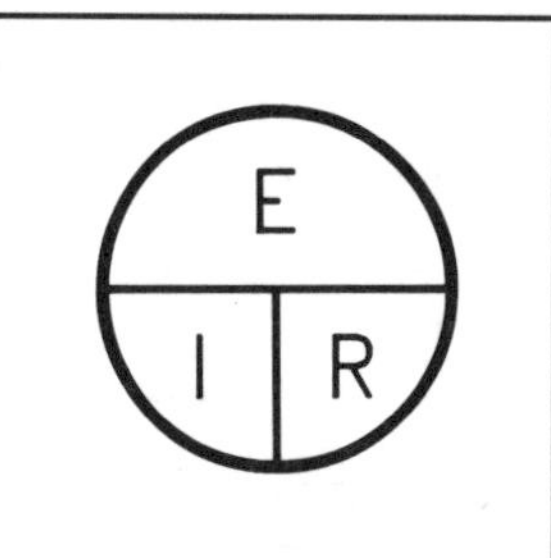

Figure 1—The Ohm's Law Circle provides a simple way to write the Ohm's Law equation for any particular problem.

PRACTICE PROBLEMS

Practice Problem 1.

The schematic diagram shows a circuit with a 33-kilohm resistor connected to a battery. There is an ammeter in the circuit to measure current. It indicates 5 milliamps of current in the circuit. What is the battery voltage?

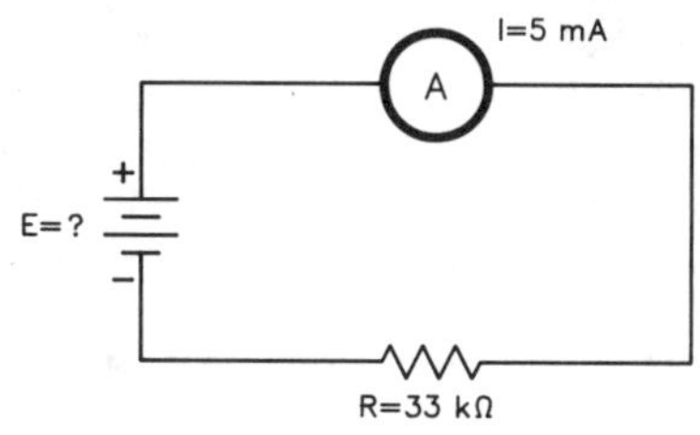

Practice Problem 2.

We use a voltmeter to measure the voltage of a battery, and find it to be 24 volts. How much current will flow through a 1.5-megohm resistor connected across the battery terminals?

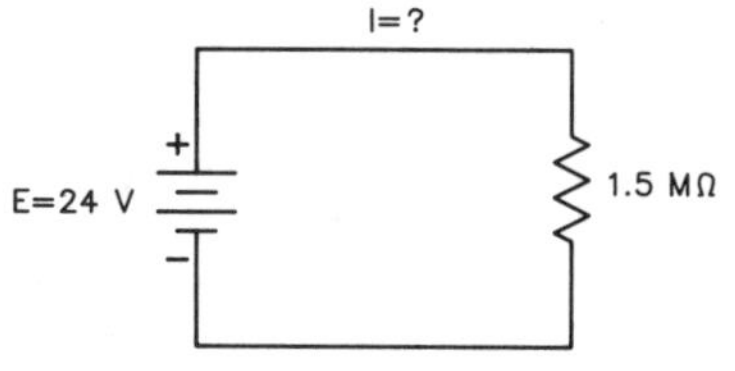

Practice Problem 3.

This schematic diagram shows an unknown value of resistance connected to a voltage supply. By connecting meters in the circuit, we are able to measure the circuit voltage and current. The voltage supply is 50 volts and there is 125 milliamps of current in the circuit. What is the resistance of the resistor in this circuit?

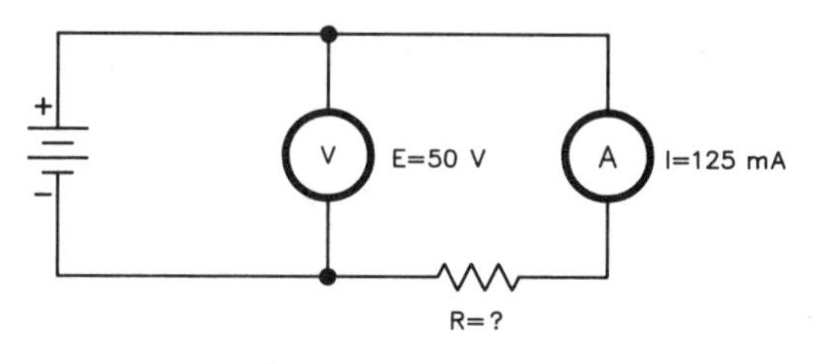

Practice Problem 4.

What voltage must you apply to a 250-ohm resistor to force 1 ampere of current through it?

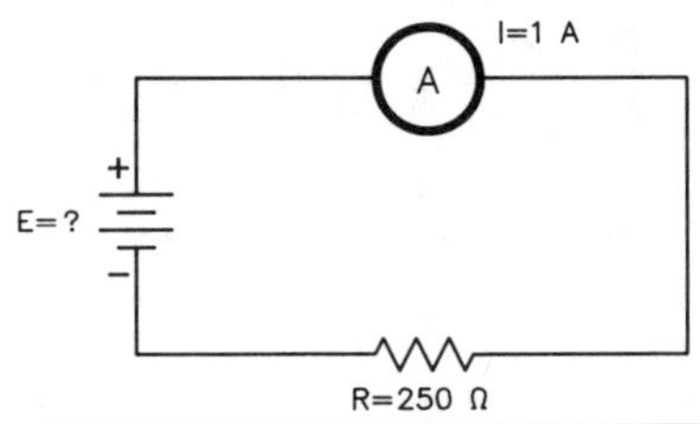

Practice Problem 5.

How much current will there be in a circuit that applies 120 volts to a 4700-ohm resistor?

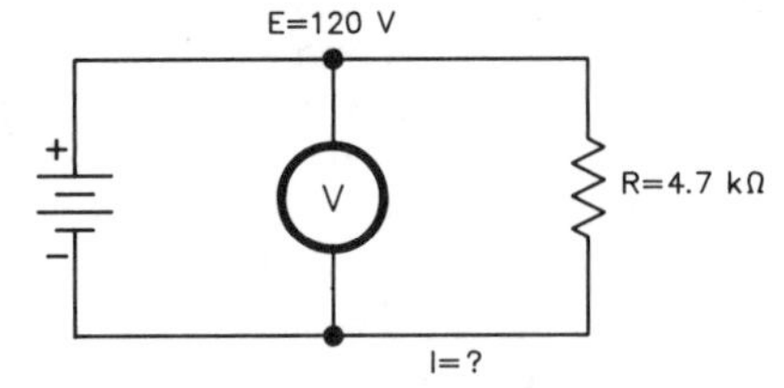

Practice Problem 6.

A 6-volt battery connects to a complicated mixture of resistors connected in series and parallel with each other. An ammeter in the circuit indicates that a current of 75 milliamps flows out of the battery. What is the total effective resistance of this complicated resistor maze?

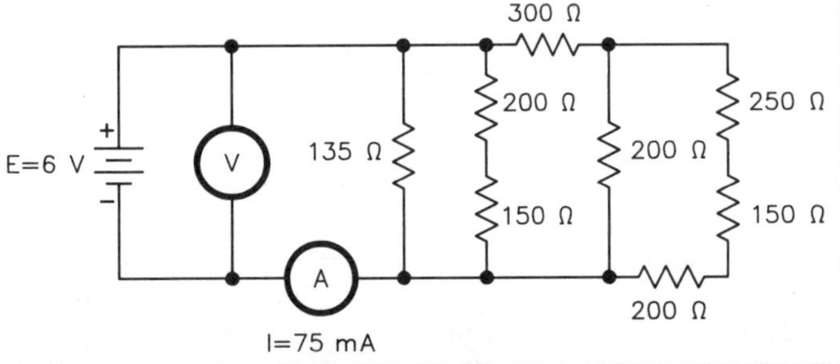

SOLUTIONS TO PRACTICE PROBLEMS

Solution to Problem 1.

Use the Ohm's Law Circle to find the equation to calculate voltage. From this, write the equation, E = I R.

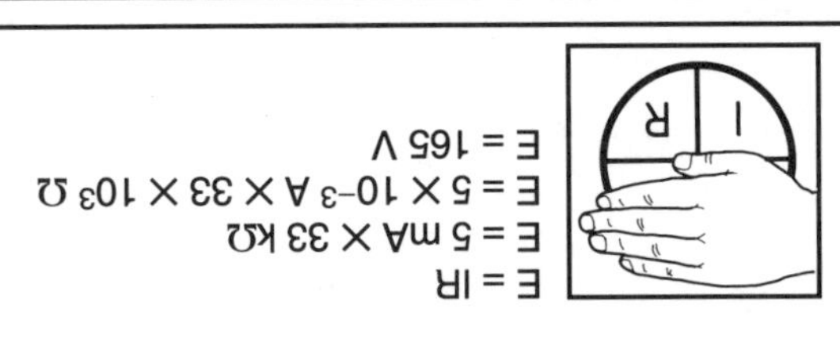

$E = IR$

$E = 5 \text{ mA} \times 33 \text{ k}\Omega$

$E = 5 \times 10^{-3} \text{ A} \times 33 \times 10^{3}\ \Omega$

$E = 165 \text{ V}$

Solution to Problem 2.

Using the Ohm's Law Circle, we write the equation to solve this problem.

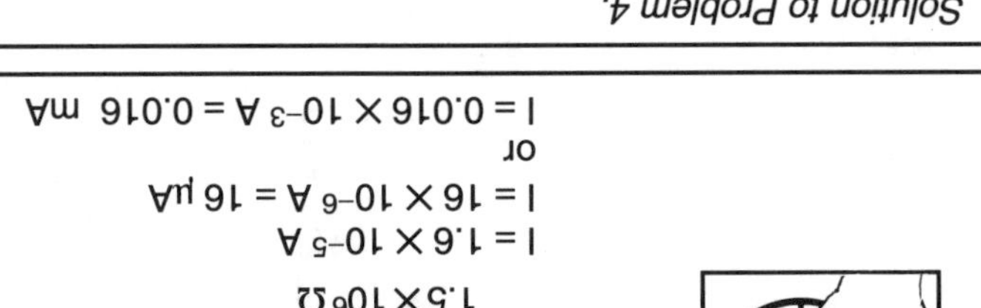

$$I = \frac{E}{R}$$

$$I = \frac{24 \text{ V}}{1.5 \text{ M}\Omega}$$

$$I = \frac{24 \text{ V}}{1.5 \times 10^{6}\ \Omega}$$

$I = 1.6 \times 10^{-5}$ A

$I = 16 \times 10^{-6} \text{ A} = 16\ \mu\text{A}$

or

$I = 0.016 \times 10^{-3} \text{ A} = 0.016 \text{ mA}$

Solution to Problem 3.

Use the Ohm's Law Circle to solve Ohm's Law for resistance.

$$R = \frac{E}{I}$$

$$R = \frac{50 \text{ V}}{125 \text{ mA}}$$

$$R = \frac{50 \text{ V}}{125 \times 10^{-3} \text{ A}}$$

$R = 400\ \Omega$

Solution to Problem 4.

By now you probably remember the basic Ohm's Law equation, solved for voltage. If not, use the Ohm's Law Circle.

$E = I R$

$E = 1 \text{ A} \cdot 250\ \Omega$

$E = 250$ V

Solution to Problem 5.

If you need to, use the Ohm's Law Circle to find the Ohm's Law equation solved for current.

$$I = \frac{E}{R}$$

$$I = \frac{120 \text{ V}}{4.7 \text{ k}\Omega}$$

$$I = \frac{120 \text{ V}}{4.7 \times 10^{3}\ \Omega}$$

$I = 0.0255$ A or $I = 25.5 \times 10^{-3} \text{ A} = 25.5$ mA

Solution to Problem 6.

Don't think of this as a trick question. The problem illustrates one important practical use for Ohm's Law. We can find the total effective resistance of a complicated maze of resistors simply by measuring the applied voltage and the current from the voltage supply. What is the Ohm's Law equation solved for resistance?

$$R = \frac{E}{I}$$

$$R = \frac{6 \text{ V}}{75 \text{ mA}}$$

$$R = \frac{6 \text{ V}}{75 \times 10^{-3} \text{ A}}$$

$R = 80\ \Omega$

Ohm's Law Applies to Circuits and Components

We have been working example problems showing how to use Ohm's Law to calculate voltage, current or resistance of an entire circuit. Can this important rule of electronics help us calculate information about more complex circuits as well? The answer is an enthusiastic YES!

Look at the circuit of Figure 1. Here we have three resistors connected in series with the battery. The voltmeter shows that the battery applies 12 volts to the circuit. The ammeter tells us that the current is 2 milliamps.

It takes some battery voltage to push electrons through each resistor in the circuit. This is similar to a water pipe that is blocked in several places. Figure 2 shows a water pipe with high water pressure at the left end. The blockage at point A prevents the full amount of water from flowing through the pipe. After forcing its way through that blockage, there is not as much pressure when the water comes to point B. Here there is another blockage, which restricts the water even more. By the time the water gets through the blockage at point C, there is only a trickle of water coming out of the pipe!

Each resistor in an electronics circuit is similar to a blockage in a water pipe. There is a voltage across each resistor, usually called a *voltage drop*. The voltage drop across a resistor represents the force or electrical potential needed to push the electrons through the resistor. (If you want a larger current—more electrons—it will take a larger voltage.)

For example, if you connect a voltmeter into the circuit, as shown in Figure 3A, the meter will read the full 12 volts of the battery. When you connect the voltmeter across R1 (in parallel with R1), as shown in Part B, the meter will read less than the full battery voltage. Part C shows that the meter also reads less than the battery voltage when you connect it between points B and C, in parallel with R2. Of course, the meter also reads something less than 12 volts when you connect it between points C and D. Part D shows the meter connected across R3 to make this measurement.

Our question now is, just how much does the voltage drop as we move across each resistor? Is the voltage drop the same across each resistor? Well, let's see how Ohm's Law can answer those questions for us. We know the value of resistance for each resistor because the diagram shows those values. We also know the current through each resistor because the same current must flow through every part of a series circuit. The ammeter also shows that the circuit current is 2 milliamps.

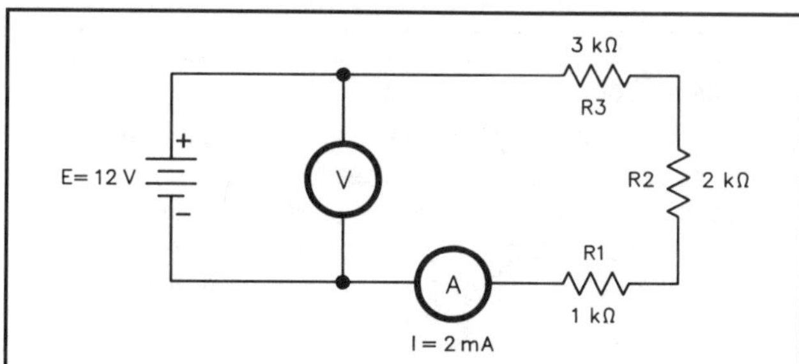

Figure 1—Ohm's Law will help us calculate the *voltage drop* across each resistor in this series circuit. The voltmeter measures the supply voltage applied to the entire circuit. The ammeter measures the amount of current flowing in each part of the circuit.

We use Ohm's Law to find the voltage drop across resistor R1:

$$E_1 = I R_1 \qquad \text{(Equation 1)}$$

$$E_1 = 2 \text{ mA} \times 1 \text{ k}\Omega$$

$$E_1 = 2 \times 10^{-3} \text{ A} \times 1 \times 10^{3} \ \Omega$$

$$E_1 = 2 \text{ V}$$

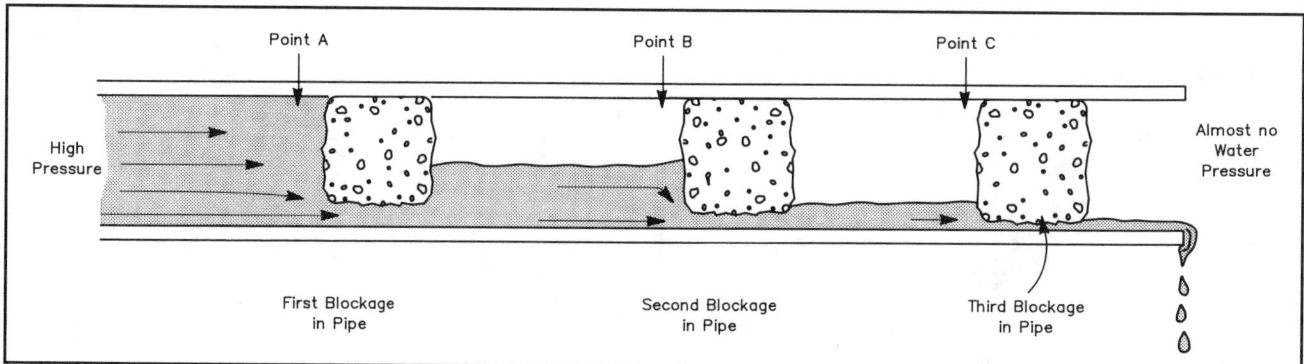

Figure 2—Resistance in an electronics circuit is similar to a blockage in a water pipe. The water forced past each blockage produces less pressure than there was before the blockage. Here we see that by the time the water gets past the blockage at point C, there is little pressure left. The water only drips out of the pipe.

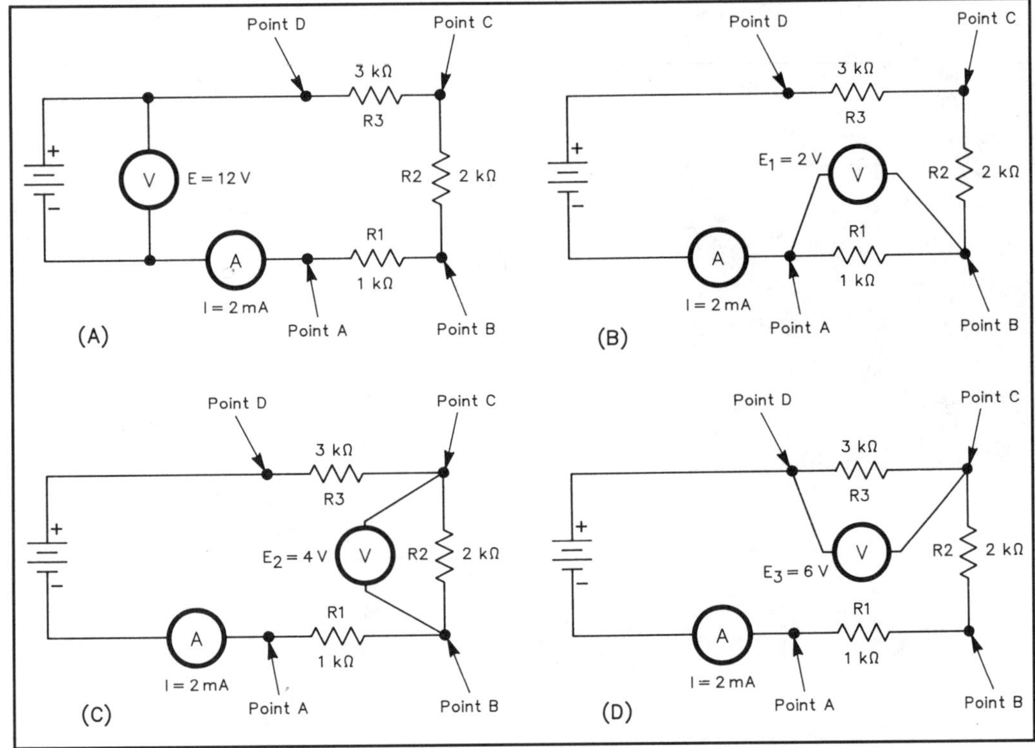

Figure 3—Part A shows a simple series circuit. A voltmeter measures the battery voltage and an ammeter measures the current through the circuit. If we connect the voltmeter between points A and B, as shown in Part B, we can measure the voltage drop across R1. (Notice that the voltmeter is in parallel with the resistor. You should always connect a voltmeter in parallel with the part of the circuit you want to measure.) In Part C, the voltmeter connects between points B and C, in parallel with R2. We are measuring the voltage drop across R2 in this example. To measure the voltage drop across R3, we connect the voltmeter between points C and D, as shown in Part D.

The voltmeter connected as shown in Figure 3B reads 2 volts, so it agrees with our calculation!

How about the voltage drop across resistor R2?

$E_2 = I\ R_2$ (Equation 2)
$E_2 = 2\text{ mA} \times 2\text{ k}\Omega$
$E_2 = 2 \times 10^{-3}\text{ A} \times 2 \times 10^3\ \Omega$
$E_2 = 4\text{ V}$

So the voltage drop across R2 is 4 volts, as the voltmeter in Figure 3C shows.

What is the voltage drop across resistor R3?

$E_3 = I\ R_3$ (Equation 3)
$E_3 = 2\text{ mA} \times 3\text{ k}\Omega$
$E_3 = 2 \times 10^{-3}\text{ A} \times 3 \times 10^3\ \Omega$
$E_3 = 6\text{ V}$

We can use Ohm's Law to calculate the voltage drop across any resistor in a series circuit. The only requirement is that we know the resistance and the current through the series circuit.

Now take a look at Figure 4. This parallel circuit presents a slightly different problem. The current through each branch of the circuit is different. If you examine the circuit carefully, however, you will realize that it applies the full battery voltage to each parallel branch. Can we then use Ohm's Law to calculate the amount of current through each branch? Of course!

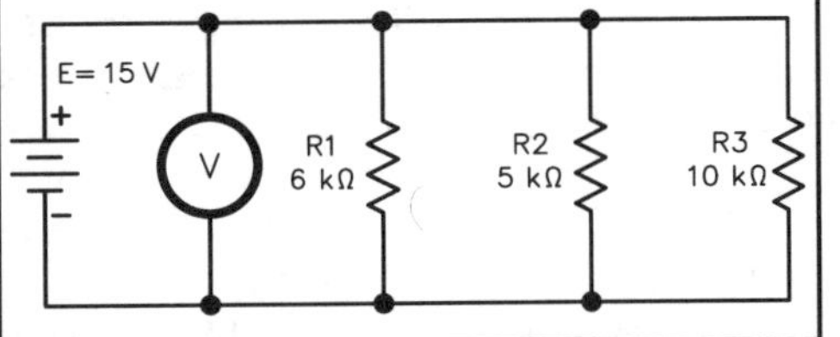

Figure 4—Here is a circuit with three resistors connected in parallel with a voltage source. Ohm's Law can help calculate the current through each branch of a parallel circuit.

Draw an Ohm's Law Circle and write the Ohm's Law equation solved for current.

$I = E / R$ (Equation 4)

Let's find the current through R1 first.

$I1 = 15\text{ V} / 6\text{ k}\Omega$
$I1 = 15\text{ V} / 6 \times 10^3\ \Omega$
$I1 = 2.5 \times 10^{-3}\text{ A} = 2.5\text{ mA}$

Now calculate the current through R2.

$I2 = 15\text{ V} / 5\text{ k}\Omega$
$I2 = 15\text{ V} / 5 \times 10^3\ \Omega$
$I2 = 3.0 \times 10^{-3}\text{ A} = 3.0\text{ mA}$

Finally, we'll calculate the current through R3.

$I3 = 15\text{ V} / 10\text{ k}\Omega$
$I3 = 15\text{ V} / 10 \times 10^3\ \Omega$
$I3 = 1.5 \times 10^{-3}\text{ A} = 1.5\text{ mA}$

Ohm's Law is important to us in our study of electronics not just because it will help us learn something about the conditions of an entire circuit. Ohm's Law is also important because we can apply it to individual components in the circuit. Later in your studies you will have to expand your understanding of Ohm's Law a bit. When we begin to study more complex circuits we'll have to add a bit more to the definition of this important rule. The result, however, is that every electronics circuit obeys Ohm's Law.

CHAPTER 13 Kirchhoff's Laws

Kirchhoff's Voltage Law

Ohm's Law can help us calculate voltage, current or resistance. You are familiar with this important electronics principle and can use it to calculate information about a circuit. There are times, however, when you can save yourself work if you know a few other rules.

Kirchhoff's Voltage Law, sometimes called *Kirchhoff's First Law*, is one of these rules. Gustav Kirchhoff, like Georg Simon Ohm, was a German scientist. In 1857, just 3 years after Ohm's death, Kirchhoff stated his two rules of electric circuits.

According to Kirchhoff's Voltage Law, *all voltage drops must equal all voltage rises around a closed loop.* There is a *voltage drop* across any circuit resistance when there is a current through it. We use Ohm's Law to calculate the amount of these voltage drops. A *voltage rise* normally refers to a voltage source, such as a battery. We add all the voltage rises and all the voltage drops. If you think the statement of Kirchhoff's First Law sounds like a mathematical equation, you are exactly right! We might write this rule as:

$$\Sigma E_{rises} = \Sigma E_{drops} \qquad \text{(Equation 1)}$$

The uppercase Greek sigma (Σ) represents the summation, or addition of all related terms. To use Kirchhoff's Law, you must first identify the voltage rises around the loop you are working with, and add them. Then you must identify the voltage drops around the loop, and add them. Do you remember how to use Ohm's Law to calculate the voltage drops?

Figure 1 shows a simple circuit with one voltage rise and four voltage drops. You must go around a complete loop of the circuit to use Kirchhoff's Voltage Law.

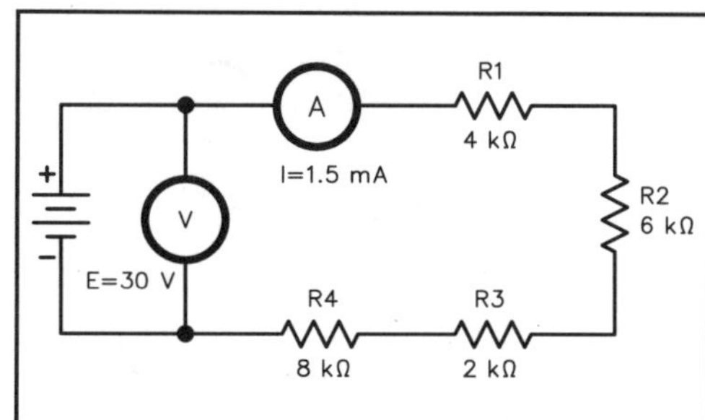

Figure 1—A simple series circuit, which forms a closed loop. We can write a Kirchhoff's Voltage Law equation by going around that loop adding all voltage drops and all voltage rises.

Let's write the Kirchhoff's First Law equation for the circuit of Figure 1.

$$\Sigma E_{rises} = \Sigma E_{drops} \qquad \text{(Equation 1)}$$

$$30\ V = I\ R1 + I\ R2 + I\ R3 + I\ R4$$

$$30\ V = 1.5\ mA \times 4\ k\Omega + 1.5\ mA \times 6\ k\Omega + 1.5\ mA \times 2\ k\Omega + 1.5\ mA \times 8\ k\Omega$$

$$30\ V = 6\ V + 9\ V + 3\ V + 12\ V$$

$$30\ V = 30\ V$$

This example shows the meaning of Kirchhoff's Voltage Law. Often, we use this law to calculate the current through a circuit. Figure 2 shows another circuit. This one has three voltage sources to include in the equation. Let's use Kirchhoff's Voltage Law to calculate the current through this circuit.

Figure 2 will help illustrate another important point about using Kirchhoff's Voltage Law. When you go around the loop, there are two possible directions, clockwise (↻) or counterclockwise (↺). Normally, go in a direction that takes you through the voltage sources from the negative to the positive terminal. Notice this is the direction we followed in the example of Figure 1A. Going through a voltage source from negative to positive results in a voltage increase, or voltage rise. If you go through a voltage source from positive to negative, you have a voltage drop! The

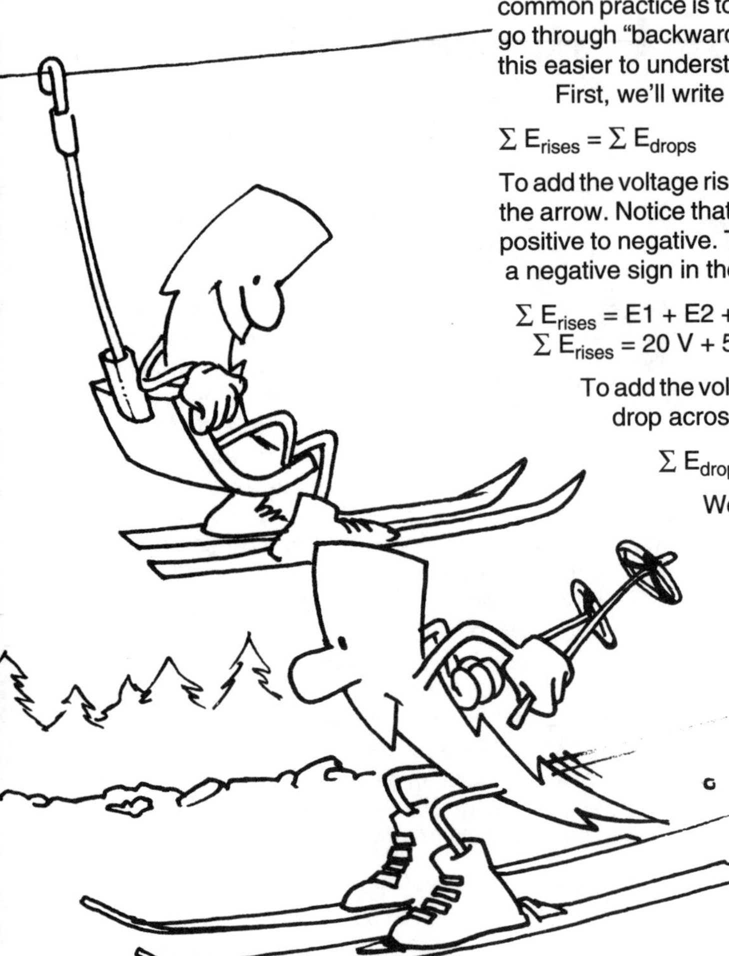

common practice is to assign a negative value to any voltage source that you go through "backwards" around the loop. The example of Figure 2 will make this easier to understand.

First, we'll write the equation for Kirchhoff's Voltage Law.

$\Sigma E_{rises} = \Sigma E_{drops}$ (Equation 1)

To add the voltage rises, proceed around the loop in the direction indicated by the arrow. Notice that when you come to E3, you go through the battery from positive to negative. That results in a voltage decrease at this point. Give E3 a negative sign in the equation.

$\Sigma E_{rises} = E1 + E2 + E3$ (Equation 2A)

$\Sigma E_{rises} = 20\text{ V} + 5\text{ V} - 10\text{ V} = 15\text{ V}$ (Equation 2B)

To add the voltage drops, we'll use Ohm's Law to calculate the voltage drop across each resistor. (E = IR)

$\Sigma E_{drops} = I\,R1 + I\,R2 + I\,R3 + I\,R4$ (Equation 3A)

We don't know the current in the circuit, (that's what we are trying to find) but we do know each of the resistances.

$\Sigma E_{drops} = I \times 200\ \Omega + I \times 800\ \Omega + I \times 900\ \Omega + I \times 500\ \Omega$

Did you notice that we multiply each resistance value by the current, I? When you multiply each value in a group by some constant value, you simplify the problem. First you can add all the numbers, and then multiply that sum by the constant value. We can rewrite Equation 3A as:

$\Sigma E_{drops} = I \times (R1 + R2 + R3 + R4)$ (Equation 3A)

$\Sigma E_{drops} = I \times (200\ \Omega + 800\ \Omega + 900\ \Omega + 500\ \Omega)$

$\Sigma E_{drops} = I \times (2400\ \Omega)$ (Equation 3B)

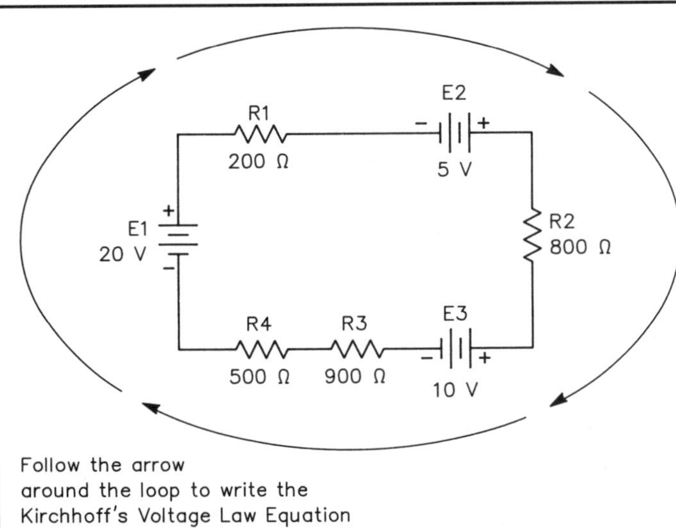

Figure 2—This diagram shows a more complicated series circuit, including three voltage sources and four resistors. With Kirchhoff's Voltage Law, we can write an equation and calculate the current in this circuit.

Now we combine Equations 2B and 3B into a single Kirchhoff's Law equation (like Equation 1). (We could have built both sides of the equation at the same time. It's often more convenient to work with one side at a time, though, and then combine them for the final solution.)

$\Sigma E_{rises} = \Sigma E_{drops}$

$15\text{ V} = I \times 2400\ \Omega$ (Equation 4)

To solve this equation for I, divide both sides of the equation by 2400 Ω.

$$\frac{15\text{ V}}{2400\ \Omega} = \frac{I \times 2400\ \Omega}{2400\ \Omega} \quad \text{(Equation 5)}$$

$I = 6.25 \times 10^{-3}\text{ A} = 6.25\text{ mA}$

You can use Kirchhoff's Voltage Law to solve problems even more complicated than this. In the next section you'll learn several ways this important electronics principle can help you simplify a complex series circuit.

Simplifying Series Circuits

Kirchhoff's Voltage Law gives us several ideas for ways to simplify series electronics circuits. Look at the circuit shown in Figure 1. It includes some voltage sources, such as batteries, connected in series, along with a string of resistors. Sometimes we might like to redraw a circuit like this to show an *equivalent circuit.* An *equivalent circuit* has the same voltage and current as the original. The equivalent circuit uses one voltage source and one resistor, however. Figure 2 shows the simplified circuit, but without values for the voltage and resistance.

We'll use Kirchhoff's Voltage Law to find these component values. Start with the basic Voltage Law equation.

$\Sigma E_{rises} = \Sigma E_{drops}$ (Equation 1)

Remember that we worked with each half of this equation separately at first, to make the job easier. To add the voltage rises, go around the circuit in a direction that takes you through most of the voltage sources from negative to positive. Notice that the order of connecting the voltages doesn't really matter. The order in which we add them doesn't make a difference, either. It doesn't even matter if there are a few resistors between some of them. Just remember to give a negative sign to any voltage source that you go through from positive to negative along the way. (These voltages subtract from the others.)

From Kirchhoff's Voltage Law you can see that voltage sources either help or restrict electrons flowing through the circuit. Sometimes you might hear the terms *aiding* and *bucking* in relation to how we connect the voltage sources. *Aiding* sources connect positive to negative, so the voltages add. *Bucking* sources connect negative to negative (or positive to positive), so the voltages subtract. You can find the total voltage of a circuit like the one shown in Figure 1 with some simple arithmetic.

$E_{total} = 20\ V + 30\ V + 10\ V - 15\ V - 15\ V$ (Equation 2)
$E_{total} = 30\ V$

This is the equivalent voltage for the single source of Figure 2.

Now let's calculate the equivalent resistance of the single resistor for Figure 2. Write an equation using Kirchhoff's Voltage Law again, adding the voltage drops around the circuit loop. Let I represent the circuit current, since we don't know the actual value.

$\Sigma E_{drops} = I \times R1 + I \times R2 + I \times R3 + I \times R4 + I \times R5 + I \times R6$ (Equation 3)

Since we are multiplying the same current, I, times each resistor value, we can rewrite this equation.

$\Sigma E_{drops} = I\ (R1 + R2 + R3 + R4 + R5 + R6)$ (Equation 4)

All we are really trying to do here is find the total equivalent resistance of the resistors in Figure 1. That one equivalent resistor must produce the same voltage drop as all the others combined. The current, I, times the equivalent resistance must equal the right side of Equation 4, then.

$I\ (R_{Total}) = I\ (R1 + R2 + R3 + R4 + R5 + R6)$ (Equation 5)

Divide both sides of this equation by I to cancel current out of the equation. Then we have a simple equation to calculate the total resistance of a string of resistors connected in series.

$R_{total} = R1 + R2 + R3 + R4 + R5 + R6$ (Equation 6)

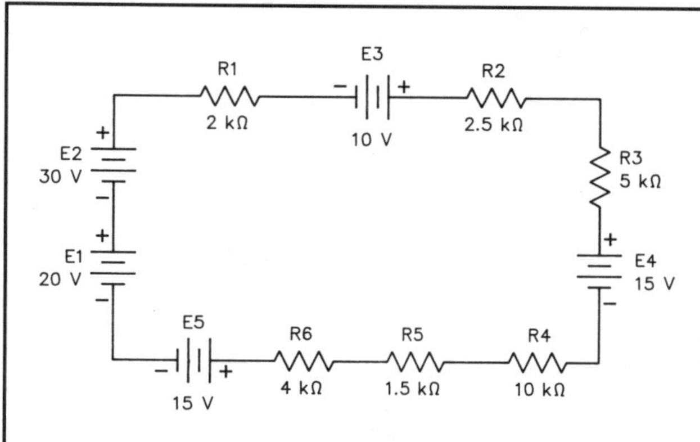

Figure 1—This circuit is a rather complex combination of resistors and voltage sources connected in series. Kirchhoff's Voltage Law can help us find a simple equivalent circuit that includes one voltage source and one resistor. (An equivalent circuit means that both circuits produce the same external effects. In this example, it means the same current will flow in both circuits.)

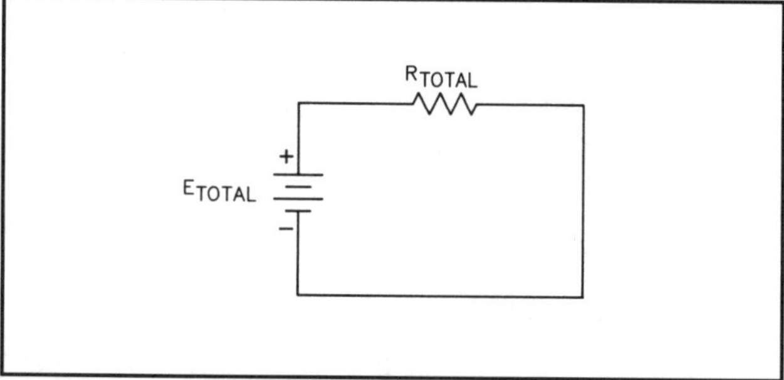

Figure 2—This diagram represents a simple circuit that is equivalent to the complex circuit of Figure 1. The text explains how to calculate the voltage and resistance values for the new circuit.

The total equivalent resistance of series-connected resistors is equal to the sum of the individual resistor values. A general equation that you can apply to any combination of series resistors, is:

$$R_{total} = R1 + R2 + R3 + ... + Rn \qquad \text{(Equation 7)}$$

where Rn represents the last resistor in the string. This equation works if there are two resistors in series or two hundred (or more) in series. This equation does not work for resistors connected in parallel, however. (We'll come to that case later.)

For the example of Figure 1, the total resistance is:

$$R_{total} = 2\ k\Omega + 2.5\ k\Omega + 5\ k\Omega + 10\ k\Omega + 1.5\ k\Omega + 4\ k\Omega$$
$$R_{total} = 25\ k\Omega$$

Going back to Equation 5, we can write the equation as:

$$I\ (R_{total}) = I \times 25\ k\Omega$$

We can also write Equation 4 as:

$$\Sigma\ E_{drops} = I \times 25\ k\Omega$$

Now we can write Equation 1 with all of the additional information we have calculated.

$$\Sigma\ E_{rises} = \Sigma\ E_{drops} \qquad \text{(Equation 1)}$$

$$30\ V = I \times 25\ k\Omega$$

Does this look like an Ohm's Law equation? It should. Solve the equation for current, I.

$$I = \frac{30\ V}{25\ k\Omega}$$

$$I = \frac{30\ V}{25 \times 10^3\ \Omega}$$

$$I = 1.2 \times 10^{-3}\ A = 1.2\ mA$$

With the help of Kirchhoff's Voltage Law, we calculated the total circuit current. You can use this method to calculate unknown information about circuits with even more voltage sources and resistances.

Kirchhoff's Current Law

K*irchhoff's Current Law*, sometimes also called *Kirchhoff's Second Law*, is an important electronics principle. Kirchhoff's Current Law provides us with a very powerful tool for studying an electronics circuit and calculating conditions in that circuit.

Before you learn what this law says, you must learn another term sometimes used to describe part of an electronics circuit. A *circuit node* is any point in a circuit where two or more conductors connect. A node is the point where two resistors connect. The point where a wire connects to a battery is also a node. Most of the time we look for nodes that are branch points in the circuit. At least three conductors connect at that point. Figure 1 points out several nodes on a circuit diagram.

Kirchhoff's Current Law states that *all currents into a circuit node equal all currents out of that node.* If you think about it, this rule should seem just as obvious as Kirchhoff's Voltage Law. After all, the current has to go somewhere. You can't just lose electrons as they flow around a circuit. No matter how obvious it is, however, this law is very important and extremely powerful in helping us solve electronics circuit problems.

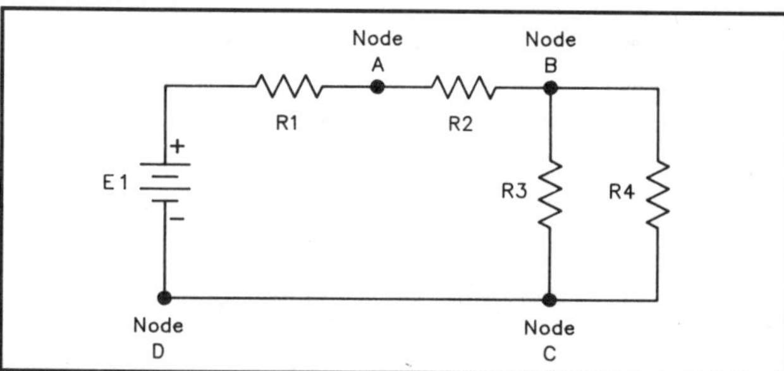

Figure 1—This circuit diagram shows several *circuit nodes.* A node is the point where two or more conductors meet. Most of the time our interest is in nodes that occur at circuit branch points. In those cases, there will be three or more conductors, such as at nodes B and C on this diagram.

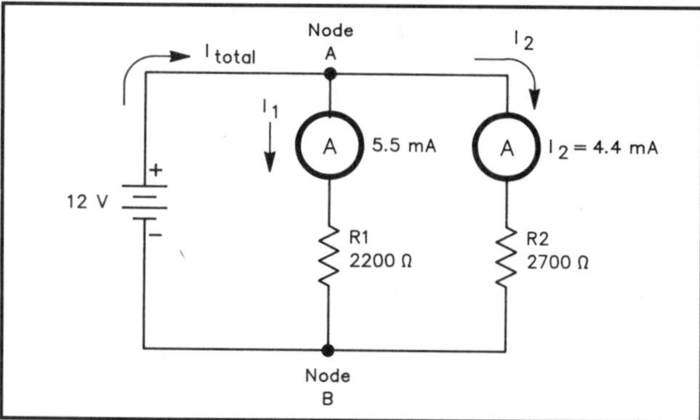

Figure 2—This simple parallel circuit includes ammeters to measure the current in each branch. There is no ammeter to measure the total battery current, however. We must use Kirchhoff's Current Law to find the total current.

Kirchhoff's Current Law indicates that we could write an equation for the current at any circuit node. Again, we use the Greek letter sigma (Σ) to represent a summation of similar values.

$$\Sigma I_{IN} = \Sigma I_{OUT} \qquad \text{(Equation 1)}$$

Figure 2 shows a simple parallel circuit. Ammeters measure the current in the two branches, but we want to know the total battery current. Kirchhoff's Current Law will help us make that calculation. There is one current flowing into node A, I_{TOTAL} and two currents flowing out, I_1 and I_2.

Remember that engineers define the direction of current to be from the positive side of the voltage source to the negative side. We normally use this *conventional current* direction when dealing with Kirchhoff's Current Law. This is just a matter of convenience, because when you divide a positive voltage (a voltage rise) by a resistance, the positive number indicates a positive current. A negative value would mean the current flows in the opposite direction.

We can write a Kirchhoff's Current Law equation at node A of Figure 2, using Equation 1. This will allow us to calculate the total current out of the battery.

$$\Sigma I_{IN} = \Sigma I_{OUT} \qquad \text{(Equation 1)}$$

$$I_{TOTAL} = I_1 + I_2 \qquad \text{(Equation 2)}$$

$$I_{TOTAL} = 5.5 \text{ mA} + 4.4 \text{ mA} = 9.9 \text{ mA}$$

You also might want to write an equation for node B, just to prove that the results are the same.

$$\Sigma I_{IN} = \Sigma I_{OUT} \qquad \text{(Equation 1)}$$

$$I_1 + I_2 = I_{TOTAL} \qquad \text{(Equation 3)}$$

$$5.5 \text{ mA} + 4.4 \text{ mA} = I_{TOTAL} = 9.9 \text{ mA}$$

This example is almost too easy. The real usefulness of Kirchhoff's Current Law is for much more complicated circuits than this. For these more complicated problems, we often have to use both of Kirchhoff's Laws. Most of the time there are several circuit conditions we want to calculate. The solution involves the technique of solving *simultaneous equations.* That term simply means that if there is more than one variable or unknown in a circuit, you will need more than one equation about circuit conditions. You'll need one equation for each unknown, and those equations must all apply to the circuit at the same time. (That's what *simultaneous* means.)

Most problems like that are beyond the *beginner's* level of this book. Let's work through one easy example, though, just so you can see the steps involved.

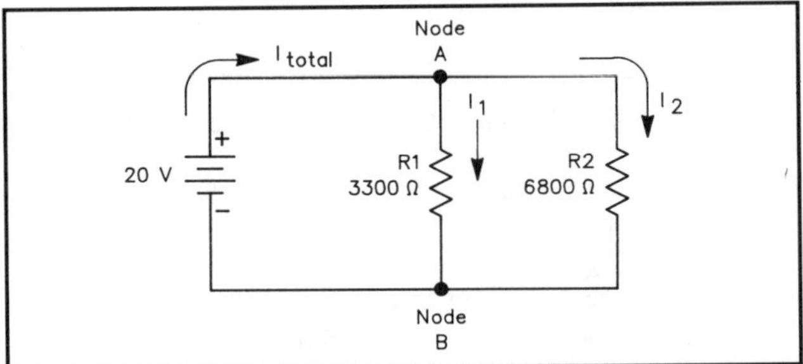

Figure 3—This diagram will help illustrate the technique of using *simultaneous equations* to solve a problem that has more than one unknown. We want to calculate the total battery current and the current in each branch of this circuit. The text explains the procedure.

Figure 3 shows a parallel circuit. We know the battery voltage and resistance values as shown on the diagram. We want to calculate the current through each branch of the circuit, and the total current drawn from the battery. That amounts to three currents that we want to calculate. We might start by writing a Kirchhoff's Current Law equation at node A.

$\Sigma I_{IN} = \Sigma I_{OUT}$ (Equation 1)
$I_{TOTAL} = I_1 + I_2$ (Equation 4)

This equation represents the three quantities we want to find, so you might wonder how it can help. Remember, with three unknowns, we'll need three simultaneous equations about this circuit. What other equations can we possibly write? There are at least two Kirchhoff's Voltage Law equations for this circuit. Figure 4 shows the same circuit as Figure 3, with two loops identified to write these equations.

$\Sigma E_{RISES} = \Sigma E_{DROPS}$ (Equation 5)

For loop 1:

$20\text{ V} = I_1 \times R_1$ (Equation 6)

For loop 2:

$20\text{ V} = I_2 \times R_2$ (Equation 7)

Now we have three different equations that represent our three unknown quantities in this circuit. There are several mathematical techniques for solving simultaneous equations. One of those, which works well with simple equations like these, is *substitution.* To use this method, we solve one or more of the equations for known quantities or in terms of the other variables, and substitute that value into the remaining equations. (This method is easier than it sounds! Let's continue with the example, to see how it's done.)

First, we'll solve Equation 6 for I_1 by dividing both sides of the equation by R_1.

$$\frac{20\text{ V}}{R_1} = \frac{I_1 \times R_1}{R_1} \quad \text{(Equation 6)}$$

$$\frac{20\text{ V}}{R_1} = I_1 \quad \text{(Equation 8)}$$

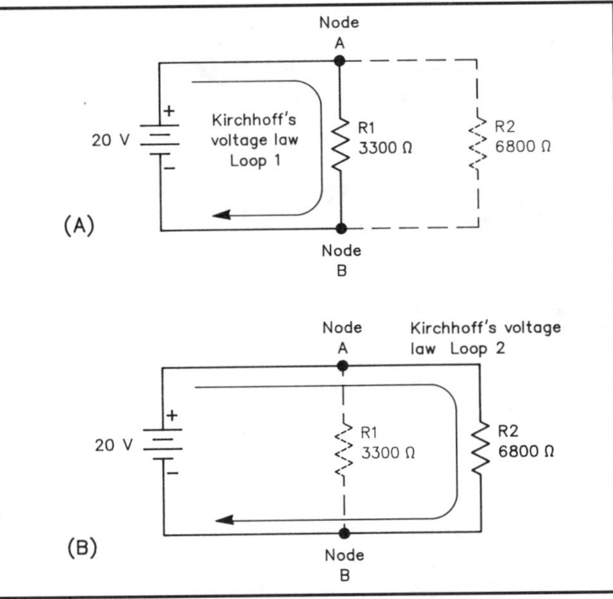

Figure 4—The two parts of this drawing duplicate the diagram of Figure 3. Part A shows one loop that we can use to write a Kirchhoff's Voltage Law equation and Part B shows a second loop. The two equations from these loops, along with a Kirchhoff's Current Law Equation, allow us to calculate the three currents in the circuit.

Since the value for R_1 is given on the diagram, we can substitute that into the equation, and calculate I_1.

$$\frac{20\text{ V}}{3300\ \Omega} = I_1$$

$6.06 \times 10^{-3}\text{ A} = 6.06\text{ mA} = I_1$

Then we'll solve Equation 7 for I_2 in the same way.

$$\frac{20\text{ V}}{R_2} = \frac{I_2 \times R_2}{R_2} \quad \text{(Equation 7)}$$

$$\frac{20\text{ V}}{R_2} = I_2 \quad \text{(Equation 9)}$$

We also know the value for R_2 from Figure 3, so we can complete this calculation.

$$\frac{20\text{ V}}{6800\ \Omega} = I_2$$

$2.94 \times 10^{-3}\text{ A} = 2.94\text{ mA} = I_2$

Now we have expressions for I_1 and I_2 that we can substitute into Equation 4.

$I_{TOTAL} = I_1 + I_2$ (Equation 4)
$I_{TOTAL} = 6.06\text{ mA} + 2.94\text{ mA}$ (Equation 10)
$I_{TOTAL} = 9.00\text{ mA}$

This example shows you that Kirchhoff's Laws can solve complex circuit problems. You can solve problems with circuits that have many parallel branches, although the math can be tricky.

Simplifying Parallel Circuits

When we studied Kirchhoff's Voltage Law, we discovered that it could help us simplify series circuits. For example, if there were several voltage sources connected in series, we could replace them with one equivalent source. We also discovered a simple equation for calculating the value of a single resistor to replace several resistors connected in series.

You might be wondering if Kirchhoff's Current Law provides a similar way to simplify parallel circuits. The answer is a very definite *yes.*

Figure 1 shows a circuit that has two batteries connected in parallel as a voltage source. What effect will this have on the circuit? Each battery supplies some current to the circuit. Without writing an equation, let's think about how Kirchhoff's Current Law applies to this circuit. Each battery supplies a current that flows into node A. The total circuit current flows out of node A. Connecting the batteries in parallel helps them last longer. It's similar to using a bigger battery when the circuit needs more current than a smaller one can supply. For example, compare a 1-1/2-volt AAA cell with a 1-1/2-volt D cell or a hobby cell. They all produce the same voltage, but the bigger batteries can supply a lot more current.

Figure 1—We can connect batteries in parallel if they have equal voltage. This increases battery life because the circuit draws less current from each battery.

You wouldn't normally connect batteries of different voltages in parallel. That might lead to a large current flowing between the batteries, overheating them and possibly causing other damage or dangerous conditions. There are times, however, when you may want to connect several batteries of the same voltage in parallel.

Take another look at Figure 1. There are two voltage sources, each having the same voltage, connected in parallel. We can simplify this circuit by replacing the two voltage sources with one that can supply the same current. Pretty simple isn't it?

Can Kirchhoff's Current Law help us combine several resistors connected in parallel and replace them with a single equivalent resistor? Again, *yes*! Let's start with the circuit of Figure 3. What is the resistance of a single resistor that could replace the three parallel resistors shown? The same total current flows from the battery. First, we'll write a Kirchhoff's Current Law equation at node A.

$$\Sigma I_{IN} = \Sigma I_{OUT} \quad \text{(Equation 1)}$$

$$I_{TOTAL} = I_1 + I_2 + I_3 \quad \text{(Equation 2)}$$

We also can write Kirchhoff's Voltage Law equations for three loops around this circuit. See Figure 4. Since each of these loops involves the battery voltage and a single resistor, the full battery voltage appears across each resistor. We can use Ohm's Law to write equations for the voltage across each resistor. Then we can solve these equations for the branch currents.

$$E = I_1 \times R_1 \quad \text{(Equation 3)}$$

Solving for I_1:

$$\frac{E}{R_1} = \frac{I_1 \times R_1}{R_1} \quad \text{(Equation 4)}$$

$$I_1 = \frac{E}{R_1} \quad \text{(Equation 5)}$$

Next, we'll find I_2.

$$E = I_2 \times R_2 \quad \text{(Equation 6)}$$

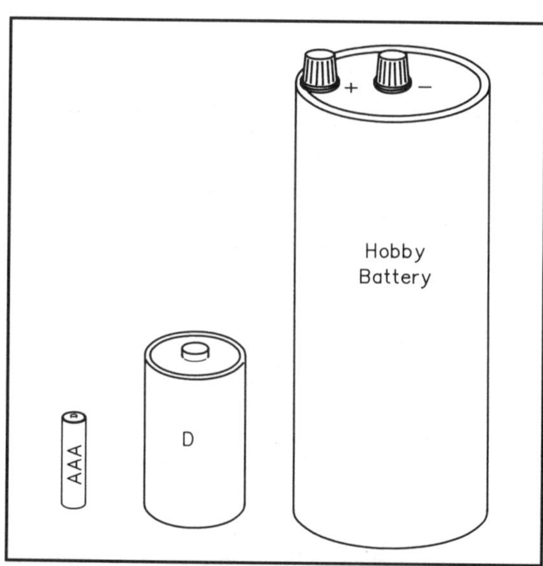

Figure 2—An AAA cell, a D cell and a hobby cell all produce 1.5 volts. The larger the battery size, however, the more current it can supply before it goes dead.

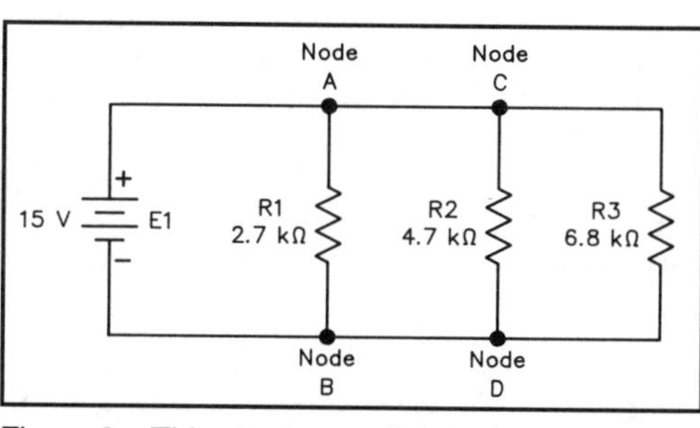

Figure 3—This simple parallel circuit has three resistors connected in parallel with the battery. We want to replace these three resistors with a single one that will have the same effect in the circuit. The text describes the procedure to follow.

$$I_2 = \frac{E}{R_2} \quad \text{(Equation 7)}$$

Finally, the third branch current.

$$E = I_3 \times R_3 \quad \text{(Equation 8)}$$

$$I_3 = \frac{E}{R_3} \quad \text{(Equation 9)}$$

Now we can substitute these branch current values into Kirchhoff's Current Law, Equation 2.

$$I_{TOTAL} = I_1 + I_2 + I_3 \quad \text{(Equation 2)}$$

$$I_{TOTAL} = \frac{E}{R_1} + \frac{E}{R_2} + \frac{E}{R_3} \quad \text{(Equation 10)}$$

The total current in this circuit is the same as the current through a single resistor equivalent to the three in parallel. With this piece of information, we can solve an Ohm's Law equation for I_{TOTAL}.

$$I_{TOTAL} = \frac{E}{R_{TOTAL}} \quad \text{(Equation 11)}$$

Substituting into Equation 10, we have:

$$\frac{E}{R_{TOTAL}} = \frac{E}{R_1} + \frac{E}{R_2} + \frac{E}{R_3} \quad \text{(Equation 12)}$$

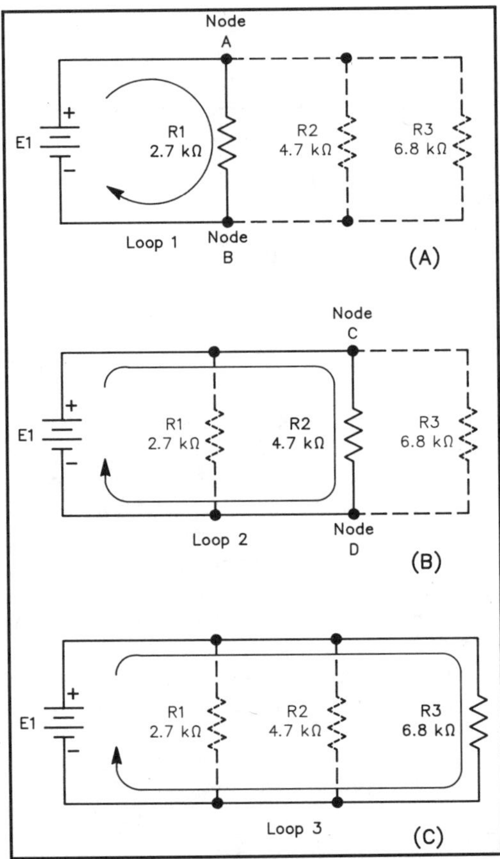

Figure 4—The three parts of this figure each shows the circuit of Figure 3. Each part shows a circuit loop that we can use with Kirchhoff's Voltage Law, to write an equation about the circuit.

By dividing every term on both sides of this equation by E, we can eliminate the voltage from our equation. Then we are left with an equation to calculate the total resistance of a string of parallel-connected resistors.

$$\frac{1}{R_{TOTAL}} = \frac{1}{R_1} + \frac{1}{R_2} + \frac{1}{R_3} \quad \text{(Equation 13)}$$

It's pretty easy to find reciprocals (one divided by a number) on a scientific calculator. You can solve this whole equation in a few simple steps.

$$\frac{1}{R_{TOTAL}} = \frac{1}{2.7\ k\Omega} + \frac{1}{4.7\ k\Omega} + \frac{1}{6.8\ k\Omega}$$

$$\frac{1}{R_{TOTAL}} = \frac{3.7 \times 10^{-4} + 2.1 \times 10^{-4} + 1.5 \times 10^{-4}}{\Omega}$$

Cross multiply by R_{TOTAL} to take it out of the denominator:

$$1\ \Omega = 7.3 \times 10^{-4}\ R_{TOTAL}$$

Now we can cross multiply to solve the equation for R_{TOTAL}:

$$\frac{1\ \Omega}{7.3 \times 10^{-4}} = R_{TOTAL}$$

$$1400\ \Omega = R_{TOTAL} \quad \text{(Equation 14)}$$

(Your calculator probably will read something like 1369.86, but we've rounded the answer off to show only two significant figures.)

If you have a string of resistors connected in parallel, you find the value of one resistor that is equivalent to the string. Just add the reciprocal of each resistor in the string, and then take the reciprocal of that result.

$$\frac{1}{R_{TOTAL}} = \frac{1}{R_1} + \frac{1}{R_2} + \frac{1}{R_3} + \ldots + \frac{1}{R_n} \quad \text{(Equation 15)}$$

It's very important for you to realize that you must add the reciprocal values of the resistances. This is *not* the same as adding the resistance values themselves! With your scientific calculator in hand, you can easily add the reciprocals of the resistances.

There are a couple of other important observations that we should make about resistors connected in parallel. Notice the answer given in Equation 14. The combined resistance of the parallel resistors in Figure 3 is only 1.4 kΩ. That's less than any resistor in the circuit! Whenever you have some resistors connected in parallel, the equivalent resistance of the combination is less than any of the individual resistors.

Equal value resistors connected in parallel present another interesting condition. Connect two equal resistors in parallel. The total resistance will be 1/2 the resistance of either one by itself. Connect three equal resistors in parallel and the total will be 1/3 the value of any one resistor. You can find the equivalent resistance of any number of equal-value resistors connected in parallel. Divide the value of a single resistor by the number of resistors in the string. This gives you a quick way to calculate the equivalent resistance of a parallel string of equal resistors.

Simplifying Series-Parallel Circuits

Kirchhoff's Voltage and Current Laws provide us with the tools to simplify electronics circuits. If a circuit includes a string of resistors connected in series, we can find a single equivalent resistor to replace the string. Likewise, if a circuit includes a string of resistors connected in parallel, we can find a single equivalent resistor to replace that string. To combine *series resistors*, we simply add the individual resistor values.

$$R_{TOTAL} = R_1 + R_2 + R_3 + ... + R_n \qquad \text{(Equation 1)}$$

To combine *parallel resistors*, we must add the *reciprocals* of the individual resistor values, and then take the reciprocal of that sum.

$$\frac{1}{R_{TOTAL}} = \frac{1}{R_1} + \frac{1}{R_2} + \frac{1}{R_3} + ... + \frac{1}{R_n} \qquad \text{(Equation 2)}$$

The best way to learn how to use these equations is through practice, so let's try a couple of examples. Figure 1 shows a circuit with five resistors connected in series. What is the equivalent resistance of a single resistor that can replace them?

Since this is a series circuit, we'll have to use Equation 1. Simply add the five resistance values to find the equivalent resistance. Sounds easy, doesn't it?

$$R_{TOTAL} = R_1 + R_2 + R_3 + ... + R_n \qquad \text{(Equation 1)}$$
$$R_{TOTAL} = 1\ k\Omega + 3.3\ k\Omega + 5.6\ k\Omega + 4.7\ k\Omega + 1.5\ k\Omega$$
$$R_{TOTAL} = 16.1\ k\Omega$$

Now let's try a parallel-circuit problem. Figure 2 shows a circuit that includes three resistors connected in parallel. What is the value of a single resistor that can replace these three? This one requires us to use Equation 2.

$$\frac{1}{R_{TOTAL}} = \frac{1}{R_1} + \frac{1}{R_2} + \frac{1}{R_3} + ... + \frac{1}{R_n} \qquad \text{(Equation 2)}$$

$$\frac{1}{R_{TOTAL}} = \frac{1}{2.2 \times 10^3\ \Omega} + \frac{1}{1.5 \times 10^3\ \Omega} + \frac{1}{2.7 \times 10^3\ \Omega}$$

$$\frac{1}{R_{TOTAL}} = \frac{4.5 \times 10^{-4} + 6.7 \times 10^{-4} + 3.7 \times 10^{-4}}{\Omega}$$

$$\frac{1}{R_{TOTAL}} = \frac{14.9 \times 10^{-4}}{\Omega}$$

Now solve this equation for R_{TOTAL} by cross multiplying the terms.

$$R_{TOTAL} = \frac{1\ \Omega}{14.9 \times 10^{-4}} = 670\ \Omega$$

This took a bit more work than the previous problem. With the help of your scientific calculator, it shouldn't have given you too much trouble, though. You'll get plenty of additional practice adding series and parallel resistors as you learn more about electronics.

You will find electronics circuits that look much more complicated than these simple series or parallel circuits. Most circuits include some combination of series and parallel connections. You'll have to use your knowledge of the simple circuits to solve these problems. Take them one step at a time, and work with part of the circuit at each step. This way, you can solve some difficult problems.

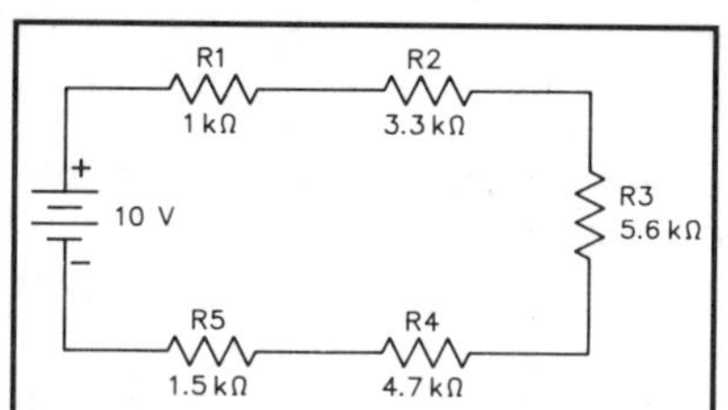

Figure 1—Find a single resistor to replace this string of resistors connected in series. The text explains how to do this problem.

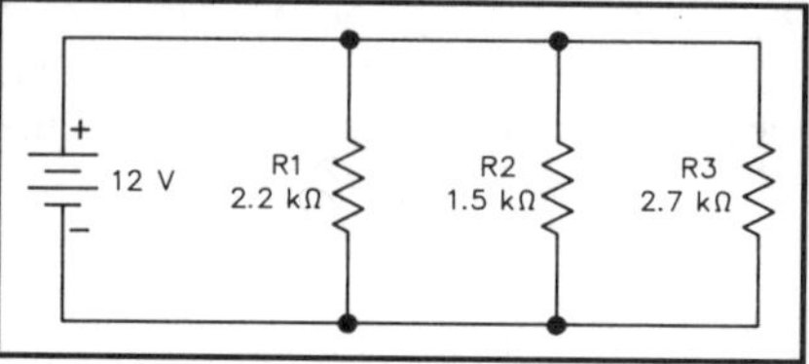

Figure 2—Find a single resistor to replace this string of resistors connected in parallel. The text explains the steps involved in solving this problem.

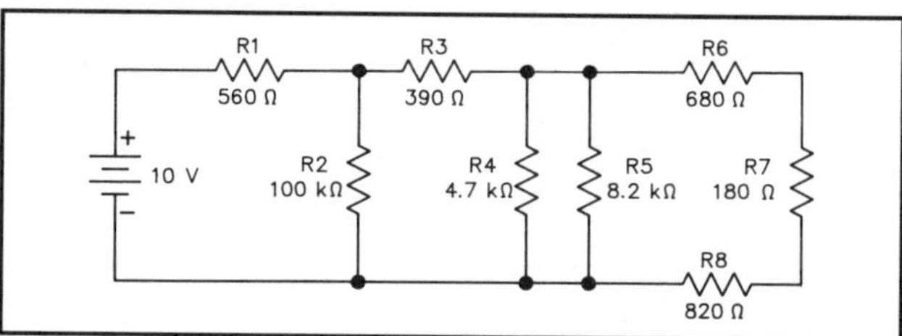

Figure 3—Here is a series/parallel circuit. You are to find a single resistor that is equivalent to the resistors in this network. The text, along with Figure 4, explains the steps involved.

Let's look at Figure 3. Here is a network of series and parallel resistors. Your job is to find the value of a single resistor that can replace the entire network. Why would you want to do that? Well, one practical reason would be to calculate the total current flowing in the circuit. Now you could write Kirchhoff's Voltage and Current equations and solve them for the total circuit current. This is a difficult task for even a circuit like the one of Figure 3, however. You would have up to seven variables, and need seven simultaneous equations to solve this problem! That's an exercise better left for an advanced student.

How will we find the equivalent resistance of this circuit? Start working at the point farthest from the voltage source. Study the circuit diagram to determine if the resistors at that point connect in series or parallel. When you look at Figure 3, you'll notice that R_6, R_7 and R_8 are in series. The first step is to combine these three resistors into one. We'll rewrite Equation 1 a bit to fit this problem.

$$R_{6\text{-}8} = R_6 + R_7 + R_8 \qquad \text{(Equation 3)}$$

where $R_{6\text{-}8}$ means the equivalent resistor that replaces resistors numbered from 6 to 8.

$R_{6\text{-}8} = 680\ \Omega + 180\ \Omega + 820\ \Omega$

$R_{6\text{-}8} = 1680\ \Omega$

Figure 4A shows the circuit of Figure 3 with this resistor replacing resistors R6, R_7 and R_8. Take a close look at this new circuit of Figure 4A. You should immediately notice that the three resistors farthest from the voltage source connect in parallel. Our next step, then, is to combine these three into a single equivalent resistance. Using Equation 2, we have:

$$\frac{1}{R_{4\text{-}8}} = \frac{1}{R_4} + \frac{1}{R_5} + \frac{1}{R_{6\text{-}8}} \qquad \text{(Equation 4)}$$

where $R_{4\text{-}8}$ represents a single resistor that is equivalent to resistors R_4 through R_8 in the original circuit.

$$\frac{1}{R_{4\text{-}8}} = \frac{1}{4.7\ \text{k}\Omega} + \frac{1}{8.2\ \text{k}\Omega} + \frac{1}{1.68\ \text{k}\Omega}$$

$$\frac{1}{R_{4\text{-}8}} = \frac{1}{4.7 \times 10^3\ \Omega} + \frac{1}{8.2 \times 10^3\ \Omega} + \frac{1}{1.68 \times 10^3\ \Omega}$$

$$\frac{1}{R_{4\text{-}8}} = \frac{2.13 \times 10^{-4} + 1.22 \times 10^{-4} + 5.95 \times 10^{-4}}{\Omega}$$

$$\frac{1}{R_{4\text{-}8}} = \frac{9.30 \times 10^{-4}}{\Omega}$$

To solve this equation for $R_{4\text{-}8}$, cross multiply the terms.

$$\frac{\Omega}{9.30 \times 10^{-4}} = R_{4\text{-}8}$$

$10.75\ \Omega = R_{4\text{-}8}$

Figure 4B shows the circuit redrawn with this new equivalent resistor. Notice that R_3 and $R_{4\text{-}8}$ are in series. We can replace those two resistors with a single value found by adding them.

$R_{3\text{-}8} = R_3 + R_{4\text{-}8}$

$R_{3\text{-}8} = 390\ \Omega + 1075\ \Omega = 1465\ \Omega$

Figure 4C includes this substitution in our circuit. It's beginning to look much simpler, isn't it? How do R_2 and $R_{3\text{-}8}$ connect? How will you find an equivalent value for them? That's right, they connect in parallel, so you'll have to add the reciprocals of their values, using Equation 2.

$$\frac{1}{R_{2\text{-}8}} = \frac{1}{R_2} + \frac{1}{R_{3\text{-}8}} \qquad \text{(Equation 5)}$$

$$\frac{1}{R_{2\text{-}8}} = \frac{1}{100\ \text{k}\Omega} + \frac{1}{1465\ \Omega}$$

$$\frac{1}{R_{2\text{-}8}} = \frac{1.00 \times 10^{-5} + 6.826 \times 10^{-4}}{\Omega}$$

$$\frac{1}{R_{2\text{-}8}} = \frac{6.926 \times 10^{-4}}{\Omega}$$

Cross multiply the terms to solve for $R_{2\text{-}8}$.

$$\frac{\Omega}{6.926 \times 10^{-4}} = R_{2\text{-}8}$$

$1444\ \Omega = R_{2\text{-}8}$

Figure 4D shows this newest circuit, one step from completion. All that's left is to combine R1 with $R_{2\text{-}8}$. Since they connect in series, you probably don't even need to write an equation. Just add the values.

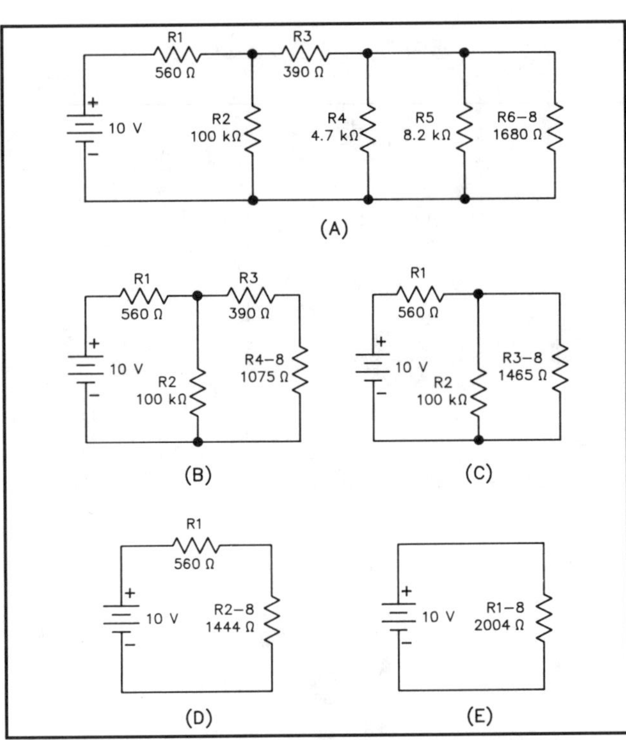

Figure 4—The parts of this figure show the steps involved in simplifying the circuit of Figure 3. Each part shows a further reduction in the number of resistors in the circuit, until, in Part E, There is only one left. This one resistor draws the same current from the voltage source as the entire circuit of Figure 3.

$R_{TOTAL} = R_1 + R_{2\text{-}8}$ (Equation 6)

$R_{TOTAL} = 560\ \Omega + 1444\ \Omega$

$R_{TOTAL} = 2004\ \Omega$

Part E of Figure 4 shows this final equivalent circuit.

By simplifying this circuit in steps, we found a single resistor that is equivalent to the whole circuit. You could now calculate the total current that flows from the voltage supply using this equivalent resistor. Ohm's Law will help you with that calculation.

$E = I \times R$ (Equation 7)

Solve this equation for I (or use an Ohm's Law Circle to find the new equation).

$I = E / R$ (Equation 8)

$I = 10\ \text{V} / 2004\ \Omega = 4.99 \times 10^{-3}\ \text{A}$

$I = 4.99\ \text{mA}$

We won't go through any more calculations here. It's not difficult to work back through the original circuit, calculating all the branch currents. Kirchhoff's Current and Voltage Laws will be a big help. Even though it was lots of work to simplify this circuit, it was the easiest way to calculate the total circuit current.

Remember, when you want to simplify a complicated electronics circuit, begin at the farthest point from the voltage source. Replace series and parallel resistor combinations one step at a time. Eventually you will find a single resistor to take the place of all those in the original circuit.

CHAPTER 14 Energy and Power

Energy is the Ability to do Work

What can electricity do for us? It gives light when electrons flow through a light bulb. Electricity can run a motor to turn a drill, spin a saw blade or run an electric hair dryer. The heating elements in an oven or stove use electricity to produce the heat needed to cook our food. Electrons flowing through the circuits inside our radios and TV sets convert radio signals into sounds and pictures. The list could go on and on.

Electric companies sell us this electricity, and the amount they charge us depends on how much electricity we use each month. So how do they decide how much to charge? They measure the electricity we use during the month. This is more than just measuring the current or voltage, since those quantities change from moment to moment. The electric company uses a meter that measures the amount of *electrical energy* you use. (A kilowatt-hour meter measures this energy.)

What is *electrical energy*, and how do we measure it? Well, as the title for this module says, *energy* is the ability to do *work*. The amount of work to be done determines how much energy is required. Since you may not be familiar with these terms, let's go over some basic definitions. We'll also look at a few examples that will help you understand the terms a little better.

To do some work, there are two requirements. There must be a force acting on an object, and that force must move the object some distance. To calculate how much work was done, multiply the amount of the force and the distance the object moves. In the metric system, the unit for work is the *joule*. (Pronounce that jool, which rhymes with pool.) Energy is a measure of how much work something

Figure 1—Hammlett helps us understand the idea of work and energy by lifting a football and holding it above his head. When he does work on the football to lift it, he also stores some energy in it, which we call *potential energy*. Later, when he drops the football, Hammlett shows us that the football's potential energy changes into *kinetic energy*, the energy of motion.

can do. It takes one joule of energy to do one joule of work. The common abbreviation for joule is J.

Figure 1 shows a simple example of work and energy. Hammlett has picked up a football, and is holding it above his head. To lift the ball, he had to exert a force equal to the force of gravity pulling down on the ball. In metric units, let's say that took a force of 1 newton. If Hammlett lifts the ball 1 metre, he does 1 joule of work. We also can say the ball has 1 joule of *potential energy* now. If Hammlett lets go, gravity will pull the ball back to the earth. *Potential energy* is energy stored in an object for later release. As the ball falls back to the ground it loses some of its potential energy because it doesn't have as far to fall anymore. (Remember, it's force times distance.) But, the ball now has *kinetic energy*, which is energy of motion.

What does all this have to do with electronics? Well, electrons can have potential energy stored in them. For example, the electrons at the negative side of a battery have potential energy. This is because the battery voltage will push and pull the electrons through a circuit when we connect it to one. (Remember, voltage is the force that moves electrons through a circuit.) Electrons also have kinetic energy, because each is in constant motion around the nucleus of an atom. After an electron goes through a resistor, there is less voltage pushing and pulling it the rest of the way through the circuit. Ohm's Law tells us there is a voltage drop across the resistor, and that drop depends on the amount of current and the resistance. The electron has less potential energy as it gets closer to the positive battery terminal. This is similar to the football, which has less potential energy as it gets closer to the ground. Some of the electron's energy changes to heat energy in the resistor. When electrons move through a light bulb, some of their energy changes into light energy. (Some of it also changes to heat energy in a light bulb.)

We start with a certain amount of energy stored in the electrons at the negative battery terminal. As we move around the circuit, we find that the electrons have less and less potential energy. We also find that the electricity has given us other forms of energy. This is an important idea. The total amount of energy doesn't change. Some energy changes from one form to another, but we still have the same total amount. Scientists call this principle *conservation of energy*. Work is done when one form of energy changes to another. The amount of work done is equal to the amount of energy converted.

Earlier, we mentioned that electric companies measure the amount of electric energy we use, so they can charge us for this electricity. Remember that we often measure the amount of electric charge (or the number of electrons) in *coulombs*. There are 6.24×10^{18} electrons in one *coulomb* of charge. One coulomb of electrons flows past a point in a circuit in one second. There is a *one-ampere* current in that circuit. Now that we've reminded you of those important definitions, we can explain what a joule of electrical energy means.

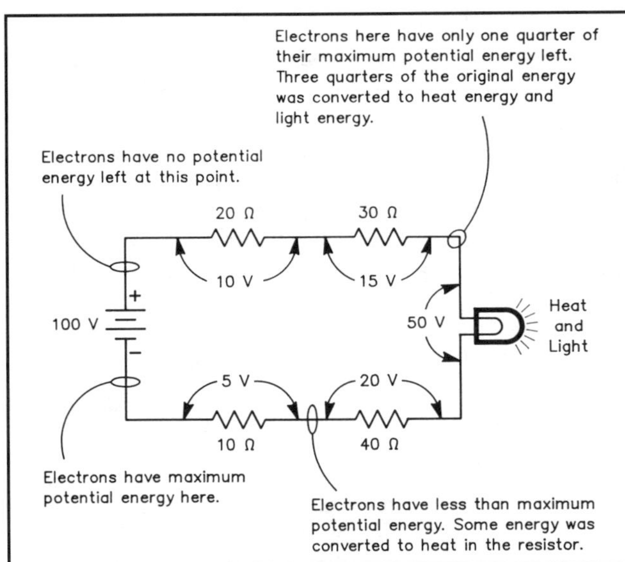

Figure 2—As we move away from the negative battery terminal, electrons have less potential energy after passing through any circuit resistance. Electrons at the negative terminal have the most potential energy and electrons at the positive terminal have zero potential energy. (Don't confuse *conventional current* direction with *electron flow* here. With conventional current, we would say the positive battery terminal has the highest potential, just the opposite of looking at the electrons.)

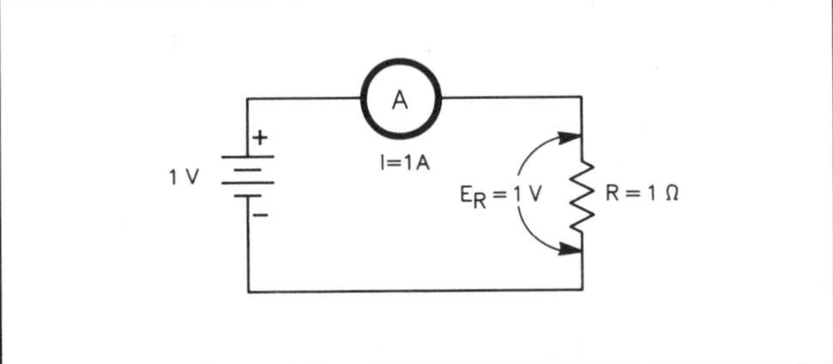

Figure 3—This simple electronics circuit shows a battery doing 1 joule of work every second. A joule is the amount of work done on a coulomb of electrons to move them through a voltage drop of 1 volt.

Suppose you have a circuit like the one shown in Figure 3. A 1-ampere current flows through the circuit. There is a voltage drop of 1 volt across the 1-ohm resistor. The battery does 1 joule of work every second as it pushes the 1-amp current through the resistor. From this, you can guess that 1 joule of electrical energy is a small amount. Too small, in fact, for the electrical companies to use. They use a larger unit, the *kilowatt-hour*, to calculate our monthly electricity bills. We'll define that unit in more detail in the next section, but for now you can recognize the metric prefix kilo (meaning thousand) and the time unit *hour*.

Power is the Rate of Doing Work, or the Rate of Using Energy

The circuit of Figure 1 uses 2 joules of energy every second. We also could say the voltage supply in this circuit does 2 joules of work every second. If we just measure the amount of energy in joules, it doesn't tell us as much about the circuit. That's why we also include a time.

Suppose we just said that the battery in the circuit of Figure 1 does 2 joules of work. We also can say that the battery in Figure 2 does 2 joules of work. Of course these circuits are much different from each other. In fact, it takes eight seconds for the battery in the circuit of Figure 2 to do 2 joules of work. (That's eight times as long as it takes the battery of Figure 1!) We might even draw a circuit that would take an hour or more to do 2 joules of work. The point is that different circuits will use energy (or do work) at different rates. The time rate of doing work provides us with some important information about a circuit.

A special term defines this rate of doing work or using energy. *Power* is the rate of doing work. If something uses energy at the rate of 1 joule per second, the power is 1 *watt* (abbreviated W). Most readers have at least heard of this term. This basic power unit is named in honor of James Watt, a scientist and inventor who built the first successful steam engine. A watt is a small unit of power, so we use the *kilowatt* (1000 W, or 1 kW) to measure large amounts of power. When you are talking about electric generating plants, you might even measure the power output in megawatts (MW).

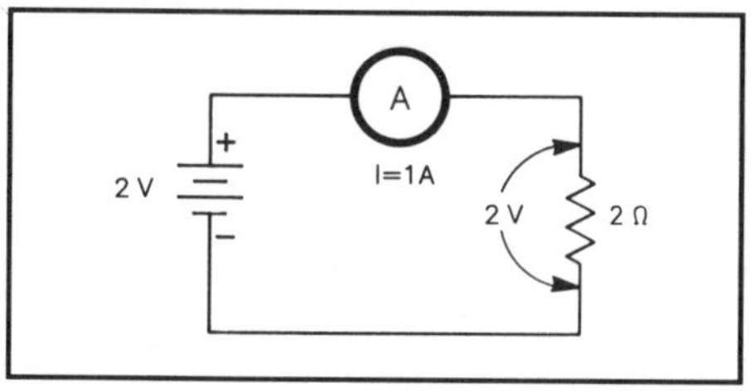

By now you are probably wondering how we measure energy and power in a practical way. The *joule* represents the energy used to move one *coulomb* of electrons through a 1-V change in *potential* (or *voltage*). A coulomb of electrons per second gives us the definition of an *ampere*. Without going through a lot of mathematics, you should begin to realize that we can define the *watt* in terms of current and voltage. We multiply the current through a circuit times the voltage drop across that circuit to get the power in the circuit. Since we can measure current and voltage easily, this gives us a good, practical way to measure power.

We can write an equation to calculate power.

$$P = I\,E \qquad \text{(Equation 1)}$$

where

P = power in watts
I = current in amperes
E = voltage in volts

This may be the easiest electronics equation to remember. Just look at the letters. Don't they spell one of your favorite kinds of dessert? Whether it's lemon meringue, cherry crumb or pumpkin, most people like some type of *PIE*. Once you've made that association, you will always remember the equation to calculate power.

Figure 1—This simple electronics circuit uses energy at the rate of 2 joules every second.

Take a look at Figure 3. How much power does the battery in this circuit provide? The battery voltage is 15 V, and the circuit current is 25 mA. Since Equation 1 calls for current in amps, we'll have to convert the 25 mA to 0.025 A. Then, using Equation 1, we have:

$$P = I\,E \qquad \text{(Equation 1)}$$

$$P = 0.025 \text{ A} \times 15 \text{ V} = 0.375 \text{ W}$$

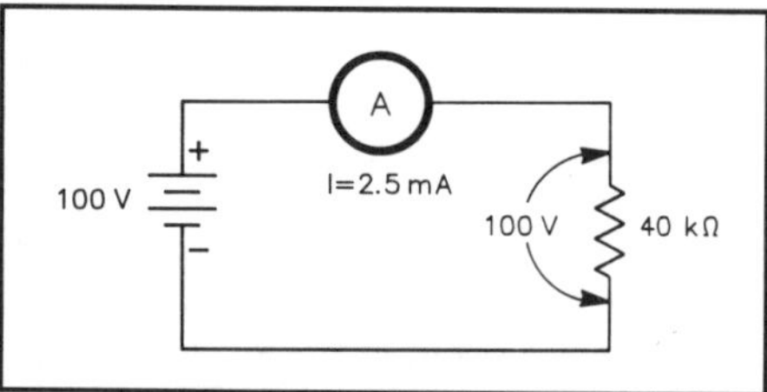

Figure 2—The voltage of this battery is much larger than the voltage of the battery shown in Figure 1. This battery supplies energy to its circuit at a much slower rate, because the resistance is much larger. In fact, this circuit only uses 1/4 J of energy every second.

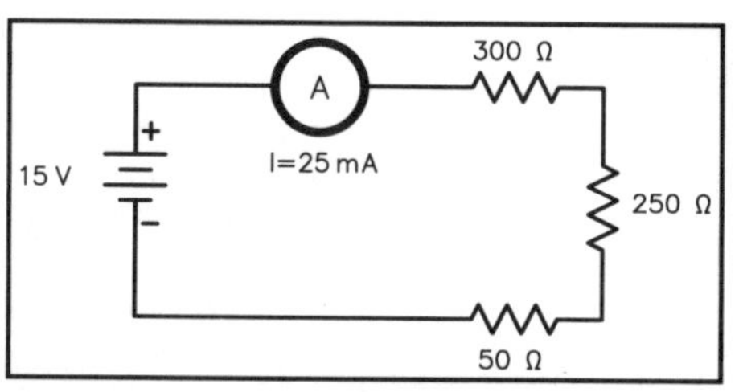

Figure 3—You can calculate the rate that a circuit uses energy (its power) by multiplying the circuit voltage times the total current.

Figure 4 shows a radio transmitter and amplifier tuned to 14.030 in the amateur 20-meter band. A meter in the feed line to the antenna measures the radio frequency current from the transmitter. The current in the feed line is 5 amps. A voltmeter measures the amplifier output as 270 V. What is the power output from this transmitter and amplifier combination?

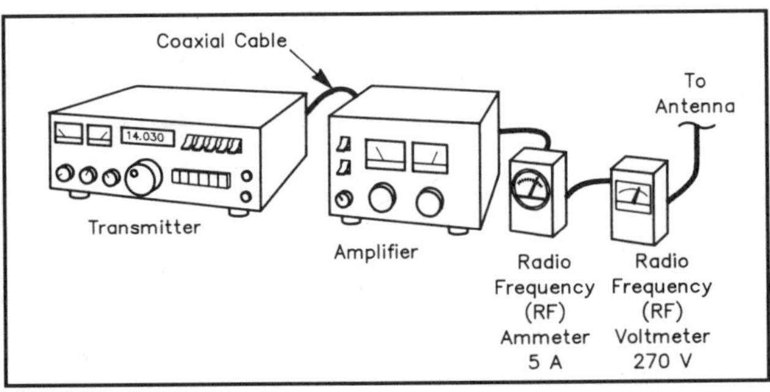

Figure 4—This Amateur Radio station includes a transmitter and power amplifier. It also includes meters to measure the radio frequency current and voltage in the antenna feed line. You can determine the output power of the amplifier by multiplying the current and voltage readings. This amplifier feeds energy to the antenna at the rate of 1350 watts.

P = I E (Equation 1)

P = 5 A × 270 V = 1350 W

You also can express this in kilowatts. The amplifier is feeding 1.35 kW to the antenna in this amateur station.

If you know power and either current or voltage in a circuit, the power equation will help you calculate the third quantity. You can solve Equation 1 for current and voltage in much the same way we solved Ohm's Law for current or resistance. We also can draw a "Power Equation Circle," as shown in Figure 5. Use the Power Circle by covering the quantity you want to find. Write the equation from the remaining parts. Be sure you can write Equations 2 and 3 from the Power Circle.

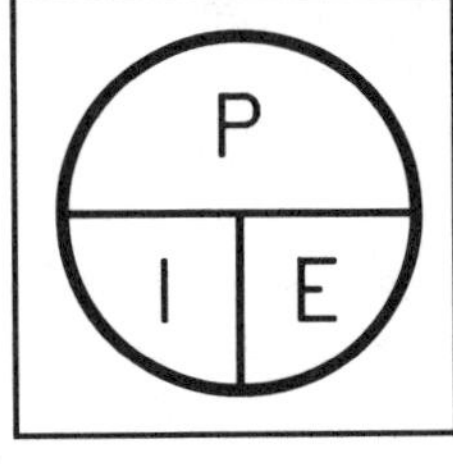

Figure 5—This "Power Circle" provides a way to solve the basic power equation for any of the three quantities in the equation. As with the Ohm's Law Circle, simply cover the part you want to calculate. The other two sections show you how to calculate it. Practice writing equations to calculate power, current and voltage from this circle.

I = P / E (Equation 2)

E = P / I (Equation 3)

Let's use a simple flashlight circuit as an example. Our flashlight uses a 6-V battery and a bulb rated at 2 W. How much current will this bulb draw from the battery? From the Power Circle, we write Equation 2.

I = P / E (Equation 2)

I = 2 W / 6 V = 0.33 A

You could use a similar procedure to find the circuit voltage if you knew power and current.

The power rating on a light bulb (or any electric appliance) describes the rate that it uses electric energy. In an incandescent light bulb, some of that energy gets converted to light energy and some gets converted to heat energy. The power rating helps you calculate the current or voltage that the bulb requires. Of course it also gives you an idea how much light the bulb will produce. Don't expect a 75-W bulb to produce 75 J of light energy every second, though! Fluorescent lights produce less heat than incandescent bulbs, so there is a lot less "wasted" energy for a given power rating.

More Power Calculations

It's an easy task to calculate the power in a circuit if you know the voltage and current. Equation 1 shows how to do this calculation.

$$P = I\,E \qquad \text{(Equation 1)}$$

Many times, you won't know both of those quantities directly, though. You may know some circuit conditions that will help you calculate power, however. Remember, Ohm's Law gives us a relationship between voltage, current and resistance. If you know the circuit voltage and resistance, you can calculate the current using Ohm's Law. Then you can calculate power from those values.

We don't have to use numbers to solve Ohm's Law for current. We can use the letter symbols, and then substitute those symbols into the power equation. By doing that there'll only be one calculation step to find power. That will save time and reduce the number of possible errors.

First, let's assume we know the voltage applied to a circuit, and the circuit resistance. From Ohm's Law, we can calculate the current.

$$E = I\,R \qquad \text{(Equation 2)}$$

so

$$I = E / R \qquad \text{(Equation 3)}$$

If we substitute this expression for current into the power equation, we'll get a single equation to calculate power under these conditions.

$$P = \frac{E}{R} \times E = \frac{E^2}{R} \qquad \text{(Equation 4)}$$

Figure 1 shows a simple parallel circuit that includes a voltage source connected to two parallel resistors. How much power does the voltage source supply to this circuit? The voltage source applies 25 volts and the two resistors have a combined equivalent resistance of 50 ohms. (Be sure you can calculate that equivalent resistance. If you need some help, refer to the Kirchhoff's Laws chapter of this book.)

Calculate the power for this circuit by using Equation 4.

$$P = \frac{E^2}{R} \qquad \text{(Equation 4)}$$

$$P = \frac{(25\text{ V})^2}{50\ \Omega} = \frac{625\text{ V}^2}{50\ \Omega}$$

$$P = 12.5\text{ W}$$

Suppose we knew the current in a circuit, but not the voltage. If we also know the resistance, can we still calculate the power? Sure. Take a look at Figure 2. Here is a series circuit in which we know the current and resistance (at least we can calculate the equivalent resistance easily). The ammeter indicates 700 mA of current, and when we combine the series resistors into one equivalent value we have 300 ohms. First, we'll find a single equation to solve for power. Using Ohm's Law, we can find the circuit voltage.

$$E = I\,R \qquad \text{(Equation 2)}$$

Next, we can substitute this expression for voltage into the power equation.

$$P = I\,E \qquad \text{(Equation 1)}$$

$$P = I \times I\,R = I^2\,R \qquad \text{(Equation 5)}$$

$$P = (0.700\text{ A})^2 \times 300\ \Omega$$

$$P = 0.490\text{ A}^2 \times 300\ \Omega = 147\text{ W}$$

Equations 4 and 5 are very important variations of the power equation. As you have learned here, you can combine Ohm's Law and the basic power equation to find these equations. They are important enough that you should memorize them along with the basic power equation, though.

Let's try a few more practice problems. Figure 3 shows another circuit with a voltage source and three

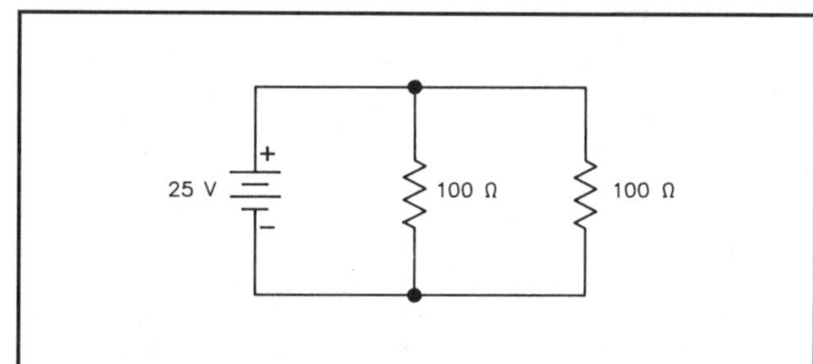

Figure 1—Find the power that this battery supplies to the circuit. You'll have to find a single equivalent resistor to replace the parallel pair first. The text explains the steps to follow to calculate power.

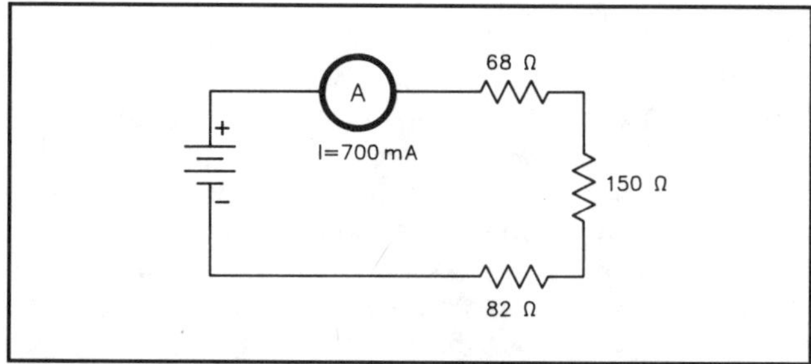

Figure 2—How much power does the battery supply to this series circuit? You should have little trouble calculating an equivalent resistance to replace these series resistors.

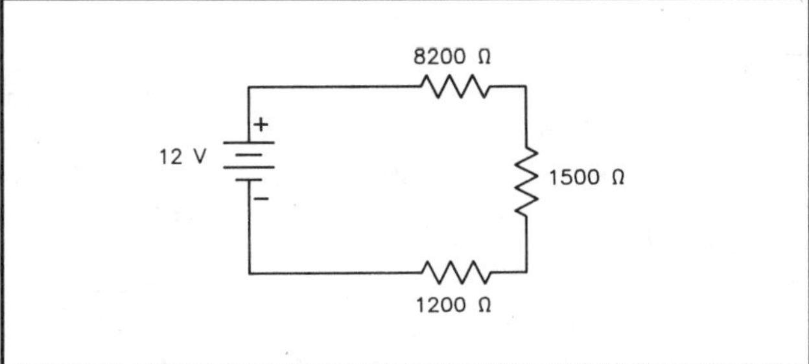

Figure 3—In this series circuit, the diagram shows the battery voltage and the values for three resistors. We must combine the resistors into a single equivalent resistance. Then we can calculate the power supplied by the battery.

resistors connected in series. The voltage supply is a 12-V battery, but the diagram doesn't give the current. The resistor values are 8200 ohms, 1500 ohms and 1200 ohms. How much power does the battery supply to the circuit? First, you'll have to calculate the equivalent resistance of the circuit. That works out to be 10.9 kΩ (10900 Ω). Now, since we know the resistance and voltage, we can use Equation 4 to calculate the battery power.

$$P = \frac{(E^2)}{R} \qquad \text{(Equation 4)}$$

$$P = \frac{(12\ \text{V})^2}{10.9 \times 10^3\ \Omega} = 0.0132\ \text{W}$$

This is a pretty small amount of power. Since the watt is a unit in the metric system, we could write the value with a common metric prefix. We can write this power as 13.2×10^{-3} W. That should help you recognize the metric prefix that we often use for small power levels.

$$13.2 \times 10^{-3}\ \text{W} = 13.2\ \text{mW}$$

Figure 4 shows one more example for you to solve. There is an ammeter to measure the circuit current, which turns out to be 1.5 milliamperes. If the resistor values are 390 kilohms and 0.150 megohms, what is the equivalent circuit resistance? What is the power, in milliwatts, of this circuit?

Since these resistors are in parallel, we'll have to add their reciprocals, and then take the reciprocal of the result. Again, be sure you can do that calculation. Round off your answer to include only 3 digits. You should get 108 kΩ. Now you are ready to use Equation 5 to calculate the circuit power.

$$P = I^2 R \qquad \text{(Equation 5)}$$

$$P = (1.5 \times 10^{-3}\ \text{A})^2 \times 108\ \text{k}\Omega$$

$$P = 2.25 \times 10^{-6}\ \text{A}^2 \times 108 \times 10^3\ \Omega$$

$$P = 0.243\ \text{W} = 243 \times 10^{-3}\ \text{W} = 243\ \text{mW}$$

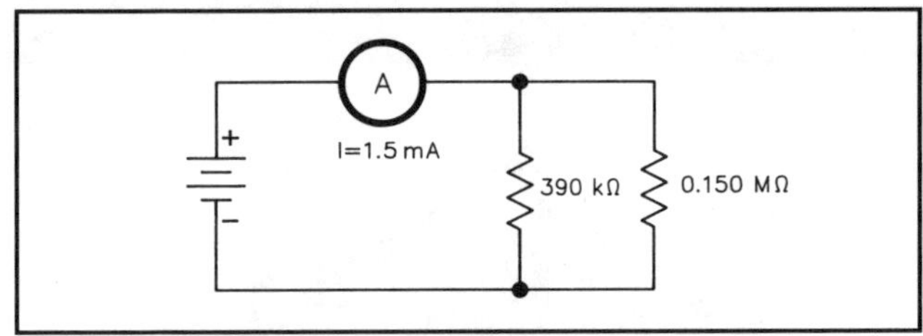

Figure 4—Find the value of a single equivalent resistor to replace the two parallel resistors in this circuit. Then calculate the power that the battery supplies to the circuit.

CHAPTER 15 Decibels

Defining the Bel

The *decibel* is one unit that you will hear used (and misused) quite often in electronics. Just what does this term mean? First we'll define the term, and then we'll take a look at some of the ways we use decibels in electronics.

You have probably recognized *deci* as the metric prefix that means one-tenth. So the unit we are really talking about here is the *bel*, and a decibel is just 1/10 of a bel. We often use a capital B to abbreviate bel. Since a lower case d is the abbreviation for deci, the proper abbreviation for a decibel is dB. The bel is named for Alexander Graham Bell. Most people remember Bell for his invention of the telephone. Bell was also very interested in working with deaf people and studying the way we hear sounds. Bell tried to invent a device that would amplify sounds, to help people with a partial hearing loss. The telephone is a result of this work.

We can hear very soft sounds, like a leaf rustling through the other leaves on a tree as it falls. We also can hear sounds that are extremely loud. A jackhammer pounding the pavement on a city street or the roar of a nearby jet engine are some examples. Some sounds can be so loud that they actually cause us pain. These painfully loud sounds can be nearly 1×10^{12} times louder than the soft rustling of the leaves or a quiet whisper, yet we can hear sounds within this full range.

Figure 1—We can hear very soft sounds, if there is not much other noise.

To make the numbers easier to work with, we often use logarithms to represent the numbers on such a wide scale. We could say that our ears have a logarithmic response to sound loudness or *intensity*. The *bel* compares the loudness of two sounds with each other. One of these sounds serves as a reference for the comparison. To calculate how many bels louder or softer the second sound is, simply divide the reference intensity into the other value. Then find the logarithm of that result.

$$\text{bels} = \log (I_1/I_0) \qquad \text{(Equation 1)}$$

where:

I_0 is the intensity (or loudness) of the reference sound

I_1 is the intensity of the sound compared to the reference

Use the quieter sound intensity (a smaller number) as the reference sound to get a positive value of bels. If you use the louder sound intensity as the reference sound, however, you get the same value but with a negative sign. So a positive value of bels indicates a sound is louder than the reference sound. A negative value of bels indicates a sound is quieter than the reference.

Painfully loud sounds can be as much as 1×10^{12} times louder than the softest sounds we can hear. Use that number as the ratio I_1/I_0 in Equation 1, and calculate the range of our hearing, in bels.

$$\text{bels} = \log (I_1/I_0) \qquad \text{(Equation 1)}$$

$$\text{bels} = \log (1 \times 10^{12}) = 12$$

We can hear sounds that differ in intensity by as much as 12 bels. Normal sounds in your home, like soft music, conversation or the TV are about 4 to 7 bels louder than the softest sounds you can hear.

Sound *intensity* is similar to sound *power*, so we can apply the bel to power levels in electronics. The bel is a rather large unit, even to compare sound intensity levels, so we normally use the decibel. It takes 10 decibels to make one bel. Therefore, the equation to compare two power levels in decibels is 10 times the equation to calculate bels.

$$\text{dB} = 10 \log (P_1/P_0) \qquad \text{(Equation 2)}$$

where

P_0 is the reference power level

P_1 is the power level compared to the reference

Let's look at an example to help you understand bels and decibels. Remember that we

Figure 2—Loud sounds can make it impossible to hear softer sounds.

use a ratio of two power levels, which means we divide one power by a reference, or comparison, power.

Suppose we measure the output power from an Amateur Radio transmitter, and find that it is 15 watts. If we use a power amplifier after the transmitter, and measure the power again, we measure 1500 W. What is the gain, or power increase provided by this amplifier? To solve this problem, we will use the 15-W power as the reference. We want to compare the amplified power with this value. Equation 2 will help us answer the question.

$$dB = 10 \log (P_1 / P_0) \qquad \text{(Equation 2)}$$
$$dB = 10 \log (1500\ W / 15\ W) = 10 \log (100)$$
$$dB = 10 \times 2 = 20$$

The amplifier provides a 20 dB increase in power. This example shows how to do the calculation. In a real Amateur Radio station, we would probably use one amplifier to go from 15 to 150 watts. Then another amplifier would increase the power from 150 to 1500 watts. Each amplifier would have a gain of 10 dB, which is a more realistic figure.

Remember that a logarithm is just the exponent of 10 that will give you the original number. You can find the log of 100 even without your calculator, because 100 is 10^2! That's one benefit to having a power ratio that turns out to be a number like 10, 100, 1000 and so on. (The logarithms for those numbers are 1, 2, 3 and 4.)

What happens if the power decreases? Well, let's look at an example and find out. We can continue with the problem above, and measure the actual power arriving at the antenna. In this station, a *long* length of coaxial cable connects the transmitter to the antenna. Because some power is lost in this cable, we measure only 150 watts at the antenna. This time we'll use the 1500-W amplifier output as our reference. We want to compare the power at the antenna with the amplifier power. Again, Equation 2 helps us answer our question.

$$dB = 10 \log (P_1/P_0) \qquad \text{(Equation 2)}$$
$$dB = 10 \log (150\ W/1500\ W) = 10 \log (0.10)$$
$$dB = 10 \times -1 = -10$$

The negative sign tells us that we have *less* power than our reference. Of course, we knew that because there was less power at the antenna than the amplifier was producing. What happened to that power? Some of the energy going through the coaxial cable changed to heat, and there may be other losses in the cable. All coaxial cables would have some loss. If you have a cable with 10 dB of loss in any reasonable length, however, it's probably no good!

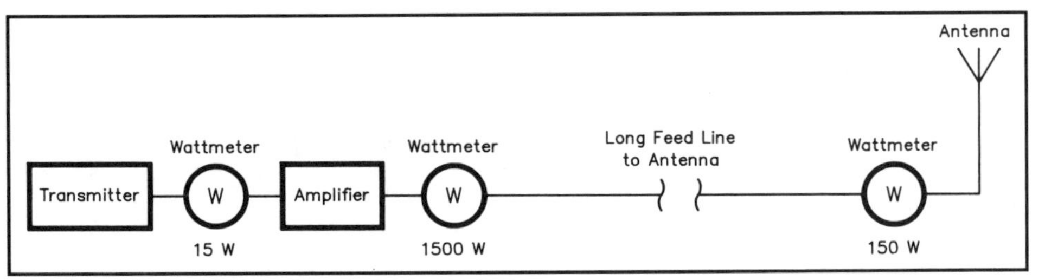

Figure 3—This diagram shows a power level measured at the output of an Amateur Radio transmitter. The diagram also shows power levels after the signal goes through a power amplifier and a long length of coaxial cable.

Decibels and Power Ratios

We often use decibels to compare power levels in electronics circuits. How do you find the number of decibels? First select the power you want to use to compare other power levels. This is your reference power, which often is the beginning power. It may be the power before the signal goes through an amplifier or through a line to an antenna. Next, divide the new power by your chosen reference power. Now find the logarithm of that power-level ratio. Finally, multiply the result by ten. A few examples will show you how easy it is to calculate decibels.

You won't even need a calculator to find the logarithm when the ratio is a number like 10, 100, 1000 and so on! Multiples of 10 less than 1, like 0.1, 0.01, 0.001 and so on are also easy. Use your calculator as we work through a few examples. You'll soon be doing logarithms like these without it, though.

log (1) = 0	**log (0.1) = –1**
log (10) = 1	**log (0.01)= –2**
log (100) = 2	**log (0.001) = –3**
log (1000) = 3	

Figure 1—This chart shows the logarithms of some common multiples of 10. After you work with these values for a while, you will begin to remember them without using a calculator.

Let's suppose you have an amateur transmitter that operates on the 2-meter band. Your transmitter has an output power of 10 watts, but you would like a little more power to use with a distant repeater. An amplifier is just what you need. After connecting your new amplifier, you measure the output power again, and find it is now 100 watts. How many dB increase is this? We'll use the 10-W signal as the reference in this case. Divide 100 by 10 to find the power ratio.

$$\text{Power ratio} = \frac{P_1}{P_0} \qquad \text{(Equation 1)}$$

where

P_0 is the reference power level

P_1 is the power level compared to the reference power

$$\text{Power ratio} = \frac{100}{10} = 10$$

Now find the logarithm of the power ratio.

log (10) = 1

Finally, multiply this result by 10

decibels = 10 × 1 = 10

Your amplifier has increased the power of your 2-meter signal by 10 dB!

Now suppose the amplifier increased your signal to 1000 watts. Choose the reference power to be 10 W again, and divide the new power by the reference.

$$\text{Power ratio} = \frac{P_1}{P_0} \qquad \text{(Equation 1)}$$

$$\text{Power ratio} = \frac{1000}{10} = 100$$

Find the logarithm of the power ratio.

log (100) = 2

Multiply this result by 10 to find the number of decibels.

decibels = 10 × 2 = 20

If we put all these steps together into a single equation, we have the definition of a decibel.

$$\text{decibels dB} = 10 \log \left(\frac{P_1}{P_0}\right) \qquad \text{(Equation 2)}$$

where

P_0 is the reference power level

P_1 is the power level compared to the reference

Use this equation to calculate the number of decibels between power levels.

You should be aware of certain power ratios, because they occur so often. For example, let's see what happens if we double a given power. Suppose we start with a circuit that has a power of 2 mW. What dB increase occurs if we double

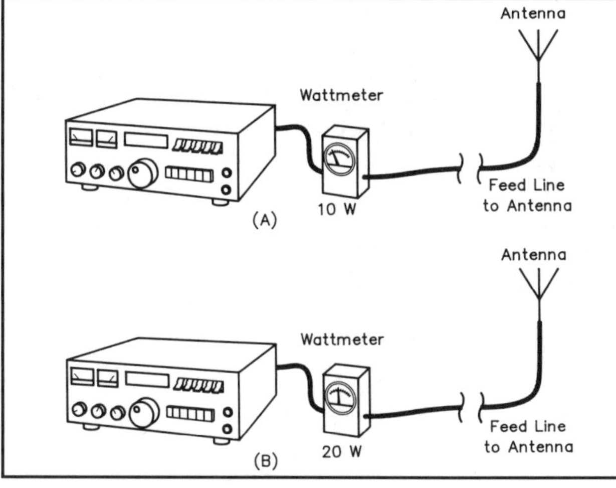

Figure 2—The output power from an Amateur Radio transmitter is 10 watts. After making some adjustments to the transmitter tuning, you measure the power again. Now you find the power has increased to 20 watts. The text describes how to calculate the decibel increase that occurred.

the power to 4 mW? We'll start with the basic definition of a decibel.

$$dB = 10 \log \left(\frac{P_1}{P_0} \right) \qquad \text{(Equation 2)}$$

$$dB = 10 \log \left(\frac{4 \text{ mW}}{2 \text{ mW}} \right) = 10 \log (2)$$

$$dB = 10 \times 0.30 = 3.0$$

When we double the power, there is a 3 dB increase. This is true no matter what the actual power levels are. Let's look at an example with higher power levels to show that the dB increase is the same.

We measure the transmitter output power at an Amateur Radio station, and find that it is 10 W. Use this power as a reference power for the station. After making some adjustments to the circuit, we measure the transmitter output power again. This time we find that the output power has increased to 20 W. What is the power increase, in dB? Equation 2 will help us answer this question.

$$dB = 10 \log \left(\frac{P_1}{P_0} \right) \qquad \text{(Equation 2)}$$

$$dB = 10 \log \left(\frac{20 \text{ W}}{10 \text{ W}} \right) = 10 \log (2.0)$$

$$dB = 10 \times 0.30 = 3.0$$

So our transmitter adjustments gave us a 3 dB increase in transmitter power.

What would happen if we used the 20-W power as the reference power?

$$dB = 10 \log \left(\frac{P_1}{P_0} \right) \qquad \text{(Equation 2)}$$

$$dB = 10 \log \left(\frac{10 \text{ W}}{20 \text{ W}} \right) = 10 \log (0.5)$$

$$dB = 10 \times -0.30 = -3.0$$

What does a negative value mean? There was a 3 dB *decrease* in power! So it is important which power level you use as the reference level.

Suppose you measured the power output from another transmitter, and found it to be 100 W. Later, after experimenting with a new circuit in the transmitter, you measure the output power as 50 W. What effect did your experiment have on the output power? What is the power change in decibels? Again, Equation 2 will help answer this question.

$$dB = 10 \log \left(\frac{P_1}{P_0} \right) \qquad \text{(Equation 2)}$$

$$dB = 10 \log \left(\frac{50 \text{ W}}{100 \text{ W}} \right) = 10 \log (0.50)$$

$$dB = 10 \times -0.30 = -3.0$$

Although the power levels in our two examples were much different, we still had a 3 dB change. This is an important point about the decibel. It compares two power levels. The number of decibels depends on the *ratio* of those levels, not on the actual power. The 3 dB value is also important, because it shows that one power level was twice the other one. Increasing a power by two gives a 3 dB increase and cutting a power in half gives a 3 dB decrease.

Whenever you multiply or divide the reference power by a factor of 2, you will have a 3 dB change in power. You might guess, then, that if you multiplied the power by 4 it would be a 6-dB increase. If you multiplied the power by 8 it would be a 9-dB increase. You would be right in both cases!

Suppose the power in part of a circuit measures 5 milliwatts and in another part of the circuit it measures 40 mW. Using the 5-mW value as the reference power, how many decibels greater is the 40-mW power?

$$dB = 10 \log \left(\frac{P_1}{P_0} \right) \qquad \text{(Equation 2)}$$

$$dB = 10 \log \left(\frac{40 \text{ mW}}{5.0 \text{ mW}} \right) = 10 \log (8.0)$$

$$dB = 10 \times 0.90 = 9.0$$

Figure 3—A simple amateur transmitter amplifies the signal from an oscillator and then feeds that signal to an antenna. It uses several amplifier stages. The input power to one of those stages is 5 milliwatts and the output from that stage is 40 milliwatts. The text describes how to calculate the gain of that amplifier stage.

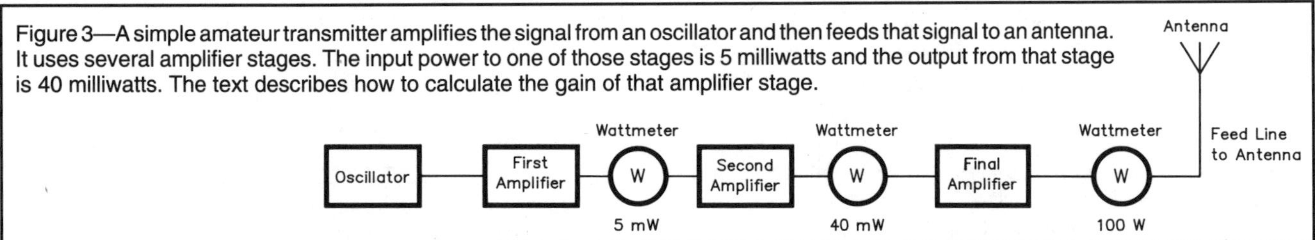

Specify the Reference Power

Any time you calculate a decibel value, you must use some power level as the reference. That's all you must do to compare two power levels, or two circuit conditions. If you increase the power two times, you have a 3 dB increase. If you increase it four times, you have a 6 dB increase, and so on. Likewise, if you cut the power in half there is a 3 dB decrease.

One circuit takes an input signal of 2 watts and amplifies that signal to 4 watts. Another circuit takes a signal of 500 watts and amplifies it to 1000 watts. Both circuits produce a signal 3-dB stronger than the input signal. What if you want to compare these two circuits? Would you say the amplifiers are the same? They are both 3-dB amplifiers, but they operate on signal levels that are very much different. Putting a 500-watt signal into the first amplifier probably would destroy the amplifier.

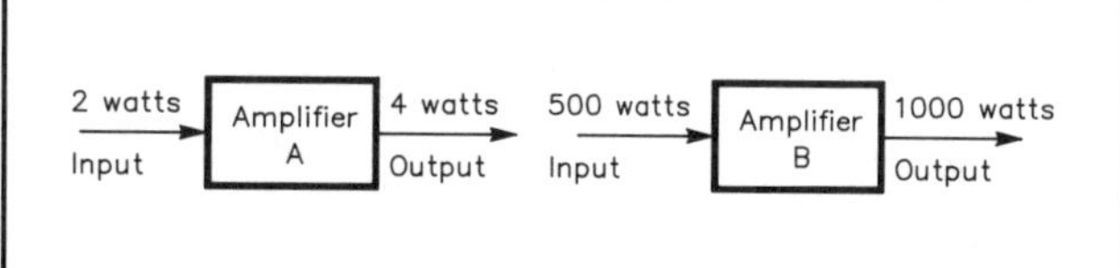

Figure 1—Here are two amplifiers. Each increases the input signal by 3 dB. They work with input signals that are very different, however. One amplifier works with signals in the range of a few watts and the other with signals of several hundred watts. Knowing this will help you select the right amplifier for a particular job.

The decibel is an important tool for studying electronics circuits. It becomes even more important, however, when we specify the reference power used for the comparison. By using the same reference level to specify power in several circuits, we have a convenient way to compare those circuits.

You will often see the abbreviation for a decibel written with another letter following the dB, such as dBm or dBW. The last letter tells us about the reference power used in the comparison. We often measure a power level in some section of a circuit. To compare this circuit with other similar circuits, we may want to use a "standard power" for a decibel calculation. You might expect that one watt would be a convenient power level for comparison. Indeed, we often use one watt as the basis for comparison. Another common "standard power" is the milliwatt, 0.001 W.

It is convenient to compare low power levels, like those found in a radio receiver, with a milliwatt. A signal specified as 3 dBm has twice as much power as a 1 mW signal. (A 3-dBm signal has a power level of 2 mW, then.) We specify a 1-mW signal as 0 dBm and a 10-mW signal as 10 dBm.

If we use 1 watt as a standard reference, we can specify power levels in dBW. (A 100-W signal from a transmitter would be 20 dBW under this system.)

Most Amateur Radio operators like to experiment with antennas. Decibels provide a convenient unit to compare several antennas. Measure the power transmitted from each antenna in a certain direction. Some hams even measure the transmitted signal power every few degrees around their antennas. Then they plot a graph, called an antenna radiation pattern. (Never mind how they do this now.) Right away you can see that we need to specify some reference to compare this power. You probably even suspect that decibels will play some role here.

A *beam antenna* will transmit a stronger signal in one direction and a weaker signal in other directions. A "Super Signal Boomer" antenna transmits a signal that is twice as strong as the signal from a "Little, Simple Bandspanner" antenna. (Of course the same transmitter power goes into each antenna.) You probably are thinking, "Wow, the 'SSB' antenna has a gain of 3 dB compared to the 'LSB'!" You are correct. But how will either of these antennas compare to a "Carefree Wire" antenna? We will have to build a "CW" antenna and measure the transmitted signal power to make a comparison. It isn't always convenient to build a new antenna and measure the radiation pattern, however. Sometimes we want to make a theoretical calculation instead. This is especially true if you are considering the purchase of a brand new "Super Signal Boomer." You certainly want to know how much better it will be than your present "Carefree Wire" antenna before you spend all that money.

We sometimes use a theoretically perfect antenna as a reference for such comparisons. This imaginary perfect

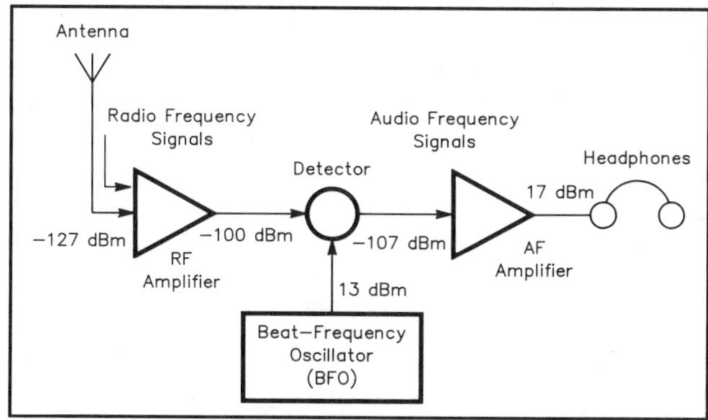

Figure 2—This block diagram of a radio receiver shows typical signal values, all referenced to a milliwatt (dBm).

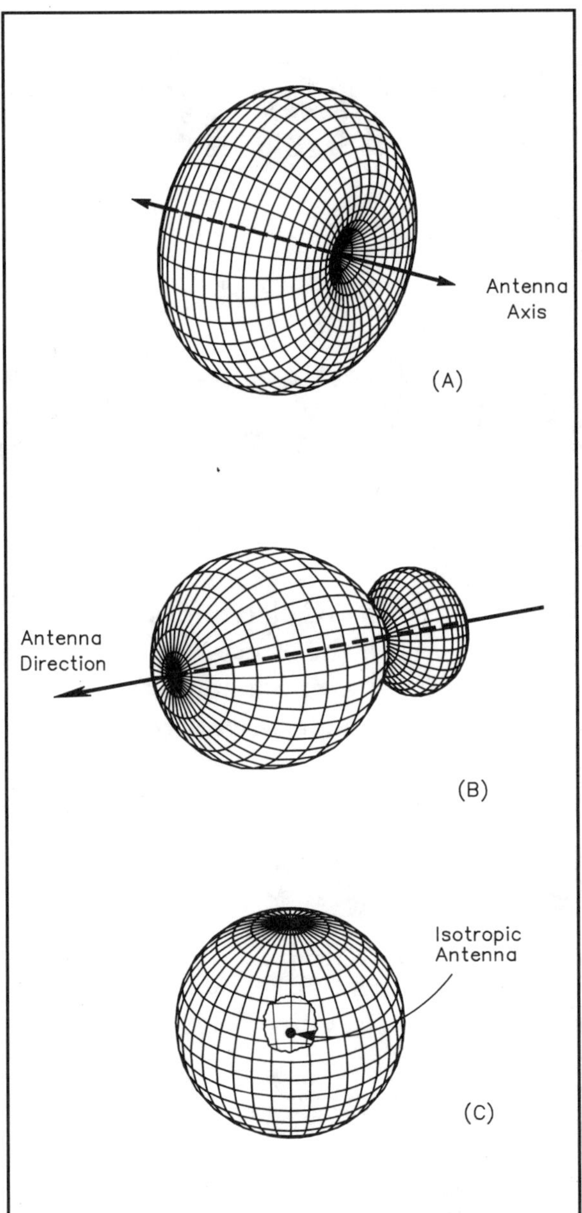

Figure 3—Part A shows the antenna radiation pattern for a simple dipole antenna. Part B shows the pattern for a beam antenna and C shows the pattern for an imaginary isotropic antenna.

antenna is a point suspended in space. Far from Earth, there is nothing nearby to interfere with the antenna as it sends the signals off to other stations. (Real antennas are affected by other objects, such as the ground, trees, buildings and mountains.) We call this imaginary perfect antenna an *isotropic antenna*. A lower-case letter i used with the decibel tells us that we are comparing the power from one antenna with the power that would be sent out from the isotropic antenna. Some antenna engineers calculated that the "Super Signal Boomer" antenna has a gain of 9 dBi. That means the "SSB" antenna transmits eight times more power in one direction than an isotropic antenna would. What is the gain of the "LSB" antenna described earlier, compared to the isotropic antenna? Well, we know the "SSB" antenna has a gain of 3 dB compared to the "LSB" antenna. The "LSB" must have a gain of 6 dBi, then.

Sometimes it is better to compare our antennas with one that is more practical than an imaginary antenna. Many hams use a half-wavelength dipole antenna, so it makes a good standard for comparison. (Don't worry about what these antennas look like or how to build them.) A lower-case letter d with the decibel tells us the reference antenna is a dipole. How does a "Triple-Loop Sender" (gain 5 dBd) compare with a "TR35B" (gain 2 dBd)? Right. The "Triple-Loop" signal is twice as strong.

Do you get the idea? We specify a standard reference, so we can compare amplifiers or antennas or other systems. As you study electronics you will see letter combinations like dBm, dBW, dBi or dBd. When you do, remember that in addition to the decibel measurement, you know the power level used as a reference.

We also use other letters with the decibel. Each describes something about the reference level used in the calculation. You'll learn about them as you study more electronics. When you see a decibel value with an extra letter, try to figure out what the reference power might be. If you aren't familiar with the letter, try asking another Amateur Radio operator. You also can look in an electronics dictionary.

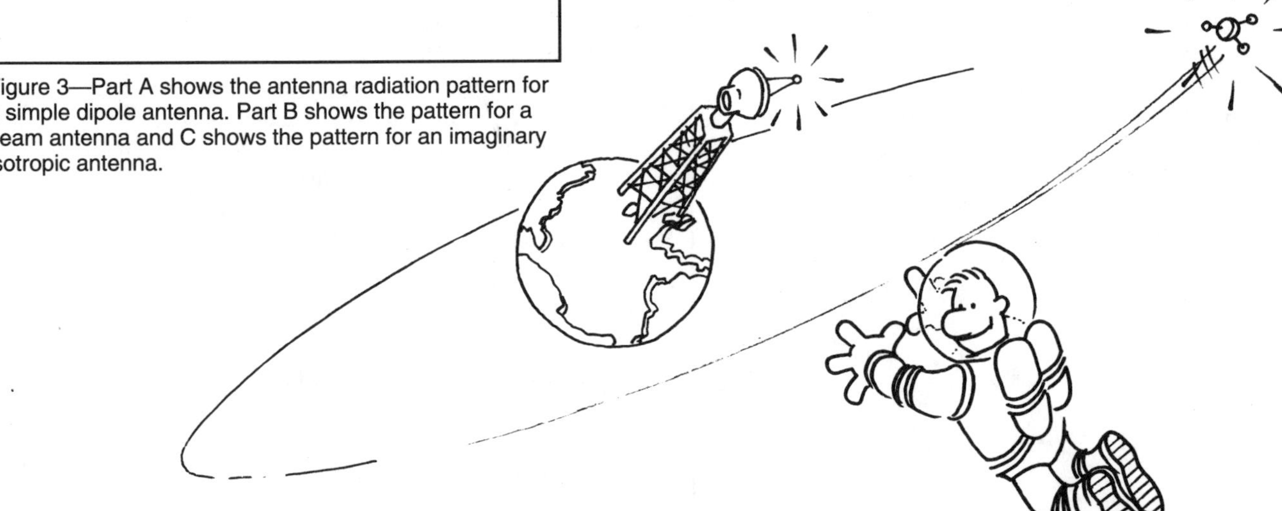

UNIT 3 Elementary AC Electronics

In the last unit we studied steady, unchanging voltages and currents. We learned to use electronics principles to calculate circuit conditions with these steady voltages applied. Now we are ready to consider the effects of changing voltages and currents.

When we studied dc electronics in the last unit, we learned about circuit resistance. Resistance is an important circuit condition when all signals are steady, unchanging ones. Most electronics circuits include some signals that are changing, however. Circuit currents may increase and decrease over time even if their direction doesn't change. Other circuits include currents that change direction continuously. Signals that change direction are called *alternating current (ac) signals.*

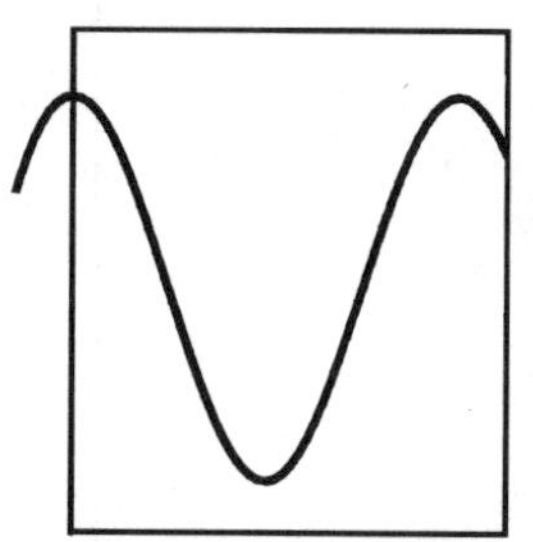

AC electronics is a study of the effects various circuit components have on sine-wave signals.

Inductors and *capacitors* are components with a special purpose in circuits that have varying signals. They help guide dc and ac signals to the proper parts of a circuit. These components oppose changes in the current going through them. We call this opposition *reactance.* Reactance provides a way to control the amount of alternating current in a circuit.

Electric and magnetic fields are important for you to understand how capacitors and inductors work. These components store energy in electric and magnetic fields. The fields change when the signal going into an inductor or capacitor changes.

Most of the signals we study in this unit will change in a regular way over time. Such signals are easier to consider than ones that vary in an irregular way. We call regularly varying ac signals *sine waves.* AC electronics is a study of the effects various circuit components have on sine-wave signals.

In This Unit You Will Learn:

- That all ac signals have a certain wavelength and frequency.
- What capacitors are and how they affect various types of signals.
- That capacitors present a special opposition to ac signals. We call this special opposition *reactance.*
- About inductors and how they affect various types of signals.
- That inductors also present opposition, or reactance, to ac signals. You also will learn the difference between inductive reactance and capacitive reactance.
- About the quality factor of capacitors and inductors, *circuit Q.*
- How transformers change the voltage and current of an input signal to larger or smaller output values.
- About the combined effects of circuit resistance and reactance, called *impedance.*
- That when a circuit's inductive reactance and capacitive reactance are equal, we say the circuit is *resonant.*

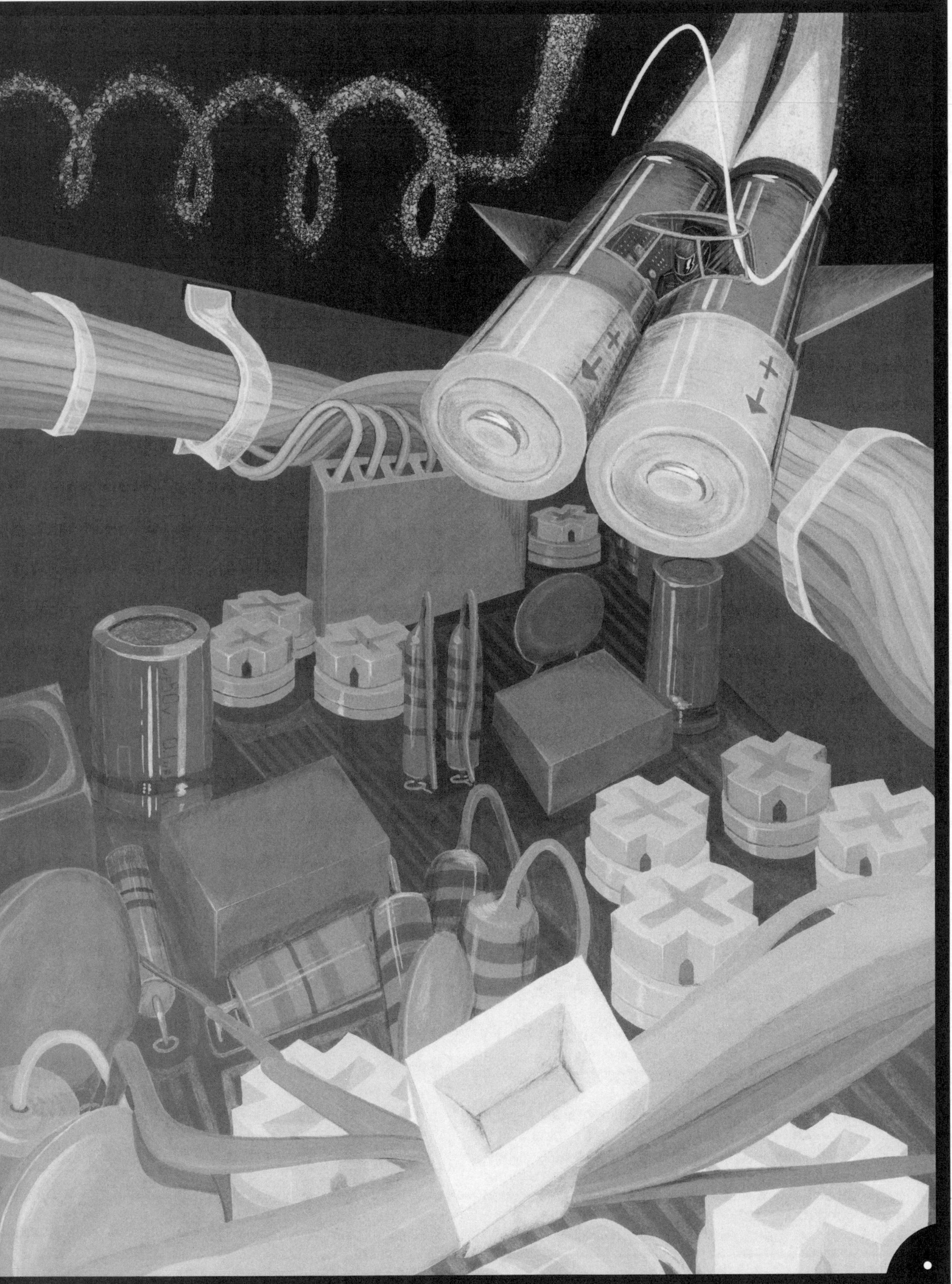

CHAPTER 16 Frequency and Wavelength

Current Goes in Only One Direction in a DC Signal

In Unit 2 you learned about many important electronics principles. You studied voltage, current and resistance. You learned there are two circuit types; series and parallel. Ohm's Law and Kirchhoff's Voltage and Current Laws helped you learn about conditions in those circuits.

We used steady, unchanging voltage sources to supply the current for those circuits. The current through the circuits didn't vary. Steady, unchanging currents make it easy for you to learn electronics principles. *Direct-current* (*dc*) supplies produce circuit currents that don't vary.

Batteries are common direct-current supplies. Figure 1 is the schematic diagram of a simple circuit. There is a battery, a switch and two resistors. *Conventional current* flows from the positive battery terminal, through the resistors and back to the negative battery terminal. The electrons move the opposite direction through the circuit.

Imagine the electric circuit is like a four-lane superhighway. All the traffic is going the same direction as your car. Most of the cars are traveling at the speed limit, so you move along at a steady speed.

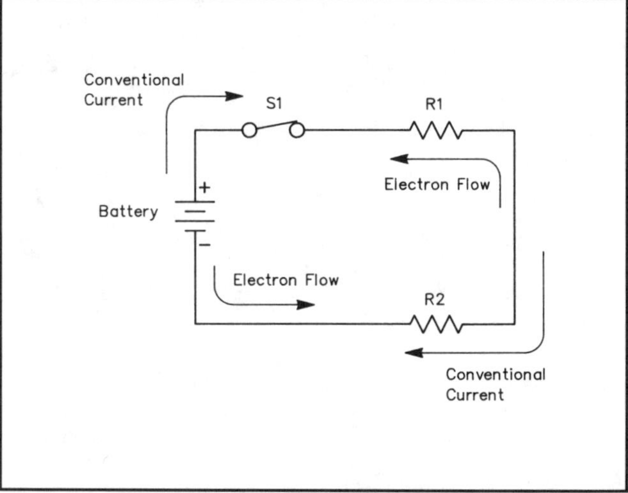

Figure 1—This simple direct-current circuit has a steady, unchanging current. Electrons flow from the negative battery terminal, through the circuit, to the positive battery terminal.

This superhighway is like a direct-current circuit. The cars are like the electrons moving through the circuit. All the electrons move in the same direction at nearly the same speed.

It is easy to study traffic patterns on our superhighway. It is also easy to study electrons moving through a simple dc circuit.

Most roads are not like our superhighway, though. Traffic speeds up and slows down depending on the amount of traffic and the road conditions. Few electronics circuits are simple dc circuits, either. To understand more complex circuits, we will use the basic principles we learned to study what happens when circuit voltage and current change. In the next several sections we will study circuits with changing voltages and currents.

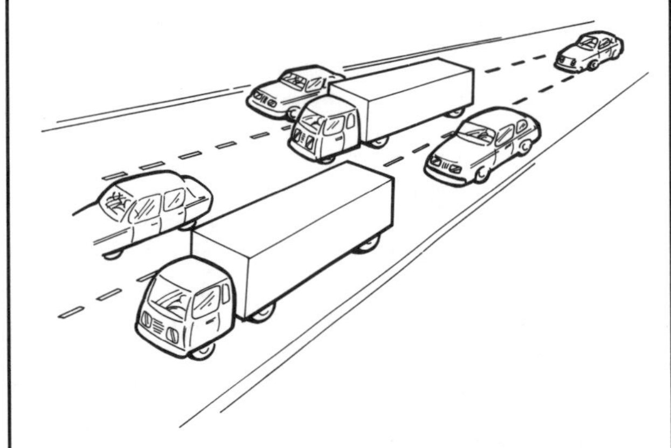

Figure 2—You can think of a simple dc circuit as being similar to a superhighway. All the cars go in one direction, at nearly the same speed.

Current Can be Constant or Changing

All the circuits we studied so far had constant currents. Most circuits have currents that do change with certain conditions, however. The simplest kind of change is to have a circuit with a switch. When we close the switch, there is current through the circuit. When we open the switch, there is no current.

Figure 1 shows a simple circuit with a switch to connect or disconnect the battery. There is a voltmeter to measure the voltage across the resistor, and an ammeter measures the current through the resistor.

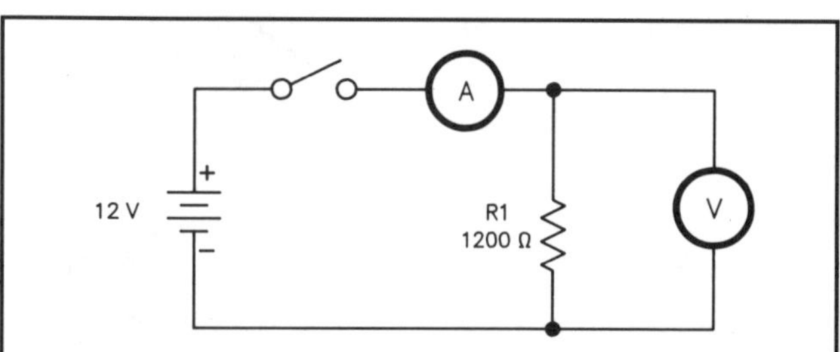

Figure 1—This circuit's current and voltage change every time you change the switch position. The voltmeter measures the voltage across the resistor and the ammeter measures the current through the resistor.

With the switch in the open position, there is no voltage across the resistor. When we close the switch, the voltage across the resistor jumps to 12 volts. When the switch is open there is no current through the circuit and when we close it the current jumps to 10 mA. This circuit has a changing current and voltage.

Now suppose we change the switch position every 10 seconds. Figure 2 graphs the voltage across the resistor and the current through the circuit. Notice the vertical axis on the left side indicates voltage and the vertical axis on the right side indicates current. Voltage and current both change each time the switch opens or closes, so this one graph shows both quantities.

When we refer to the *waveform* of a current or voltage, we mean the shape of the signal voltage or current. The Figure 2 graph shows a *square-wave signal.*

Have you ever had a ride on a highway with toll

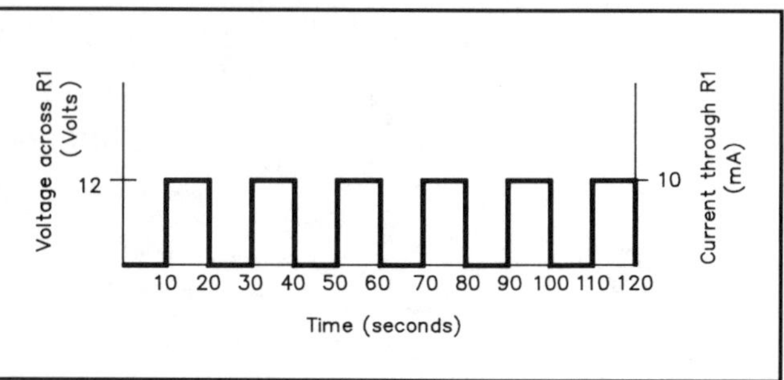

Figure 2—Suppose we change the switch position in Figure 1 every 10 seconds. The horizontal or X axis shows time, marked off in 10-second intervals. The left-side vertical or Y axis shows the voltage across R1, which the voltmeter indicates. The right-side Y axis shows the current through R1, which the ammeter indicates.

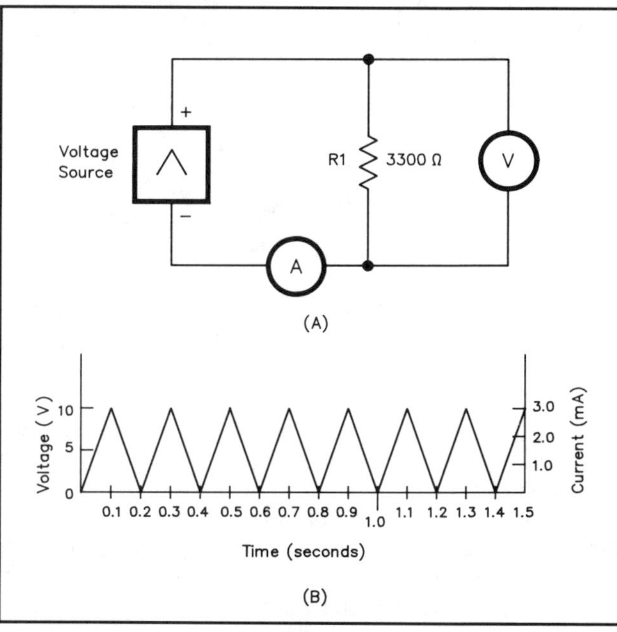

Figure 3—Part A shows a circuit that has a voltage source that gradually increases and decreases at regular intervals. This voltage source produces a *triangle waveform.* Part B shows the voltage and current of this triangle waveform.

booths? On some toll roads you have to stop at a toll booth every few miles to pay a small fee. You can think of such roads as similar to the Figure 1 circuit. Between toll booths, the circuit is complete, and you drive down the highway. When you come to a toll booth, though, the switch opens and traffic must stop.

Your car won't stop instantly, however. You need some time to apply the brakes and slow the car. Neither will it go from 0 to 55 miles per hour in an instant. It will take at least a few seconds to get back up to travel speed.

Some circuits increase the voltage and current gradually. The voltage and current also may decrease gradually. Figure 3A shows a circuit with a voltage source that increases and decreases gradually. We drew this voltage source as a block instead of using the common battery symbol. Part B shows the output voltage and current from this voltage source. The output signal looks like a series of triangles. Are you surprised to learn that we call this type of signal a *triangle waveform*?

There are many ways that a voltage or current can change. Some waveforms are simple, like those we studied here. Others may be complex, following no apparent pattern. As long as the electrons move in one direction through the circuit, we say it is a direct-current signal.

If Current Changes Direction, It Becomes an AC Signal

In the two previous sections we used examples of a superhighway to study dc electronics circuits. We'll use another highway example in this section to study electric currents that change *direction*.

Instead of riding in a car, imagine you are the toll collector on a bridge. Your toll plaza stands at the eastern end of the bridge. There are three lanes going each direction to cross your bridge.

Figure 1 shows your toll plaza. There are nine toll booths and ten lanes through the plaza. As a toll collector, you control a red X or a green O above the lanes at your booth. Figure 1 shows the plaza with four lanes open coming from the West. (This is from the bottom of the picture.) There are six lanes open coming from the East. Most of the traffic is from the East now, so the center lanes are open from that direction.

A few hours later most of the traffic comes from the West. You suddenly realize that traffic is backed up on the bridge because the cars can't get through the four open lanes fast enough. There is almost no traffic coming from the East, though, so those six lanes are empty.

You are in the center toll booth, so you decide to change the lights over the two lanes on either side of your booth. First you change the green Os to red Xs on the westbound side. Then you change the red Xs to green Os on the eastbound side. Immediately, some cars move out of the crowded lanes and toward the two new lanes you just opened. Now there are six lanes open for the eastbound direction and only four open lanes for those heading West.

Traffic in the two center lanes changed direction when you switched the lights. Cars can move through the lane in only one direction at a time. You were able to change the traffic-flow direction by switching the lights.

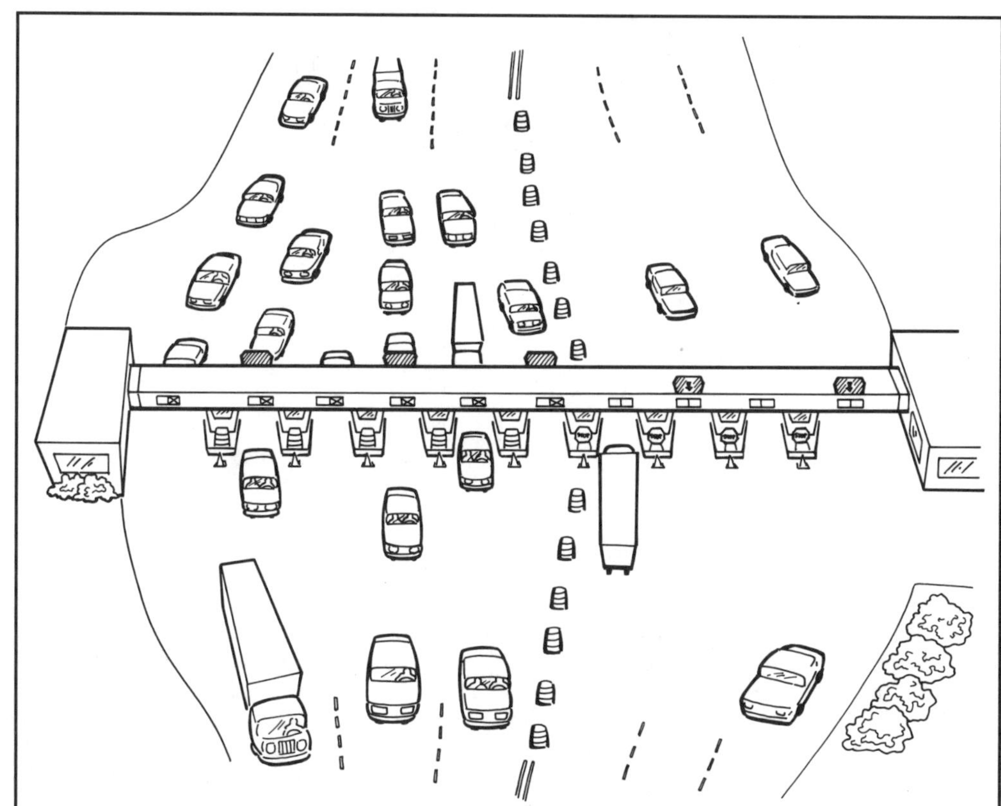

Figure 1—The operation of this toll plaza at a bridge will help you understand electronics circuits with voltage and current that change direction. Cars approach the toll plaza from the bottom when they are driving East. Those cars just crossed the bridge. As they approach the toll plaza they see red Xs and green Os. They can go through the lanes with green Os. Those lanes with red Xs are open to traffic from the opposite direction. Cars approach the toll plaza from the top when they are driving West. As shown, there are six lanes open for cars driving West. There are only four lanes open for cars driving East. Later in the day, the center two lanes close to westbound traffic and open to eastbound traffic. Cars go through those lanes in the opposite direction. The direction of traffic through those two lanes reverses, depending on how much traffic there is going each way.

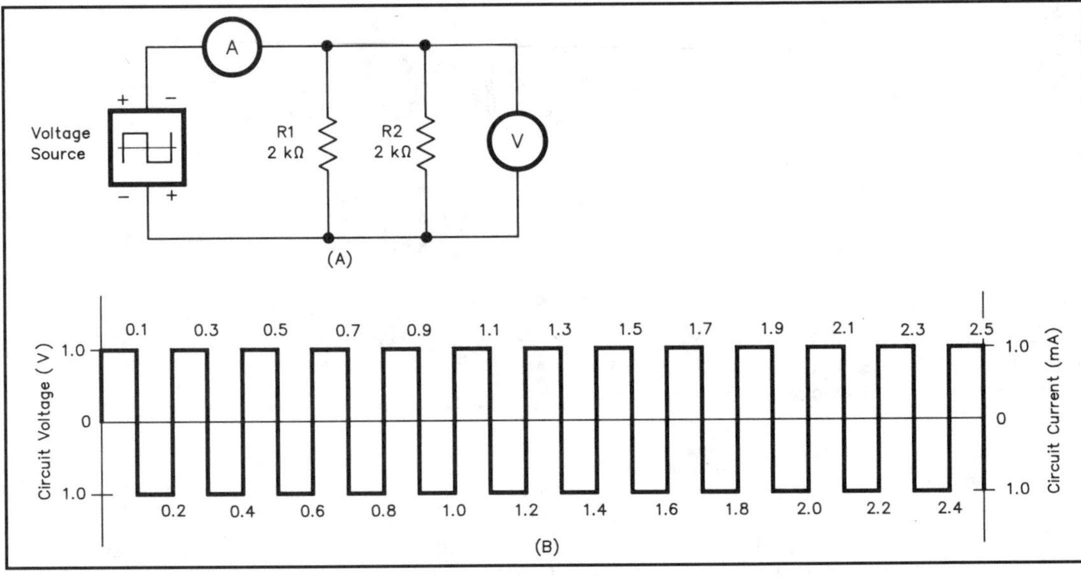

Figure 2—This circuit includes a voltage source that changes polarity ten times each second. There are two resistors in parallel. A voltmeter and an ammeter measure circuit conditions. The graph shown at B represents the voltage across the resistors and the total circuit current.

Some electronics circuits are like this. Electrons can move through the circuit in one direction. A very short time later they can change direction and move through the circuit the other way. When electrons flow first in one direction and then in the other, we say there is an *alternating current.* We abbreviate alternating current as *ac.*

Figure 2A shows a circuit with an alternating-current signal. This is a simple circuit that includes a voltage source and two resistors. We connected the resistors in parallel this time. We drew the voltage source as a box with two terminals. Notice that when the top terminal is positive, the bottom one is negative. When the top terminal becomes negative, the bottom one becomes positive.

Figure 2B shows a graph of the voltage across the resistors and the current through them. The voltage polarity reverses ten times each second. When the voltage polarity reverses, the current also reverses.

Let's make the top of the supply positive and the bottom negative. This is a positive voltage on the graph of Figure 2B. The *conventional current* flows clockwise through the circuit, from positive to negative. We drew this as a positive current on Figure 2B.

One tenth of a second later the voltage polarity reverses. Now the top of the supply is negative and the bottom is positive. This is a negative voltage on the graph. The current also reverses, so it moves in a counterclockwise direction, still from positive to negative.

You must learn some important definitions about alternating current circuits. You will hear these terms frequently as you continue your study of electronics.

The *waveform* of an ac signal refers to the shape of the voltage or current. Our voltage source in Figure 2 produces a waveform with a square (or rectangular) shape. We say it is a *square-wave signal.*

Many ac signals have a regular repeating shape. Look at Figure 2B. Start when the signal is positive, changes to negative and then becomes positive again. This represents one complete *cycle* of the wave.

The time it takes to make one cycle is the wave's *period.* We measure period in time, so we use a capital T to abbreviate period. The Figure 2B wave changes polarity every 0.1 second. That means it takes 0.2 second to make a complete cycle from positive to negative and back to positive. The period of the wave in this example is 0.2 second.

The *frequency* of a wave is the number of complete cycles the wave makes in one second. Frequency is the reciprocal of period. Do you remember how to find a number's reciprocal? You divide the number into one. Your calculator probably has a key labeled 1/X to make it easy to find reciprocals.

> *The basic unit of frequency is the hertz. This unit is named in honor of Heinrich Rudolf Hertz (1857 - 1894).*
> *This German physicist was the first person to demonstrate the generation and reception of radio waves.*

We measure frequency in hertz, abbreviated Hz. This is in honor of Heinrich Hertz, a German scientist who experimented with radio waves. You can combine any of the metric prefixes with this basic unit. A kilohertz (kHz) is 1000 Hz, or 1000 complete cycles in one second. A frequency of 5 megahertz (MHz) is 5 million cycles per second.

$$f = 1 \text{ cycle} / T \text{ (in seconds)} \qquad \text{(Equation 1)}$$

where f is frequency in hertz.

Now let's use Equation 1 to calculate the frequency of the wave in Figure 2.

$$f = 1 \text{ cycle} / 0.2 \text{ s} = 5 \text{ cycles} / \text{s} = 5 \text{ Hz}$$

This square wave has a period of 0.2 second and a frequency of 5 Hz.

Smoothly Varying AC is a Sine Wave

When you studied dc signals you learned that the voltage levels sometimes change gradually rather than abruptly. Voltage and current polarities also can change gradually. Many waveforms are possible with alternating current signals. Instead of a square wave, you might find that a circuit has a triangle-shaped waveform. You might even find a signal that has a sawtooth-shaped waveform. Figure 1 shows a few possible ac waveforms.

An ac waveform that you will frequently find in electronics is the *sine wave.* A sine-wave voltage varies gradually between its peak positive and peak negative values. There are no sudden changes or "corners" on a sine wave. Figure 2 shows a sine-wave voltage or current graph. We marked one complete cycle on this graph, starting when the wave crossed the zero line going positive. Notice that the cycle doesn't end the next time the waveform crosses the zero line. At that point the wave has completed its positive half cycle and is ready to begin the negative half cycle.

We marked another complete cycle starting when the voltage or current is at the peak positive value. It doesn't matter where you start to measure a cycle, as long as you measure to that same point the next time it occurs.

Alternating current signals travel through conductors in much the same way as dc signals do. Alternating signals also produce *electromagnetic radiation.* We often refer to electromagnetic radiation as *radio waves.* Visible light, ultraviolet radiation, X-rays, gamma rays and other types of radiation also are electromagnetic radiation.

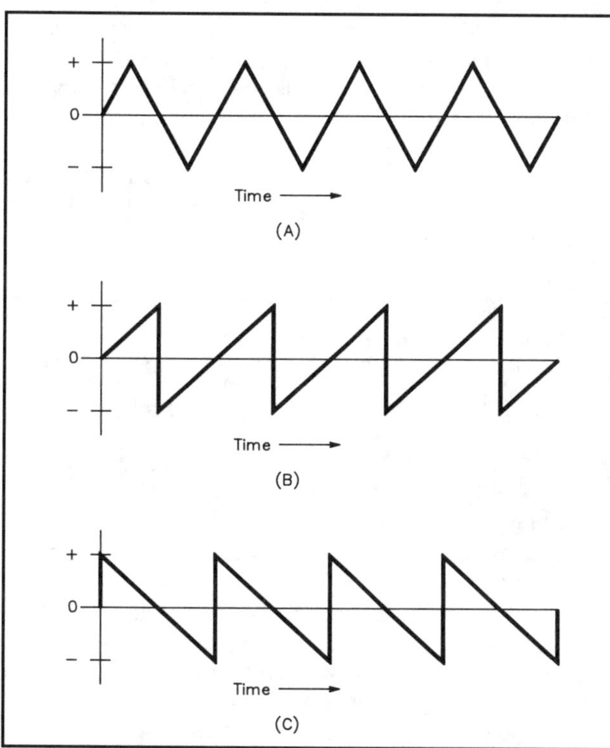

Figure 1—These graphs show a few types of alternating-current waveforms. The signal at Part A is a triangle waveform. Part B shows a sawtooth or ramp waveform. The voltage increases gradually from its peak negative value. When it reaches its peak positive value the voltage (or current) suddenly drops to the peak negative value. Then the voltage or current begins to increase gradually again. The signal at C is a reverse sawtooth. In this example, the voltage increases suddenly to its peak positive value. Then it decreases gradually to its peak negative value.

From our earlier discussion of frequency, you should realize that alternating currents and voltages can change direction at almost any rate imaginable. Some signals have low frequencies. The power company supplies electricity to your house as 60-Hz ac. Other signals have higher frequencies. Radio signals, for example, can alternate at more than several million hertz.

If you know the frequency of an ac signal, you can use that frequency to describe the signal. Your toaster uses 60-Hz power, for example. Your favorite AM Broadcast station may have a frequency of 1080 kHz. Perhaps you like to operate on the 222-MHz Amateur Radio band. All these frequencies describe particular signals.

Wavelength is another quality that we associate with every ac signal. As the name implies, wavelength refers to the distance the wave travels through space in a single cycle. We use the lower-case Greek lambda (λ) to represent wavelength. All electromagnetic radiation travels through space at the speed of light, 300,000,000 metres per second. You can write that in scientific notation as 3.00×10^8 m/s.

The faster a signal alternates the less distance it can travel during one cycle. There is an equation that relates a

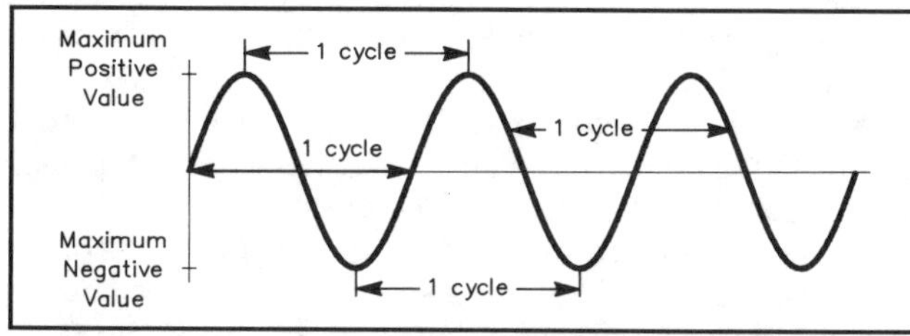

Figure 2—This graph shows a sine-wave alternating current waveform. This is one of the most common ac signals. One cycle is marked off on the waveform starting at several places. Notice that it really doesn't matter where you start measuring a cycle. Wherever you start, you must measure to the same position on the waveform the next time it occurs.

signal's frequency and wavelength to the speed of light.

$c = f\lambda$ (Equation 1)

where:

c is the speed of light, 3.00×10^8 m/s
f is the frequency of the wave in hertz
λ is the wavelength of the wave in metres

You can draw an equation circle for this equation. The equation circle will help you solve the equation for frequency or wavelength. Use the equation circle of Figure 3 to write Equations 2 and 3.

$f = c / \lambda$ (Equation 2)

$\lambda = c / f$ (Equation 3)

From these equations you should realize that as the frequency increases the wavelength becomes shorter. As the frequency decreases the wavelength gets longer.

Use Equation 3 to find the wavelength of a 7.125-MHz radio signal. (Remember to change the frequency in megahertz to frequency in hertz.)

$\lambda = c / f$ (Equation 3)

$$\lambda = \frac{3.00 \times 10^8 \text{ m/s}}{7.125 \times 10^6 \text{ Hz}} = 42.1 \text{ metres}$$

As another example, use Equation 2 to find the frequency of a radio signal that has a wavelength of 80.5 metres.

$f = c / \lambda$ (Equation 2)

$$f = \frac{3.00 \times 10^8 \text{ m/s}}{80.5 \text{ metres}} = 3.727 \times 10^6 \text{ Hz} = 3.727 \text{ MHz}$$

You should be familiar with several other alternating current and voltage measurements. The ac-waveform graphs show the voltage and current change with time. For all the waveforms we looked at, the waveform is positive for half its cycle and negative for half its cycle. What good is this? Doesn't the negative part cancel the effects of the positive part? Do the electrons just move back and forth in the wire, never really going anywhere? You have probably asked these questions as we began our study of ac signals.

If you connect a dc voltmeter to an ac signal, the meter will read 0. Yes, half the time the needle tries to move positive, and half the time it tries to swing negative. So we can't use a dc voltmeter to measure ac voltage. (We can't use a dc ammeter to measure ac current for the same reason.)

Meters that measure ac voltage and current use a diode to change the ac signal to a dc signal. We say the diode *rectifies* the ac signal when it converts it to a dc signal. (You will learn more about diodes and how they rectify ac signals in Unit 4.) The manufacturer adjusts the meter scale to read the proper values.

Most ac meters read *effective* voltage or current. Start by putting some direct current through a resistor. Let's use 1 amp, for example. Measure the heat produced in the resistor. Then put an alternating current through the resistor. Adjust the current until it produces the same heating effect. Then you have an *effective* alternating current of

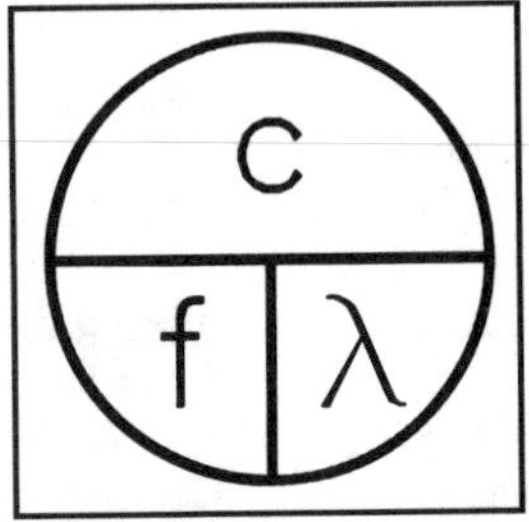

Figure 3—This equation circle represents the relationship between the speed, frequency and wavelength of electromagnetic radiation, or radio waves. Cover the quantity you want to calculate and the rest of the circle shows how to do that calculation.

1 amp. (We also can make this measurement with voltage, to determine the effective ac voltage.)

Another name for effective voltage or current is *RMS voltage* or *current*. This name comes from the mathematical method of determining the value. The letters stand for *root mean square*. *Mean* is a mathematical term for average. The mathematics of the calculation are beyond the scope of this book. You will hear the term RMS voltage or RMS current used frequently in relation to ac signals.

Every ac waveform has a different RMS value compared to the peak voltage or current. You should be familiar with the values for a sine wave, because you will see them often. The RMS value of a sine-wave voltage or current is 0.707 times the peak value. If a sine-wave voltage reaches a positive peak of 100 volts it has an RMS voltage of 70.7 volts. (This signal also will reach a negative peak of 100 volts.) This sine-wave signal has a peak-to-peak value of 200 volts. Figure 4 shows the graph of a sine-wave signal with the peak, peak-to-peak and RMS values indicated.

RMS value = $0.707 \times$ peak value (Equation 4)

This equation applies to both voltage and current of a sine wave.

Most meters respond to the *average value* of the rectified ac signal. For a sine wave, this average value is equal to 0.637 times the peak value.

Average value = $0.637 \times$ peak value (Equation 5)

This equation also applies to both voltage and current of a sine wave.

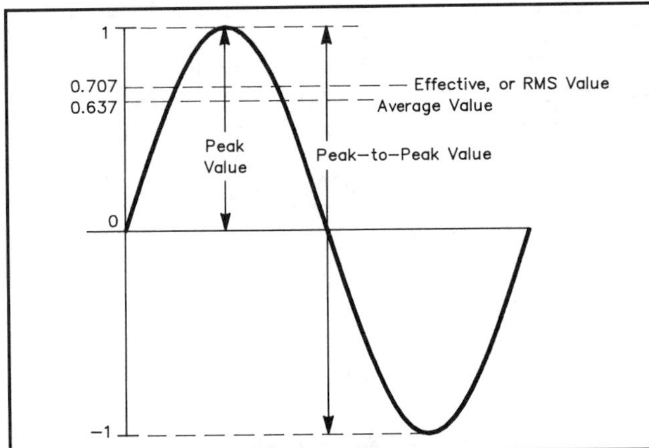

Figure 4—This sine wave shows the relationship between peak, peak-to-peak, RMS and average current and voltage values. The RMS and average values shown here apply only to sine-wave signals.

Sine Waves and a Rotating Wheel

The sine wave is an important ac-signal waveform. It is the simplest waveform to study, because the voltage or current changes smoothly. You can combine sine waves that have different frequencies and amplitudes to produce more complex waveforms.

Figure 1 is a drawing of a sine wave. The horizontal, or X, axis represents time. The vertical, or Y, axis represents voltage. (It also could represent current.) The graph shows how the voltage changes with time. You can make two observations about this graph. First, the wave repeats at regular intervals. Second, there are no sudden changes or corners on the graph.

Let's conduct a simple experiment to help you understand the shape of a sine wave. Imagine a wheel, such as a bicycle wheel, a car wheel or even a Ferris wheel. Concentrate on a single spot on your wheel. Watch that spot as the wheel spins. Figure 2 shows a bicycle wheel spinning clockwise. The motion of your spot repeats every time the wheel goes around. That motion is also very smooth. There are no sudden changes in the motion. The drawing shows the dot position at the start of a rotation. Shaded dots indicate the position after the wheel rotates 90°, 180° and 270°.

If you can, find a bicycle and turn it upside down so you can spin one wheel. Tie a string or ribbon around the tire so you can watch a single spot on the wheel. Spin the wheel, then stand back and watch it for a few seconds. (Stand so you are looking at the bicycle from the front or back, watching the wheel from the edge.)

It is difficult to show motion on the pages of this book.

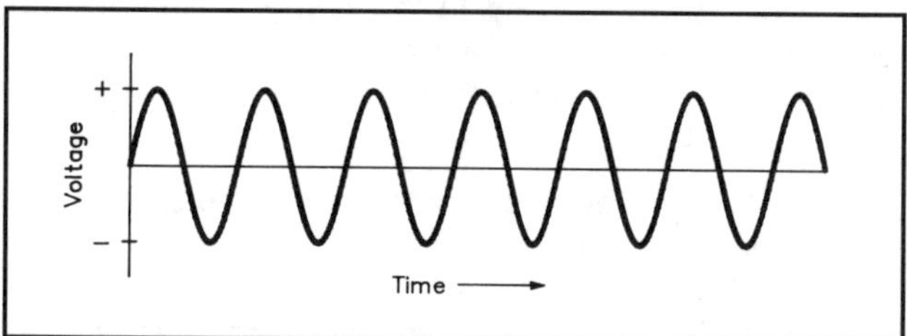

Figure 1—This graph shows how a sine-wave voltage changes with time. (The graph also could represent the current.) The voltage varies between the peak positive and peak negative values. Those voltage variations repeat at a regular rate.

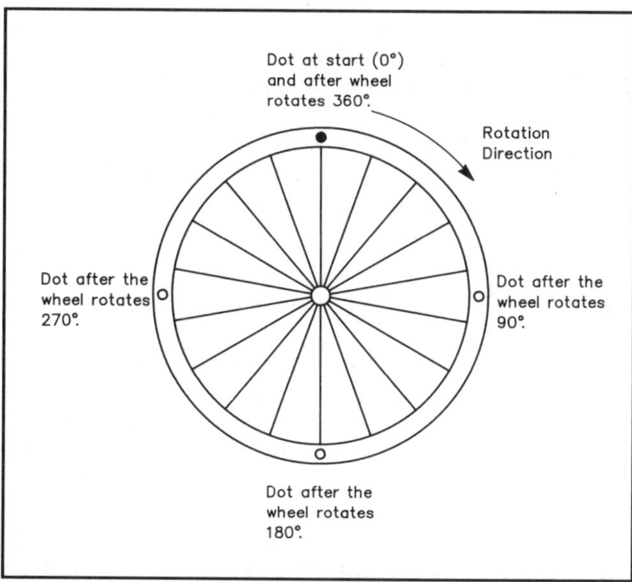

Figure 2—This spinning bicycle wheel will help you understand the shape of a sine wave. The dot at the top of the wheel indicates the start of a rotation. Shaded dots indicate the position after the wheel rotates 90°, 180° and 270°. When the wheel rotates 360°, the spot is back at the top.

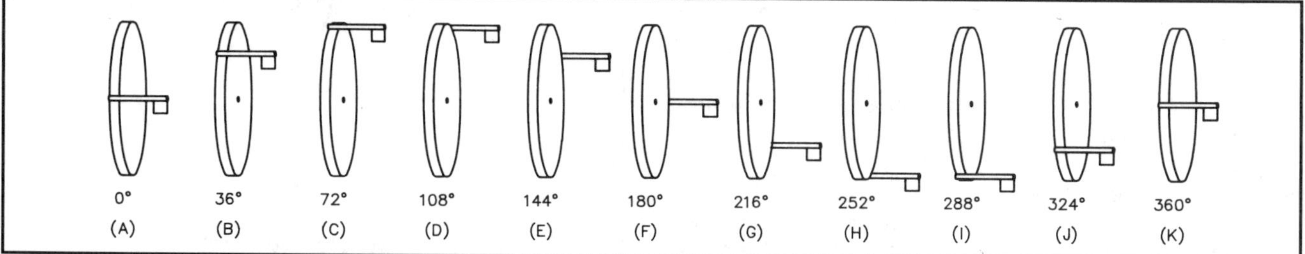

Figure 3—This series of drawings represent snapshots of a rotating wheel. The flag will help you visualize the rotation. The drawings are 0.1 second apart. The wheel rotates 36° between each snapshot. Notice that the flag appears to move up and down, as it rotates away from you. It takes longer for the flag to go over the top or under the bottom than it does to move across the middle of the wheel.

To help you imagine the wheel spinning, Figure 3 is a series of "snapshots," freezing the wheel's motion. Figure 3 shows the wheel rolling away from you. We attached a flag to the wheel, so there is a marker for you to watch as the wheel spins. The flag, attached to the wheel's front edge, moves up and then down the back.

The wheel rotates 36° between each picture. We marked the rotation angles, measured from 0° at A, under each wheel. We took these pictures 0.1 second apart. The wheel makes one complete rotation in 1 second.

Study the vertical motion of the flag. Can you see it moving up and down? This motion is similar to the voltage variations of a sine wave.

The wheel rotates at a constant speed. It doesn't speed up or slow down while we watch it. The same time goes by between each snapshot. The vertical movement between snapshots does change, though. Notice the vertical motion seems to slow down and stop as it changes direction at the top and bottom positions. The flag moves quickly past the center points.

Table 1
Values of the Sine Function for Angles Listed in Figure 3

Angle in Degrees	*Sine Function*
0°	0.000
36°	0.588
72°	0.951
108°	0.951
144°	0.588
180°	0.000
216°	−0.588
252°	−0.951
288°	−0.951
324°	−0.588
360°	0.000

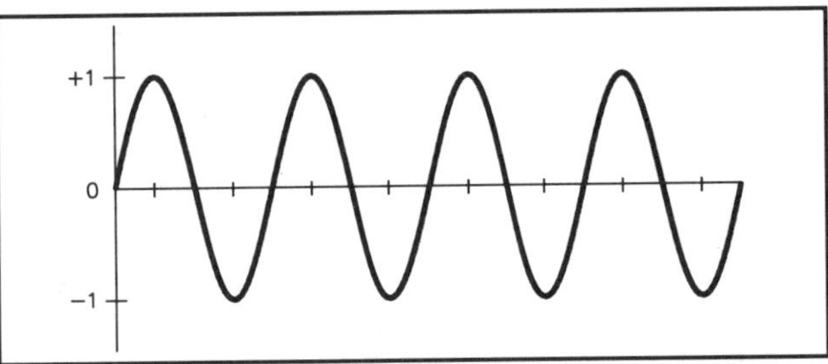

Figure 4—This drawing represents the motion of the flag on the Figure 3 wheel, as the wheel moves from left to right in front of you.

Now imagine the wheel moving from the left to the right as it spins. Trace the flag movement onto the wall in the background. Figure 4 shows the result. This drawing looks exactly like the sine-wave graphs we've been drawing!

Let the center line represent 0 volts. The flag goes between +1 and −1 volts, then. With this scale you can easily calculate the position of the flag at any point of the wheel's rotation. Just find the sine of the wheel's rotation angle. Use your calculator to find the sine of each angle shown in Figure 3. Compare your answers with the results listed in Table 1.

Rotation Speed Relates to Wave Frequency

In the last section we studied the rotation of a wheel to help you understand the sine wave. We will continue that analogy in this section. Imagine two wheels with flags attached, like we used in the last section. See Figure 1. The first wheel makes one complete turn every second. The second wheel spins twice as fast, so it makes two turns every second. Now imagine that both wheels move left to right in front of you. Both wheels move across the room at the same speed. Figure 2 shows two graphs that picture this motion.

Notice the graph at Part B moves up and down two times in the same distance the graph at Part A moves up and down once. This is because the second wheel spins twice as fast as the first.

What if we had another wheel spinning ten times faster than the first wheel? Figure 2C shows a graph of this motion. Again, this graph goes up and down ten times while Part A goes up and down once.

The wheel's rotation speed relates to the sine-wave *frequency*. *Frequency* of a wave relates to how many complete voltage or current variations the wave goes through each second. We measure frequency in *hertz*, abbreviated *Hz*. When the

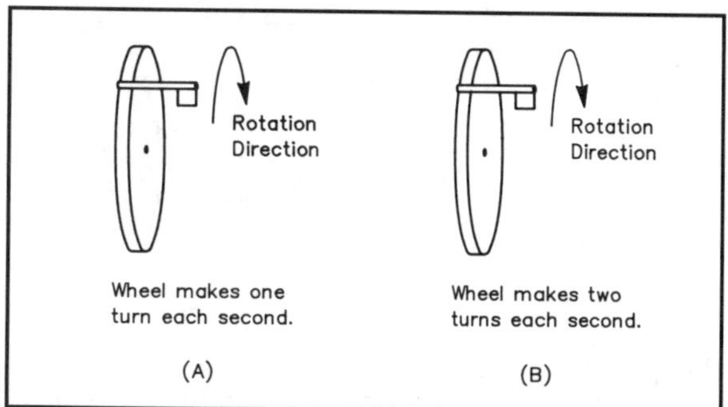

Figure 1—Two wheels spinning at different rates produce sine waves with different frequencies. The wheel at A makes one complete turn each second. The wheel at B makes two complete turns each second.

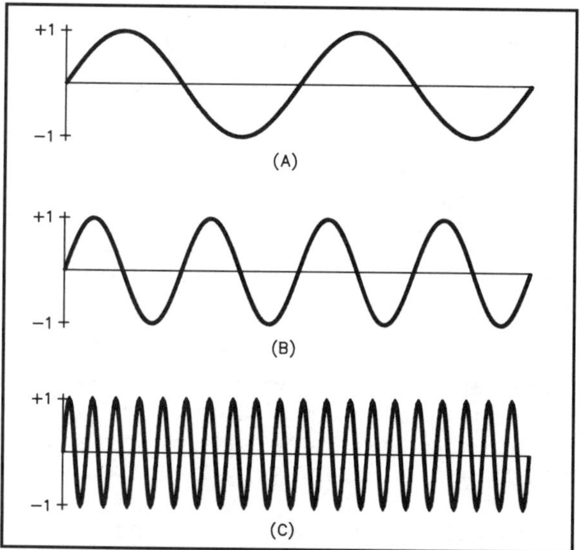

Figure 2—When the two wheels of Figure 1 move across the room in front of you they trace out the motion shown in Parts A and B. The Part B sine wave has a frequency twice as large as the Part A wave. Part C shows the sine wave produced by a wheel spinning ten times faster than the one from Part A.

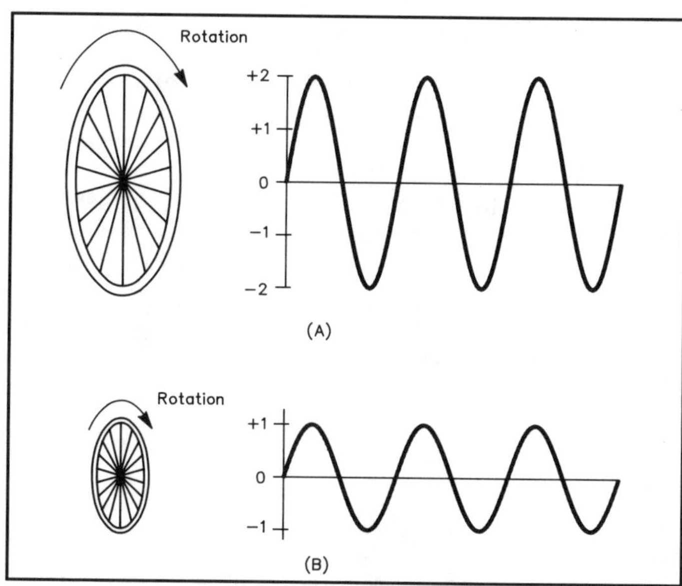

Figure 3—Part A shows a wheel with a diameter of 1.0 metre. Part B shows a wheel that has a 0.5-metre diameter. When the wheels move across in front of you they produce the sine waves shown at C and D. The amplitude of the wave at C is twice as large as the amplitude of the wave at D.

wheel spins faster it produces a higher-frequency sine wave. Suppose Figure 2A shows a sine wave with a frequency of 1 hertz. Then Figure 2B is a sine wave with a frequency of 2 hertz and Figure 2C has a frequency of 10 Hz.

All wheels don't have the same diameter. What if we use two wheels with different diameters? The graphs produced by these wheels won't have the same *amplitude*, or positive and negative peak values. Figure 3 shows two wheels and the sine waves they trace. The Part A sine wave has twice the amplitude of the Part B wave. This could represent a wave that has twice the peak voltage or current as another wave. These two waves have the same frequency.

The Figure 3 sine waves start at the same time. When two or more waves reach the same point in their motion at the same time, we say the waves are *in phase. Phase* refers to the relative wave positions.

Waves aren't always in step, however. Figure 4 shows two waves with the same frequency but which don't begin at the same time. These waves are *out of phase.*

We measure the position of a wave in degrees, related to the wheel rotation. It takes 360° to make one complete cycle, just as the wheel rotates 360° to make one complete turn.

We also measure phase as an angle. The two waves shown in Figure 4 are 90° out of phase. The second wave reaches the zero point as the first one comes to its 90° point. Figure 5 shows two waves that are 180° out of phase.

The *phase angle* between two or more waves can be any angle. We commonly use angles between 0° and 180°, however. If you find a phase angle greater than 180° you can subtract 180° from it. Think about our wheel analogy again. Two points on the wheel can never be farther apart than half way around the wheel.

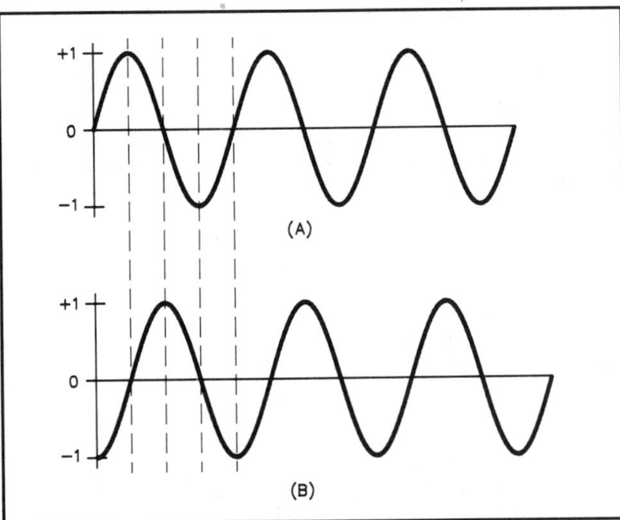

Figure 4—These two sine waves have the same frequency. They don't both begin at the same time, however. These waves are 90° out of phase. The wave at A reaches its maximum positive value when the wave at B crosses the zero point. The wave at A has gone through one quarter of its cycle (90°) when the wave at B begins.

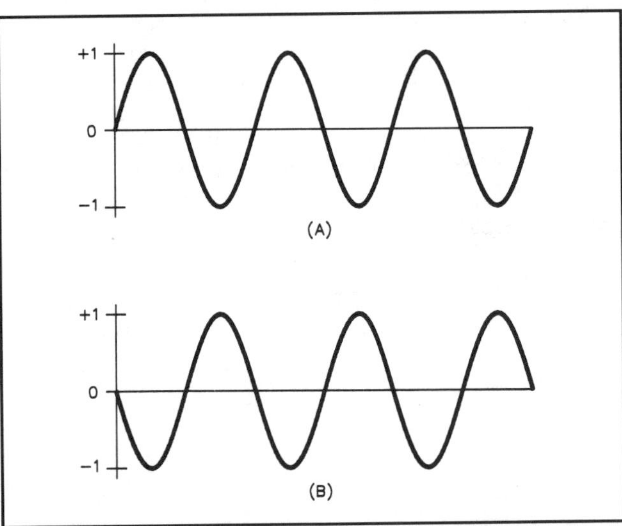

Figure 5—These sine waves have the same frequency, but they are 180° out of phase. The first one crosses the zero point going toward its positive maximum when the second wave crosses zero going toward its negative maximum. Notice that when one wave reaches its positive maximum value the other wave reaches its negative maximum.

CHAPTER 17 Capacitors

Stored Charge Produces an Electric Field

In this chapter you will learn about capacitors. Capacitor sounds a bit like *capacity*. You might want to look up capacity in a dictionary. Capacity is the ability to hold, receive or accommodate. This gives us a good beginning definition for a *capacitor*. Capacitors hold electrical energy. As you learned earlier, *energy* is the ability to do work.

You are familiar with several devices that store energy. Batteries store chemical energy. They produce electricity, or electrical energy, when the chemicals react under the proper conditions. A dam on a river stores water. The energy of that water pressure can run a generator to produce electricity.

You may wonder how capacitors store electrical energy. Figure 1 shows a circuit that stores electricity in a capacitor. Notice the capacitor schematic diagram looks like two parallel conductors with a space between them. This schematic symbol accurately pictures capacitor construction.

When you close the switch, electrons leave the negative battery terminal. The positive battery terminal attracts the electrons, and they move through the circuit. Use Ohm's Law to calculate the current when you first close the switch. Do you remember Ohm's Law? That's right, voltage equals current times resistance. Solve this equation for current, then use the voltage and resistance from Figure 1.

$$I = \frac{E}{R} \qquad \text{(Equation 1)}$$

$$I = \frac{20 \text{ volts}}{1000 \text{ ohms}} = 0.02 \text{ amperes} = 20 \text{ mA}$$

This current can't continue for long, however. The electrons reach the capacitor, but they can't get through it because there is an insulating layer between the conductors. Two things happen here. First, electrons move onto the conductor surface. These electrons exert an electric force to repel electrons from the opposite conductor. Second, the positive battery terminal attracts electrons from the second conductor. As electrons move off the second conductor, it has a positive charge.

Figure 2 shows the capacitor conductors, or plates, as the charge begins to accumulate. The negative charge on the first capacitor plate also repels any more electrons that try to reach it. At first there are few electrons and this force is small. The battery pushes with a stronger force, so electrons continue to accumulate. As more electrons pile up, however, the force becomes stronger. Eventually, the repelling force of electrons on the capacitor plate equals the battery force, and no more electrons flow onto the capacitor.

While electrons are piling up on one capacitor plate, the battery is pulling electrons off the other plate. Here, as the battery pulls electrons off that plate it becomes more difficult to pull electrons away. Again, the forces reach a balance, and the battery can't pull any more electrons off the capacitor plate.

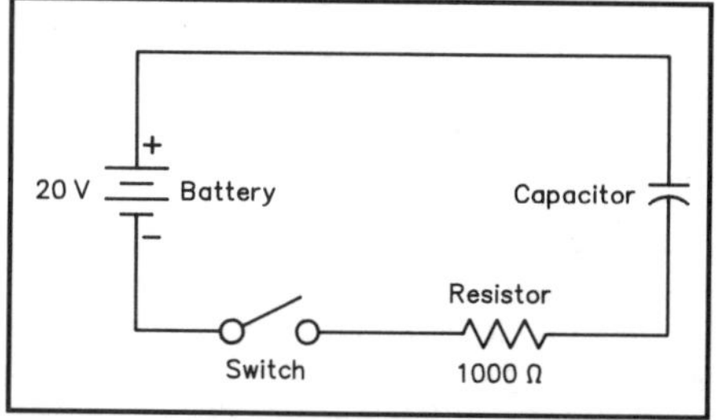

Figure 1—This circuit includes a battery, a switch, a resistor and a capacitor. The text explains how this circuit stores electrical energy in the capacitor.

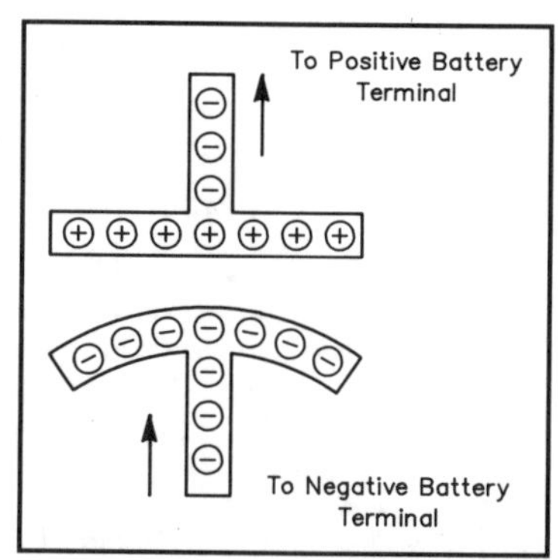

Figure 2—One capacitor plate receives a negative charge and the other one receives a positive charge. Eventually no more electrons can flow onto the negative side and no more can leave the positive side.

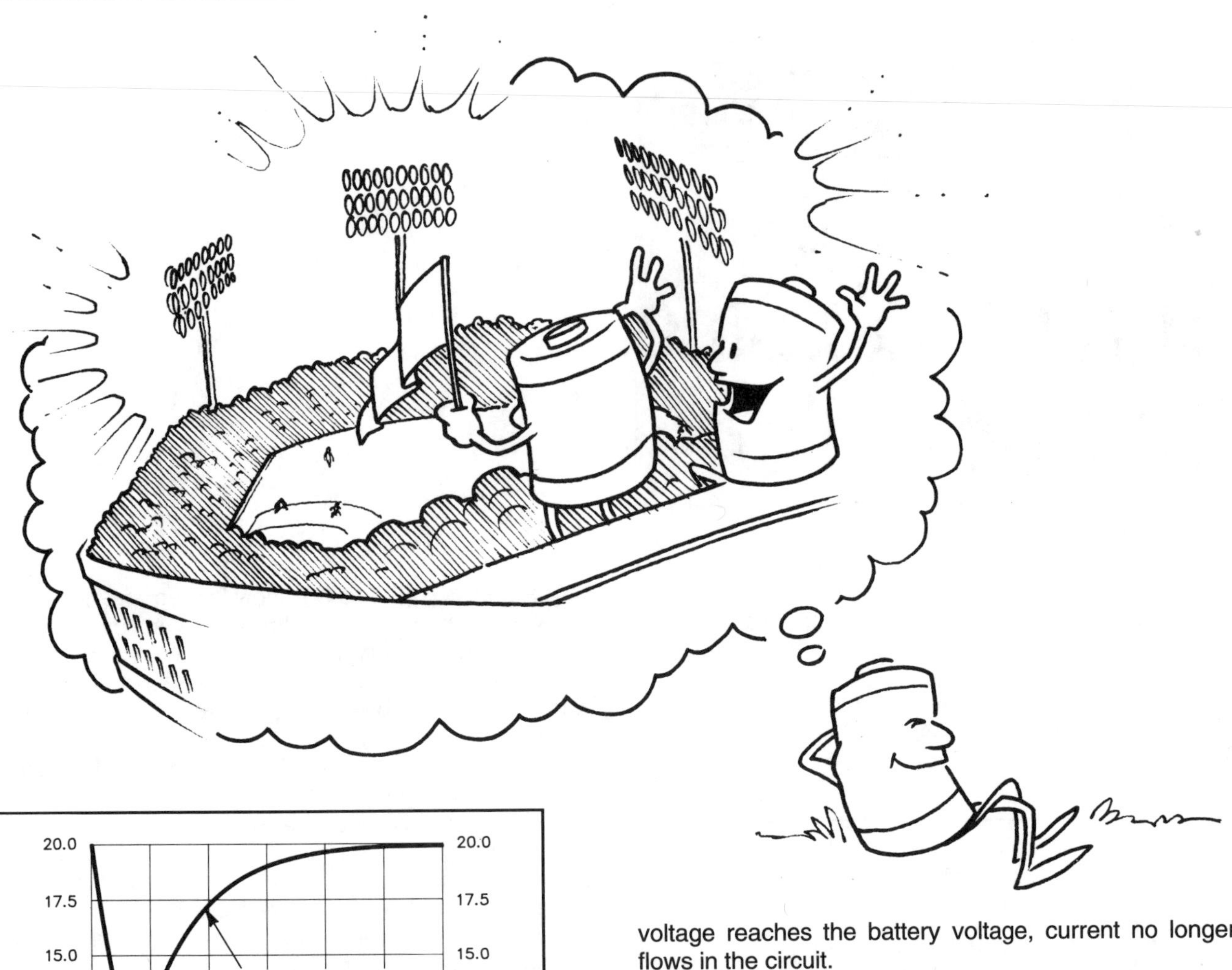

Figure 3—This graph shows the current through the Figure 1 circuit. It also shows the voltage across the capacitor.

Figure 3 shows a graph of the circuit current. When you first close the switch the current is 20 mA. Almost immediately the current decreases, though.

Figure 3 also has a graph of the voltage across the capacitor plates. When you first close the switch there is no voltage on the plates. As the electric charge builds up, there is a voltage across them, however. When that voltage reaches the battery voltage, current no longer flows in the circuit.

The time, current and voltages shown on Figure 3 apply only to this circuit. These numbers depend on the battery voltage, circuit resistance and the capacitor used in the circuit. Other combinations result in different numbers on the graph. The graph *shapes* don't change, however.

Do you remember there is an *electric field* between two electrically charged objects? This is similar to the magnetic field between two magnets. As charge builds up on the capacitor plates, then, there is an electric field between the plates.

This electric field between the capacitor plates represents stored electrical energy. The capacitor will store this energy as long as the charge remains on the capacitor plates.

Suppose we remove the capacitor from the circuit. A perfect capacitor would not lose any charge. Of course no capacitor is perfect, so some of the charge will leak through the insulation between the capacitor plates. The capacitor must have wire leads to attach it to a circuit, so some charge always leaks off the capacitor into the air surrounding the capacitor leads.

You should be careful with capacitors that are in circuits. They can hold a dangerous charge long after you turn off the circuit. If you accidentally contact the leads of such a capacitor, you can receive an electrical shock.

Increasing Plate Surface Area Strengthens the Electric Field

A capacitor has two conducting surfaces separated by an insulator. A battery or other voltage source puts extra electrons on one surface, leaving a negative charge. The voltage source removes electrons from the other surface, leaving a positive charge. The electric charge on these surfaces, or plates, produces an electric field.

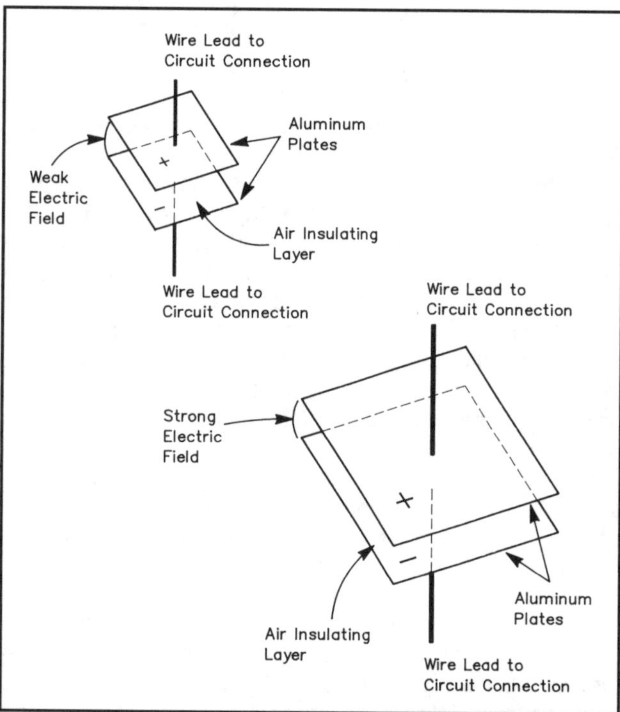

Figure 1—The capacitor shown at A has a small plate surface area. It holds a small electric charge and there is a weak electric field between the capacitor plates. Capacitor B is much larger than capacitor A. When the same voltage connects to each capacitor, capacitor B holds a larger electric charge. The electric field between capacitor B's plates is stronger than the electric field between capacitor A's plates.

Suppose you connect a capacitor to a voltage source. The capacitor charges until the voltage across the plates equals the supply voltage. You can't put any more charge on the capacitor, unless you increase the supply voltage. A larger voltage stores a larger charge, and thus more energy.

Do you keep a bottle of water in your refrigerator? Sometimes nothing quenches thirst better than a glass of ice-cold water. Have you ever reached for the water bottle only to find it nearly empty? That can be frustrating, especially on a hot summer day. Perhaps you need a water bottle that has a larger *capacity*. Instead of a 1 litre bottle you might want a 2 litre bottle or even a 4 litre bottle. Larger bottles hold more water.

Does this give you any ideas about how to store more energy in an electronics *capacitor*? What do you suppose would happen if the capacitor plates were larger? Yes, the negative plate could hold more electrons. The positive plate also would give up more electrons, to have a larger positive charge.

Figure 1 shows two capacitors made from aluminum plates. A layer of air separates the plates, and insulates them so no electrons can flow between them. The capacitor shown at A has small plates, and only stores a small charge. The electric field between the plates of this capacitor is weak. It doesn't store much electrical energy.

Figure 1B shows a capacitor with much larger plates. Let's imagine they are ten times larger than the plates of capacitor A. Capacitor B can hold ten times more charge than capacitor A. Capacitor B's electric field is ten times stronger, too.

Ceramic disk capacitors have two metal plates with a thin ceramic insulating layer between them. Small wire leads attach to each plate, so you can connect the capacitor into a circuit. Figure 2 shows a cut-away view of a ceramic disk capacitor.

The insulating layer material and thickness determine the capacitor's highest safe operating voltage. If you connect a capacitor to a voltage that is too high for the

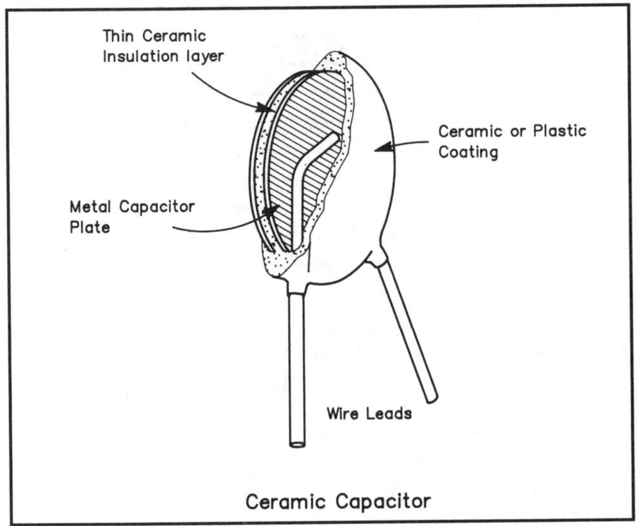

insulating layer, a spark will jump through the insulation. This will destroy the capacitor.

Suppose you find two ceramic disk capacitors with the same voltage rating. One disk has a larger diameter than the other. Which capacitor is likely to have the higher capacitance rating? That's right. The capacitor with the larger diameter probably has a higher capacitance. It will hold a larger electric charge and will store more electrical energy.

Figure 2—This is a cut-away view of a ceramic disk capacitor. The metal plates have a thin ceramic insulating layer between them. A small wire lead attaches to each plate, so you can connect the capacitor to a circuit. This assembly then has a ceramic or plastic coating applied to protect the capacitor from moisture and other harmful effects.

Increasing the Number of Plates Increases Plate Surface Area

Larger capacitor plates hold more electric charge. This produces a stronger electric field and stores more energy. How large can we make the capacitor plates, though? Can you imagine capacitors made with plates 50 centimetres on a side? That sure would limit how small we could make electronics circuits.

We don't want a capacitor that takes up the area of a kitchen table top. How can we make a capacitor with more plate surface area? Suppose we fold the two plates in half. Fit one side of each plate between the two halves of the other plate. Figure 1 shows a capacitor made this way.

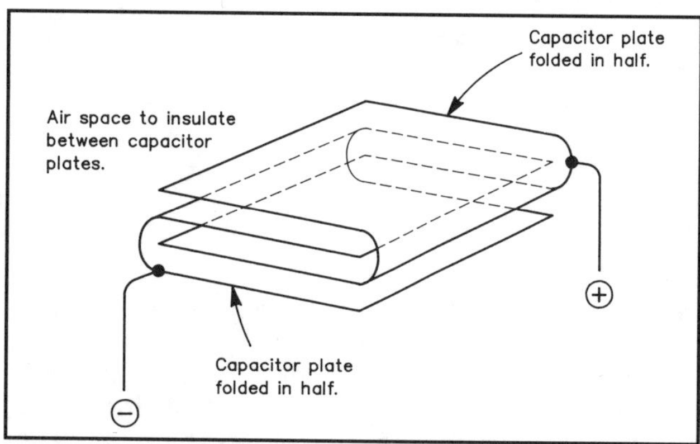

Figure 1—This capacitor has two plates folded in half to save space.

It will be difficult to fold the plates more than once because there is no way to fit them together after you fold them. You can cut the large plates into smaller pieces, though. Then you can connect half the pieces along one edge and half along the other edge. You can put the two stacks together, then, like Figure 2 shows.

This is a much more practical construction method than using two large plates. Many capacitor types use this stacked-plate construction. It is difficult to see the construction when you look at a capacitor, though. Most capacitors have a protective coating, so you can't see inside.

Many electronics circuits require adjustable capacitors. You may change the *capacitance*, or the amount of charge a capacitor can hold when you tune a radio receiver.

One common variable capacitor type uses one set of plates that rotates and another set of fixed plates. The rotating plates change the overlapping area. When the plates completely mesh, the capacitor has its maximum value. When the plates have no overlapping area, the capacitor has its minimum value.

Figure 3 shows such a variable capacitor. A layer of air separates the rows of capacitor plates, preventing electrons from flowing between them.

You can find these "air variable capacitors" in many sizes. Some have a few small plates. Others have larger plates, and usually more of them.

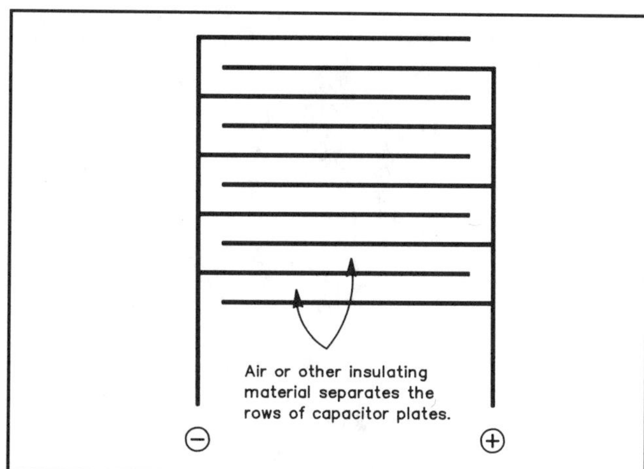

Figure 2—This capacitor consists of two stacks of plates each connected along one edge. Air or another insulating material separates the two sets of plates.

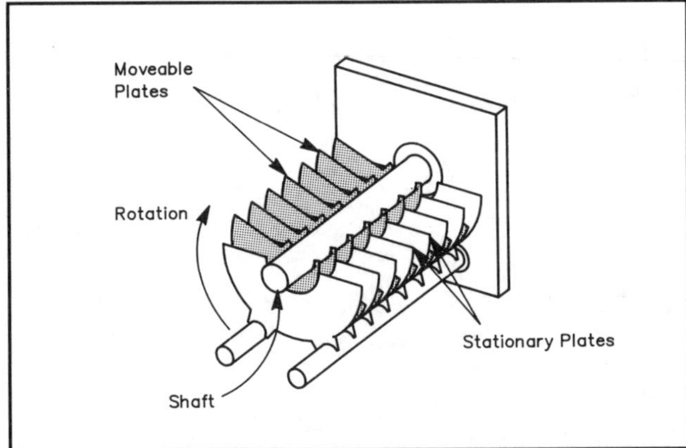

Figure 3—This variable capacitor has two sets of plates. One set can rotate, while the other set cannot move. The rotating plates turn in and out of the fixed plates, changing the overlapping surface area.

There is another way to reduce the capacitor's package size. If you make the plates from a metal foil, you can roll the plates and insulating material into a cylinder. This only works if you use a solid material to insulate between the plates, of course.

You can make a capacitor using this construction technique. You'll need some aluminum foil and waxed paper. The size of the aluminum-foil capacitor plates isn't critical. Tear 4 square pieces of waxed paper. Make their length equal the roll's width. Then tear 4 pieces of aluminum foil. Make the foil about an inch smaller than the waxed paper in both dimensions. Leave a foil tab on each foil piece. These will connect alternate foil layers. When you assemble your capacitor, alternate the direction of the tabs, so the first and third layers connect on one side and the second and fourth layers connect on the other side.

Figure 4 shows a "sandwich" of several aluminum-foil and waxed-paper layers. The odd-numbered foil layers connect to form one side of the capacitor. Even-numbered foil layers connect to form the other side. Waxed-paper layers separate each foil layer. "Alligator clip leads" connect to each side of the capacitor.

Now you can roll your capacitor, starting at the end opposite the clip leads. A few small pieces of tape will hold the capacitor in a cylindrical shape. Figure 4B shows the rolled capacitor.

Now you can perform an experiment with your capacitor. Connect one lead to a 6- or 12-volt battery's negative terminal and the other lead to the positive terminal. Now connect a voltmeter across the capacitor leads. The voltmeter should register the battery voltage.

Disconnect the battery and watch the voltmeter. The reading will gradually decrease. The voltmeter drains off the stored energy because it needs some current to register the voltage.

Charge your capacitor again, but don't connect the voltmeter. Disconnect the battery and wait a minute or two.

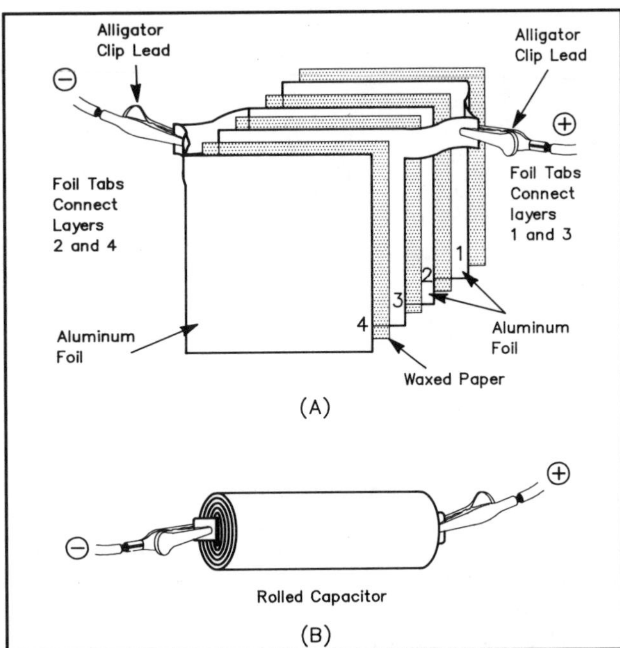

Figure 4—You can make an experimental capacitor. Part A shows a "sandwich" of aluminum-foil layers with waxed paper between them. Part B shows how to roll the capacitor into a cylinder.

Then connect the voltmeter. Remember which side went to the positive battery terminal, and connect the voltmeter the same way. Otherwise the meter needle will try to swing the wrong way, and you could damage the meter.

Do not connect your capacitor across the ac mains. This would be extremely dangerous. To be safe, only connect your capacitor to batteries. Don't use an ac-operated power supply.

Decreasing Plate Separation Strengthens the Electric Field

Figure 1 shows a single positive charge and a single negative charge. These charges are far apart, and there is a very weak electric field between them. Now let's move the charges closer together. As they move toward each other, the electric field becomes stronger. Remember that opposite electric charges attract. When the charges are closer together there is a stronger attraction force between them.

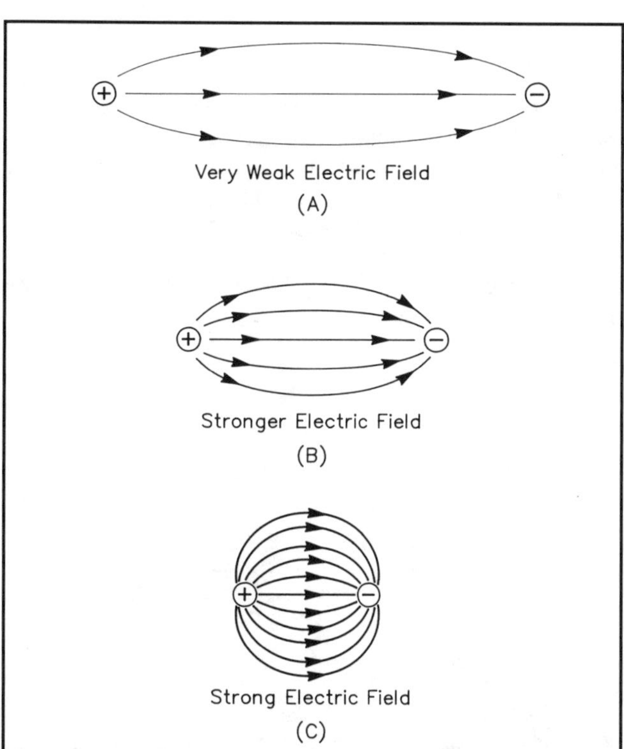

Figure 1—A positive and negative electric charge gradually move closer together. In Part A, they are far apart and exert a weak force of attraction on each other. There is a weak electric field between them. We moved the charges closer together in Part B. Now the attraction force is stronger, and the electric field between them is stronger. At C, the charges are close together. Now there is a strong attraction force between them. There is also a strong electric field.

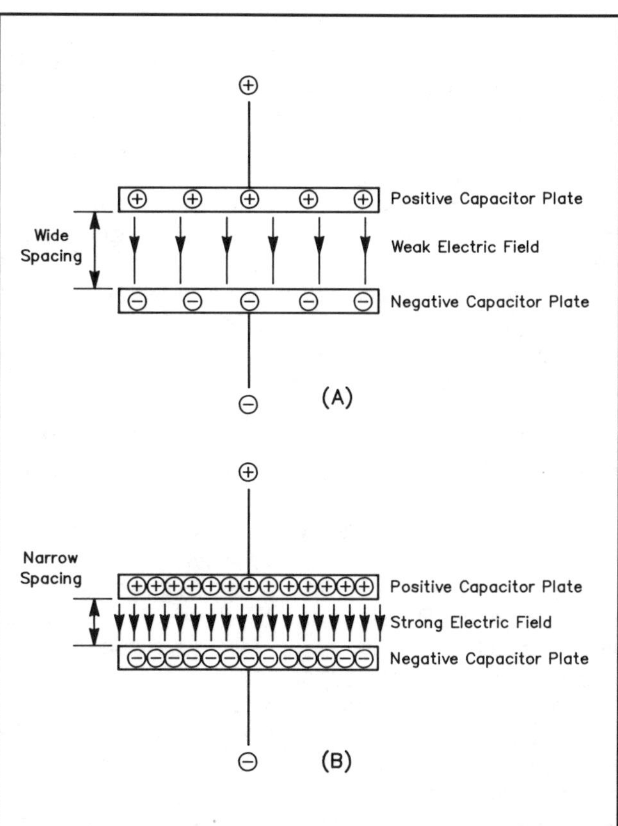

Figure 2—The plates of capacitor A have a wide spacing. We charged this capacitor by connecting it to a battery, so there is some positive and negative charge on the plates. The plate spacing is wide, so the electric field between the plates is weak. At B, we move the capacitor plates closer together. The plates have the same dimensions. We connected the new capacitor to the same battery we used to charge the first capacitor. This time there are more electrons on the negative capacitor plate. The positive charge is also larger. We increased the electric-field strength by moving the plates closer together.

The capacitor plates in Figure 2A are far apart. The extra electrons on the negative plate exert a weak force to repel electrons from the positive plate. (The negative side also exerts a weak force to attract more positive charge onto the positive capacitor plate.)

Now let's move the capacitor plates closer together. Figure 2B shows a much stronger electric field between the plates. There are more electrons on the negative plate. There is more positive charge on the positive capacitor plate.

The plates are the same size for both capacitors. We connected the same voltage source to the capacitors in each case as well. The only variable for the two capacitors in Figure 2 is the plate spacing.

How close can we safely move our capacitor's plates? Remember that the insulation between the plates keeps the electrons from jumping across to the positive plate. Air is a good insulator, as long as the voltage doesn't become too large. Lightning is an example of an electric spark jumping through air, however. Even at low voltages, the electrons will jump across the gap when the plates move too close.

There is a trade-off between the safe operating voltage and increased capacitance achieved by moving the plates closer together. This is why most capacitors use insulators other than air. Many materials can withstand higher voltages than air. A thin layer of mica or ceramic material is a good capacitor insulator.

Figure 3 shows the construction of a mica compression variable capacitor. Several sets of plates have a thin mica wafer separating them. A screw passes through a hole in the center of the assembly and threads into the ceramic capacitor base. As you turn the screw into the base, it compresses the capacitor layers. This brings the plates closer together. As you learned earlier, the plates hold more charge when they are closer together. This produces a stronger electric field.

Other capacitors use a construction method similar to the one you used to make your experimental capacitor in the last section. Solid insulating materials allow the plates to be close together, increasing the amount of charge the capacitor can hold. Rolling the assembly into a cylinder reduces package size.

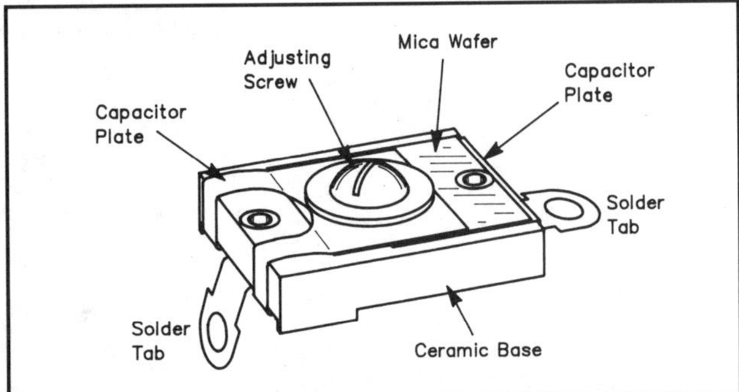

Figure 3—A mica compression capacitor uses thin mica wafers between the capacitor plates. A screw passes through the plates and mica, threading into the ceramic base. As you turn the screw into the case it presses the capacitor plates closer together. This increases the capacitance, producing a stronger electric field between the plates.

Dielectric Constants and the Electric Field

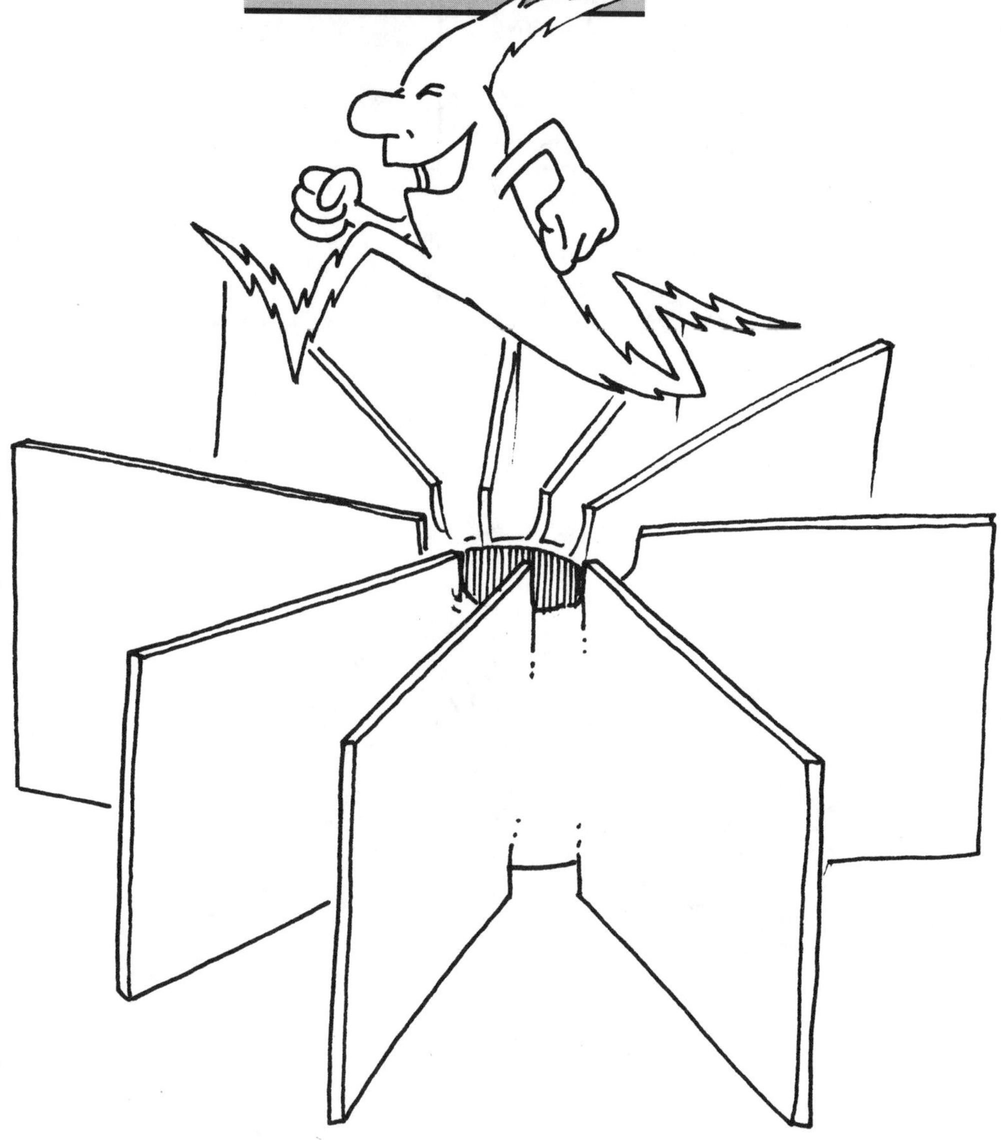

A capacitor does not conduct direct-current electricity. The insulation between the capacitor plates blocks the flow of electrons. You learned earlier there is a short current pulse when you first connect the capacitor to a voltage source. The capacitor quickly charges to the supply voltage, and then the current stops.

We call the insulation used in a capacitor a *dielectric* because the capacitor does not conduct *di*rect *electric* current. An *air dielectric capacitor* uses air to insulate the capacitor plates. Some common capacitor dielectrics are mica, polystyrene plastic, paper, ceramic and aluminum oxide.

Breakdown voltage is an important dielectric rating. You usually will see breakdown voltage expressed as a number of volts per mil of dielectric thickness. A *mil* is a thousandth of an inch (0.001 inch).

Air has a breakdown voltage of 21 volts per mil. Suppose you plan to use the capacitor in a circuit that has a 20-volt supply. That means you will need at least 0.001-inch spacing between the capacitor plates. (You probably will want more spacing to allow a safety factor.) What if you want to use the capacitor in a circuit that has 200 volts? You'll need a spacing of 10 mils, or 0.01 inch.

Most capacitor-dielectric materials have higher breakdown-voltage ratings than air. Mica, for example, has a breakdown-voltage rating of 3800 to 5600 volts per mil. Obviously, a 1-mil thickness of mica is sufficient for most applications.

You should always check the voltage rating of a capacitor before you install it in a circuit. Normally, low-voltage capacitors are physically smaller than higher-voltage units with the same capacitance rating. If a capacitor cannot withstand the voltage applied to it, a spark probably will jump through the insulation. This will destroy the capacitor, and may damage other circuit components as well.

Another important dielectric rating is the *dielectric constant*. This is a measure of how much energy you can store in the insulation between the capacitor plates. Another way to say this is that the dielectric constant indicates the electric-field strength.

Usually when you see a dielectric constant listed, it will be the *relative dielectric constant*. This means it is given in comparison to air. Engineers give air a relative dielectric constant of 1.0.

Porcelain is one type of ceramic used in capacitors. The relative dielectric constant for porcelain is between 5.1 and 5.9. What does this mean? Suppose you make a ceramic capacitor and an air-dielectric capacitor. Both have the same plate size and spacing. The porcelain-dielectric capacitor will have a capacitance that is 5.1 to 5.9 times greater than the air-dielectric one.

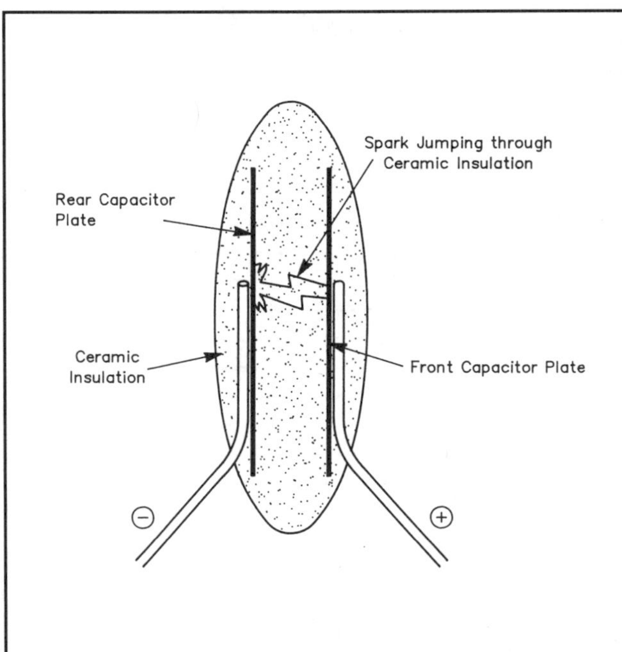

Figure 1—When the voltage is large enough, a spark will jump through any insulating material. A thicker insulator requires a higher voltage to make the spark jump.

Table 1
Dielectric Constants and Breakdown Voltages

Material	*Relative Dielectric Constant**	*Breakdown Voltage***
Air	1.0	21
Bakelite	4.4-5.4	240
Bakelite, mica filled	4.7	325-375
Formica	4.6-4.9	450
Glass, window	7.6-8	200-250
Glass, Pyrex	4.8	335
Mica, ruby	5.4	3800-5600
Mycalex	7.4	250
Paper, Royalgrey	3.0	200
Polyethylene	2.3	1200
Polystyrene	2.6	500-700
Porcelain	5.1-5.9	40-100
Quartz, fused	3.8	1000
Teflon	2.1	1000-2000

*At 1 megahertz
**In volts per mil (0.001 inch)

Table 1 lists some common dielectric materials. The Table gives relative dielectric constants and breakdown-voltage ratings for these materials. The breakdown voltages are in volts per mil of thickness.

These relative dielectric constants are measured at 1 megahertz. In general, dielectric constants decrease as you increase the operating frequency. The change is normally small, however.

Factors that Determine Capacitance

In this chapter you have been studying capacitors. When you connect a capacitor to a voltage source, an electric charge builds up on the capacitor plates. Extra electrons build up on one plate, giving it a negative charge. The other plate loses electrons, so it takes a positive charge. These opposite charges, separated by an insulating material, produce an electric field. Capacitors store electrical energy in the form of an electric field.

In this chapter's previous sections we discussed several ways to increase the amount of charge a capacitor holds. In this section we will summarize those factors. First, however, let's discuss the units we use to measure this stored charge. *Capacitance* is the measure of a capacitor's ability to store electric charge.

The basic unit of capacitance is the *farad*. Scientists named the farad for Michael Faraday. Faraday was a British scientist during the early 1800s. Faraday had an interest in static electric fields, such as the field in a charged capacitor. We use a capital letter F to abbreviate farad.

The farad is too large a unit for practical capacitor measurements. A capacitor with a 1-farad capacitance would be physically very large. Microfarads (10^{-6}), abbreviated µF, and picofarads (10^{-12}), abbreviated pF, are more practical measurement units. You also may see capacitance values given in nanofarads (10^{-9}), abbreviated nF. You should recognize these metric prefixes.

Three main factors determine capacitance. The first of these is the plate surface area. Capacitance varies directly with plate surface area. You can double the capacitance value by doubling the capacitor's plate surface area. Figure 1 shows a capacitor with a small surface area and another one with a large surface area.

Remember you can increase the plate surface area by adding more capacitor plates. Figure 2 shows alternate plates connecting to opposite capacitor terminals. Many capacitors use this multiple-plate construction technique.

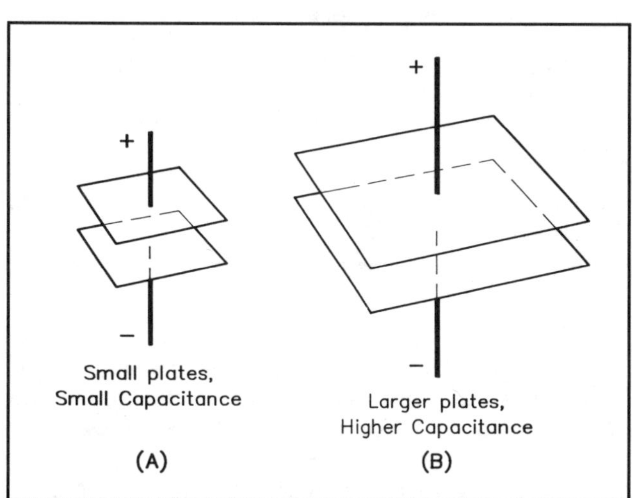

Figure 1—Capacitance varies directly with plate surface area. If all other factors are the same, a capacitor with more plate surface area will have a larger capacitance.

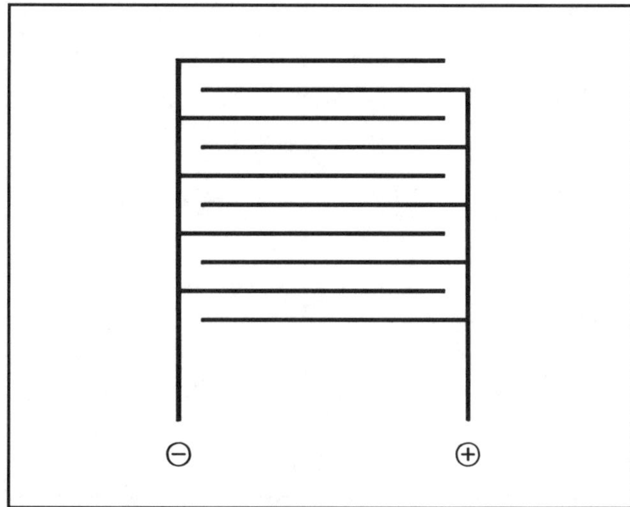

Figure 2—You can connect several sets of plates to produce a capacitor with more surface area.

Larger plates produce physically larger capacitors. Stacking many smaller plates also increases a capacitor's size. Manufacturers roll the stack of plates and insulating material into a cylinder for some capacitor types.

The second important factor affecting capacitance rating is the distance between the plates. Capacitance varies inversely with the distance between plate surfaces. The capacitance increases when the plates are closer together. Figure 3 shows capacitors with the same plate surface area, but with different plate spacing.

The third important factor determining capacitance is the dielectric constant of the insulating material. An insulating material with a higher dielectric constant produces a higher capacitance rating. (This assumes the same plate surface area and spacings.)

Figure 4 shows two capacitors. Both have the same plate surface area and spacing. Air is the dielectric in the first capacitor and mica is the dielectric in the second one. Mica's dielectric constant is 5.4 times greater than air's dielectric constant. The mica capacitor will have 5.4 times more capacitance than the air-dielectric capacitor. In addition, the mica capacitor can withstand much higher voltages.

To summarize, there are three ways to increase capacitance. You can increase the plate surface area, either by increasing the size of one pair of plates or by increasing the number of plates. You can reduce the spacing between the capacitor plates, by using a thinner layer of insulating material. You can use a better insulator between the capacitor plates.

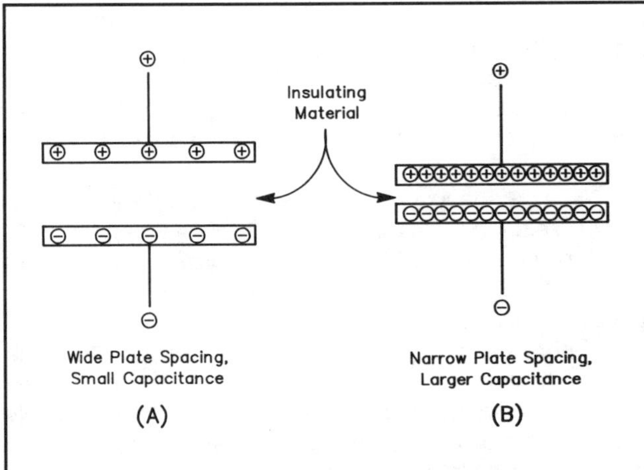

Figure 3—Capacitance varies inversely with the distance between the plate surfaces. If all other factors are the same, a capacitor with plates closer together will have a larger capacitance.

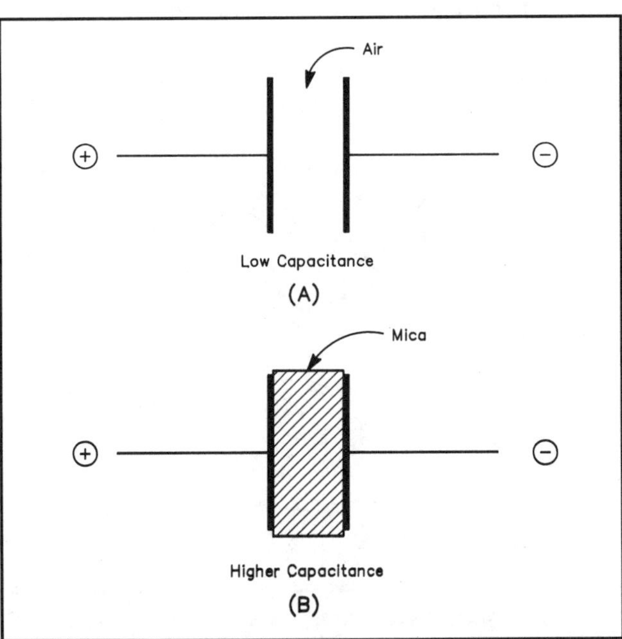

Figure 4—Capacitance depends on the dielectric constant of the insulating material.

Practical Capacitors

You can buy capacitors made with a variety of dielectrics. Each dielectric lends itself to certain construction techniques. There are many trade-offs with regard to cost, breakdown-voltage rating and size. Some dielectric materials are less effective at higher operating frequencies, so you want to avoid those for radio-frequency circuits. Operating-temperature changes have a significant effect on the capacitance value of some capacitor types.

Selecting the proper capacitor for a specific job isn't difficult. You do have to understand the effects of some trade-offs, however. For example, an inexpensive ceramic capacitor may do fine as an audio-bypass capacitor. You would not use a normal ceramic capacitor in a radio-frequency oscillator circuit.

This section describes the common capacitor construction methods and dielectric materials. The information serves as an introduction. It will help you understand why a 10 microfarad paper capacitor may not be a good substitute for a 10 microfarad tantalum capacitor.

You may see the breakdown-voltage rating given as working-volts dc (WVDC). This is the highest direct voltage you can safely connect to the capacitor. Select a capacitor with a breakdown voltage at least two times larger than the highest voltage you expect it to endure. This provides a safety margin.

Mica capacitors consist of metal-foil strips separated by thin mica layers. Figure 1 shows this construction.

Figure 1—Mica capacitors have thin sheets of mica separating metal-foil capacitor plates. Alternate foil layers connect to the capacitor leads. A ceramic coating protects the assembly from dirt and moisture.

Alternate plates connect to each electrode. A plastic or ceramic coating seals the capacitor.

Mica has a breakdown-voltage rating between 3800 and 5600 volts per mil. This is why we use mica capacitors in transmitters and high-power amplifiers. Their ability to withstand high voltages is important.

Mica capacitors have good temperature stability. Their capacitance does not change much as the temperature changes.

Typical capacitance values for mica capacitors range from 1 picofarad to 0.1 microfarad. Breakdown-voltage ratings as high as 35,000 are possible.

Ceramic capacitors have a metal film on both sides of a thin ceramic disc. Wire leads attach to the metal film. Figure 2 shows this construction. The capacitor has a protective plastic or ceramic covering.

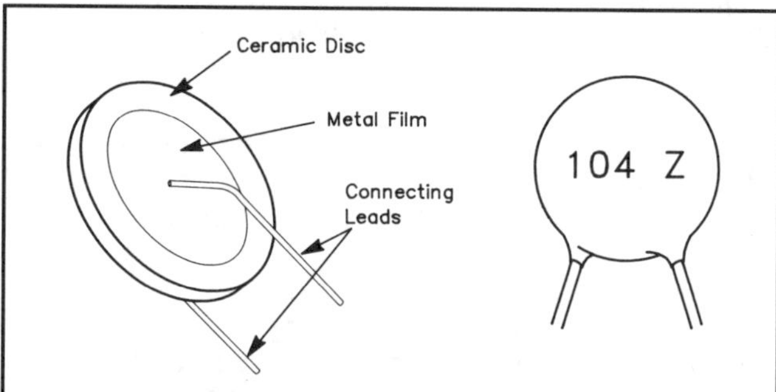

Figure 2—This drawing shows the construction of a disc ceramic capacitor. A metal film deposited on both sides of a small ceramic disc forms the electrodes. Wires attach to the metal, and the assembly gets a protective plastic or ceramic coating.

Ceramic capacitors are inexpensive and easy to make. Many electronics circuits use ceramic capacitors. You can't use them in every application, however.

The capacitance of ordinary ceramic capacitors changes when the temperature changes. You can't use them in a circuit requiring a capacitance that doesn't change with temperature.

Some capacitors have a *negative temperature coefficient*. This means their value decreases when the temperature goes up. Others have a *positive temperature coefficient*. Their capacitance increases when the temperature goes up.

There are special ceramic capacitors that don't change value with temperature changes. These *NP0 capacitors* have a zero temperature coefficient. The NP0 stands for "negative-positive zero." An NP0 capacitor has neither a negative temperature coefficient nor a positive one. The capacitance of an NP0 capacitor remains nearly unchanged over a wide temperature range.

You can buy ceramic capacitors with values from 1 picofarad (1 pF) to 0.1 microfarad (0.1 μF). Ceramic

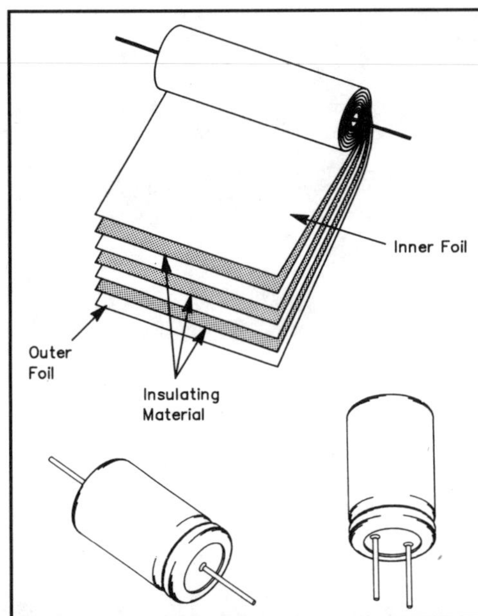

Figure 3—Several capacitor types use the same construction method as the paper capacitor. An insulating material between metal-foil plates forms the capacitor. Roll the assembly into a cylinder and finish it with a protective coating. Paper capacitors use an insulating paper as the dielectric. Plastic-film capacitors use mylar or polystyrene. Chemical-soaked paper goes between the aluminum-foil layers of an aluminum electrolytic capacitor.

capacitors with breakdown-voltage ratings up to 1000 volts are common.

Paper capacitors consist of alternate layers of metal foil and insulating paper. Wire leads attach to the two sets of metal-foil plates. Then the manufacturer rolls the layers into a cylinder, as Figure 3 shows.

The capacitor may have a plastic covering, or it may have a wax coating. This outer layer protects the capacitor from dirt and moisture.

The capacitor may have a stripe around one end. This shows the lead that attaches to the outer metal-foil plate. You can connect this end to the circuit ground, so the outer foil shields the capacitor from radio-frequency energy.

Capacitance values of paper capacitors range from about 500 pF to about 50 μF. They come with voltage ratings up to about 600 WVDC.

Paper capacitors are generally inexpensive. They have a larger size for a given value than other capacitor types, and this makes them impractical for some uses.

Plastic-film capacitors are similar in construction to paper capacitors. Thin sheets of mylar or polystyrene serve as the insulating layers. The plastic material gives the capacitors a high voltage rating in a physically small package. Mylar and polystyrene capacitors have good temperature stability. Typical values range from 5 pF to 0.47 μF.

Aluminum electrolytic capacitors also use a similar construction technique. Sheets of aluminum foil have a layer of paper soaked in a chemical solution between them. The rolled assembly goes into a protective casing, usually a metal can.

The chemical causes a reaction when you apply electricity, so we call the chemical an *electrolyte*. This is where we get the name *electrolytic*.

When you apply a voltage to the capacitor it causes a chemical reaction on the positive plate surface. This produces a thin aluminum-oxide layer, which forms the capacitor dielectric. Electrolytic capacitors have a high capacitance value in a small package because of this thin dielectric layer.

One lead of an electrolytic capacitor always has a + or a – sign clearly marked. You must observe this polarity when you connect the capacitor into a circuit. If you connect an aluminum electrolytic capacitor with the wrong polarity, a gas will form inside the capacitor. This may cause the capacitor to explode. At the very least, you will destroy an electrolytic capacitor by connecting it with reverse polarity.

Electrolytic capacitors are available in capacitance values from 1 μF to 100,000 μF (0.1 farad). Some of them have voltage ratings of 400 V or more. Electrolytic capacitors with high capacitance values and/or high voltage ratings are physically large.

Tantalum capacitors are another form of electrolytic capacitor. These are much smaller than aluminum electrolytic capacitors for a given value. They usually have the shape of a water drop.

The *anode*, or positive capacitor plate, is a small tantalum pellet. A layer of manganese dioxide forms the solid electrolyte, or chemical, which produces an oxide layer on the outside of the tantalum pellet. This oxide layer serves as the dielectric. Layers of carbon and silver form the *cathode* or negative capacitor plate.

An epoxy coating gives the capacitor its characteristic shape. This is why we often call them "tear-drop capacitors." Figure 4 shows the construction of a tantalum electrolytic capacitor.

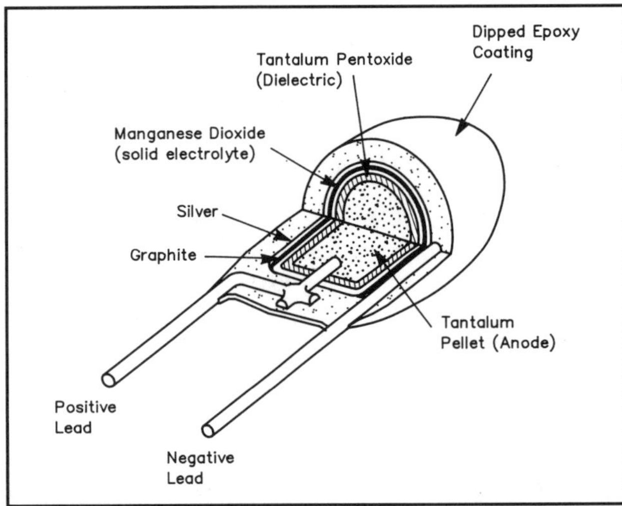

Figure 4—Tantalum capacitors, sometimes called "tear-drop capacitors" use a small tantalum pellet as the positive plate. This drawing shows the construction of these capacitors.

CHAPTER 18
Capacitive Reactance

Capacitors Oppose a Change in Applied Voltage

A constant direct voltage applied to a capacitor puts an electric charge on the capacitor. As the charge increases, there is less current into the capacitor. When the voltage across the capacitor plates equals the supply voltage, the current stops. Figure 1A shows a simple circuit to charge a capacitor. With the switch in position 1, there is current through R1 to charge the capacitor. Figure 1B is a graph of the circuit current and the capacitor voltage.

Now let's move the switch to position 2, as Figure 2A shows. This disconnects the voltage supply and puts R2 across the capacitor terminals. The capacitor returns the energy stored in it by forcing current through R2. As the capacitor returns its stored energy to the circuit it loses charge. The voltage across the capacitor decreases. Ohm's Law tells us the current through R2 decreases as the voltage decreases. Figure 2B is a graph of the current through R2 and the capacitor voltage after we change the switch position.

Can you guess what happens if we connect an alternating voltage to our capacitor? Figure 3 shows a simple circuit with a sine-wave voltage supply. We'll make the voltage applied to the top capacitor plate positive for the first sine-wave half cycle. As the sine-wave voltage increases from zero there is a sudden rush of current as the capacitor begins to charge. That current tapers off as the charge increases.

The sine-wave voltage reaches a maximum, and begins to decrease to zero again. When the voltage begins to decrease, the capacitor begins to return its stored energy to the circuit. It is interesting to realize that the current direction changes when the voltage begins to

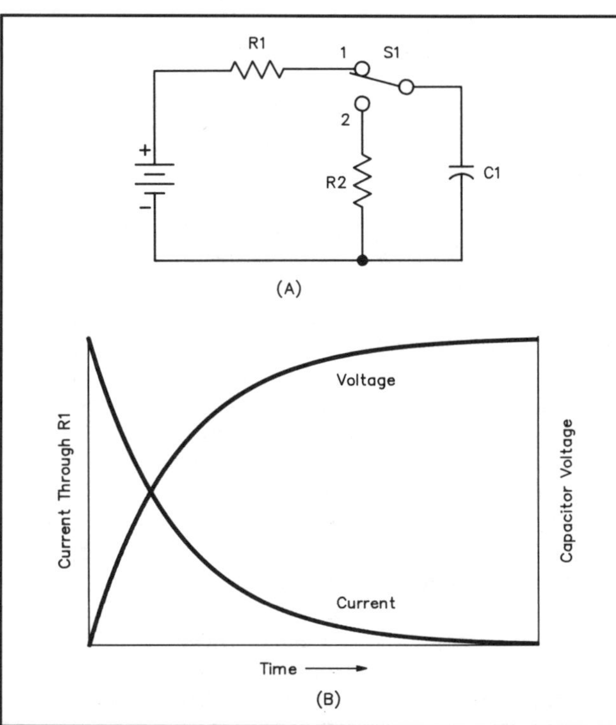

Figure 1—We can charge a capacitor using the circuit shown at Part A, when the switch is in position 1. Part B is a graph of the circuit current and voltage across the capacitor plates as it charges. Notice there is no voltage on the capacitor to begin, and the current is large. As charge builds up on the capacitor plates, however, the voltage increases and the current decreases. When the voltage across the capacitor plates equals the battery voltage the current stops.

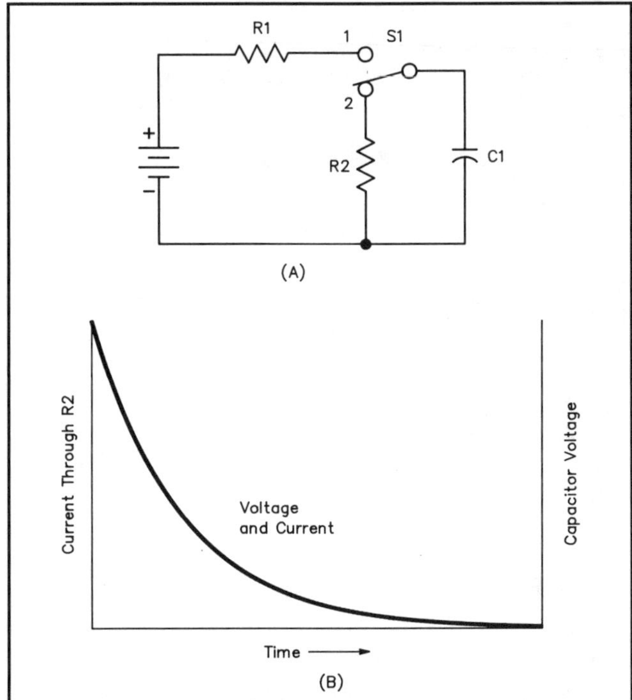

Figure 2—Part A shows the circuit of Figure 1A with the switch moved to position 2. This removes the battery from the circuit and connects R2 across the capacitor terminals. The Part B graph shows the current through R2 and the voltage across the capacitor terminals. The capacitor releases its stored energy.

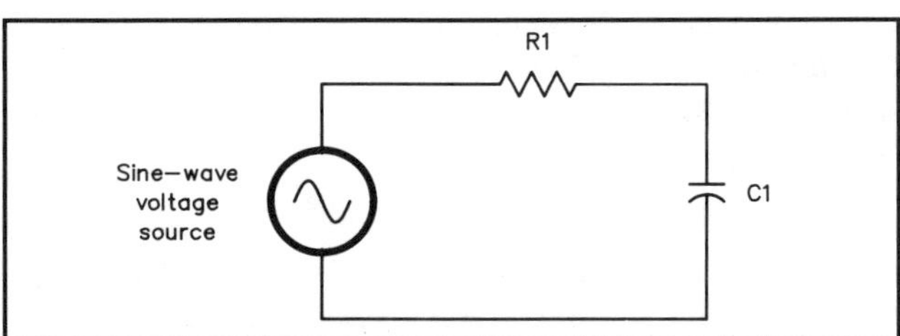

Figure 3—We can apply an alternating voltage to a capacitor and observe the circuit current and capacitor voltage with this circuit. The text describes the results as the applied voltage goes through one sine-wave cycle.

decrease. Electrons were moving *onto* the bottom capacitor plate of Figure 3 as the capacitor was charging. Now electrons move *off* the bottom plate, because the capacitor is returning energy to the circuit.

What happens during the next sine-wave half cycle? The voltage polarity reverses, so the voltage applied to the top capacitor plate is negative. Electrons continue to move *off* the bottom plate, and *onto* the top plate. Soon the capacitor returns all the original charge to the circuit, and it begins to charge in the opposite direction. As the capacitor charge increases the current decreases.

After the sine-wave voltage reaches its maximum negative value, the voltage begins to decrease to zero again. Now the capacitor returns its stored energy to the circuit again. Electrons flow *off* the top plate. The current changes direction again.

Figure 4 graphs the sine-wave voltage source and capacitor current. We used a solid line to represent the voltage and a dashed line to represent the current. There are no numbers on the vertical or horizontal scales because we are not trying to show specific values. The actual voltage, current and time values depend on the resistor and capacitor values, the applied voltage and the sine-wave frequency.

As you study Figure 4, notice that capacitor current is a maximum at 0°. The applied voltage doesn't reach a maximum until 90° on the graph. This describes the *phase* relationship between the alternating current *through* a capacitor and the voltage *across* it. The alternating voltage reaches every point on its waveform 90° after the current. Sometimes we say the voltage across a capacitor *lags* the current through it. Another way of saying this is the current through a capacitor *leads* the voltage across it. In either case the *phase angle* between the voltage and current is 90°.

It appears that capacitors don't like the applied voltage to change. They react to a voltage change as to oppose that change. When the voltage is increasing, they take energy from the voltage supply. You could view this as an attempt to prevent the voltage from increasing. When the voltage is decreasing, the capacitor returns stored energy to the circuit. Think of this action as working to prevent the voltage from decreasing.

Capacitors *react* to voltage changes, trying to prevent the change. We call this opposition to voltage change *reactance*. This opposition to voltage (and current) change in a capacitor is similar to the opposition to current of a resistor. In fact, we measure this opposition in *ohms*.

In the next section you will learn how to calculate *capacitive reactance*. You also will learn how the applied-signal *frequency* affects this opposition to voltage and current changes.

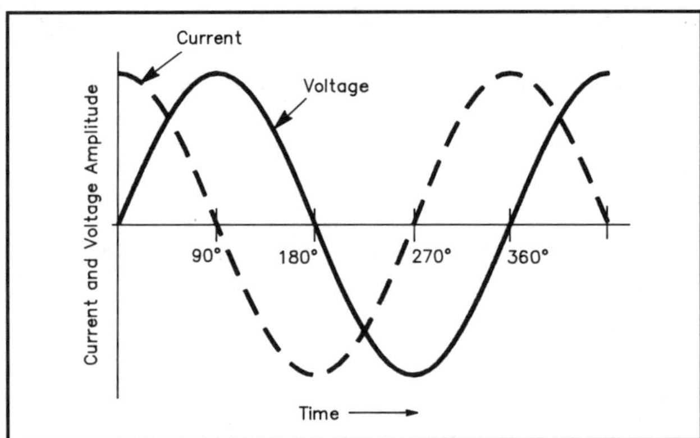

Figure 4—The solid line shows the voltage applied to the capacitor plates of Figure 3. This is a sine-wave voltage that starts at zero. It applies a positive voltage to the top capacitor plate during the first half cycle. During the second half cycle the positive voltage goes to the bottom capacitor plate.

There is More Opposition at Lower Frequencies

A discharged, or "empty," capacitor acts like a short circuit when you first connect voltage to it. There is a large current as the capacitor begins to charge. As the voltage builds up, the current through the capacitor will drop toward zero. Capacitors block direct current.

Capacitive reactance is the opposition capacitors have to current through them. Since capacitors won't pass a steady direct current, you could say they have a large reactance to dc.

Look at Figure 1. We applied a 10-hertz alternating voltage to a 1-microfarad capacitor. The voltage source applies 100-volts RMS to the circuit. The ac ammeter shows a 6.28-milliamp current through the circuit. How much reactance does the capacitor have?

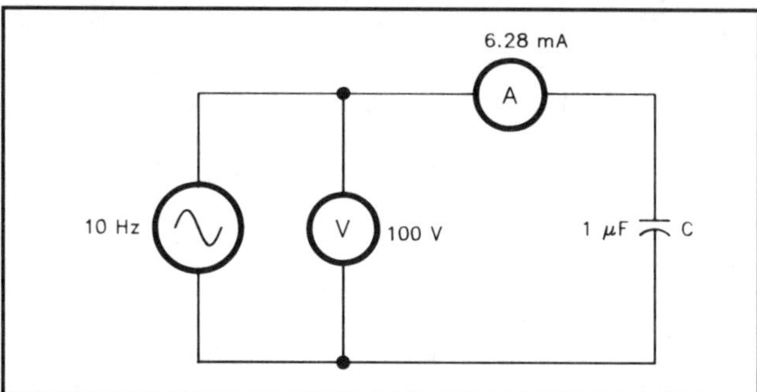

Figure 1—We'll use this circuit for a thought experiment about capacitive reactance.

Does this sound a bit like an Ohm's Law problem? It should. You know a circuit voltage and current, and want to know the opposition to that current. Use Ohm's Law to calculate the reactance in this problem instead of resistance.

Since this is a "thought experiment," we assume the components are *ideal*. That means there is no resistance in the capacitor. The wires don't have any resistance either. *Real* components always have some resistance. The math is more difficult if a circuit has both resistance and reactance. We won't try any problems like that here. You can learn how to handle the more difficult math after you learn the basic principles.

We use X to represent reactance. A subscript C reminds us we are working with a capacitor in this problem. Capacitive reactance is X_c

$$E = I\,X_c \qquad \text{(Equation 1)}$$

Solve this equation for X_c, or use the equation circuit given in Figure 2.

$X_c = E / I$
$X_c = 100 \text{ volts} / 6.28 \text{ mA}$
$X_c = 15{,}924 \text{ ohms}$

(Don't forget that 6.28 mA is 6.28×10^{-3} amperes.)

Let's change our circuit for the next part of our experiment. Replace the 1-μF capacitor with a 2-μF unit. This time the ammeter reads 12.56 mA. What is the new capacitor's reactance? Solve Equation 1 for X_c again.

$X_c = E / I$
$X_c = 100 \text{ volts} / 12.56 \text{ mA} = 7962 \text{ ohms}$

What did you learn from this experiment? When we double the capacitance value, the reactance becomes 1/2 the original value. What do you expect the reactance to be if we used a 3-μF capacitor? If you said 5308 ohms, you are correct. Capacitive reactance is inversely related to capacitance. Larger capacitance values produce smaller reactances.

Now let's make another change to our circuit. This time we will connect a 100-Hz voltage source. Change back to the original 1-μF capacitor. The voltage supply still produces 100 volts. This time the ammeter reads 62.8 mA. Calculate the reactance of the 1-μF capacitor with the new voltage source.

$X_c = 100 \text{ volts} / 62.8 \text{ mA} = 1592.4 \text{ ohms}$

The new signal frequency is ten times the original frequency. Did you notice the current is ten times larger than the first example? Were you surprised to find the reactance is ten times smaller? This part of the experiment taught you that reactance is inversely related to frequency. Higher-frequency signals produce smaller reactances, for a given capacitor.

Do you think this is a reasonable conclusion? Let's think about what happens when we apply higher-frequency signals to a capacitor.

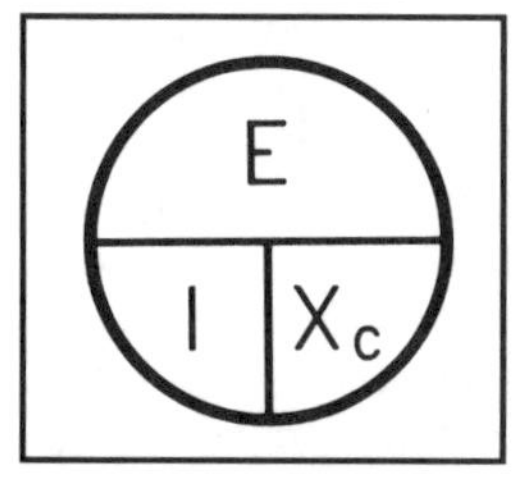

Figure 2—Use this equation circle to solve Equation 1 for capacitive reactance, X_c.

You know that capacitors have a high reactance to dc signals. When you connect a low-frequency ac signal, the capacitor charges to the full signal voltage. The capacitor takes as much charge as it can hold, for that voltage.

Now begin to increase the signal frequency. What if the signal voltage reaches its peak, and begins to decrease before the

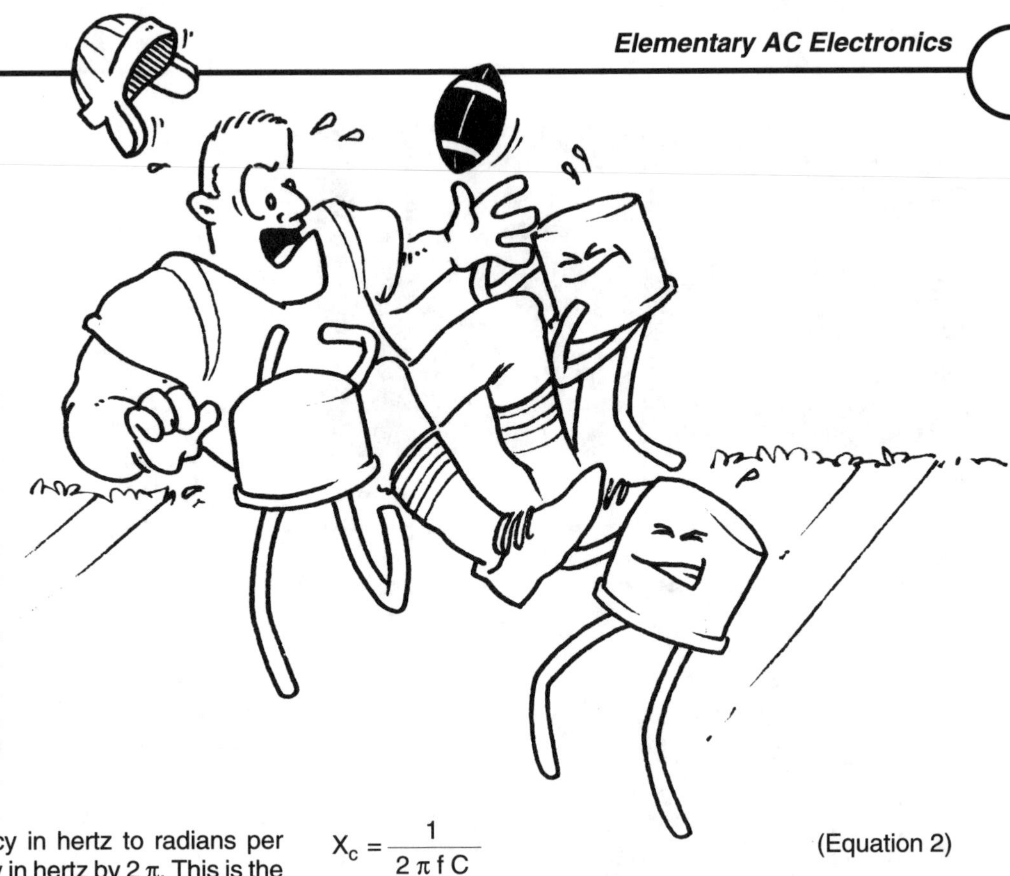

capacitor takes a full charge? If the capacitor doesn't "fill up" with charge, it won't have as much opposition. As the frequency increases even more, the capacitor has less opposition. Figure 3 is a graph of reactance and frequency. Reactance is highest at dc, and decreases quickly.

We found two factors that affect reactance in this section. When either capacitance or signal frequency increases, the reactance decreases. We can write an equation to calculate capacitive reactance, based on these two factors.

We normally express frequency as a number of waveform cycles per second, measured in hertz. We also can relate each cycle to the rotation of a wheel. Then we express frequency in radians per second. You can convert any frequency in hertz to radians per second. Just multiply frequency in hertz by 2 π. This is the number of radians in one cycle. Equation 2 shows how to calculate capacitive reactance.

$$X_c = \frac{1}{2\pi f C} \qquad \text{(Equation 2)}$$

where:

f is the frequency in hertz
C is the capacitance in farads

If you know a frequency in kilohertz or megahertz, you must write it in hertz for this equation. If you know the capacitance in microfarads or picofarads, you must convert those values to farads.

Calculate the reactance of a 1-µF capacitor with a 10-Hz signal applied. This is the example from Figure 1.

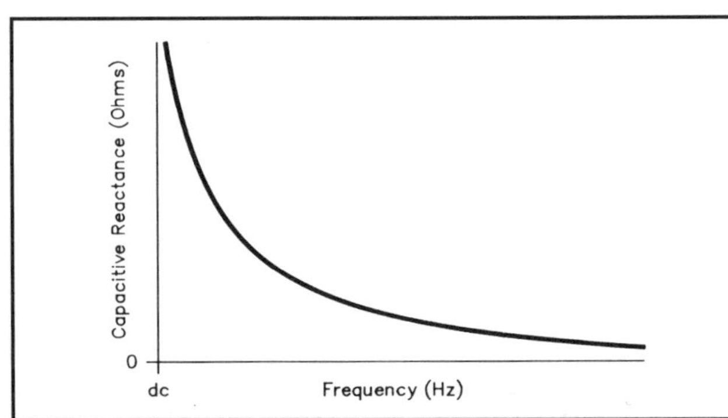

Figure 3—This graph shows how capacitive reactance varies with signal frequency. At dc, the reactance is very large. At high frequencies the reactance is very small.

$$X_c = \frac{1}{2\pi f C} \qquad \text{(Equation 2)}$$

$$X_c = \frac{1}{2 \times 3.14 \times 10\ \text{Hz} \times 1 \times 10^{-6}\ \text{F}}$$

$$X_c = \frac{1}{6.28 \times 10 \times 10^{-6}} = \frac{1}{6.28 \times 10^{-5}}$$

X_c = 15,924 ohms

You shouldn't be surprised to discover this is the same answer we found using the voltage and current in Ohm's Law earlier. Let's try another problem. Find the reactance of a 2-µF capacitor with a 100-Hz supply. (This is not the same as the problem earlier in this section.)

$$X_c = \frac{1}{2\pi f C} \qquad \text{(Equation 2)}$$

$$X_c = \frac{1}{2 \times 3.14 \times 100\ \text{Hz} \times 2 \times 10^{-6}\ \text{F}}$$

$$X_c = \frac{1}{6.28 \times 200 \times 10^{-6}} = \frac{1}{1.256 \times 10^{-3}}$$

X_c = 796.2 ohms

You can practice a few more capacitive reactance problems. Try calculating the reactance for the other circuits we studied at the beginning of this section. Your answers should agree with the ones we found there. You might even want to calculate the reactance of a few capacitance values for frequencies of 1 kilohertz and 1 megahertz. With practice (and your trusty scientific calculator) you'll soon be calculating capacitive reactance values with ease.

CHAPTER 19 Inductors

Electric Current Produces a Magnetic Field

When electrons move they produce a magnetic field. An electric current flowing through a wire produces a magnetic field. The magnetic field extends outward from the wire. It is stronger close to the wire and gets weaker as you move away from the wire. Figure 1A pictures this magnetic field around a wire as a series of concentric circles. Each larger circle represents a weaker part of the field. Figure 1B shows the magnetic field as tiny bar magnets around the wire.

You can't see a magnetic field, but you can observe its effects. One way to observe a magnetic field is with a compass. Strip the insulation from the ends of a piece of wire that is about 1 metre long. Set a D-cell battery on one end of the wire. Lay a compass on the edge of a table and hold the wire close to it. Touch the other end briefly to the battery top. Watch the compass closely. You should see the needle move slightly when you touch the wire to the battery. The compass needle will move back to North when you remove the wire.

Don't hold the wire on the battery for more than a few seconds at a time. Otherwise the battery may overheat. Then there is a danger that it could explode. You will quickly kill the battery with this experiment, so only touch the wire to the battery a few times.

If the compass does not seem to move, you may need

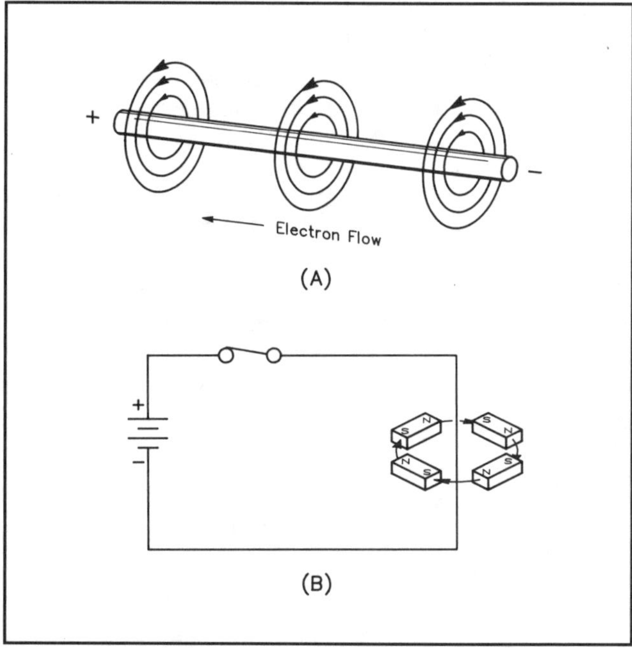

Figure 1—Electrons produce a magnetic field when they move. Part A shows a series of circles surrounding the wire to represent the magnetic field. Part B represents this magnetic field with several small bar magnets around the wire.

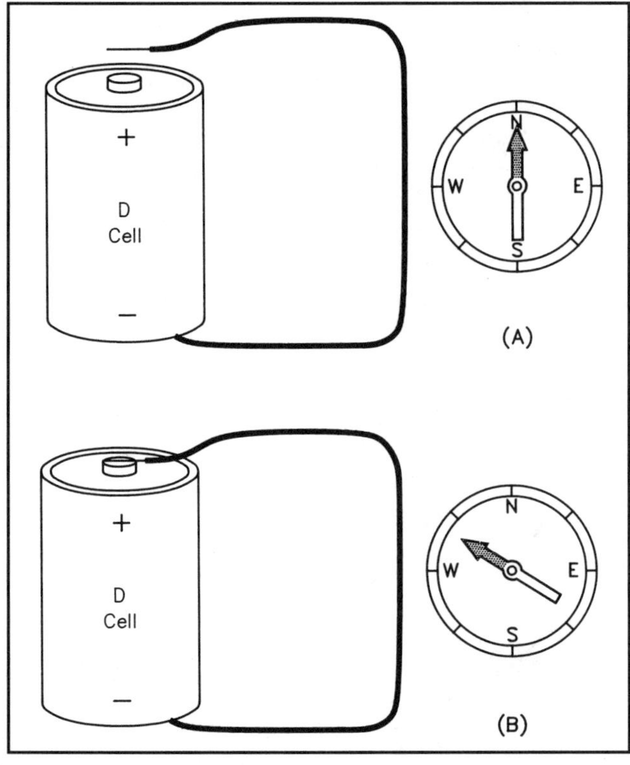

a new battery. It also is possible that you don't have the compass close enough to the wire. Figure 2 shows how the compass needle moves when you touch the wire to the battery.

This relationship between electricity and magnetism is very important. You may have built a simple electromagnet sometime. Most electrical meters operate because a small electric current creates an electromagnet. This electromagnet turns because there is a permanent magnet around it. A pointer attached to the electromagnet moves across a scale. Voltmeters, ammeters and ohmmeters all operate on this principle.

Electric motors turn because of the magnetic field produced when an electric current moves through the motor's wires. Even the existence of radio signals depends on the relationship between an electric field and a magnetic field.

Figure 2—These drawings show a D-cell battery, a wire and a compass. The compass needle moves slightly when you touch the wire ends to the battery.

Winding the Conductor into a Series of Loops Strengthens the Magnetic Field

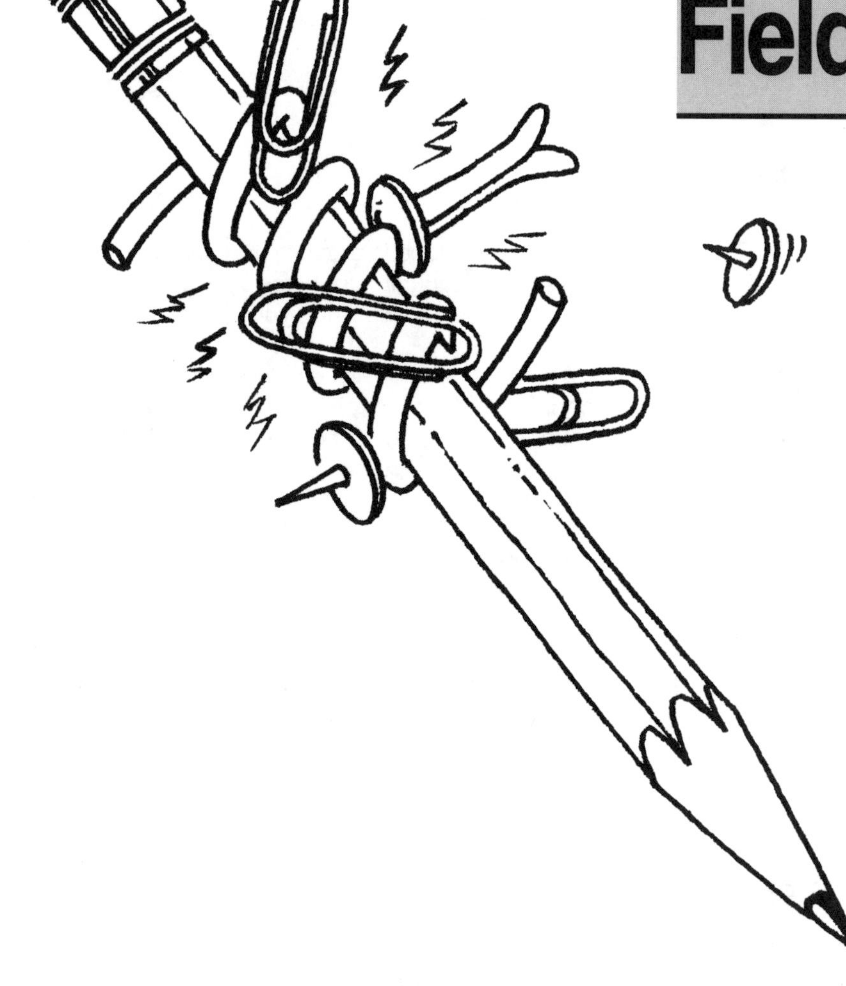

In the last section you learned that an electric current produces a magnetic field around the wire. The magnetic force surrounding a straight wire is usually weak. Larger currents produce stronger magnetic fields. It takes a current of many amperes to produce a strong magnetic field with a straight wire.

What if you bend the wire into a circle, and make several turns next to each other? The weak magnetic field around each piece of wire adds to the force of the pieces on either side of it. Figure 1 shows a piece of wire wound into several turns. The magnetic-force lines wrap through the center of the cylinder, and around the outside. The magnetic field through the coil is much stronger than the individual field of a straight wire.

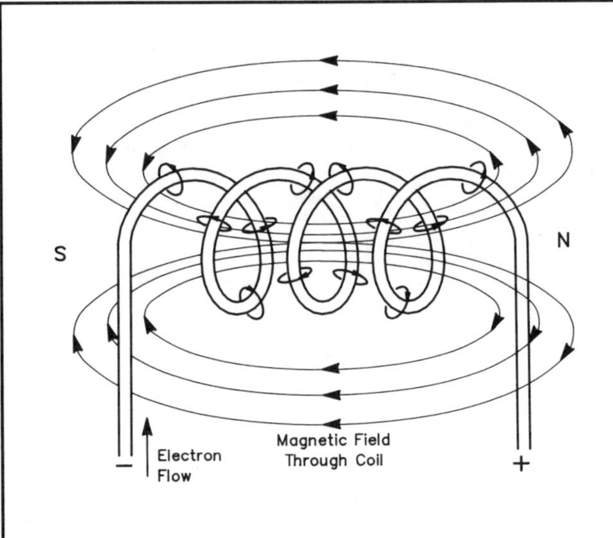

Figure 1—Electric current through a wire coil produces a stronger magnetic field than the same current through a straight wire. The lines of force around each turn reinforce the field around the other turns.

You probably remember building an electromagnet sometime. If you have never made an electromagnet, this is a good time to try it.

For this electromagnet you will need a pencil or plastic stick-type pen. You also will need a coil of small-diameter, enamel-coated copper wire. Leave about 20 centimetres of wire free on one end. Begin winding wire around the pencil, as Figure 2 shows. Wind each turn against the previous one, but don't overlap the wire. Continue winding until you have covered about 10 cm of the pencil. Leave another 20 cm of wire at the end, and then cut the wire. Clean the enamel insulation off both ends of your magnet's wire.

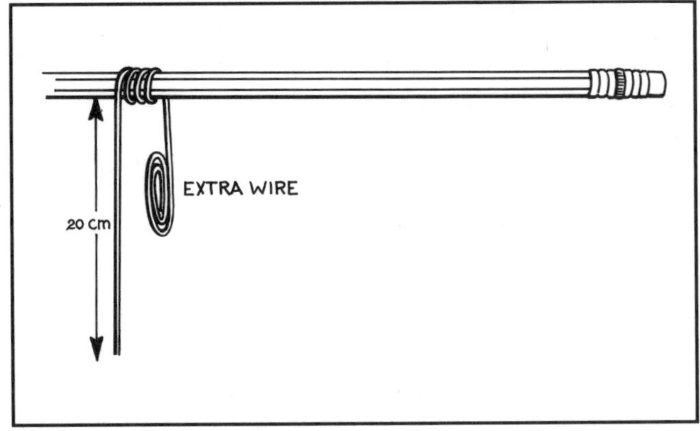

Figure 2—You can make a simple electromagnet by winding a coil of wire around a pencil or other cylindrical form. The text explains how to wind the wire.

To complete your magnet, touch the wire ends to opposite sides of a D-cell battery. Now try to pick up some paper clips by bringing one coil end close to a pile of paper clips. You may have to hold one paper clip into the coil center. That paper clip becomes magnetized, and it will attract other clips. See Figure 3.

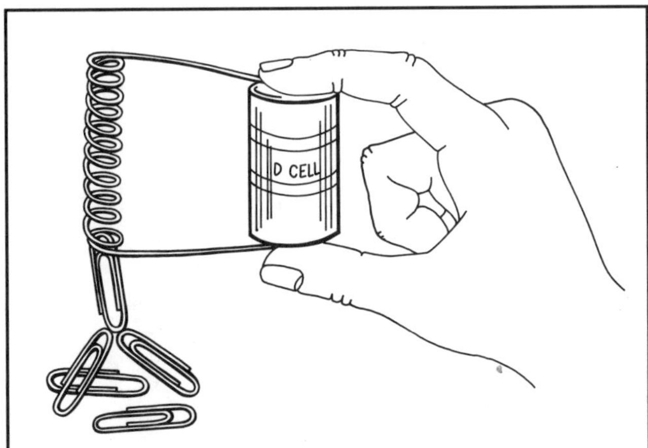

Figure 3—A paper clip held into the wire coil becomes magnetized. Then it picks up more paper clips.

How many paper clips will your magnet hold? Do you think your magnet would be stronger if you connected it to a 3-volt battery instead of the 1.5-volt battery? What do you think would happen if you wound a longer coil of wire?

Increasing the battery voltage increases the current through the wire. This makes a stronger magnetic field. Increasing the number of wire turns on the pencil also makes a stronger magnet. You must be careful to wind the coil in only one direction, however. If you wind back over the top of your first layer, the direction of the magnetic field reverses. In that case, the two layers oppose each other. Your magnet would seem weaker instead of stronger.

We call a coil of wire used in an electronics circuit an *inductor*. An electric current through an inductor creates a magnetic field. You will learn more about inductors and the magnetic fields they create in the rest of this chapter and the next one.

Winding the Coil Around a Magnetic Material Strengthens the Magnetic Field

In the last section we described how to make an electromagnet by winding some wire around a pencil. Have you ever made an electromagnet like that? Chances are, you used a nail or an iron bar other times you made an electromagnet.

You might want to make such an electromagnet now. Use a large nail or iron rod. Wind the enamel-coated copper wire around the nail as you did with the pencil in the previous section. Then connect your new magnet to a D-cell battery. How many paper clips will this magnet pick up? Probably many more than the coil without a nail in the last section.

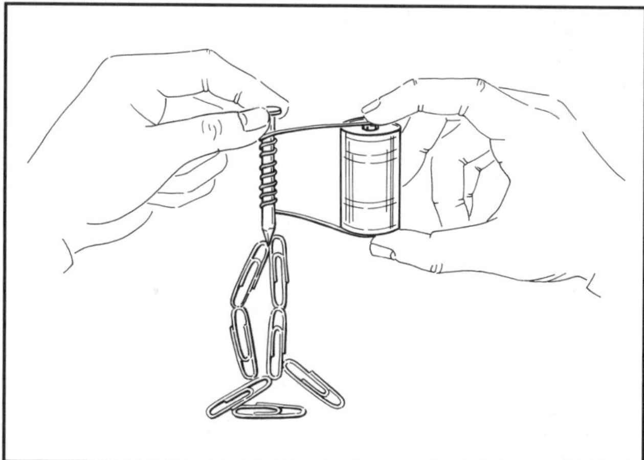

Figure 1—A coil wound around a nail or iron rod produces a strong electromagnet.

You just learned one way to increase an inductor's magnetic field strength. Winding the coil around an iron core produces a stronger magnetic field.

What other materials can we use for the center of an inductor or coil? There are many choices. There are many materials made especially for use as inductor cores. Iron powder mixed with clay and other compounds forms one type of core. The manufacturer forms the material into rods or other shapes, and then bakes them to harden the material. Manufacturers prepare special mixtures for different applications. Two general categories of these products are the *powdered iron* and *ferrite* materials. *Ferrite* comes from the Latin word for iron. Chemists often refer to iron compounds as *ferrous* materials.

Ferrite materials have other metal alloys mixed with the iron. Manufacturers select various alloys for the mixture to provide specific characteristics. You can select a mixture to provide the best performance over a certain frequency range. Iron and powdered iron materials normally cover the audio-frequency range. Various ferrite mixtures cover certain portions of the radio-frequency range.

Figure 2 shows two common inductor types. You will find the air-core inductors in many radio transmitters and

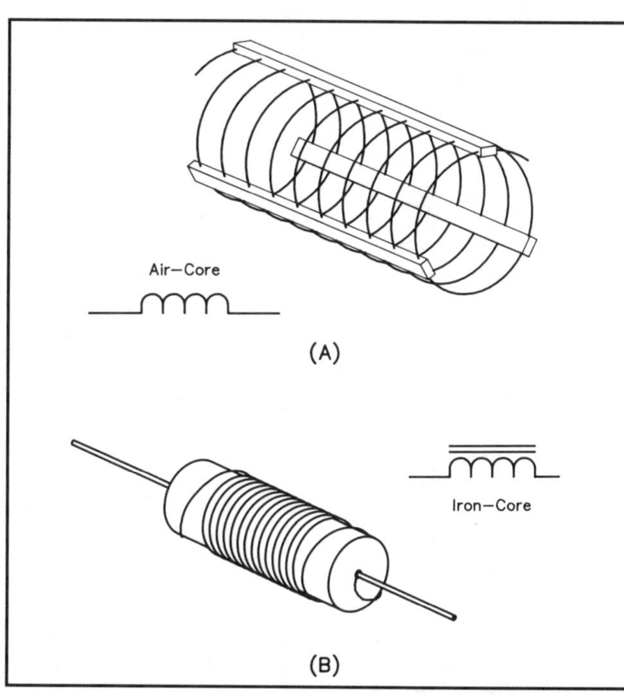

Figure 2—Plastic strips hold the wires securely in place on one common type of air-core inductor. Iron-core inductors take many forms. The one shown at B has a wire wound on a powdered-iron rod. Heavier wires embedded into the core form leads to connect the inductor into a circuit.

power amplifiers. There are iron-core inductors in many electronics circuits. Figure 2 also includes the schematic-diagram symbols used for these inductors.

Many transmitter and receiver circuits use a special type of inductor core. A *toroid* is an inductor wound on a core that curves into a donut shape. *Toroid cores* are available in many powdered-iron and ferrite core materials. Figure 3 shows a coil wound on a toroid core.

Remember the magnetic field produced by an inductor goes through the center of the coil. The magnetic field from a straight coil curves back around outside the core, because magnetic-field lines form a closed loop.

A magnetic field inside a toroid core doesn't have to loop around outside the core, however. Since the toroid forms a donut, closed back on itself, the field stays completely inside the core. This is one big advantage of toroid inductors.

Many electronics circuits require an inductor with a magnetic field that you can change. There are several ways to make a *variable* or adjustable *inductor*. One method is to attach a lead to an air-core inductor. See Figure 4A. You can attach this lead to any point in the coil, selecting the proper number of turns for an application. Radio frequency power amplifier circuits often use a multi-position switch to select coil tap points. You select a different tap point for each band on which the amplifier is to operate.

You also can make an adjustable inductor with a powdered-iron or ferrite core that moves into or out of the coil. You can wind the coil on a form made of ceramic or other material. Then a *slug* or piece of ferrite core material threads into the ceramic form. A screw adjustment allows you to adjust the position of the material in the coil. This changes the magnetic-field strength. Figure 4B shows this type of variable inductor.

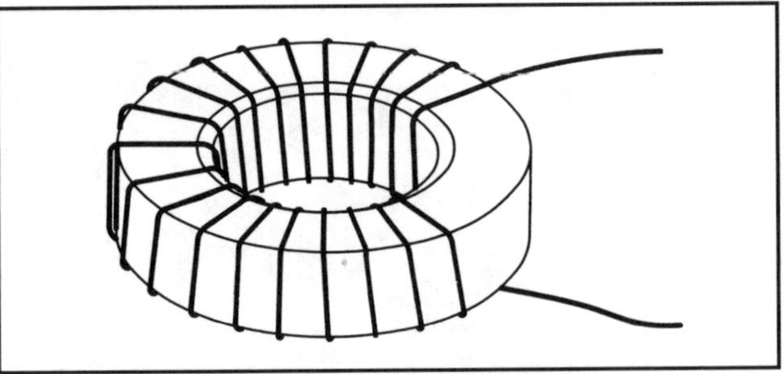

Figure 3—This coil has a toroid (donut-shaped) core. This coil produces a magnetic field that remains completely inside the core.

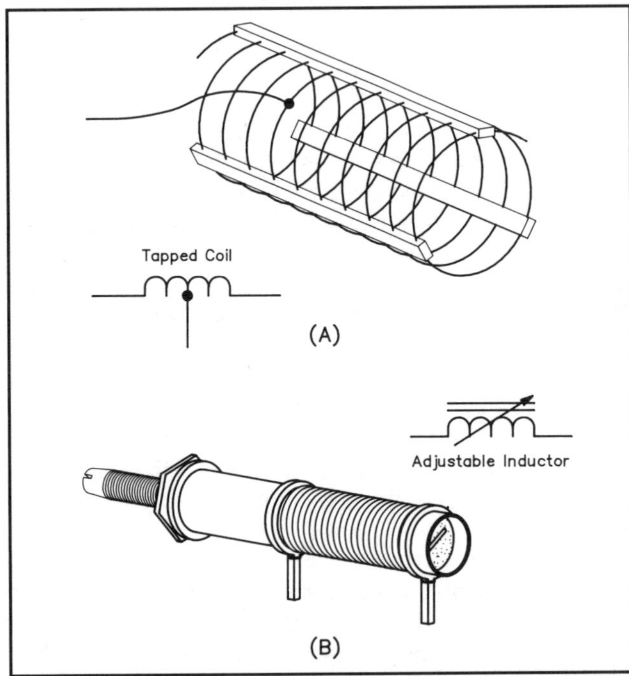

Figure 4—Part A shows how you can attach a lead to an air-core inductor. This allows you to select the number of turns required for your application. Part B shows another type of adjustable inductor. A form of ceramic or other material holds the wire coil. A core of powdered iron or ferrite material threads into the form. You can set the position of the ferrite material inside the coil to adjust the magnetic field.

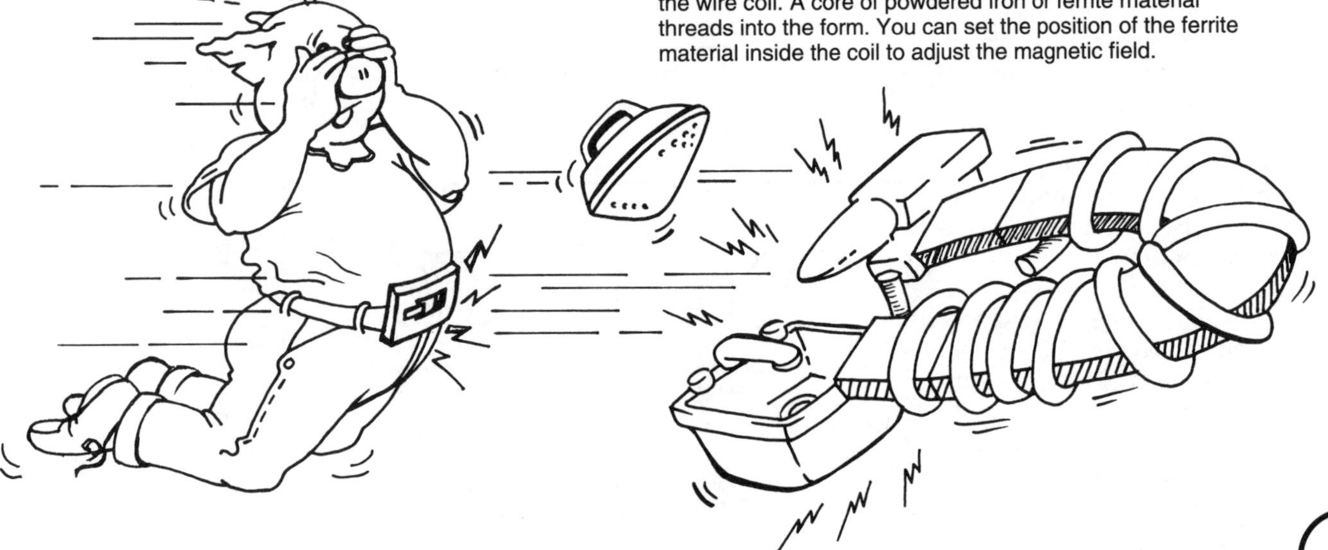

Factors that Determine Inductance

In this chapter you have been studying inductors. You learned that when electrons move through a conductor they produce a magnetic field. By winding the conductor into a coil, we produced a stronger magnetic field. You also learned that you can wind coils on iron and other magnetic materials. Magnetic materials placed in the center of an inductor also increase the magnetic-field strength.

In the next section you will discover why we use the name *inductor* to describe a coil in an electronics circuit. The *inductance* of a coil is a property that relates to the magnetic-field strength produced by the coil. We use a capital L to represent inductance. A coil with more inductance will produce a stronger magnetic field than one with less inductance.

The basic unit of inductance is the *henry*, abbreviated *H*. This name is in honor of the American physicist Joseph Henry. He discovered a relationship between electricity and magnetism during the early 1800s.

The *henry* is usually too large for practical inductance measurements. A 1-henry inductor is physically very large. We use millihenrys (mH, 10^{-3}) and microhenrys (μH, 10^{-6}) for most measurements. You should be familiar with these metric prefixes.

Four main factors help determine a coil's inductance. Changing any of these factors changes the coil's inductance.

Inductance depends on the number of wire turns used to wind the coil. Wind more turns to increase the inductance. Take off turns to decrease the inductance. Figure 1 compares the inductance of two coils made with different numbers of turns.

Inductance depends on the inductor's length, or the spacing between turns. Figure 2 shows two inductors with the same number of turns. The first inductor's turns have a wide spacing. This coil is 5 cm long. The second inductor's turns are close together. The second coil is only 1 cm long. Which one do you think has more inductance? Yes, the close-spaced inductor has a larger inductance value.

The third factor that helps determine inductance is the coil diameter. You may see coil diameter called *cross-*

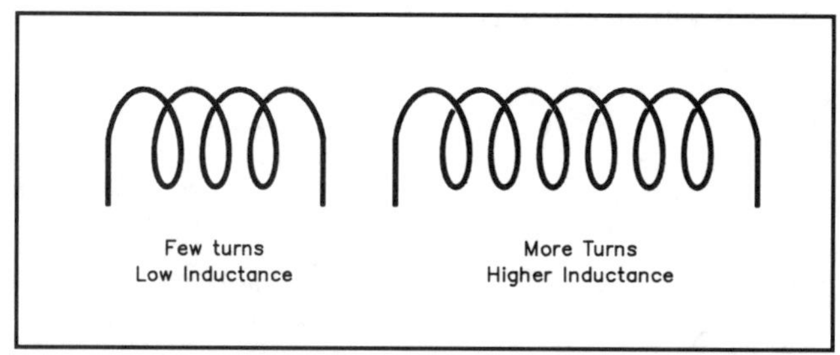

Figure 1—A coil's inductance depends on the number of turns used to make the coil.

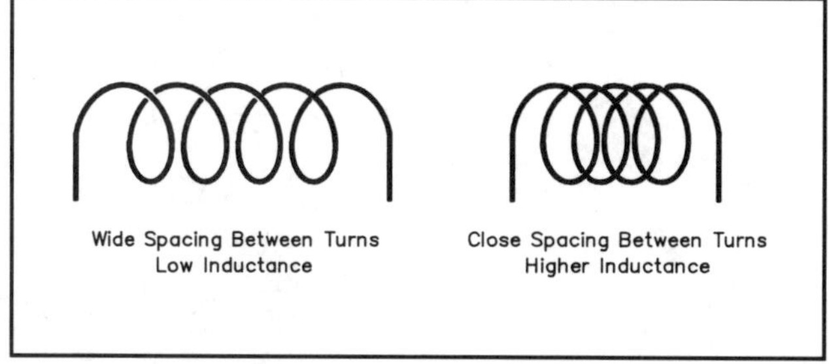

Figure 2—The length of a coil, or the spacing between turns, affects the inductance.

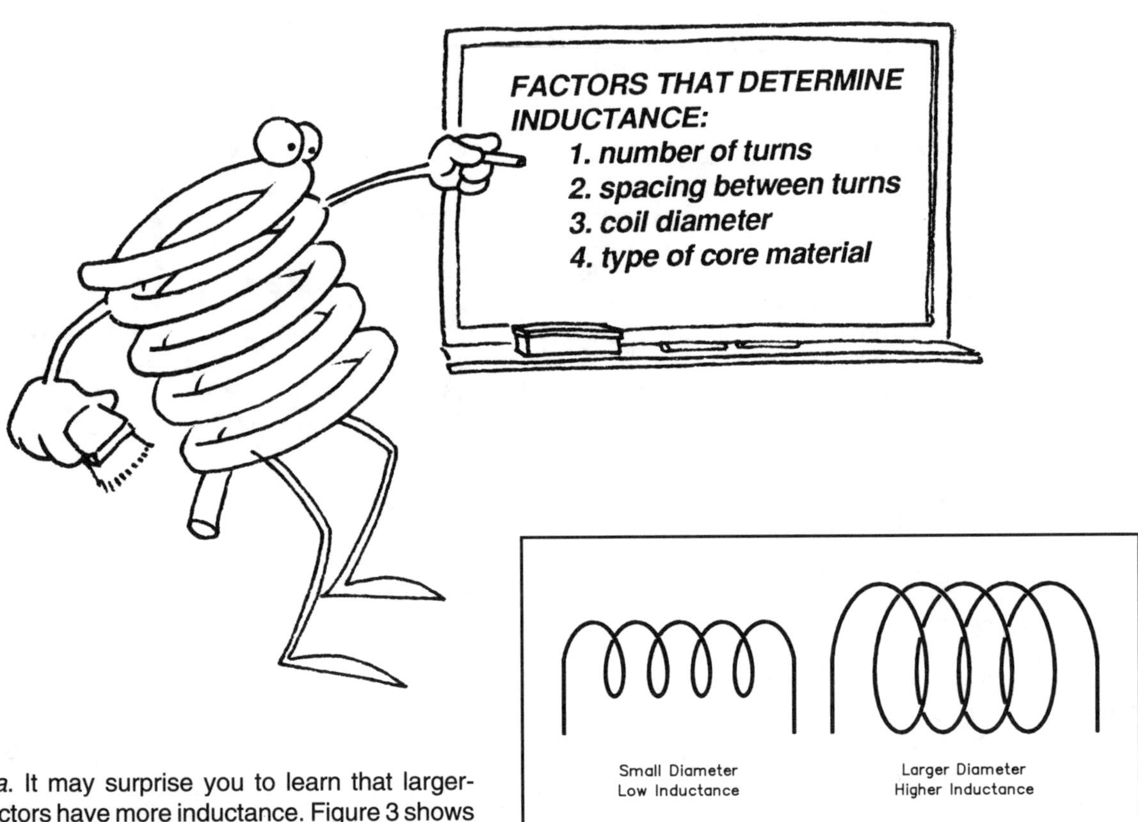

Figure 3—An inductor's value depends on the coil diameter.

sectional area. It may surprise you to learn that larger-diameter inductors have more inductance. Figure 3 shows two inductors. They have the same number of turns, and the spacing between turns is the same. The first inductor has a small diameter and the second one has a larger diameter. The second inductor has more inductance than the first one.

The fourth factor affecting inductance is the type of core material in the center of the coil. *Permeability* is a measure of how easily a magnetic field goes through a material. Permeability also tells us how much stronger the magnetic field will be with that material inside the coil.

Permeability is usually given as *relative permeability*. This means the value compares the material with the permeability of air.

One ferrite material has a relative permeability of 75. How much more inductance will a coil wound on this material have than an identical air-core coil? Did you say 75 times more? That's right! Another material with a relative permeability of 25 will produce an inductance 25 times larger than the same air-core coil.

Figure 4 shows three identical coils. One has an air core, one has a powdered-iron rod in the center and the other has a soft iron rod. (*Soft* means the iron hasn't been tempered to harden it. Most nails are soft iron.) This figure illustrates the effects of core material on inductance.

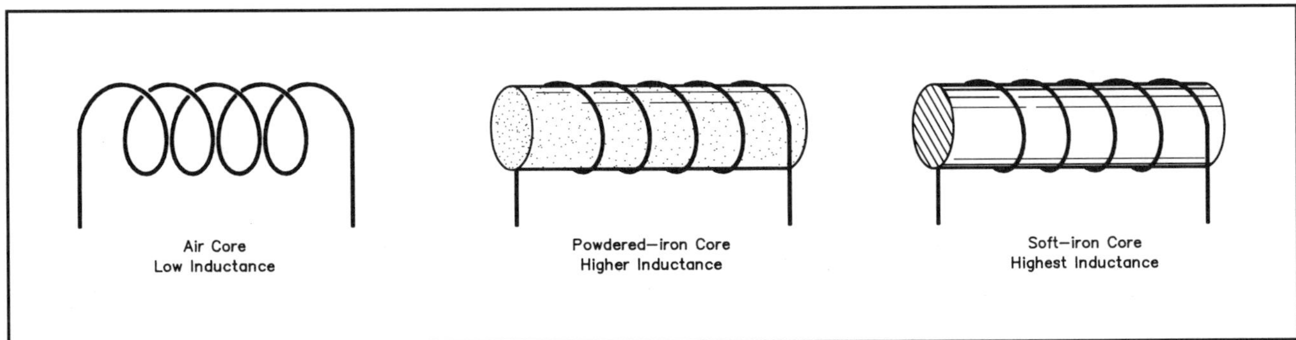

Figure 4—The core material, or material in the center of the inductor, helps determine the inductance value.

Inductors Store Electrical Energy in a Magnetic Field

What purpose does a magnetic field have in an electronics circuit? You may be wondering about this. It is a good question. After all, we haven't discussed any other effects that an inductor might have on a circuit. An inductor's primary purpose results from a changing current rather than from a steady, direct current. This section gives a partial answer to the question. The next few chapters will give you a clearer picture of the role inductors play in electronics.

A piece of wire does not produce a magnetic field by itself. An electric current flowing through it does produce a magnetic field, though. There is no magnetic field before you connect the wire to a voltage source. When you do connect it to a voltage, electrons begin to flow. The magnetic field gradually increases. (Well, okay, the magnetic field is there almost immediately. If you could study it in slow motion, however, you would see the magnetic field increase gradually.)

Figure 1 shows an inductor connected to a battery. The battery supplies a direct voltage that doesn't change. This voltage produces a steady, direct current. The magnetic field reaches a strength determined by the amount of current and the coil's inductance. This magnetic field represents stored energy. Once the magnetic field reaches the strength set by the current and inductance, it doesn't change, as long as the steady current continues.

Imagine you are watching the electrons move through the wire. We can make an interesting observation here. Let's open the switch, breaking the circuit.

When you disconnect the voltage source, the magnetic field collapses. This happens quickly, but again, if you could watch it slowly, you would see the field decrease gradually.

The electrons don't stop immediately when you remove the voltage. Instead, the number of moving electrons gradually decreases as the magnetic field collapses. The magnetic field returns the energy stored in it by continuing to move electrons through the wire.

If the electrons continue to flow, even for a short time, there must be a voltage to push them. (This conclusion results from Ohm's Law.) We removed the battery from the circuit, so it doesn't provide a voltage any longer. The inductor is the only part left in the circuit, so it must produce

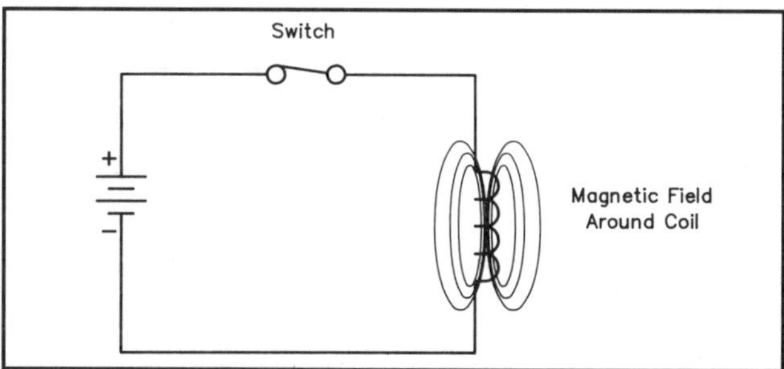

Figure 1—The magnetic field around an inductor increases from zero when you first connect it to a voltage source. The field strength depends on the current and inductance value. The magnetic field reaches a steady strength with a steady direct current through the inductor.

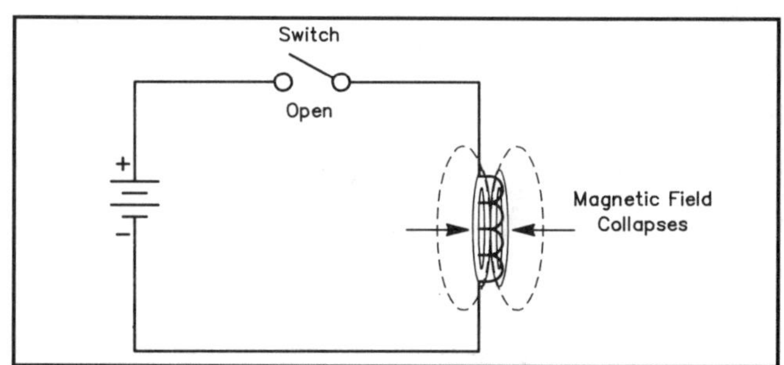

Figure 2—When we open the switch, the magnetic field begins to collapse. The field returns the energy stored in it by inducing a voltage across the coil, which tries to keep the electrons flowing.

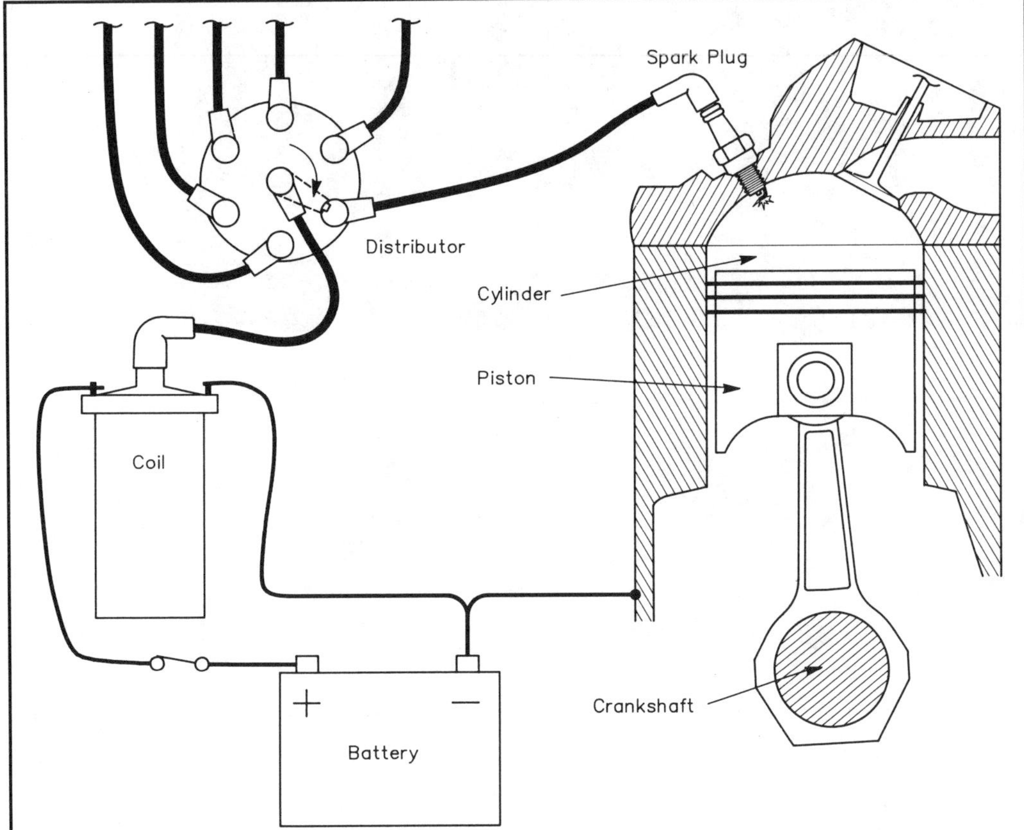

Figure 3—A gasoline engine depends on a large induced voltage to create a spark that explodes the gasoline. There is a strong magnetic field through the induction coil before the circuit is broken. This creates a large induced voltage across the coil, and that voltage causes a spark to jump across the spark plug.

a voltage that keeps the electrons moving while the magnetic field collapses.

The collapsing magnetic field *induces* a voltage across the inductor, trying to maintain the current. The dictionary definition of *induce* is "to move by persuasion or influence." The collapsing magnetic field persuades some electrons to continue moving through the coil.

You might ask, "If a collapsing magnetic field induces a voltage across an inductor, does an increasing field also induce a voltage?" Yes, it does. In that case the voltage tries to oppose the current, however. So the induced voltage has the opposite polarity to the applied voltage when the magnetic-field strength is increasing. This is why the current increases gradually when you first connect the battery.

The magnitude of the induced voltage depends on how rapidly the current changes. Suppose you have a strong magnetic field built up in a coil. If you suddenly break the circuit, there is a large induced voltage across the coil. This large voltage tries to maintain the current through the coil. The voltage can be many times larger than the original applied voltage. It is common to have a spark jump across the switch contacts as they open.

This is exactly the principle used in the induction coil of a gasoline engine. A large current flows through the coil. At the proper instant a set of contacts open, inducing a much larger voltage than the battery applies. This induced voltage causes a spark to jump across the gap in the spark plug. The spark ignites the gasoline in the cylinder, and this explosion drives the piston to turn the engine. See Figure 3.

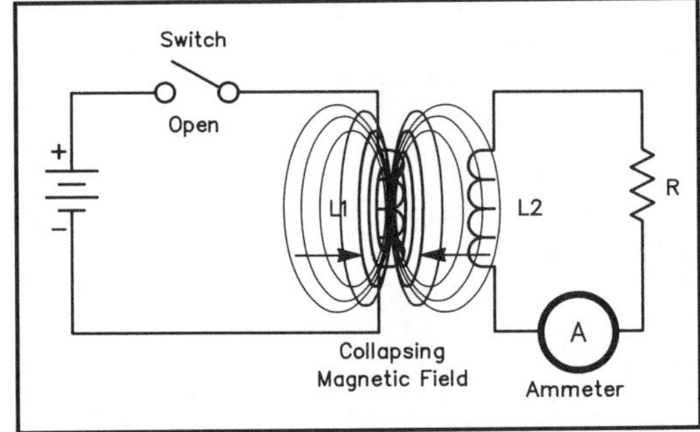

Figure 4—The magnetic field from one inductor can induce a voltage across other inductors.

Figure 4 shows two inductors. A battery connects to the first one, but the second one has only a resistor connected across its terminals. The second inductor is close to the first one. Now suppose we open the switch, disconnecting the first inductor from its circuit. A sensitive ammeter will show a small current in the second circuit when we open the switch.

This experiment is another example of a collapsing magnetic field inducing a voltage. This time the induced voltage was across another coil, however. The magnetic field of one inductor can induce a voltage across other inductors positioned close to the first one.

CHAPTER 20 Inductive Reactance

Inductors Oppose Any Change in Current

Figure 1 shows a circuit with an inductor and a battery that supplies a steady direct voltage. There also is a variable resistor in series with the inductor. Since the inductor is just a piece of wire, it has very little resistance to this direct current. You can use Ohm's Law to calculate the circuit current. What is the current if we set the resistor for 5000 ohms?

$$E = I R \qquad \text{(Equation 1)}$$

You can easily solve this equation for the current.

$$I = E / R \qquad \text{(Equation 2)}$$

$I = 12\ V / 5000\ \Omega$
$I = 2.4 \times 10^{-3}$ amperes = 2.4 mA

There is a magnetic field around the inductor with this current through the circuit. As long as the current doesn't change, the magnetic field remains constant.

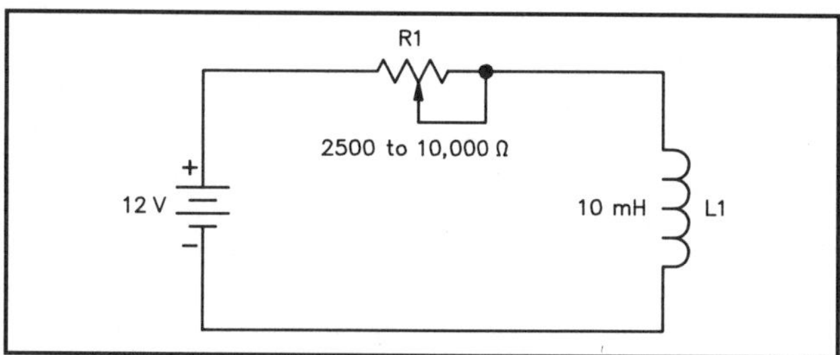

Figure 1—We'll use this circuit to learn about the effects of changing current on an inductor. Adjusting the variable resistor changes the circuit current. A capital L represents an inductor on schematic diagrams and in equations.

Suppose we increase the variable-resistor setting? With a resistance of 10,000 ohms, the circuit current decreases to 1.2 mA. (Can you prove that using Ohm's Law?) This smaller current also means the magnetic-field strength must decrease.

When we adjust the resistor, the current begins to decrease. The inductor *reacts* to this change by returning some of the energy stored in its magnetic field. This returned energy produces a voltage across the inductor that tries to keep the same current flowing through it. The current does decrease, however, as the magnetic-field strength decreases. The inductor prevents the change from occurring instantly. The new current maintains a smaller magnetic field.

What do you think will happen if we decrease the resistance? What will the current be if the new resistance value is 2500 ohms? Use Ohm's Law and Equation 2 to calculate this new current. You should calculate 4.8 mA.

The larger current creates a stronger magnetic field. Once again the inductor reacts to prevent this current change, however. There is a voltage across the inductor in a direction opposite to the battery voltage. This *back emf* as we often call it, works to prevent the current increase. Remember that *emf* is short for *electromotive force*, which is another name for *voltage*. *Back emf*, then, means a voltage produced across an inductor by a current change.

Figure 2 is a graph of the circuit current for this example. The current begins at 2.4 mA, and remains steady until we increase the resistance. Then the current decreases to 1.2 mA. Notice that this change does not occur instantly. The current decreases gradually, until it

reaches the new value. When we decrease the resistance, the current increases to 4.8 mA. Once again, this change occurs gradually.

The resistance and the inductance determine how quickly the current reaches its new value. We won't show you how to do this calculation in this book. *The ARRL Handbook* contains more information about calculating the values used to plot the Figure 2 graph. For now, you should remember that the current through an inductor changes gradually.

When an inductor *reacts* to current changes, it works against that current change. Changes in the amount of energy stored in the magnetic field cause this opposition.

You should notice some similarities between inductors and capacitors. Capacitors react to prevent changes in the amount of electrical energy they store. The applied voltage determines the amount of charge on a capacitor. A capacitor reacts to a decreased voltage by supplying current. It reacts to an increased voltage by trying to block more current.

You can think of an inductor as the opposite of a capacitor. Inductors store energy in a magnetic field instead of an electric field. Inductors produce a voltage that tries to prevent current changes.

We use the term *reactance* to describe the opposition of an inductor to changes in current. We use the same term to describe the opposition of a capacitor to changes in current. We distinguish the two with the terms *inductive reactance* and *capacitive reactance*. A capital X represents reactance in equations. Subscripts C and L indicate capacitive reactance (X_C) and inductive reactance (X_L).

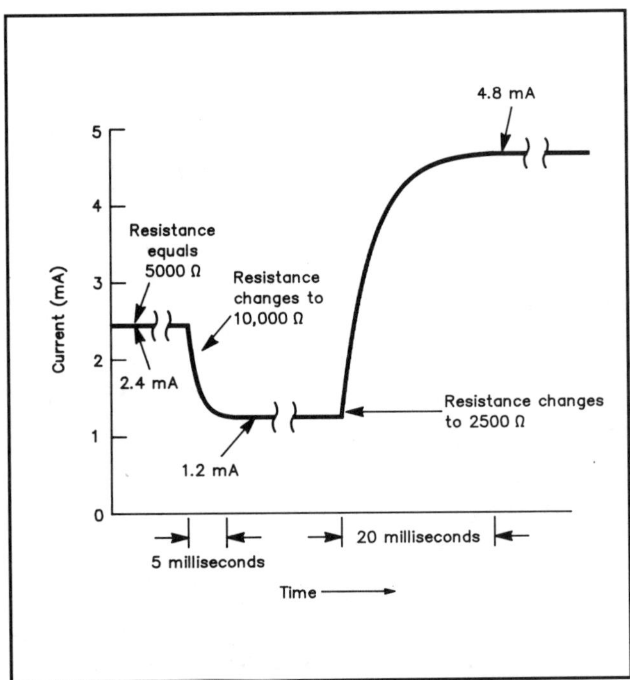

Figure 2—This graph shows the current through the Figure 1 inductor. In this experiment, the variable resistance is 5000 Ω to start. At this setting, the current is 2.4 mA. Then we increase the resistance to 10,000 Ω and the current decreases to 1.2 mA. Finally, we decrease the resistance to 2500 Ω. The current increases to 4.8 mA.

There is More Opposition at Higher Frequencies

An inductor consists of a piece of wire wound to form a cylinder. Wire has a low resistance. When you connect a dc voltage source to the inductor, there is a large current. Many circuits have a resistor in series with the inductor to limit the current.

The current through an inductor won't change instantly. It quickly builds up to the value set by the series resistance, however. Inductors act like a short circuit to direct current.

Inductive reactance is the opposition inductors have to current changes. Since inductors act like a short circuit at dc, we say they have zero reactance at dc.

Consider the effects of applying a low-frequency sine-wave voltage to an inductor. Figure 1 shows a simple circuit with a 100-hertz alternating voltage applied to a 10-millihenry inductor. The ammeter shows a 0.239-ampere current through the circuit with 1.5 volts rms from the voltage source. How much reactance does the inductor have?

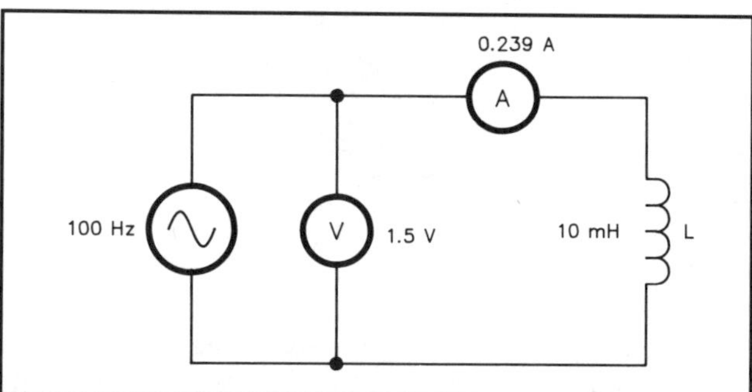

Figure 1—We'll use this circuit for a thought experiment about inductive reactance.

Does this sound a bit like another Ohm's Law problem? You know a circuit voltage and current, and want to find the opposition to that current. Use Ohm's Law to calculate the *reactance* in this problem instead of *resistance*.

Since this is a "thought experiment," we assume the inductor is *ideal.* That means the inductor's wire has no resistance. The other wires in the circuit don't have any resistance either. *Real* components always have some resistance. The math is more difficult if a circuit has both resistance and reactance. We won't try any problems like that here. You can learn how to handle the more difficult math after you learn the basic principles.

We use X to represent reactance. A subscript L reminds us we are working with an inductor in this problem. Inductive reactance is X_L.

$$E = I\,X_L \qquad \text{(Equation 1)}$$

Solve this equation for X_L, or use the equation circle given in Figure 2.

$X_L = E / I$
$X_L = 1.5 \text{ volts} / 0.239 \text{ A}$
$X_L = 6.28 \text{ ohms}$

Let's change our circuit for the next part of our experiment. Replace the 10-mH inductor with a 50-mH unit. This time the ammeter reads 47.8 mA. What is the new inductor's reactance? Solve Equation 1 for X_L again. (Remember that 47.8 mA is 47.8×10^{-3} amperes.)

$X_L = E / I$
$X_L = 1.5 \text{ volts} / 47.8 \times 10^{-3} \text{ A} = 31.4 \text{ ohms}$

What did you learn from this experiment? We made the inductance value five times larger and the reactance increased to five times larger than the original value. What do you expect the reactance to be if we use a 100 mH inductor? If you said 62.8 ohms, you are correct. Inductive reactance is directly related to inductance. Larger inductance values produce larger reactances.

Now let's make another change to our circuit. This time we will connect a 1000-Hz voltage source. Change back to the original 10-mH inductor. The voltage supply still produces 1.5 volts. This time the ammeter reads 23.9 mA. Calculate the reactance of the 10-mH inductor with the new voltage source.

$X_L = 1.5 \text{ volts} / 23.9 \text{ mA} = 62.8 \text{ ohms}$

The new signal frequency is ten times the original frequency. Did you notice the current is ten times larger than the first example? Were you surprised to find the reactance is ten times larger? This part of the experiment taught you that reactance is directly related to frequency. Higher frequency signals produce larger inductive reactances, for a given inductor.

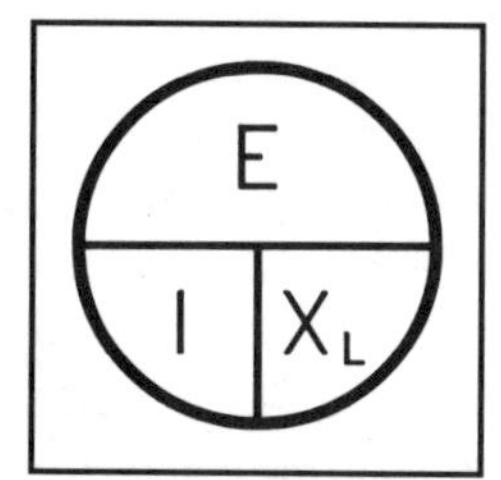

Figure 2—Use this equation circle to solve Equation 1 for inductive reactance, X_L.

Do you think this is a reasonable conclusion? Let's think about what happens when we apply higher-frequency signals to an inductor.

You know that inductors act

as short circuits to dc signals. When you connect a low-frequency ac signal, the inductor stores energy in a magnetic field. The inductor reacts to changing currents. If the current changes slowly, the inductor has almost no reaction trying to prevent that change. A low-frequency ac signal doesn't produce much reactance from an inductor.

Now begin to increase the signal frequency. It takes less time for the signal to go through one cycle at a higher frequency. You can conclude that the current and voltage change more rapidly. The inductor has a stronger reaction to this rapid current change. As the frequency increases, the inductor has more opposition to the current changes. Inductive reactance increases as the frequency increases. Figure 3 is a graph of inductive reactance and frequency. Reactance is zero at dc, and increases as the frequency goes up.

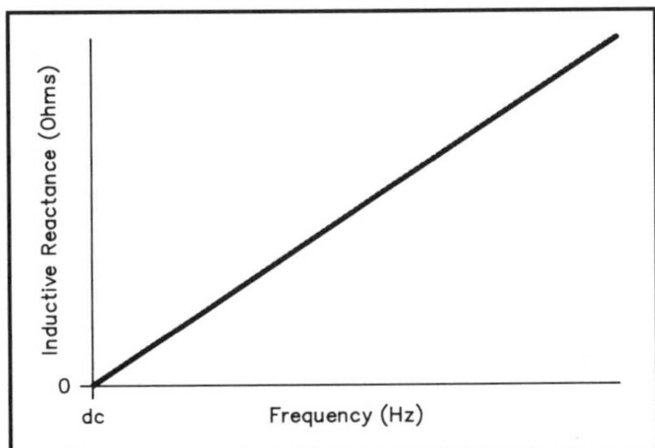

Figure 3—This graph shows how inductive reactance varies with signal frequency. At dc, the reactance is zero. At high frequencies the reactance is very large.

We found two factors that affect reactance in this section. When either inductance or signal frequency increases, the reactance increases. We can write an equation to calculate inductive reactance, based on these two factors.

We normally express frequency as a number of waveform cycles per second, measured in hertz. We also can relate each cycle to the rotation of a wheel. Then we express frequency in radians per second. You can convert any frequency in hertz to radians per second. Just multiply frequency in hertz by 2 π. This is the number of radians in one cycle. Equation 2 shows how to calculate inductive reactance.

$$X_L = 2 \pi f L \qquad \text{(Equation 2)}$$

where:

f is the frequency in hertz
L is the inductance in henrys

If you know a frequency in kilohertz or megahertz, you must write it in hertz for this equation. If you know the inductance in millihenrys or microhenrys, you must convert those values to henrys.

Calculate the reactance of a 10-mH inductor with a 100-Hz signal applied. This is the example from Figure 1.

$$X_L = 2 \pi f L \qquad \text{(Equation 2)}$$
$$X_L = 2 \times 3.14 \times 100 \text{ Hz} \times 10 \times 10^{-3} \text{ H}$$
$$X_L = 6.28 \times 1000 \times 10^{-3} = 6.28 \times 1.000 \text{ ohms}$$
$$X_L = 6.28 \text{ ohms}$$

You shouldn't be surprised to discover this is the same answer we found using the voltage and current in Ohm's Law earlier. Let's try another problem. Find the reactance of a 200-μH inductor with a 1000-Hz supply. (This is not the same as the problem earlier in this section.)

$$X_L = 2 \pi f L \qquad \text{(Equation 2)}$$
$$X_L = 2 \times 3.14 \times 1000 \text{ Hz} \times 200 \times 10^{-6} \text{ H}$$
$$X_L = 6.28 \times 2.00 \times 10^{-1} \text{ ohms} = 1.256 \text{ ohms}$$

You can practice a few more inductive reactance problems. Try calculating the reactance for the other circuits we studied at the beginning of this section. Your answers should agree with the ones we found there. You might even want to calculate the reactance of a few inductance values for frequencies of 1 kilohertz and 1 megahertz. With practice (and your trusty scientific calculator) you'll soon be calculating inductive reactance values with ease.

CHAPTER 21 Circuit Q—The Quality Factor of Components

Circuit Reactance and Resistance Determine Q

Ideal components show only one type of circuit performance. It doesn't matter if the circuit has a steady direct voltage or an alternating signal source. Wires connect parts of a circuit, but have no resistance. When you connect a resistor into a circuit it never behaves like a capacitor or an inductor. Ideal capacitors have capacitance but no resistance or inductance. Ideal inductors only show inductance. They have no resistance or capacitance.

Of course, real components don't meet these ideals. There is some resistance in the wires that connect circuit components. We often ignore this added resistance, however.

For example, Figure 1A shows a circuit that has three 1000-ohm resistors connected in series. Suppose the circuit wiring has a resistance of 0.1 ohm. Figure 1B shows the same circuit with an additional 0.1-ohm resistor added in series with the others. We labeled the new resistor R_W because it represents the wire resistance. How much will this extra resistance change the circuit current you calculate with Ohm's Law? Not much! (As a practice exercise, you should add the resistances of Figure 1A and use Ohm's Law to calculate current. Then add the resistances of Figure 1B and calculate that current.)

Since capacitors include wire leads to connect them into a circuit, they have some resistance. The electrical energy stored in a capacitor may heat the dielectric, or insulating material between the capacitor plates. The electronics circuit can't use the electrical energy that gets converted to heat energy. The energy is lost to the elec-

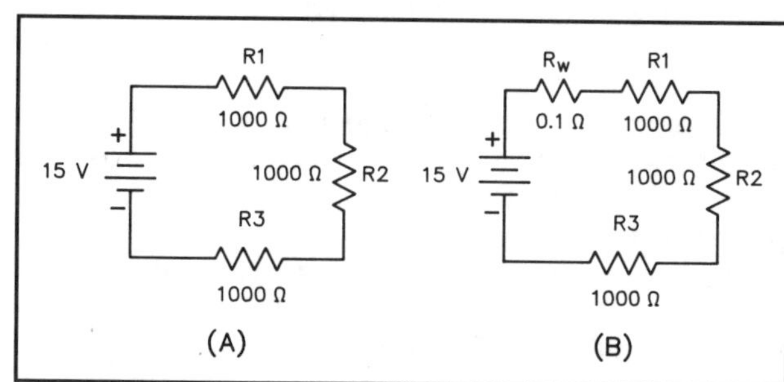

Figure 1—Part A shows three 1000-ohm resistors connected in series to a 15-V battery. Part B shows the same circuit, but with an additional resistor. R_W represents the resistance of the wires connecting the parts of this circuit. As an exercise, you should use Ohm's Law to calculate the current for both circuits.

tronics circuit, then. This is another way real capacitors vary from ideal components.

We refer to the lead resistance and dielectric loss as a resistance. You can't measure this resistance by connecting an ohmmeter across the capacitor, however. The lead resistance is too small to measure with most common ohmmeters. The insulating material will cause the ohmmeter to measure an infinite (too large to measure) resistance. The ohmmeter indicates an open circuit when you connect it between the two capacitor leads.

The capacitor's wire leads have some inductance. Inductive reactance is small at low frequencies and increases as the signal frequency increases. This is why inductance becomes more important when we connect higher-frequency signals to the capacitor.

Figure 2 shows a circuit with a capacitor and an ac signal. We included a resistor and an inductor in the circuit. We drew these with dashed lines to show they represent the resistance and inductance of the real capacitor leads.

Most inductors consist of a length of wire wound into a coil. That wire has some resistance, so real inductors have resistance. Each turn of an inductor's coil is close to the turns on either side. When an alternating signal goes through the inductor, these adjacent turns act like small capacitors. Figure 3 shows a circuit with an ac signal applied to an inductor. We drew a resistor and a capacitor in the circuit with dashed lines to represent the resistance and capacitance of a real inductor.

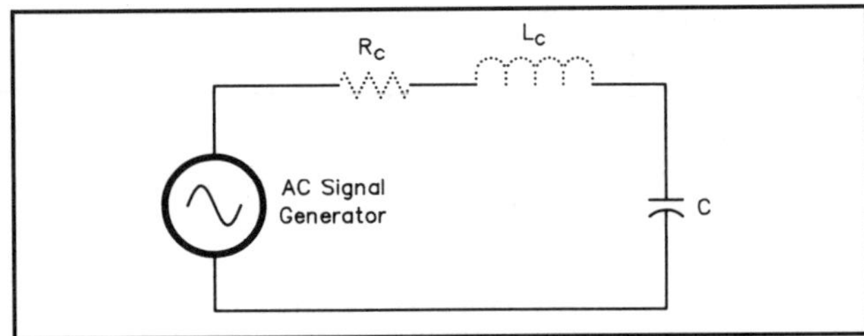

Figure 2—This diagram includes the small resistance and inductance that are part of a real capacitor. We drew the resistor and inductor with dashed lines to indicate they are part of the capacitor. They are not physical components in the circuit.

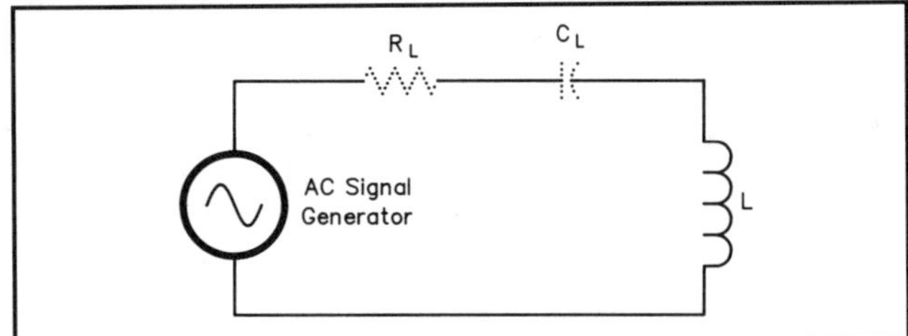

Figure 3—Real inductors include a small amount of capacitance and resistance. The diagram shows a resistor and a capacitor drawn with dashed lines. This indicates they are part of the inductor. They are not physical components in the circuit.

We frequently ignore the small inductance of a capacitor and the small capacitance of an inductor. You must consider these limitations of real components for some circuits, however. They become important in radio-frequency circuits that operate in the very-high frequency (VHF) range and higher frequencies. We will ignore these effects for the rest of this book, however.

We often must consider the wire resistance of these components. We usually want the resistance to be much smaller than the reactance of a capacitor or inductor. In general, better quality capacitors and inductors have a smaller resistance.

We can assign a number to a coil or capacitor to indicate that component's relative merits. This number represents a *quality factor*. We call the number *Q*. An entire circuit also can have a Q value. A circuit Q value gives an idea of how close to the ideal that circuit performs.

We define *Q* as the ratio of reactance to resistance. This definition applies to both inductors and capacitors.

The Q of a real capacitor, C, is equal to the capacitive reactance divided by the resistance. Equation 1 states this definition mathematically.

$$Q = \frac{X_C}{R} \qquad \text{(Equation 1)}$$

The Q of a real inductor, L, is equal to the inductive reactance divided by the resistance. Equation 2 is a mathematical statement of this definition.

$$Q = \frac{X_L}{R} \qquad \text{(Equation 2)}$$

The first thing you should notice about these equations is that smaller resistance values give larger Q values. Larger Q values indicate better-quality components.

You should remember that reactance varies with frequency. Capacitive reactance is largest at low frequency and decreases as the frequency increases. Inductive reactance is smallest at low frequency and increases as the frequency increases. How will this affect Q? Let's consider some examples.

Figure 4 shows a 25-µF capacitor that has 0.01 ohm of lead resistance. Suppose we connect a 40-Hz signal to this capacitor. What is the capacitor's Q? First calculate the capacitive reactance.

$$X_C = \frac{1}{2 \pi f C} \qquad \text{(Equation 3)}$$

$$X_C = \frac{1}{6.28 \times 40 \text{ Hz} \times 25 \times 10^{-6} \text{ F}}$$

$$X_C = \frac{1}{6.28 \times 10^{-3}} = 159\ \Omega$$

Now you can use Equation 1 to calculate Q

$$Q = \frac{X_C}{R} = \frac{159\ \Omega}{0.01\ \Omega} = 15900$$

What if we connect this capacitor to a 400-Hz signal? Use Equation 3 to calculate the new reactance.

$$X = \frac{1}{2 \pi f C} = \frac{1}{6.28 \times 400 \text{ Hz} \times 25 \times 10^{-6} \text{ F}}$$

$$X_C = \frac{1}{0.628} = 15.9\ \Omega$$

Use this reactance to calculate the new Q value.

$$Q = \frac{X_C}{R} = \frac{15.9\ \Omega}{0.01\ \Omega} = 1590$$

The reactance at the higher frequency is 10 times smaller, so the Q at that frequency is also 10 times smaller. When you apply a higher-frequency signal to a capacitor the reactance and Q both become smaller.

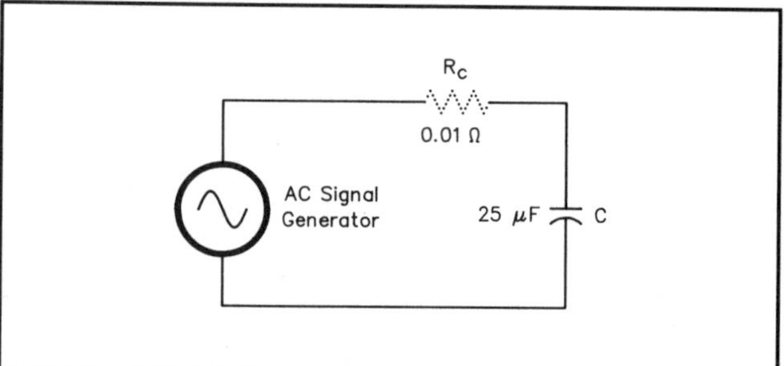

Figure 4—This 25-microfarad capacitor has 0.01 ohms of resistance. The ac signal generator applies a 40-hertz signal to the capacitor. Later the signal generator applies a 400-hertz signal. The text shows how to calculate the reactance and Q for these conditions.

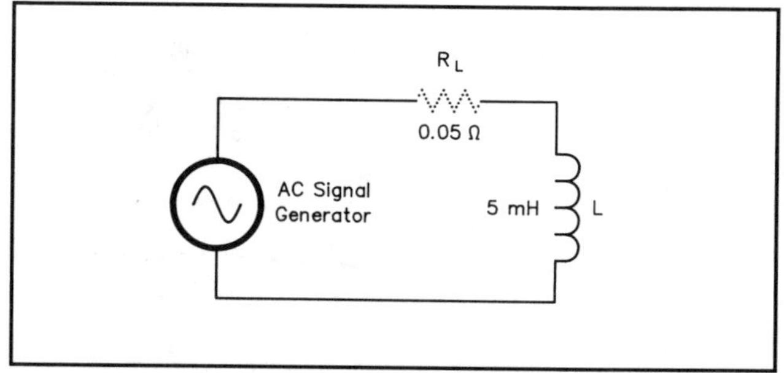

Figure 5—This 5-millihenry inductor has 0.05 ohms of resistance. The ac signal generator first applies a 40-hertz signal, and then a 400-Hz signal. The text explains how to calculate the reactance and Q for these applied signals.

The 5-millihenry inductor shown in Figure 5 has 0.05 ohm of resistance. What is the Q of this inductor when we apply a 40-hertz signal? First calculate the inductive reactance.

$$X_L = 2 \pi f L \qquad \text{(Equation 4)}$$

$$X_L = 2 \times 3.14 \times 40 \text{ Hz} \times 5 \times 10^{-3} \text{ henrys}$$

$$X_L = 6.28 \times 200 \times 10^{-3} = 1.256 \text{ ohms}$$

Next, we will use Equation 2 to calculate the inductor Q.

$$Q = \frac{X_L}{R} \qquad \text{(Equation 2)}$$

$$Q = \frac{1.256 \text{ ohms}}{0.05 \text{ ohm}} = 25.12$$

What is this inductor's Q with a 400-Hz signal applied? Verify that the inductive reactance at this frequency is 12.56 ohms. Then use Equation 2 to calculate Q again.

$$Q = \frac{X_L}{R} = \frac{12.56 \text{ ohms}}{0.05 \text{ ohm}} = 251.2$$

What can you conclude about the Q of an inductor? That's right, when you increase the frequency, the reactance and Q increase.

High Q Means Low Series Component Resistance

An ideal inductor or capacitor doesn't add extra resistance to a circuit. Real components always add some extra resistance, though. Good-quality real components will only add a small amount of resistance. Why not always use the highest-quality components available? Then we wouldn't have to worry about the inductor or capacitor quality factor, or *Q*.

There are many trade-offs to consider when selecting components. What is the circuit function? What purpose does the component serve in the circuit? What will be the frequency of signals applied to the component? Answers to these questions (and others) will help you select the proper components for the job. Other considerations are component cost, size and construction methods.

Your circuit may not require the best components available. Suppose you have a power supply that converts 120 V ac to 12 V dc. The output from this supply will have a 60 or 120-hertz ripple unless you add a filter capacitor. (Figure 1 shows how this capacitor smooths the ripples.) You needn't worry about the high-frequency characteristics of the capacitor in this case. You shouldn't be too concerned about a small energy loss in the capacitor either.

A capacitor that you plan to use in the radio-frequency portion of a VHF radio receiver must meet other requirements. This capacitor must have good high-frequency characteristics. Any signal loss in this circuit should be minimized.

Aluminum electrolytic capacitors are popular for power-supply filters. They pack a large amount of capacitance into a small package. While all capacitors have relatively high Q, aluminum electrolytic capacitors have a lower Q than some other types. This is because they have higher losses, especially at high frequencies.

Mica and ceramic capacitors are more popular for higher-frequency applications. It is not practical to make ceramic or mica capacitors with enough capacitance for use in power-supply filters. Such capacitors would be physically too large. These dielectric materials have much lower loss than electrolytics, however. Mica and ceramic dielectrics form high-Q capacitors.

How can we reduce an inductor's wire resistance? Larger-diameter wire has less resistance than small-diameter wire. Winding the coil on a larger-diameter form and putting more space between the turns also reduces losses. Sometimes you don't have room in the circuit for a very large coil, however. If the application doesn't require

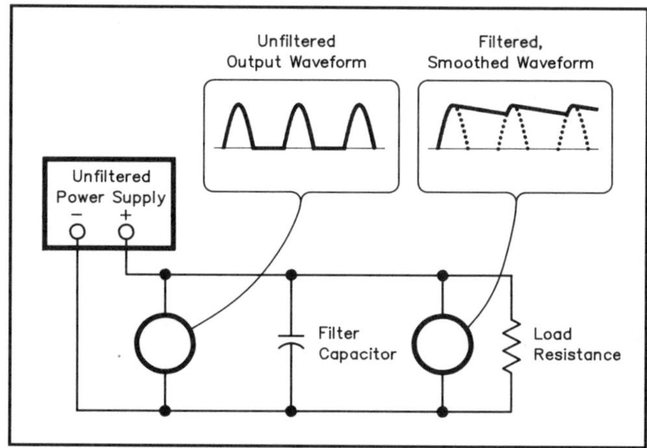

Figure 1—A power supply converts the ac mains voltage to a dc voltage for use with an electronics circuit. The output signal has *ripples* or variations left over from the ac voltage. A filter capacitor smooths these ripples to provide a steady dc output.

a high-Q coil, you can use smaller wire and a small-diameter coil form.

You may put an inductor in series with the power-supply lead to an RF amplifier. This inductor carries a small direct current from the power supply. At dc, the inductor has a very small reactance. It also has the wire resistance. The inductor has a very large reactance at the amplifier's radio frequency, though. It blocks any RF current from flowing back to the power supply. Sometimes we call such a coil a *choke*, because it chokes off high-frequency signals while allowing lower-frequency signals to pass through. Figure 2 shows a circuit with such a coil.

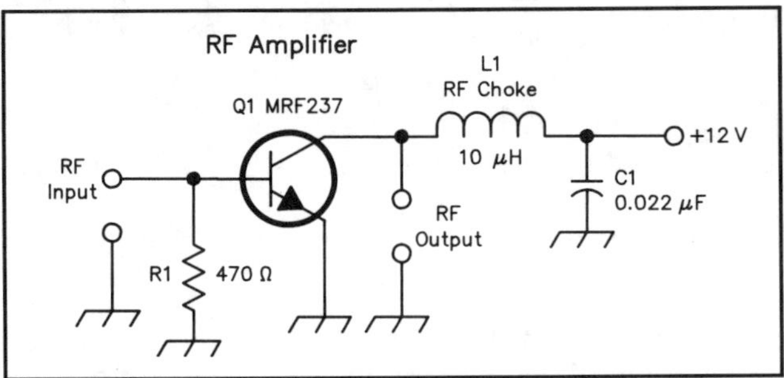

Figure 2—An *RF choke* blocks the radio-frequency signal from following the power-supply leads back into the supply. The coil's high reactance at the radio's operating frequency "chokes off" the RF. At dc, however, the inductor has only the small wire resistance and no reactance. The direct current from the power supply travels through the inductor to supply the radio with the current it needs.

You don't need a high-Q coil to serve as a choke. You just use a power supply that can supply the current needed by the amplifier and any losses in the coil.

You do want a high-Q inductor for the output network of a high-power RF amplifier. You don't want to lose any of your hard-earned RF power in the coil. The output inductor must be able to carry several amperes of RF current for a 1500-watt amplifier. If the inductor wires have too much resistance, the losses will be large.

Some high-power amplifiers use copper tubing as the conductor for this coil. The tubing's large diameter minimizes any conductor resistance, producing a high-Q inductor.

Any high-Q device or circuit has low resistance. High Q also means the component has low losses of other kinds. High-Q capacitors use low-loss dielectric materials.

A high-Q component will not lose much of the energy stored in it. Q relates the energy stored in a capacitor or inductor to the energy lost in the component. A high-Q component stores much more energy during each waveform cycle than it loses.

Adding Resistance Reduces Circuit Q

Most of our discussion about Q focused on low-loss inductors and capacitors. When you add these components in a circuit, there is nearly always some additional resistance. This resistance may be in series with the inductor or capacitor, or it may be in parallel with the components.

How do you suppose this additional resistance affects the circuit Q? From this module's title, you should guess that additional resistance decreases the Q. Why does this happen, though?

Look at Figure 1. The inductor shown here has a small amount of resistance, represented by R_L. In addition, there is another resistor in series with the inductor. You must add the two resistance values to calculate the circuit Q. We included the inductive reactance on the diagram, so you don't have to calculate that value. Use Equation 1 to calculate the circuit Q.

$$Q = \frac{X}{R} \qquad \text{(Equation 1)}$$

$$Q = \frac{4500 \text{ ohms}}{201 \text{ ohms}} = 22.4$$

Let's calculate the Q of the inductor alone, without the additional circuit resistance. We can use Equation 1 to calculate the inductor Q, also.

$$Q = \frac{X}{R} \qquad \text{(Equation 1)}$$

$$Q = \frac{4500 \text{ ohms}}{1 \text{ ohm}} = 4500$$

Adding even a small amount of resistance in series with the inductor reduces the Q significantly. Any current through a resistor converts some electrical energy to heat energy. This heat represents electrical energy that is lost to the circuit. So the additional resistance increases the energy loss, and that is what reduces the Q.

Figure 2 shows the Figure 1 inductor, with its 1 ohm of wire resistance. This time we placed a 2-megohm resistor in parallel with the inductor (and its wire resistance). Before we jump into the mathematics of calculating this circuit's Q, let's conduct a brief thought experiment.

The 2-MΩ parallel resistor provides an alternate path for the signal current. The resistor will convert some of the signal to heat, representing energy lost to the circuit. As with the series circuit, we should expect this additional resistance to reduce the circuit Q.

What do you think will happen if we replace the 2-MΩ resistor with a smaller-value unit? As we decrease the parallel resistance, it becomes easier for more signal current to flow through the resistor. As more signal flows through the resistor there is more energy lost.

We can conclude, then, that adding resistance in parallel with an inductor (or capacitor) increases losses and decreases Q. This is the same result we discovered for the series circuit. You should notice, however, that *smaller* resistance values added in parallel cause this effect.

This brings us to the mathematical part. The equation we used so far to calculate Q gave smaller values as the resistance increased. We need an equation that gives the opposite effect for this parallel circuit, however. Equation 2 gives the result we want.

$$Q = \frac{R}{X} \qquad \text{(Equation 2)}$$

What is the Q of the Figure 2 circuit? Since most of the circuit loss is from the parallel resistor, we can simplify the problem by ignoring the inductor loss. (We can use this simplification as long as the resistor losses are at least ten times greater than the inductor losses. That is the case in this example.) Divide the parallel resistance by the inductor reactance, as Equation 2 tells us.

$$Q = \frac{R}{X} \qquad \text{(Equation 2)}$$

$$Q = \frac{2 \times 10^6 \text{ ohms}}{4500 \text{ ohms}}$$

$$Q = 444$$

What would be the circuit Q if we placed the 200-ohm resistor from Figure 1 in parallel with the inductor? Use Equation 2 to calculate the Q.

$$Q = \frac{R}{X} \qquad \text{(Equation 2)}$$

$$Q = \frac{200 \text{ ohms}}{4500 \text{ ohms}} = 0.0444$$

Wow! That really decreases the circuit Q.

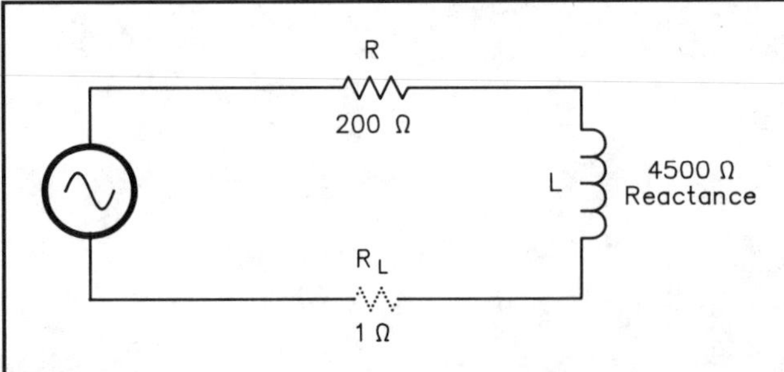

Figure 1—The inductor in this circuit has 1 ohm of resistance in the wire used to make it. R_L represents this resistance on the diagram. There is a 200-ohm resistor in series with the inductor. The text explains how to calculate the circuit Q with this additional resistor.

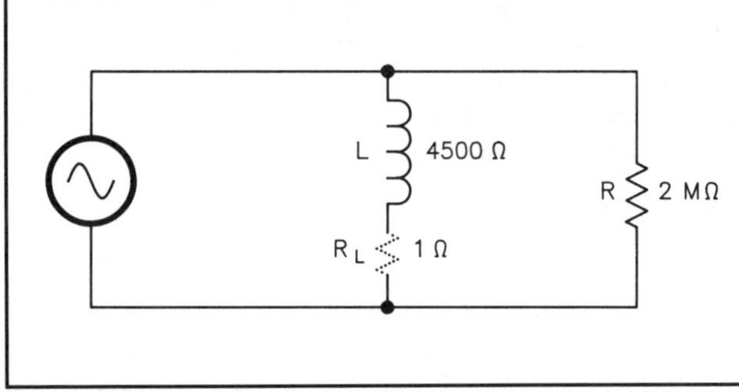

Figure 2—In this circuit we placed a 2-megohm resistor in parallel with the inductor from Figure 1. The text explains how to calculate the circuit Q with a resistor in parallel with the inductor.

CHAPTER 22 Transformers

The Magnetic Field from One Coil Can Cause a Current in Another Coil

An electric current through an inductor produces a magnetic field around the inductor. There is no magnetic field without a current. When the current begins to flow, the magnetic field "grows" around the inductor. As the current increases, the magnetic-field strength increases. While there is a steady direct current through the inductor, the magnetic field remains constant. If the current decreases, so does the magnetic field. When the current stops, the magnetic field collapses to zero.

Figure 1 shows a circuit that applies an alternating-current, sine-wave signal to an inductor. This current changes constantly. When the current changes, the magnetic-field strength also changes.

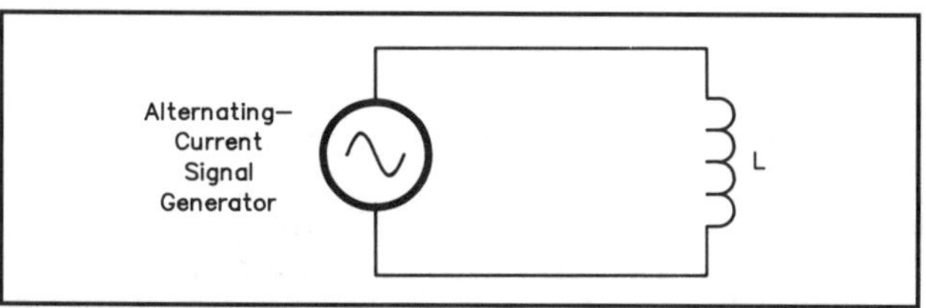

Figure 1—An alternating-current signal source creates a changing magnetic field through an inductor.

When the ac signal increases from zero, the magnetic-field strength increases. The magnetic field decreases when the signal decreases to zero again. What happens when the ac signal's current direction changes? The magnetic field polarity also changes. Figure 2 shows the magnetic field around a coil. Notice that when the current direction reverses, the North and South magnetic poles also reverse. You can visualize the magnetic-field strength as having a sine-wave pattern just like the applied signal current.

A changing electric current produces a changing magnetic field. Can a changing magnetic field also produce a changing electric current? Yes, it can! This is the principle of an electric generator.

How can we produce a changing magnetic field? One way is to move the magnet. Figure 3 shows a horseshoe magnet moving into and out of a coil of wire. The wire connects to a very sensitive ammeter. A current moving one direction through the coil moves the meter needle to the right. When the current moves in the opposite direction the needle moves to the left.

Any movement between a magnetic field and the turns of an inductor induces a voltage across the inductor.

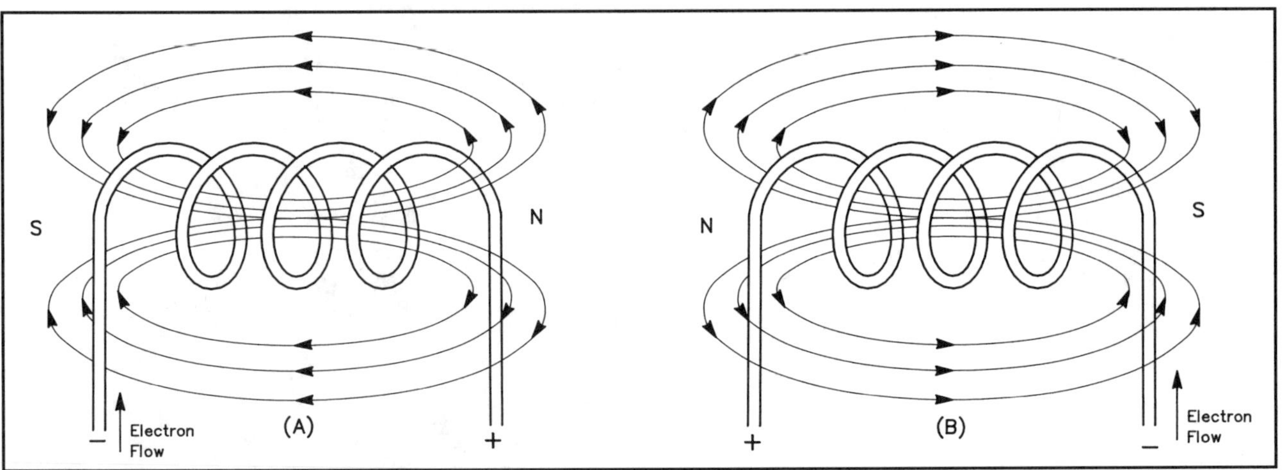

Figure 2—When the current direction changes, the polarity of the magnetic field through an inductor also changes. In Part A, the electrons flow into the wire on the left. The North magnetic pole is to the right of the coil. In Part B the electrons flow into the wire on the right. This time the North magnetic pole is to the coil's left end.

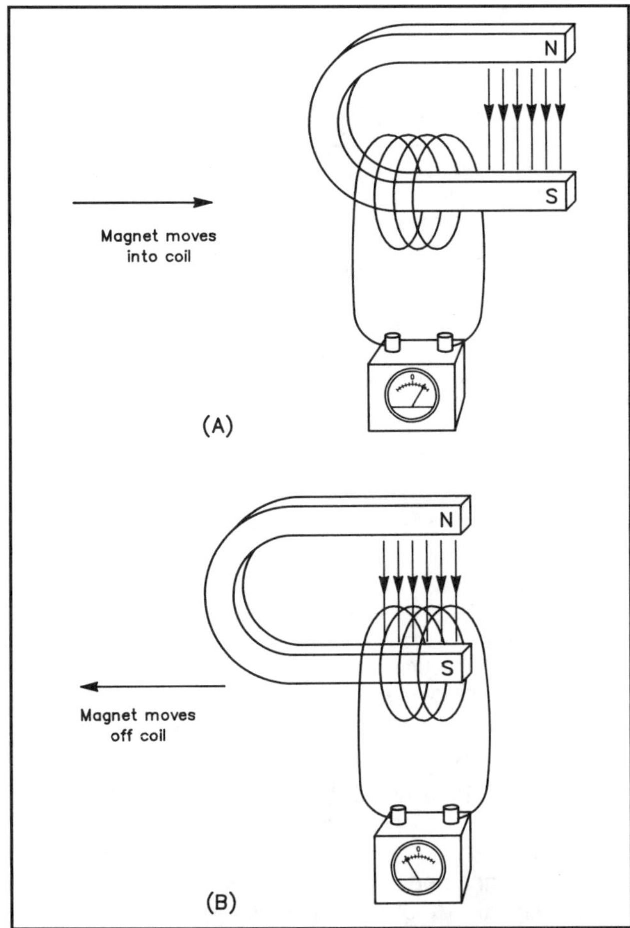

Figure 3—Moving a wire coil through a magnetic field induces a voltage in the coil. This voltage produces a current through the wire. The voltage polarity and current direction change when the direction of coil movement changes.

The coil can move through the field from a permanent magnet or a permanent magnet can move through the coil. An alternating-current signal through an inductor also produces a magnetic field that moves. The magnetic field grows, collapses and reverses direction with the current.

The changing magnetic field from one coil can induce a voltage across another coil. Figure 4 shows two inductors that are physically close to each other. An ac signal through the left coil creates a changing magnetic field that moves through the coil on the right.

When the magnetic field from one coil cuts through another coil, we say the coils have *mutual inductance*. We also sometimes call inductors with mutual inductance *linked* or *coupled* inductors.

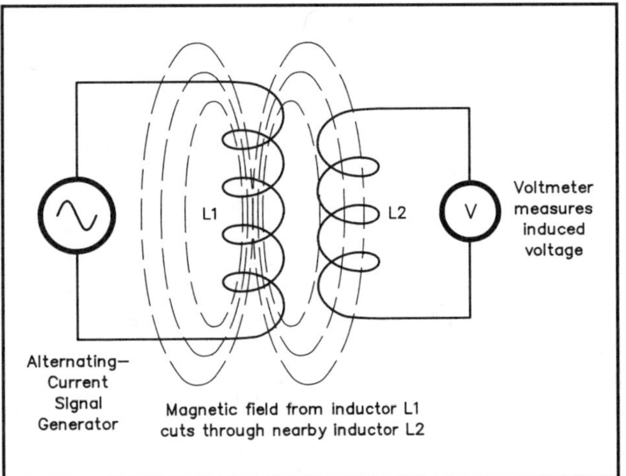

Figure 4—The changing magnetic field from inductor L1 cuts through the turns of inductor L2. The magnetic field from L1 induces a voltage across L2, producing a current through it.

Transformers Can Increase or Decrease Voltage Levels

When the changing magnetic field from one inductor cuts through the turns of another inductor, the coils have mutual inductance. The magnetic field induces a voltage across the second inductor.

Some circuits require that you minimize any mutual inductance. Induced voltages can disrupt circuit operation.

Other circuits depend on mutual inductance for proper operation. This section concentrates on the desired effects of mutual inductance.

Figure 1 shows two coils arranged so a changing current through one induces a voltage in the other. We call this combination a *transformer*. Transformers can have more than two coils.

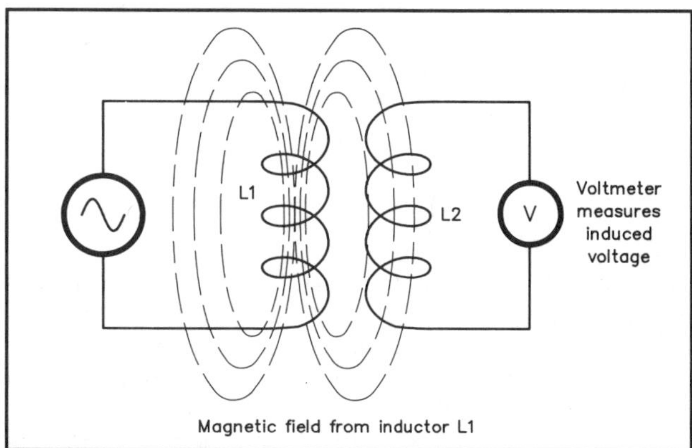

Figure 1—A current through inductor L1 produces a magnetic field that cuts through the turns of inductor L2. The changing magnetic field across L2 produces a voltage. This combination of inductors forms a transformer.

One inductor has the input signal applied across it. This is the transformer *primary* winding. The transformer primary winding induces a voltage across one or more *secondary* windings. These secondary windings provide an output voltage to other parts of the circuit.

A transformer transfers energy from one part of a circuit to another part. There is no direct connection between the two parts of the circuit, however.

We usually want to maximize the mutual inductance between the transformer primary and secondary inductors. This ensures good energy transfer between the windings. The two inductors must be physically close to each other. Most of the primary winding's magnetic field must cut through the secondary coil. We usually wind the transformer inductors on a single form, or core. The core is often made from iron, powdered iron or ferrite materials.

Figure 2 shows a transformer primary winding connected to a battery. Batteries provide a source of direct current. A resistor connects across the secondary winding. The solid lines between the two coils indicate that the transformer has an iron core.

Can you describe the voltage across the secondary? How much current do you think will flow through the resistor? Remember that only a changing magnetic field will induce a voltage across the secondary. Did you answer there is no voltage across the secondary and no current through the resistor? Good.

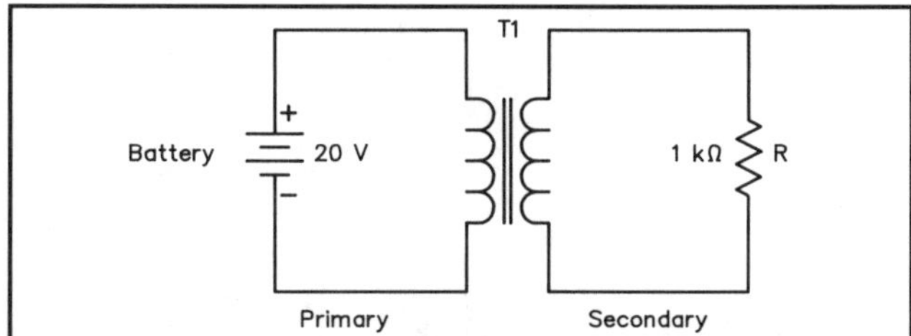

Figure 2—The primary side of this transformer connects to a 20-volt battery. There is a 1-kΩ resistor across the secondary winding. The solid lines between the inductors indicate the windings of this transformer are on an iron core. There is no voltage across the secondary winding, and no current through the resistor. Transformers require a changing primary voltage to produce a secondary voltage.

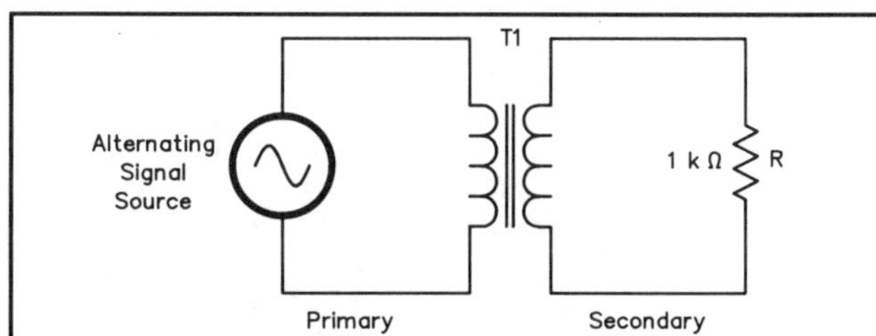

Figure 3—When we connect the Figure 2 transformer to an alternating signal source there is a voltage across the secondary. This induced voltage produces a current through the resistor.

We connected the transformer to an alternating-current signal in Figure 3. This time there is a voltage induced across the secondary, so a current will flow through the resistor.

The small "power cubes" that power many electronics devices contain transformers. These cubes plug directly into a wall outlet and have a wire that plugs into a radio, telephone, tape player or other device. The 120-V house current goes to the cube's primary winding. The secondary output voltage is usually some lower voltage, such as 9 or 12 volts.

We can make a transformer that will produce any desired output voltage. Some transformers produce an output voltage that is lower than the input voltage. Other transformers produce a higher output voltage than their input voltage.

Figure 4—Small "cube transformers" power many electronics devices, such as this portable short-wave receiver.

Some transformers have more than one secondary winding. These transformers produce several output voltages from a single input voltage. You may even find transformers that produce one voltage higher than the input and another that is lower than the input.

Transformers come in many shapes and sizes. Larger transformers generally produce higher voltages or use heavier wire to carry more current. AC power transformers often have a core made from a stack of thin sheet-iron layers. A metal "can" or covering protects the wires. Transformers designed for use with radio-frequency signals often have powdered-iron or ferrite cores. Toroidal, or donut-shaped cores are popular for RF transformers.

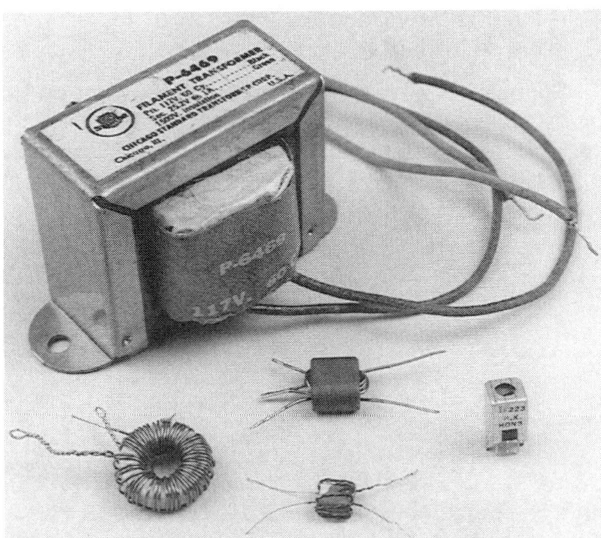

Figure 5—This photo shows a variety of transformers.

Turns Ratio Determines Voltage Transformation

The magnetic field strength around an inductor depends on several factors. The field is stronger when the voltage applied to the inductor is larger, and there is more current through the inductor. The field is also stronger when there are more turns of wire in the coil. Winding an inductor around a core of iron, powdered iron or ferrite material also makes the magnetic field stronger.

Figure 1 shows an ideal transformer with two identical windings. There are the same number of turns on the primary as there are on the secondary. An ideal transformer transfers to the secondary all the energy put into its primary winding. It does not lose any of the energy in wire resistance or losses in the transformer core.

In Figure 1 we connected the transformer to the 120-V house current. This primary voltage produces 120 volts on the secondary winding.

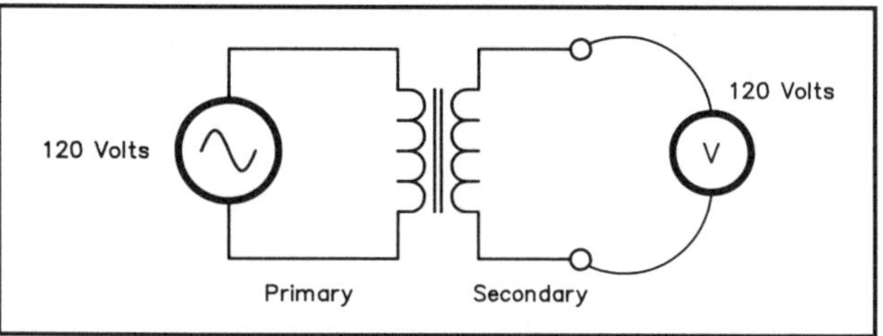

Figure 1—This transformer has identical primary and secondary windings. With 120 V ac applied to the primary, 120 V ac appears across the secondary.

The voltage induced by a changing magnetic field is proportional to the number of turns on the inductor. The magnetic field induces a larger voltage across an inductor that has more turns. An inductor with fewer turns has a smaller induced voltage.

Figure 2 has a transformer with two secondary windings. The primary winding has 1000 turns. Secondary 1 has 500 turns and secondary 2 has 250 turns. Let's connect this transformer's primary winding to 120 volts and measure the voltage across each secondary. Did you notice that secondary 1 has only half as many turns as the primary? You should expect to measure 60 volts across this winding. How many volts do you think we will measure across secondary 2? It has half as many turns as secondary 1 and one fourth as many as the primary. Therefore, we will measure 30 volts across secondary 2.

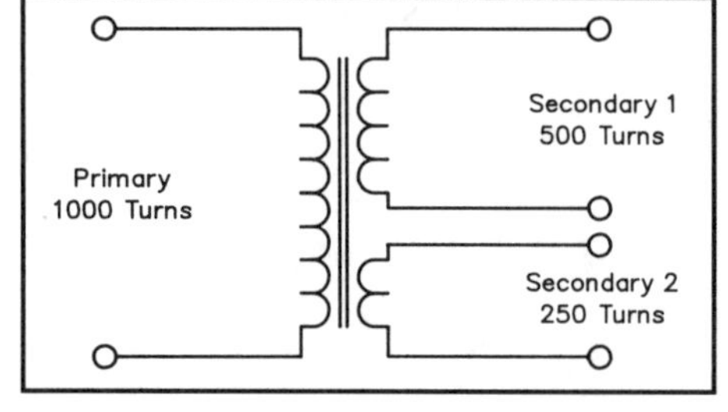

Figure 2—Here is a transformer with two secondary windings. Secondary 1 has twice as many turns as secondary 2. The voltage induced across secondary 1 will be twice as large as the voltage across secondary 2.

The previous example demonstrates how easy it is to calculate a transformer's secondary output voltage. We only need to know the *primary-to-secondary turns ratio.* The number of turns on the primary divided by the number of turns on the secondary gives us the primary-to-secondary turns ratio. The turns ratio is the same as the input voltage divided by the output voltage. We can write this statement as an equation, and use the equation to calculate the output voltage when we know the input voltage.

$$\frac{N_p}{N_s} = \frac{E_p}{E_s} \qquad \text{(Equation 1)}$$

where:

N_p is the number of turns on the primary winding
N_s is the number of turns on the secondary winding
E_p is the primary voltage
E_s is the secondary voltage

Suppose we have a transformer with 500 turns on the primary winding and 400 turns on the secondary winding. What will the secondary voltage be when a 120-V ac supply connects to the primary? Substitute these numbers into Equation 1, and solve for E_s.

$$\frac{N_p}{N_s} = \frac{E_p}{E_s} \qquad \text{(Equation 1)}$$

$$\frac{500}{400} = \frac{120 \text{ V}}{E_s}$$

$$1.25 = \frac{120 \text{ V}}{E_s}$$

(The easiest way to solve this equation for E_s is to multiply both sides of the equation by E_s. That cancels E_s on the right side, and puts the E_s on the left side. Then divide both sides by 1.25, and you have the answer.)

$$E_s \times 1.25 = \frac{120\ V \times E_s}{E_s}$$

$$\frac{E_s \times 1.25}{1.25} = \frac{120\ V}{1.25}$$

$$E_s = \frac{120\ V}{1.25} = 96\ V$$

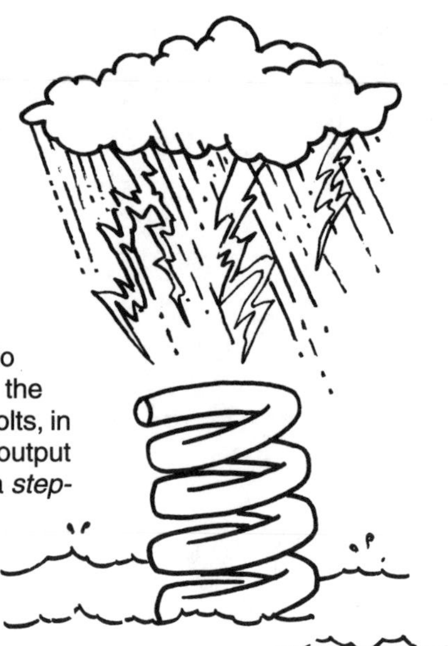

The secondary voltage from this transformer is 96 volts when the primary connects to 120 volts. Suppose 12 volts ac connects to the primary. The secondary output would be 9.6 volts, in that case. Can you calculate this value? The output voltage is less than the input voltage. This is a *step-down transformer* because it reduces the input voltage, or steps it down.

Let's try another example. Suppose we find a transformer that has 500 turns on the primary winding and 4000 turns on the secondary winding. What is the turns ratio of this transformer? What is the secondary voltage if the primary connects to 120 V ac? We can calculate the turns ratio from the left side of Equation 1.

$$\frac{N_p}{N_s} = \frac{500}{4000} = 0.125$$

The turns ratio is less than 1, which tells us this is a *step-up transformer*. A step-up transformer increases, or steps up, the input voltage to a higher level. Equation 1 will help us calculate the secondary voltage.

$$\frac{N_p}{N_s} = \frac{E_p}{E_s} \qquad \text{(Equation 1)}$$

$$0.125 = \frac{120\ V}{E_s}$$

(Solve this equation for E_s, similar to the way we solved the equation in the earlier example.)

$$E_s = \frac{120\ V}{0.125} = 960\ V$$

Let's try one more example. Suppose you find a bargain transformer at your local Electronics Parts Palace. The store manager says the box and papers that came with the transformer are missing. You aren't sure what the secondary voltage is, but you can't pass up a bargain, so you bring it home.

When you examine the transformer, you find a small P near one pair of wires and a small S near the other pair. You decide these markings indicate the intended primary and secondary windings. Figure 3 shows how you connect this transformer to the ac mains. Your test leads include 1 amp fuses in both leads, and you carefully insulate all the connections before you plug in the cord. When you measure the output voltage, you find 25 volts across the secondary winding. What is the turns ratio of your transformer?

Again, we use Equation 1. We don't know how many turns there are on either the primary or the secondary, so we can only calculate the ratio of primary to secondary turns.

$$\frac{N_p}{N_s} = \frac{E_p}{E_s} \qquad \text{(Equation 1)}$$

$$\frac{N_p}{N_s} = \frac{120\ V}{25\ V} = 4.8$$

There are 4.8 times as many turns on the primary as there are on the secondary. Congratulations! You may be able to use your bargain transformer as part of a low-voltage power supply project.

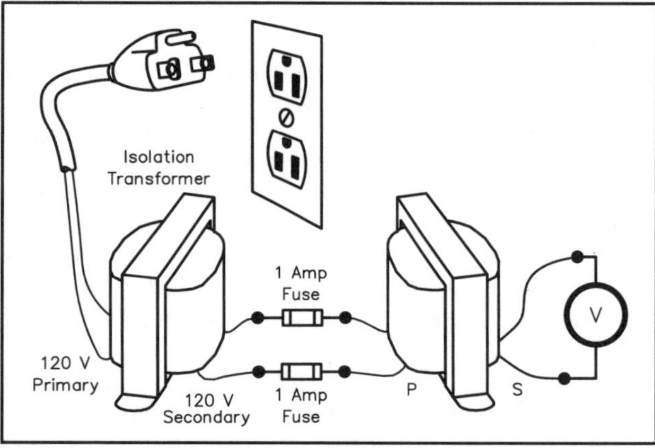

Figure 3—This test setup shows how you might test an unknown transformer to measure the output voltage. The fuses provide some protection against a short-circuit in the transformer. Be sure you insulate all connections to prevent accidental contact with the 120-V house current. You should not attempt a measurement like this until you are very familiar with proper test procedures and safety practices.

Transformers Change Current Levels, Too

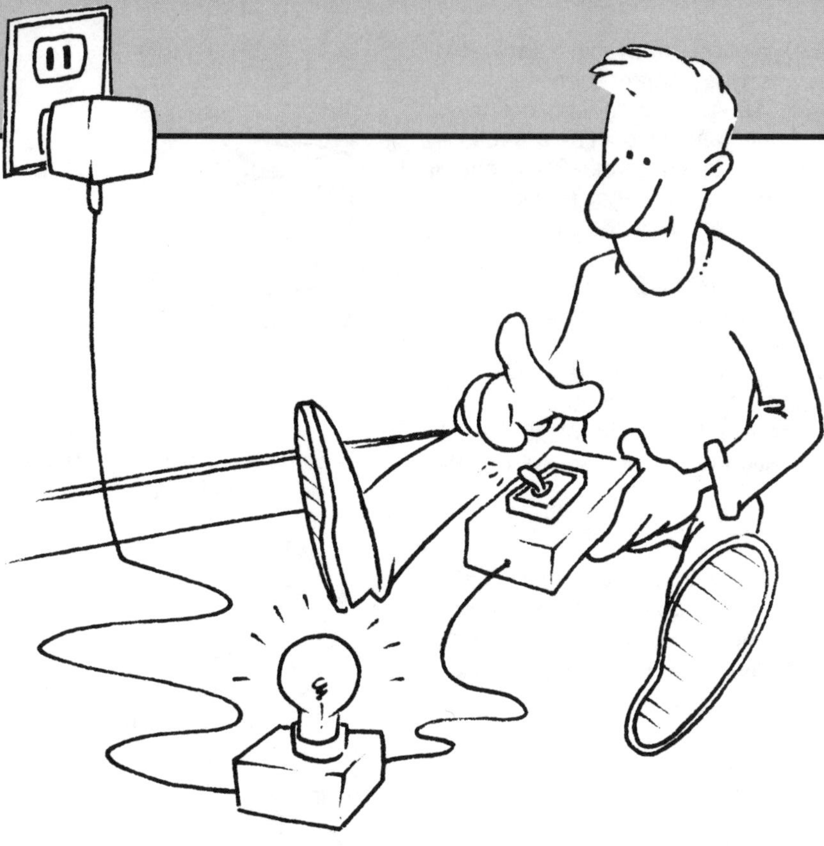

You can change from just about any ac voltage to any other voltage using a transformer. Transformers transfer electrical energy from the primary winding to the secondary winding. If the secondary winding has more turns than the primary winding, you have a *step-up transformer.* The secondary voltage is greater than the primary voltage. If the secondary winding has fewer turns than the primary, you have a *step-down transformer.* In this case the secondary voltage is less than the primary voltage.

The law of conservation of energy is a fundamental natural law that all transformers must obey. According to this law, the amount of energy in a system must remain constant. We can neither create nor destroy energy. We can change energy from one form to another.

This law tells us we can't get more energy out of the transformer secondary than we put into the primary. A *real* transformer will change some of the input energy into heat in the wires and core material. It takes some energy to overcome the wire resistance, which shows up as heat energy in the wire. It also takes some energy to magnetize the core material, and this appears as heat energy in the transformer core. The electrical energy converted to heat represents a *loss* of electrical energy, but not *lost energy.*

Power measures the rate of using energy. We calculate power by multiplying current times voltage. Let's use power to consider conservation of energy with a transformer.

We can calculate a transformer's input power by measuring the primary voltage and current. Equation 1 is a mathematical expression for transformer input power.

$$P_{in} = I_p \times E_p \qquad \text{(Equation 1)}$$

where:

P_{in} is the input or primary circuit power
I_p is the primary current
E_p is the primary voltage

Equation 2, which is similar to Equation 1, shows us how to calculate the transformer output power. The subscripts on these symbols represent output power and secondary current and voltage.

$$P_{out} = I_s \times E_s \qquad \text{(Equation 2)}$$

We will simplify the analysis by assuming we have a perfect, or *ideal*, transformer. There is no wire resistance, and it doesn't take any energy to magnetize the core material. Therefore, there is no electrical energy lost in the transformer. The law of conservation of energy tells us that the input power must be equal to the output power. Equation 1 must equal Equation 2.

$$P_{in} = P_{out} \qquad \text{(Equation 3)}$$

We also can write this equation using the voltage and current terms.

$$I_p \times E_p = I_s \times E_s \qquad \text{(Equation 4)}$$

Of course, *real* transformers will have some resistance, and it will take some energy to magnetize the core material. They will always convert some of the input energy to heat, then. The output power will always be a little less than the input energy for a real transformer.

Look at Equation 4 again. The products of voltage and current must remain equal. A step-up transformer increases the voltage, so the available current must decrease. A step-down transformer decreases the voltage, which means the secondary current is larger than the primary current. Voltage and current are *inversely related.* An increase of one results in a decrease of the other.

Let's rewrite Equation 4, putting the voltage terms on one side of the equal sign and the current terms on the other side. Cross multiply the terms to rewrite the equation.

$$\frac{E_p}{E_s} = \frac{I_s}{I_p} \qquad \text{(Equation 5)}$$

You may remember that the ratio of primary voltage divided by secondary voltage equals the number of turns on the primary divided by the number of turns on the secondary. Equation 6 shows this mathematically.

$$\frac{E_p}{E_s} = \frac{N_p}{N_s} \qquad \text{(Equation 6)}$$

Now we have a way to compare the transformer turns ratio with the current. Replace the primary-to-secondary voltage ratio in Equation 5 with the expression for the primary-to-secondary turns ratio from Equation 6.

$$\frac{N_p}{N_s} = \frac{I_s}{I_p} \qquad \text{(Equation 7)}$$

Notice the primary and secondary currents have the opposite order from the number of turns in the windings. This is an *inverse* relationship. The winding with more turns must carry a smaller current. This indicates that you should use larger-diameter wire for the winding with fewer turns. You can use smaller-diameter wire for the winding with more turns.

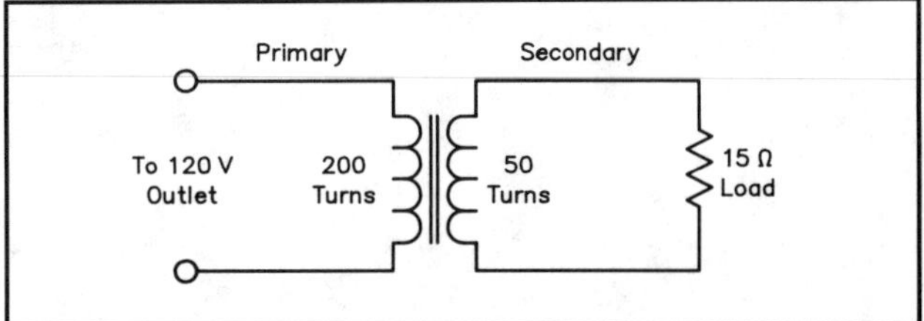

Figure 1—This transformer changes the 120-V house current to 30 volts. As the text explains, the transformer draws 0.5 A from the 120-V outlet. The secondary circuit draws 2 amps.

The wire size determines how much current the transformer can safely handle. Remember that smaller-diameter wire has more resistance. This is a design trade-off you must be aware of when you select a transformer for a particular application. A transformer made with smaller wire will be physically smaller (and less expensive). One made with larger wire can carry more current. So you must decide how much current the secondary will require, and allow a safety factor. Then select a transformer rated to handle the desired current.

Figure 1 shows a step-down transformer connected to a load that draws 2 amperes. This transformer has 200 turns on the primary winding and 50 turns on the secondary winding. How much current will the primary draw from the wall outlet? Use Equation 7 to calculate the current.

$$\frac{N_p}{N_s} = \frac{I_s}{I_p} \qquad \text{(Equation 7)}$$

$$\frac{200}{50} = 4 = \frac{2\text{ A}}{I_p}$$

Cross multiply to solve this equation for the primary current, I_p.

$$I_p = \frac{2\text{ A}}{4} = 0.5\text{ A}$$

Another transformer has a turns ratio of 5.5 to 1. That means the primary winding has 5.5 times as many turns as the secondary. How much current can you draw from the secondary, if the transformer primary can handle 5 amperes? Again, we will use Equation 7 to calculate this current.

$$\frac{N_p}{N_s} = \frac{I_s}{I_p} \qquad \text{(Equation 7)}$$

$$5.5 = \frac{I_s}{5\text{ A}}$$

$$5.5 \times 5\text{ A} = I_s = 27.5\text{ A}$$

These examples show the important relationship between a transformer's primary and secondary voltages and currents. Transformers have many applications in electronics circuits. We usually think of transformers as devices that change between voltage levels. The current available from a transformer also changes.

CHAPTER 23 Impedance

Ohm's Law Applies to AC Circuits, Too

Figure 1 shows a 12-volt battery connected to a 1200-ohm resistor. We can use Ohm's Law to calculate the current that flows in this circuit. Equation 1 gives Ohm's Law as a mathematical equation.

$$E = I\,R \qquad \text{(Equation 1)}$$

where:

E is the voltage applied to the circuit
I is the current through the circuit
R is the circuit resistance, or opposition to current

We want to calculate the current, so we must solve this equation for I.

$$I = \frac{E}{R} \qquad \text{(Equation 2)}$$

$$I = \frac{12\text{ V}}{1200\ \Omega} = 0.01\text{ A}$$

You might want to write this current in milliamperes instead of amperes.

$$I = 10\text{ mA}$$

Now, let's replace the Figure 1 battery with a sine-wave alternating-voltage source. Figure 2 shows this new circuit. The sine-wave generator produces an effective output of 12 volts. Remember that another name for the *effective value* of an ac waveform is *rms*. The signal generator in this problem produces a 12-V rms output. Figure 3 is a graph of the waveform, with labels to identify the peak, peak-to-peak and rms values. Notice the peak value is 1.414 times the rms value, or 16.97 volts. The peak-to-peak value is twice the peak value, 33.94 volts.

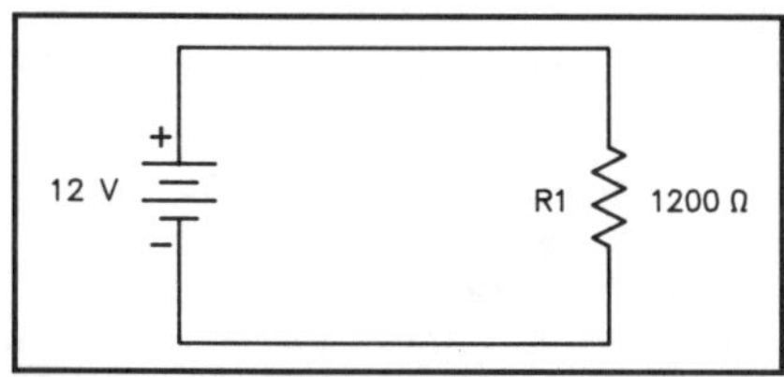

Figure 1—We use Ohm's Law to calculate the current through this dc circuit.

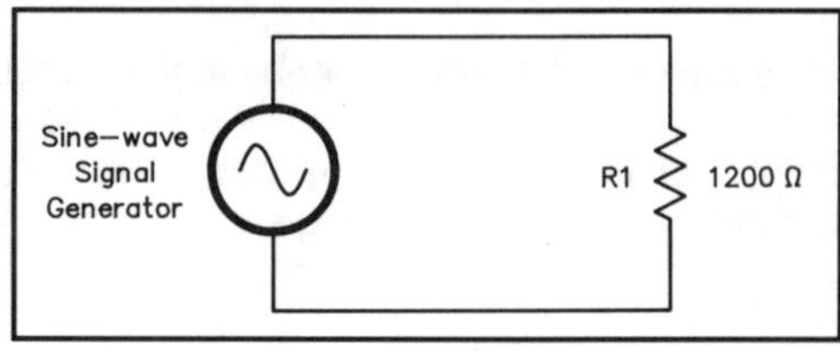

Figure 2—We can calculate the alternating current through a resistor using Ohm's Law. Specify rms, peak or peak-to-peak voltage to calculate the corresponding current value.

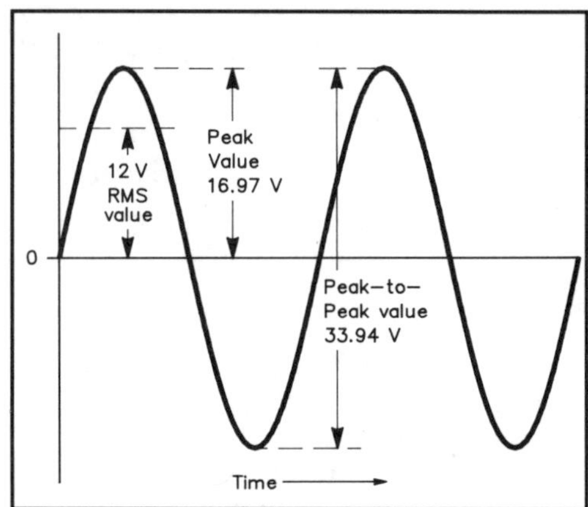

Figure 3—Labels on this sine-wave graph point out the rms, peak and peak-to-peak values.

Which voltage should we use to calculate the circuit current? Well, we can use any of the values, as long as we specify the measurement used to make the calculation. Let's use Ohm's Law to calculate the rms current, peak current and peak-to-peak current through this circuit. Equation 2 gives current.

$$I = \frac{E}{R} \qquad \text{(Equation 2)}$$

$$I_{rms} = \frac{E_{rms}}{R} = \frac{12\text{ V rms}}{1200\ \Omega}$$

$$I_{rms} = 0.01\text{ A rms} = 10\text{ mA rms}$$

$$I_{peak} = \frac{E_{peak}}{R} = \frac{16.97\text{ V peak}}{1200\ \Omega}$$

$$I_{peak} = 0.01414\text{ A peak} = 14.14\text{ mA peak}$$

$$I_{peak\text{-}to\text{-}peak} = \frac{E_{peak\text{-}to\text{-}peak}}{R} = \frac{33.94\text{ V peak-to-peak}}{1200\ \Omega}$$

$$I_{peak\text{-}to\text{-}peak} = 0.02828\text{ A peak-to-peak}$$

$$I_{peak\text{-}to\text{-}peak} = 28.28\text{ mA peak-to-peak}$$

Were you surprised to discover the peak current is 1.414 times the rms current? You could probably have predicted the peak-to-peak current value without even using Ohm's Law, once you calculated the peak current.

Ohm's Law works with ac signals just as it does with dc voltages and currents. You must be careful to specify whether the value you calculate is an rms or peak value, however. Most of the time we work with rms values of ac signals. It isn't

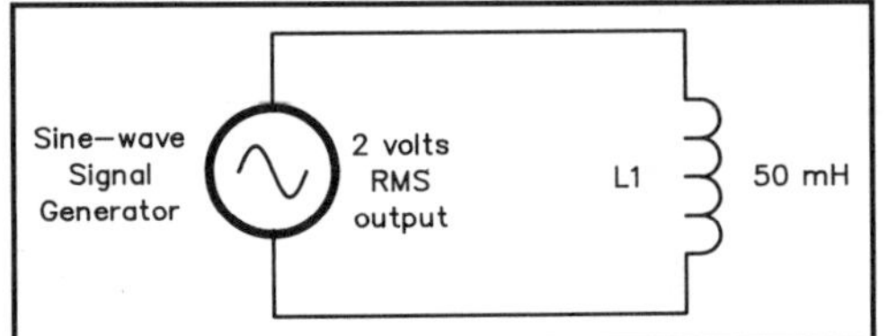

Figure 4— This ac signal generator supplies a 2-volt sine-wave output signal to the inductor. The text explains how to calculate the circuit current using Ohm's Law. Inductive reactance is a measure of the opposition to current through the inductor.

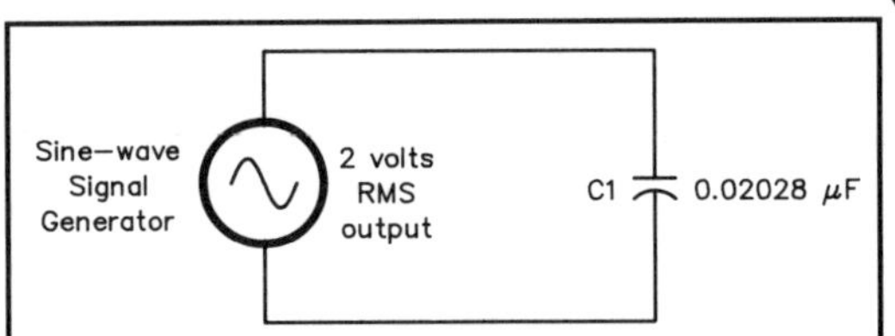

Figure 5— This ac signal generator produces a 2-volt sine-wave signal, which we apply to a capacitor. Capacitive reactance is a measure of the opposition to current through a capacitor. The text explains how to calculate the circuit current using Ohm's Law.

always necessary to write rms, then. When we don't specify a measurement type, it is understood to be an rms measurement.

Resistors aren't the only devices in an electronics circuit that oppose current. Inductors and capacitors also oppose alternating currents. This opposition depends on the frequency of the ac signal voltage. We call the opposition to current *reactance.* Equation 3 will remind you how to calculate *inductive reactance,* X_L.

$$X_L = 2 \pi f L \qquad \text{(Equation 3)}$$

where:

π is a constant approximately equal to 3.14
f is the signal frequency, in hertz
L is the inductance, in henrys

If you aren't familiar with this equation, you should review the chapters on inductors and inductive reactance.

Figure 4 shows a circuit that has a sine-wave signal generator and a 50 millihenry inductor. Let's set the output frequency to 5000 hertz. The generator produces a 2-volt output signal. What is the current through this circuit? We can use Ohm's Law to answer this question, but first we'll have to calculate the inductive reactance. Since inductive reactance is the opposition an inductor has to ac, this will take the place of resistance in Ohm's Law.

Use Equation 3 to calculate the inductive reactance.

$$X_L = 2 \pi f L \qquad \text{(Equation 3)}$$

$$X_L = 2 \times 3.14 \times 5000 \text{ Hz} \times 50 \times 10^{-3} \text{ H}$$

$$X_L = 1570 \ \Omega$$

Now use this reactance instead of resistance in Equation 2, Ohm's Law.

$$I = \frac{E}{X_L} \qquad \text{(Equation 2)}$$

$$I = \frac{2 \text{ V}}{1570 \ \Omega} = 1.27 \times 10^{-3} \text{ A}$$

$$I = 1.27 \text{ mA}$$

What will happen if we reduce the generator frequency to 2500 hertz? The inductive reactance will have a new value, and that will cause the current to change. Let's calculate this new inductive reactance and the new current. (The voltage output from our generator stays at 2 volts.)

$$X_L = 2 \pi f L = 2 \times 3.14 \times 2500 \text{ Hz} \times 50 \times 10^{-3} \text{ H}$$

$$X_L = 785 \ \Omega$$

$$I = \frac{E}{X_L} = \frac{2 \text{ V}}{785 \ \Omega} = 2.55 \times 10^{-3} \text{ A}$$

$$I = 2.55 \text{ mA}$$

Figure 5 shows a 0.02028-μF capacitor connected to our signal generator. We set the generator frequency to 5000 Hz again. The output voltage is 2. What is the current through this circuit?

Capacitive reactance is a capacitor's opposition to alternating current. Equation 4 will remind you how to calculate capacitive reactance, X_C.

$$X_C = \frac{1}{2 \pi f C} \qquad \text{(Equation 4)}$$

where:

π is a constant approximately equal to 3.14
f is the signal frequency, in hertz
C is the capacitance, in farads

Capacitive reactance takes the place of resistance in Ohm's Law to calculate the current in the Figure 5 circuit. Before we can use Equation 2 to calculate the current, we must calculate the capacitive reactance.

$$X_C = \frac{1}{2 \times 3.14 \times 5000 \text{ Hz} \times 0.02028 \times 10^{-6} \text{ F}}$$

$$X_C = \frac{1}{6.368 \times 10^{-4}} = 1570 \ \Omega$$

Now we can use Equation 2 to calculate the current in our circuit.

$$I = \frac{E}{X_C} \qquad \text{(Equation 2)}$$

$$I = \frac{2 \text{ V}}{1570 \ \Omega} = 1.27 \times 10^{-3} \text{ A}$$

$$I = 1.27 \text{ mA}$$

Suppose the generator frequency changes to 2500 Hz. As an additional practice exercise you should prove the current through the revised circuit becomes 0.637 mA. (Hint: You must calculate the new capacitive reactance first.)

Inductive reactance and capacitive reactance are not the same as resistance. We use them in place of resistance in Ohm's Law to calculate the current through an inductor or capacitor. We use the term *impedance* to mean either resistance or reactance, or some combination of both.

Imagine a circuit that includes a resistor and inductor or capacitor in series with a signal generator. Resistance and reactance are not the same, so you can't just add the values like you would with two resistors. When we talk about circuit *impedance*, then, we use a general term that may mean resistance, reactance or some combination of these quantities. You will learn more about circuit impedance, and calculations involving impedance, in the next few sections.

Voltage and Current Peaks May not Occur at the Same Time

Ohm's Law relates voltage, current and resistance in circuits that include resistors and voltage sources. The voltage sources can supply direct or alternating current.

Ohm's Law also relates voltage, current and reactance in circuits that include only voltage sources and capacitors or inductors. Most circuits that include capacitors or inductors will have alternating voltage sources.

Do you remember what happens if you feed a dc signal through a capacitor? There is current for a short time, until the capacitor charges to the full applied voltage. Then the current stops. There is a large current at first, but it quickly decreases as the capacitor charges. We won't try to calculate this changing direct current. The math is a bit beyond the scope of this book. So we will only consider alternating current through capacitors.

Ideal inductors present no opposition to steady direct current. An inductor is a short circuit to a steady direct current. Only resistance of the wire used to make the inductor limits the current in this case. We will only calculate the current through inductors with an alternating voltage source applied in this book.

Suppose we apply an alternating voltage to a resistor, as the Figure 1A circuit shows. The sine-wave signal begins to increase from 0 volts. The current through the resistor increases when the voltage increases. When the applied signal reaches its maximum voltage, the current also reaches maximum value. Figure 1B shows the current and voltage waveforms for this circuit.

These signals have the same frequency and reach their positive peak values at the same time. We say the waves are *in phase* if they meet those conditions. *Phase* refers to time, such as the time between two events that repeat at regular intervals.

Each ac cycle takes exactly the same time as any other cycle of the same frequency. We relate one cycle to a single rotation of a wheel, or an angle of 360°. Figure 2 shows how the various positions of a sine-wave signal relate to one rotation of a wheel. By using an angle related to wheel rotation we can make phase measurements that don't depend on the actual wave frequency.

When current flows through a resistor, the resistor converts some of the electrical energy into heat energy. Capacitors and inductors store electrical energy when current flows through them. Ideal inductors and capacitors don't convert any electrical energy to heat energy, however. (Real inductors and capacitors do convert some energy to heat, but this should be a small amount. We'll ignore it for now.)

One result of this energy storage is that the voltage and current waveforms are not *in phase*. Sometimes we say these waves are *out of phase*.

Figure 3 illustrates two waveforms that are out of phase. The waveform at Part A reaches its positive peak at the same time as the Part B waveform reaches its negative peak. In this example, the waveforms are 1/2 cycle different, or 180° out of phase. Notice that we labeled the horizontal, or X axis in degrees instead of time.

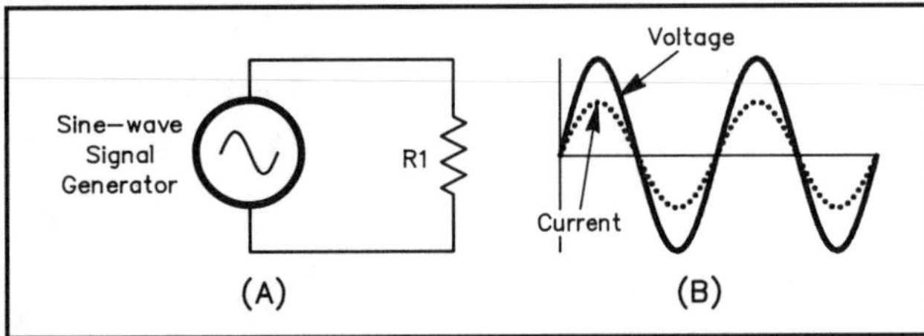

Figure 1—This simple circuit illustrates the phase relationship between the alternating voltage applied to a resistor and the current through the resistor. Part B shows the voltage and current waveforms. Notice the waveforms are in phase.

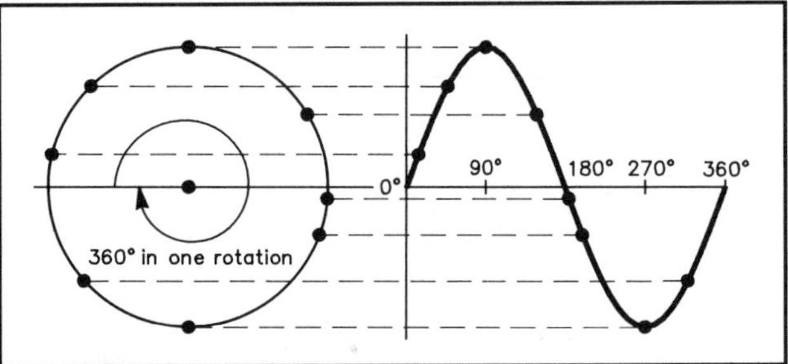

Figure 2—We often relate each cycle of a waveform to a single rotation of a wheel. Each point along the waveform corresponds to a rotation angle on the wheel. There are 360° in one complete cycle. We use an angle measurement to indicate the phase between two waveforms.

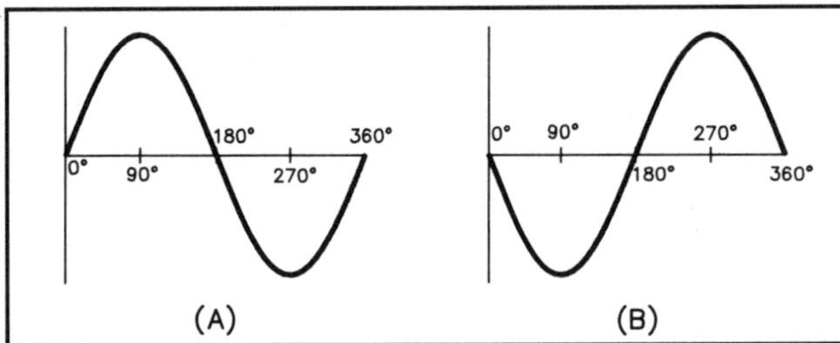

Figure 3—These two waveforms appear to be exact opposites. When the waveform at A reaches its positive peak the waveform at B reaches its negative peak. We say the two waves are 180° out of phase.

You can measure a phase angle between any two waveforms of the same frequency. The waveforms can have any phase relationship between 0° and 360°. The waveforms are in phase only if the phase angle is 0° (or 360°).

The phase angle between the voltage and current in a circuit is 0° if the circuit only includes voltage sources and resistors. If the circuit includes capacitors or inductors, the phase angle will be larger than 0°. The phase angle depends on the amount of reactance and the amount of resistance in the circuit. Later in this chapter you will learn how to calculate the phase angle.

Voltage and Current in a Capacitor

Capacitive reactance is the opposition a capacitor has to alternating current flowing through it. At low frequencies the reactance is large, and it decreases as the frequency increases. When we apply an alternating voltage across a capacitor, the increasing voltage stores energy as an electric field. The capacitor returns that energy to the circuit when the voltage decreases.

Figure 1 shows a circuit with an ac signal generator and a capacitor. We can replace resistance with reactance in Ohm's Law to calculate current or voltage in a circuit like this one.

Let's review what happens when we apply an alternating voltage to a capacitor. We'll start with the voltage at zero, as it increases toward its maximum value. There is no charge stored on the capacitor, so there is a large current to charge it. As electrons build up on the negative capacitor plate, this charge tries to prevent more electrons from moving onto the plate. The current decreases as a result. By the time the voltage increases to its maximum value, the capacitor charges to this full voltage, and the current stops.

Next the applied voltage begins to decrease. The capacitor voltage is larger than the applied voltage, so electrons move off the negative capacitor plate. This changes the current direction. As the applied voltage decreases to zero, the capacitor current increases, draining all the charge off the capacitor plates. Figure 2 is a graph showing these voltage and current relationships.

The second half of the applied-voltage cycle repeats the first half, but with opposite polarities. As the voltage increases toward the negative peak, there is a large current to charge the capacitor. This current drops to zero when the voltage reaches its negative peak. When the voltage begins to decrease toward zero again, the current direction reverses. The current increases to its maximum value when the voltage reaches zero. Figure 3 graphs two complete voltage and current cycles.

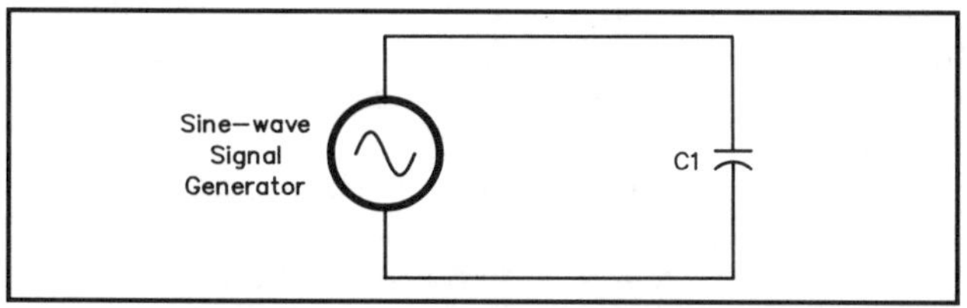

Figure 1—We will use this circuit to study the voltage and current relationship in a capacitor. When the circuit includes only voltage sources and capacitors, we can use capacitive reactance instead of resistance in Ohm's Law.

What is the phase relationship between the voltage across a capacitor and the current through it? Study Figure 3. Notice when the voltage waveform is at zero, beginning its positive half cycle, the current waveform is already at its positive maximum. When the voltage waveform reaches its positive peak the current waveform has already dropped to zero. By the time the current waveform reaches its negative peak the voltage waveform has only decreased

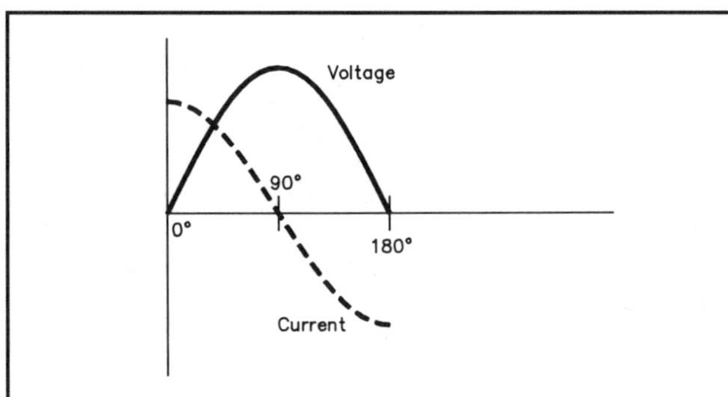

Figure 2—This graph shows how the current through a capacitor varies during the positive half cycle of applied voltage. We drew the voltage waveform as a solid line and the current waveform as a dashed line. We did not include a voltage or current amplitude scale on the graph. The two waveforms have different amplitudes so it is easier for you to distinguish the two curves.

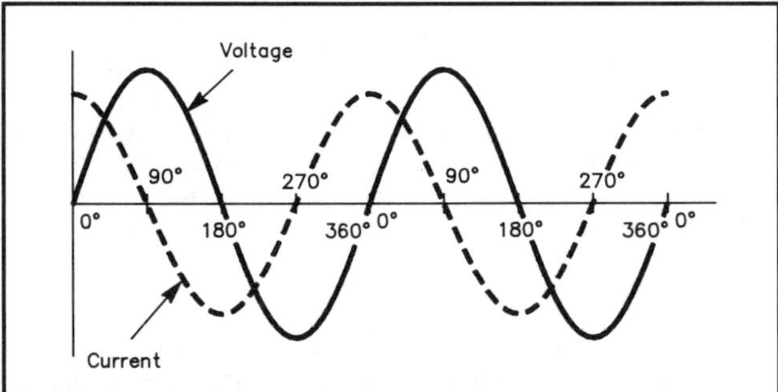

Figure 3—This graph shows how the current varies during two complete cycles of the voltage waveform. The solid and dashed lines and different amplitudes will help you distinguish between the voltage and current waveforms.

to zero from its positive peak.

These waveforms are 90° out of phase. Measure the angle between the positive peaks or between the zero crossing points as the waveforms go from positive to negative. You will measure the same angle between the negative peaks of both waveforms. You also can measure the phase angle between the zero crossings as the waves go from negative to positive. Do you have to measure phase angle only at these points? No, you can measure phase angle from any point on a waveform, as long as you measure to the same point on the other waveform. The waveform peaks and zero crossings provide convenient reference points, however.

Did you notice that the current waveform reaches each point 90° before the voltage waveform? Figure 3 shows the current waveform first reaches its peak positive value at 0°. The voltage waveform doesn't reach its peak positive value until 90°. The current waveform reaches its peak negative value at 180°. The voltage waveform doesn't reach its peak negative point until 270°; 90° later than the current.

From these observations, you can understand why we say the current *leads* the voltage in a capacitor. We sometimes also say the voltage *lags* the current. Both statements mean the same thing. This is an important relationship.

Earlier we mentioned that the applied voltage stores electrical energy in an electric field. This field starts at zero when the voltage is zero. The field strength increases as the voltage increases, and reaches a maximum when the voltage reaches its peak value. The electric field decreases as the energy returns to the circuit when the voltage decreases. So the electric field has a waveform that matches the applied voltage waveform. The electric field and the applied voltage are in phase.

Voltage and Current in an Inductor

Inductors oppose alternating current. We call this opposition inductive reactance. Inductive reactance is low at low frequencies and increases as the frequency increases. When alternating current flows through an inductor, the inductor stores some electrical energy as a magnetic field around the inductor.

The circuit of Figure 1 includes an alternating current source and an inductor. We can replace resistance with inductive reactance in Ohm's Law to calculate current or voltage in a circuit like this.

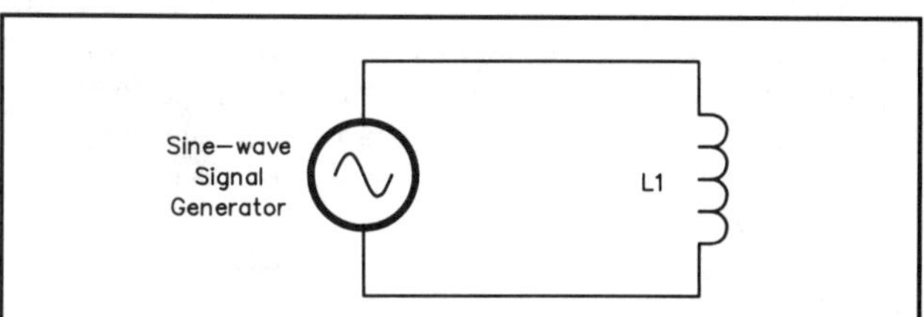

Figure 1—This alternating signal generator feeds current through the inductor. This ideal circuit does not include any resistance. The inductor stores electrical energy as a magnetic field as the current through it increases. The inductor returns all this energy to the circuit as the current decreases. The process repeats each half cycle.

Let's put an alternating current through an ideal inductor and review what happens. We'll start when the current is zero, just beginning to increase toward its positive peak value. The increasing current causes a magnetic field to build up around the inductor, storing energy. The current increases rapidly at this point, so the magnetic field strength also increases rapidly.

When the current through the coil reaches a positive peak and begins to decrease, the energy stored in the inductor reaches a maximum. As the current decreases, the magnetic field collapses, returning its energy to the circuit. When the current crosses zero the magnetic field returns all its energy, and the field strength also becomes zero.

Next the current direction reverses, and the current increases toward its peak negative value. The magnetic field direction also reverses, storing energy again. The magnetic field strength continues to increase until the current reaches its peak negative value. Then the current decreases again, and the inductor returns the energy stored in it as the magnetic field collapses.

Figure 2 is a graph of the current through an inductor and the magnetic field strength around the inductor. The magnetic field around an inductor is in phase with the current through the inductor. Similar points on both waveforms occur at the same time.

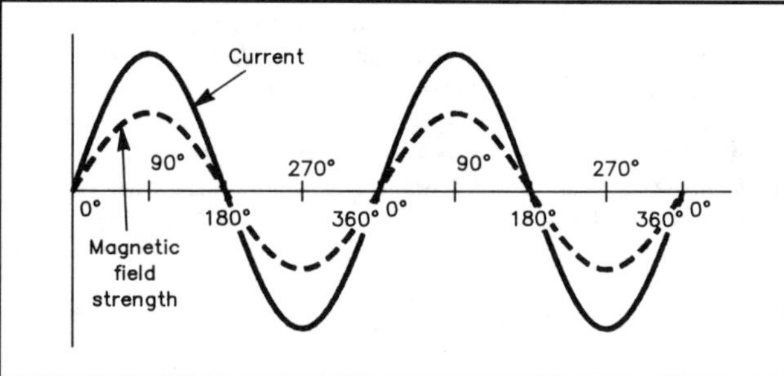

Figure 2—This graph shows that the magnetic field strength around an inductor increases when the current increases. The magnetic field decreases when the current decreases. If the current reverses direction, the magnetic field direction also reverses. The magnetic field is in phase with the current through the inductor.

Remember that a changing magnetic field induces a voltage across the inductor. The voltage depends on how fast the magnetic field changes. It is largest when the magnetic field changes fastest. A slowly changing magnetic field induces a small voltage. If the magnetic field doesn't change, it doesn't induce a voltage.

The current changes quickly when it starts at zero and begins to flow in either direction. The magnetic field also is changing the most at this point. This is when the induced voltage has its maximum value. When the current (and magnetic field) reaches its maximum in either direction, there is a brief instant when it doesn't change. At this instant the induced voltage is zero.

The induced voltage always has a polarity that opposes the changing current. The induced voltage tries to maintain a steady, unchanging current. This means that when the current is in the positive direction, the induced voltage has a negative polarity!

Figure 3 graphs the current through an inductor and the voltage induced across the inductor. Notice that as the current increases toward its peak *positive* value the induced voltage is decreasing from its peak *negative* value. The induced voltage becomes zero when the current reaches its peak value. As the current decreases from this peak value, the induced voltage becomes positive. (The positive induced voltage tries to prevent the current from decreasing.) As you can see from Figure 3, the induced voltage *lags* the current by 90°. (We also can say the current *leads* the induced voltage by 90°.)

You probably noticed that we haven't mentioned the *applied voltage* yet. You know the current source applies a voltage to the inductor to force current through it. The applied voltage is not in phase with the current through the inductor. The current produces the magnetic field around the inductor. The changing magnetic field induces the *back emf*, as we sometimes call it.

The applied voltage tries to produce current through the inductor and the induced voltage tries to prevent it. This means the applied voltage and the induced voltage are trying to accomplish opposite goals. If the two voltages are opposite each other, that means they are 180° out of phase! Figure 4 graphs current, applied voltage and induced voltage for an inductor.

Let's look at the phase relationship between the *applied voltage* and the current through an inductor. This is the relationship we're most often interested in when we discuss the phase relationship between voltage and current for an inductor.

Figure 4 shows the applied voltage reaching its peak positive value at 0°. The current reaches its peak positive value at 90°. You also can see that the voltage reaches its peak negative value at 180° and the current reaches its peak negative value at 270°. This means the applied voltage is 90° ahead of the current. The applied voltage *leads* the current through an inductor by 90°. We also can say the current *lags* the applied voltage by 90°.

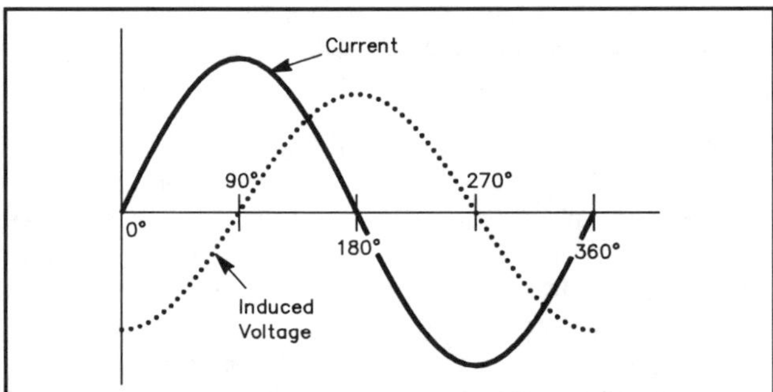

Figure 3—A heavy solid line on this graph represents the current through an inductor. A dotted line represents the voltage induced across the inductor by the changing magnetic field. Notice the induced voltage lags the current by 90°. (The current leads the induced voltage by 90°.)

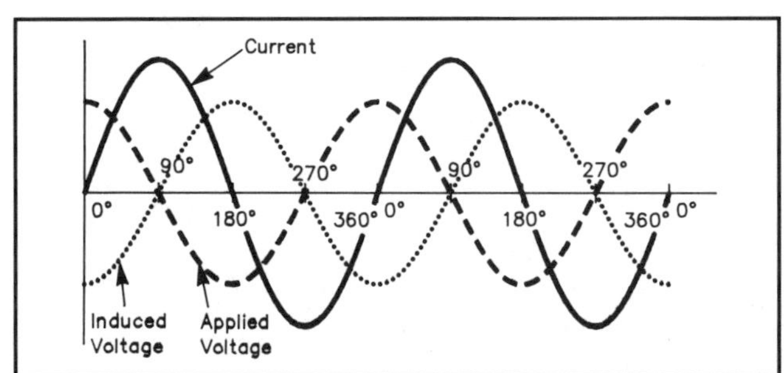

Figure 4—Here we add a dashed line to the Figure 3 graph. This new line represents the applied voltage across an inductor. The applied voltage and the induced voltage are 180° out of phase. The applied voltage leads the current by 90°. We also can say the current lags the applied voltage by 90°.

Circuit Reactance with Inductors and Capacitors

When we apply an alternating voltage to a capacitor there is a current that leads the voltage by 90°. An alternating voltage applied to an inductor produces a current that lags the voltage by 90°. An ac voltage applied to a resistor creates a current that is in phase with the applied voltage.

Figure 1 is a circuit that includes an alternating voltage source, a capacitor and an inductor. Let's see what we can determine about the voltage applied to each

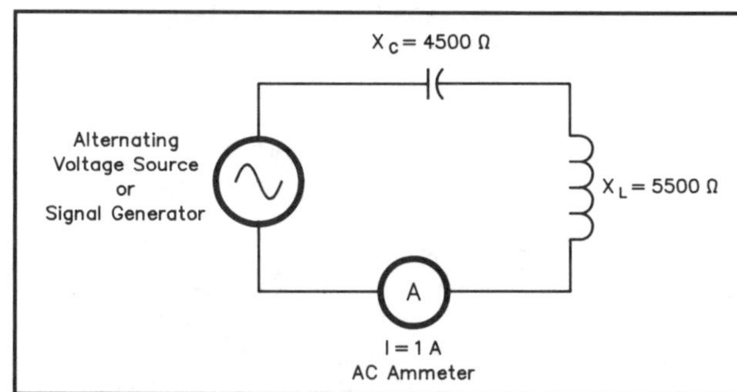

Figure 1—We'll use this circuit to study the combined effects of capacitive reactance and inductive reactance on a circuit. The reactance values apply to one output frequency from the generator. Although we don't know what that frequency is, we don't need that value for our calculations. The ac ammeter measures a 1-ampere current through the circuit.

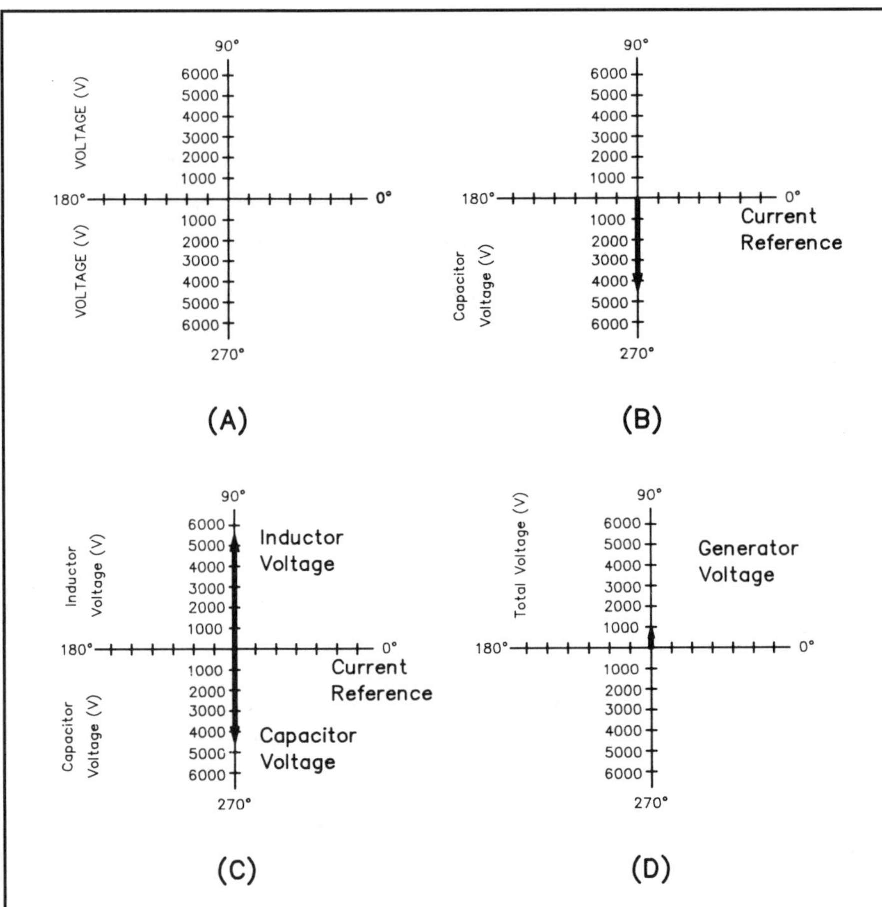

Figure 2—This graph shows how we can represent series circuit voltages through inductors and capacitors. We measure angles starting from the right side of the X axis, and go counterclockwise. We drew the capacitor voltage along the 270° line because this voltage lags the circuit current by 90°. We use the 90° line for the inductor voltage, indicating this voltage leads the circuit current by 90°. We find the total generator voltage by subtracting the capacitor voltage from the inductor voltage. Part D shows this total combined voltage as 1000 volts.

component and the current through it.

Is the Figure 1 circuit a *series circuit* or a *parallel circuit*? All the current that leaves the signal generator flows through each component to get back to the generator. There are no branches for the current to select an alternate path. Therefore, this is a series circuit.

The same current flows through each part of a series circuit. From this information, we conclude that the current through the capacitor is in phase with the current through the inductor. The voltages across each component are not in phase, however. In fact, the voltages are 180° out of phase! How do we know this? Figure 2 will help you understand.

We begin drawing Figure 2 by forming a set of X-Y coordinates, as shown in Part A. We also mark the coordinates as a set of polar coordinates, with 0° to the right, and 90° to the top. (If you are unfamiliar with polar coordinates, review that section in the Math Skills Unit of this book.)

Next we mark the capacitor and inductor voltages on our graph. We'll calculate these voltages using Ohm's Law and the component reactance. For this example, we'll assume the ac ammeter reads 1 ampere. Figure 1 lists the reactances of the inductor and capacitor. Remember this represents the reactance at one frequency, the signal generator's output frequency. Both reactances will be different at other frequencies.

Let's calculate the capacitor voltage first. A 1-ampere current flows through a 4500-ohm reactance. Multiply these values to calculate the voltage.

$$E = I R \qquad \text{(Equation 1)}$$

$$E = I X_C = 1 \text{ A} \times 4500 \ \Omega$$

$$E = 4500 \text{ volts}$$

The voltage across a capacitor lags the current by 90°. If we let 0° represent the current, we must draw the capacitor voltage at 270°. That puts the capacitor voltage 90° behind the current. Figure 2B shows the capacitor voltage added to our graph.

Next we'll calculate the inductor voltage. The same 1-ampere current flows through the inductor. Figure 1 shows the inductive reactance as 5500 ohms. Once again we multiply these values in Ohm's Law to calculate the voltage applied to the inductor.

$$E = I R \qquad \text{(Equation 1)}$$

$$E = I X_L = 1 \text{ A} \times 5500 \ \Omega$$

$$E = 5500 \text{ volts}$$

The voltage applied to an inductor leads the current through the inductor by 90°. So we'll draw the inductor

voltage at 90° on our graph. Figure 2C shows the inductor voltage added to our graph.

The inductor and capacitor voltages are 180° out of phase. When one reaches its positive peak value the other reaches its negative peak value. We can combine these two voltages to find the total applied voltage. This total voltage is the voltage the signal generator applies to the circuit. The inductor voltage goes in the positive Y direction on our graph. The capacitor voltage goes in the negative Y direction. To combine these values, we subtract! The total applied voltage is 5500 volts minus 4500 volts, or 1000 volts. How can the inductor and capacitor voltages be larger than the total applied voltage? That's one interesting property of reactance.

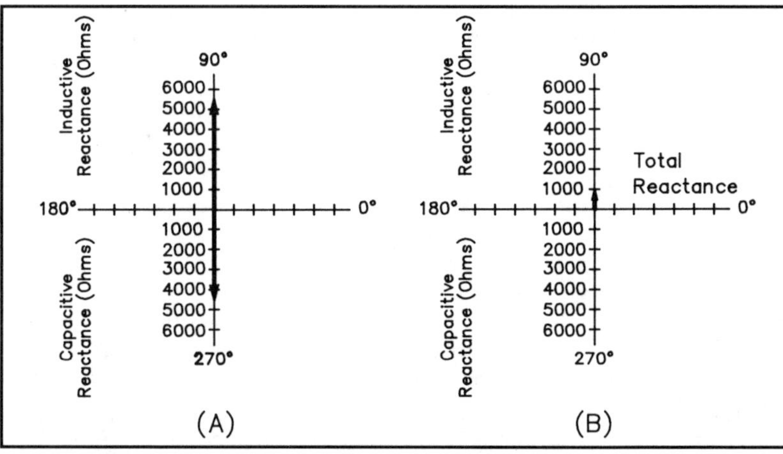

Figure 3—We often represent series circuit reactances on a graph similar to the one from Figure 2. Inductive reactance is along the positive Y axis, at 90°. Capacitive reactance is along the negative Y axis, at 270°.

The capacitor stores energy in its electric field and then returns it to the circuit. The inductor stores the energy from the capacitor in its magnetic field and returns it to the circuit later. So the capacitor and inductor hand the energy back and forth between them. If there is no resistance (like our ideal circuit of Figure 1), this process will continue forever.

If you think about the mathematics for a moment, you should realize that we can find the applied voltage without calculating the inductor and capacitor voltages first. To do this, we simply combine the two reactances first, and then use this combined value in Ohm's Law.

$$E = I X \qquad \text{(Equation 2)}$$

where X is the combined circuit reactance

$$X = X_L - X_C$$

$$X = 5500\ \Omega - 4500\ \Omega = 1000\ \Omega$$

$$E = 1\text{ A} \times 1000\ \Omega = 1000\text{ V}$$

Why do we subtract capacitive reactance from inductive reactance? Why not subtract inductive reactance from capacitive reactance? Did we just select the larger value to start? No. Take another look at Figure 2D. We drew the capacitor voltage in the negative Y direction because this voltage lags the current by 90°.

Whenever we connect capacitors and inductors in *series*, we draw the capacitor voltage in the negative Y

direction. We draw the inductor voltage in the positive Y direction.

We often use a similar graph to represent resistance and reactance on a graph. We draw resistance along the 0° line, or in the positive X direction. Inductive reactance is along the 90° line or the positive Y direction. Capacitive reactance is along the 270° line or the negative Y direction. See Figure 3.

The total reactance is in the positive direction. This means there is more inductive reactance in the circuit than capacitive reactance. It also means the total voltage leads the circuit current by 90°.

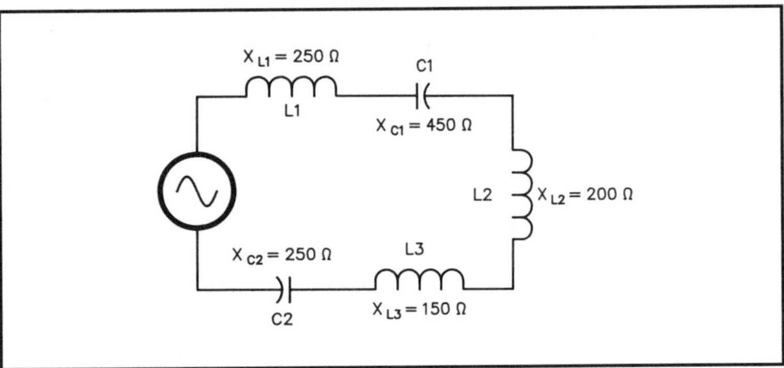

Figure 4—Circuits often have more than one inductor or one capacitor. We can find the total series circuit reactance by adding all the inductive reactances and subtracting all the capacitive reactances.

What if you have a circuit that includes several inductors and capacitors? Easy! Just add all the inductive reactances and draw the result at 90° on a graph. Then add all the capacitive reactances and draw that result at 270° on the graph. Subtract the total capacitive reactance from the total inductive reactance to find the overall reactance.

What is the total reactance of the circuit shown in Figure 4? First add the reactances of the three inductors.

$X_{L1} = 250\ \Omega$
$X_{L2} = 200\ \Omega$
$X_{L3} = 150\ \Omega$

$X_{L\ TOTAL} = 600\ \Omega$

Then add the reactances of the two capacitors.

$X_{C1} = 450\ \Omega$
$X_{C2} = 250\ \Omega$

$X_{C\ TOTAL} = 700\ \Omega$

Now combine these two values.

$$X_{TOTAL} = X_{L\ TOTAL} - X_{C\ TOTAL} \qquad \text{(Equation 3)}$$

$$X_{TOTAL} = 600\ \Omega - 700\ \Omega = -100\ \Omega$$

This time there is more capacitive reactance than inductive reactance. The total reactance is capacitive. It also means the applied voltage lags the circuit current by 90°.

We only discussed series circuits in this section. The methods you learned to find the total voltage and circuit reactance don't apply directly to parallel circuits. You will learn about the voltage, current and reactance of parallel circuits later in this chapter.

Voltage and Current in an RL Circuit

We have been studying ideal circuits with only a voltage source and inductors and capacitors. You learned how to calculate the circuit reactance. You calculated the voltage across an inductor or a capacitor, and determined the phase relationship between the voltage and current. None of these circuits included resistance.

All real circuits include some resistance. There is always some resistance in the wires used to make an inductor or connect a capacitor to the circuit. Every real part has some resistance. In this section you will learn how to calculate the voltage applied to a real inductor. You also will learn how to measure the phase angle between the current through an inductor and the voltage across it.

Figure 1 shows an ideal inductor and a resistor in series. The resistor could represent the resistance that is part of a real inductor. In this example we will treat the resistance as a separate component, however.

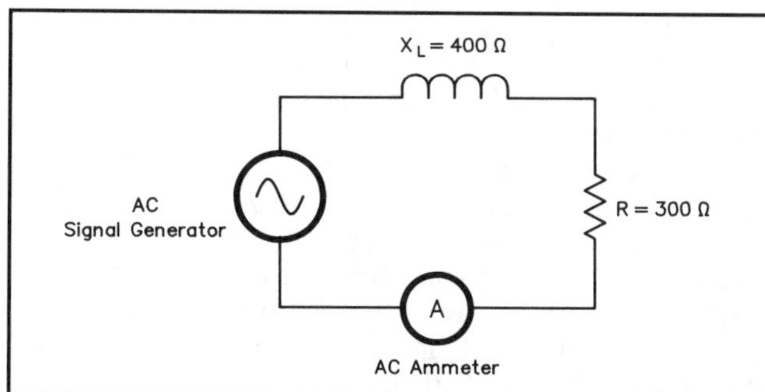

Figure 1—Real inductors always have some resistance. This circuit shows an ideal inductor in series with a resistor. This one resistor includes the resistance of any resistors in the circuit. It also includes the resistance of the wires and other components.

Suppose we look at the ammeter and find that it reads 20 mA. Since this is a series circuit, the same 20-mA current flows through every part. What is the generator output voltage?

To answer this question, we'll have to calculate the voltage across the resistor and the voltage across the inductor separately. Then we will combine these values. We can't add the voltages the way we normally add numbers, however, because the two voltages are not in phase.

We'll use Ohm's Law to calculate the voltages. Are you surprised by that?

$E_R = I\,R$ (Equation 1)

$E_R = 20 \times 10^{-3}\ A \times 300\ \Omega$
$E_R = 6\ V$

The voltage drop across the resistor is 6 volts.

$E_L = I\,X_L$ (Equation 2)

$E_L = 20 \times 10^{-3}\ A \times 400\ \Omega$
$E_L = 8\ V$

The inductor in this circuit has 8 volts across it.

Now let's see how to combine these two voltages. First, we'll make a graph to show the phase relationships between the current and the voltages. Figure 2A shows a graph with the axes labeled. Notice the angle measurements marked on the X and Y axes. We'll use the 0° mark as the current reference line.

The voltage across the resistor is in phase with the current. We will draw the resistor voltage along the 0° axis, as Figure 2B shows.

What is the phase relationship between the current and voltage for an inductor? Chances are, you're having a little trouble remembering these phase relationships. Here is a little phrase to help you remember. (The word for a memory aid like this is *mnemonic*. You pronounce this word "ni-mon-ik." This is an English word, not an electronics term!)

Whenever you want to remember the phase relationship between current and voltage through inductors, just think of your friend *ELI*. We use the letter L to represent an inductor. The letter E represents voltage and I represents current in Ohm's Law. So whenever you think of ELI you will see that the voltage comes before current with an inductor. E comes before I with an L. So voltage *leads* current through an inductor.

Your friend ELI will even help you remember the phase between current and voltage for a capacitor. That's because ELI is an ICE man. So I comes before E with a C. This second part of the mnemonic will come in handy in the next section.

Okay, back to our problem. ELI tells us the voltage leads the current through an inductor. So we'll draw the inductor voltage along the 90° line on our graph, as Figure 2B shows.

That was the easy part. Now let's go on to the more difficult step. Using the two voltages we drew on Figure 2B as a starting point, draw a complete rectangle, as Figure 2C shows. If you are using graph paper, just follow the vertical line representing 6 on the 0° axis for one line. The second line follows the horizontal line representing 8 on the 90° axis. We're just making a sketch here, though, so just make a rough rectangle. Complete your drawing by adding the dashed line that goes from the origin (where the two axes cross) to the opposite corner. Follow these steps carefully. Don't try to take any shortcuts here. There are none that work!

When your drawing looks like Figure 2C, you are ready to calculate the total applied voltage. This is where your knowledge of trigonometry becomes important. Are you familiar with the terms *sine*, *cosine* and *tangent*? If not, you should review the trigonometry chapter of the Math Skills Unit in this book now.

Either half of the rectangle forms a 90° (right) triangle. We know the two sides that include the 90° angle, and want to find the *hypotenuse* (the side opposite the right angle). Let's find the angle marked A on Figure 2C.

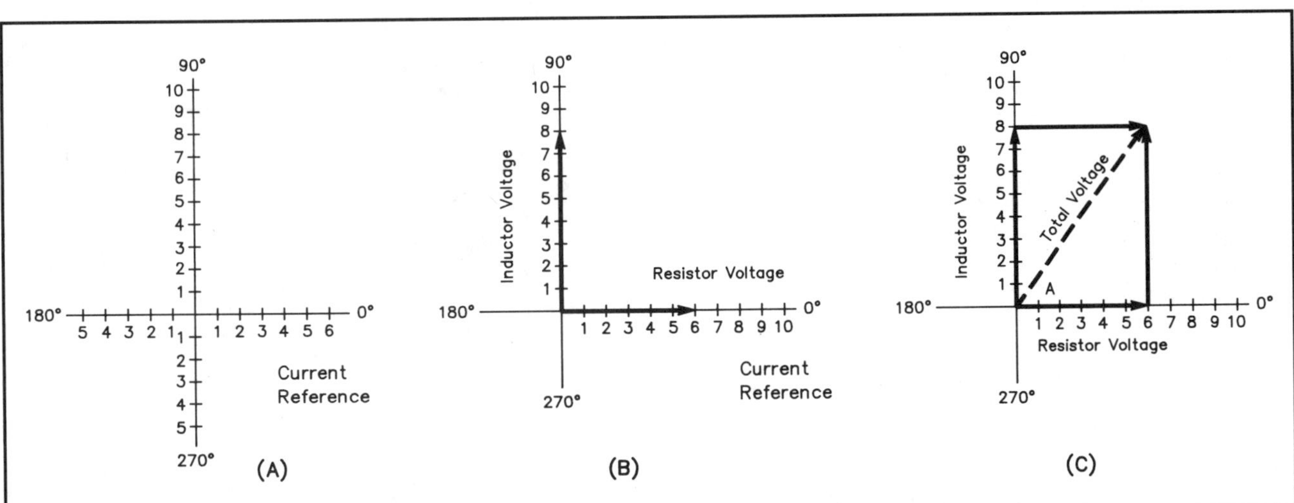

Figure 2—This drawing shows how we can represent the voltages across a resistor and an inductor on a graph. We find the total applied voltage by making a rectangle as Part C shows, and then drawing in the diagonal line from the point where the axes cross. The text explains how to calculate the total applied voltage using this drawing.

We'll use the tangent function to find angle A. Tangent is the ratio of the *side opposite* the angle divided by the *side adjacent* to the angle.

$$\tan(A) = \frac{\text{opposite}}{\text{adjacent}} \qquad \text{(Equation 3)}$$

$$\tan(A) = \frac{8\ V}{6\ V} = 1.333$$

Now find the inverse tangent on your calculator.

$A = \tan^{-1}(1.333) = 53.12°$

Finally, we'll use this angle and the sine function to find the hypotenuse.

$$\sin(A) = \frac{\text{opposite}}{\text{hypotenuse}} \qquad \text{(Equation 4)}$$

Solve this equation for the hypotenuse, which represents the total voltage applied to our circuit.

$$\text{hypotenuse} = \frac{\text{opposite}}{\sin\ (A)}$$

$$E_{TOTAL} = \frac{8\ V}{\sin(53.12°)} = \frac{8\ V}{0.800}$$

$E_{TOTAL} = 10\ V$

What is the phase angle between the circuit current and this total voltage? Angle A represents this phase angle. We calculated the angle earlier as part of our solution. The applied voltage leads the circuit current by 53.12°.

You may want to try several other solutions to this problem. For example, you can solve for the hypotenuse, or voltage, using the two known triangle sides and the Pythagorean Theorem. (This involves squaring each side, adding the results and then taking the square root.) You also can use other combinations of trigonometry functions to reach the same result.

Sometimes we want to know the total combined opposition to current for a circuit like Figure 1. We use the term *impedance* to mean the combined opposition of resistance and reactance. Just as with the voltages, we can combine resistance and reactance using a graph and trigonometry.

When working with impedance, we place the resistance along the 0° line and the inductive reactance along the 90° line. This is the same position we use for the voltage across these components. Figure 3 is a graph of the resistance and inductive reactance for our problem. The diagonal line represents the total impedance for the Figure 1 circuit.

We follow the same procedure to calculate the impedance as we used to calculate the total voltage. First use the tangent function to calculate angle A. Then solve the sine equation for the hypotenuse of the triangle to find the impedance value.

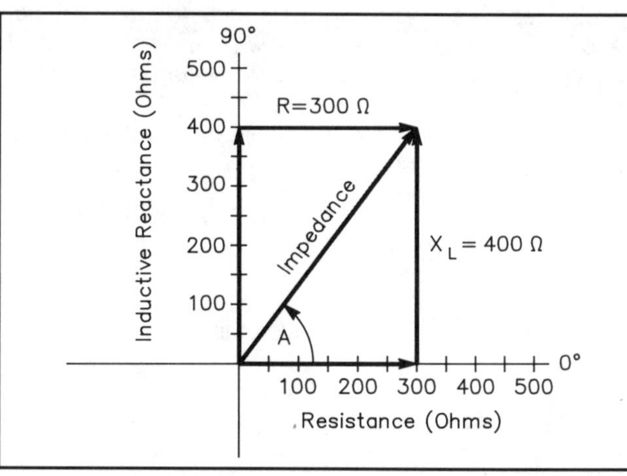

Figure 3—We can use a graph like the one from Figure 2 to represent circuit resistance and reactance. From the triangle formed with this rectangle we can calculate the circuit impedance. The text explains the steps involved in this calculation.

$$\tan(A) = \frac{\text{opposite}}{\text{adjacent}} \qquad \text{(Equation 3)}$$

$$\tan(A) = \frac{400\ \Omega}{300\ \Omega} = 1.333$$

Now find the inverse tangent on your calculator.

$A = \tan^{-1}(1.333) = 53.12°$

Notice the *phase angle* for this impedance is the same as the phase angle between the applied voltage and the circuit current.

$$\text{hypotenuse} = \frac{\text{opposite}}{\sin(A)}$$

$$\text{Impedance} = \frac{400\ \Omega}{\sin(53.12°)} = \frac{400\ \Omega}{0.800}$$

$\text{Impedance} = 500\ \Omega$

We often use a capital Z to represent impedance on circuit diagrams and in calculations. So the impedance, Z, for this circuit is 500 Ω at 53.12°. We must include the angle when we specify an impedance, because this is an important part of the measurement. A similar value at a different angle would *not* give the same circuit results.

As a final calculation, let's use this circuit impedance in Ohm's Law to find the applied voltage. Remember we measured a current of 20 mA through the circuit. Equation 5 is Ohm's Law using impedance instead of resistance.

$$E = I\,Z \qquad \text{(Equation 5)}$$

$E = 20 \times 10^{-3}\ \text{A} \times 500\ \Omega$

$E = 10$ volts

Are you surprised to discover this is the same voltage we calculated earlier? We found two ways to calculate the total applied voltage. First, we combined the voltage across the resistor and the voltage across the inductor. We found the same result by calculating the circuit impedance and using that in Ohm's Law with the total circuit current.

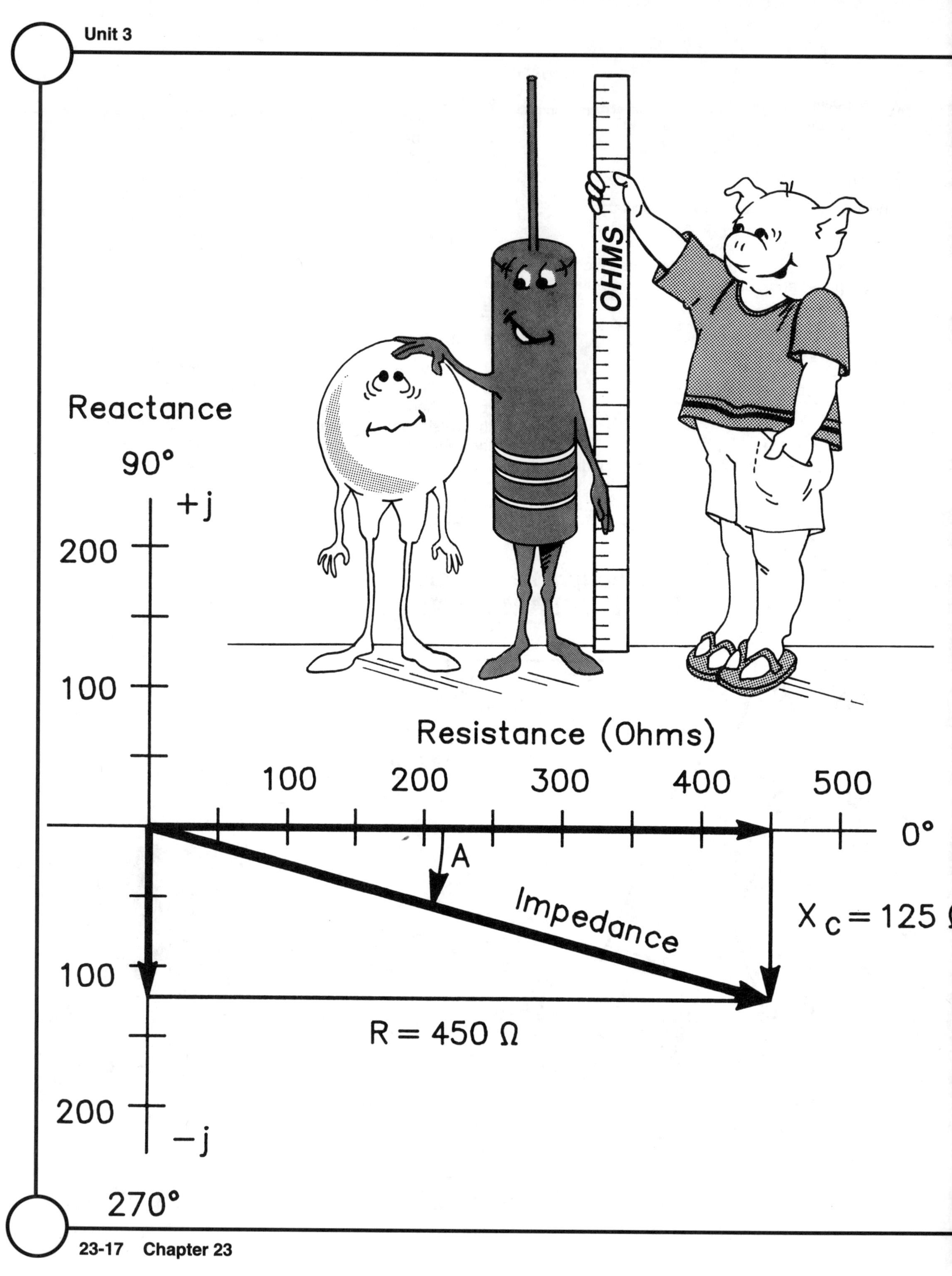
OHMS
Reactance
90°
+j
200
100
Resistance (Ohms)
100
200
300
400
500
0°
A
Impedance
X C = 125 Ω
100
R = 450 Ω
200
−j
270°

Voltage and Current in an RC Circuit

All real capacitors have some resistance. The wires that connect them to a circuit have resistance, and the capacitor plates have some resistance. There also is some energy lost in the insulation between the capacitor plates. In this section we'll learn about circuits with real capacitors, by adding a resistor in series with an ideal capacitor.

Figure 1 shows a circuit with a single ideal capacitor and one resistor. This resistor includes any resistance in the capacitor leads and other circuit wiring. A sine-wave signal generator supplies a 50-mA current to the circuit. We'll calculate the total applied voltage using Ohm's Law.

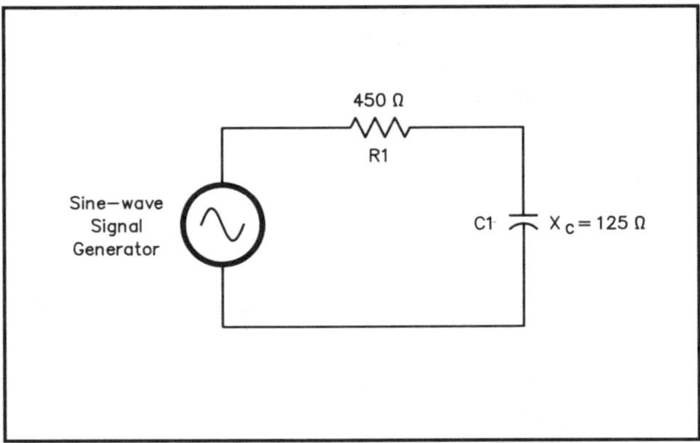

Figure 1—We'll use this circuit to study the effects of real components in a circuit with capacitors. There is always some resistance in a real circuit. The text explains how to calculate the total applied voltage from the signal generator. We also calculate the circuit impedance.

Since this is a series circuit, the same 50-mA current flows through each part of the circuit. We can calculate the voltage drop across the 450-ohm resistor with Ohm's Law. We substitute capacitive reactance in place of resistance to calculate the voltage across the capacitor. Figure 1 shows this capacitive reactance as 125 ohms.

Let's find the voltage across the resistor first. Ohm's Law tells us to multiply the current times the resistance.

$$E_R = I\,R \qquad \text{(Equation 1)}$$

$$E_R = 50 \times 10^{-3}\text{ A} \times 450\ \Omega$$
$$E_R = 22.5\text{ V}$$

A similar equation gives us the voltage across the capacitor. Simply replace R with X_C in Equation 1.

$$E_C = I\,X_C \qquad \text{(Equation 2)}$$

$$E_C = 50 \times 10^{-3}\text{ A} \times 125\ \Omega$$
$$E_C = 6.25\text{ V}$$

Do you suppose we can just add these voltages to get the total applied voltage? Of course not. They aren't in phase. We can only add voltages that are in phase. We must combine the resistor voltage and the capacitor voltage using a graph and trigonometry. This is similar to the process we used with circuits including resistors and inductors.

Figure 2A is a graph like we used with the resistor/inductor circuits. Once again, the voltage across the resistor is in phase with the current. We graph the resistor voltage along the 0° line because we use current as a reference. We added this resistor voltage to the graph in Figure 2B.

What is the phase relationship between the current through a capacitor and the voltage across it? Remember our friend ELI, the ICE man? This mnemonic reminds us that current (I) *leads* voltage (E) through a capacitor (C). (We also can say the voltage *lags* the current.)

Figure 2C shows how we add the capacitor voltage to our graph. Why did we draw this voltage along the 270° line? Since the voltage across a capacitor lags the current by 90°, it also lags the resistor voltage.

Using the resistance and reactance values for alternate sides, draw a rectangle on the graph. Then draw a diagonal line starting at the origin of our graph. Figure 2D shows the completed rectangle, with the diagonal line representing the total applied voltage.

We'll calculate angle A using the tangent function from trigonometry. Then we'll use the sine function to calculate the total combined voltage.

$$\tan(A) = \frac{\text{opposite}}{\text{adjacent}} \qquad \text{(Equation 3)}$$

$$\tan(A) = \frac{6.25\text{ V}}{22.5\text{ V}} = 0.278$$

Next, use your calculator to find the inverse tangent of 0.278.

$$A = \tan^{-1}(0.278)$$
$$A = 15.54^\circ$$

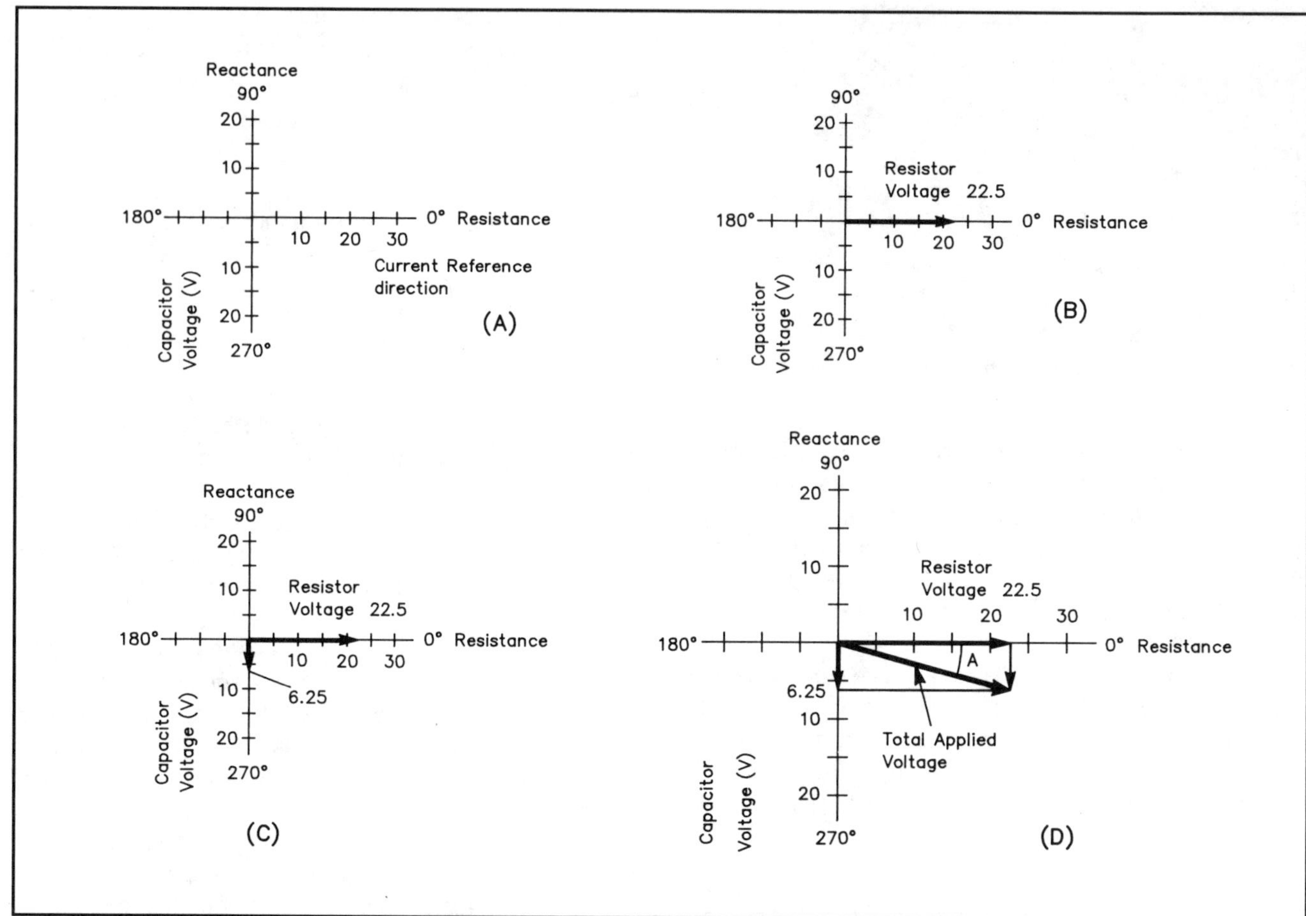

Figure 2—This diagram shows how we construct a set of X-Y coordinates to graph resistor voltage and capacitor voltage. Part A shows the coordinates, with degree markings to indicate phase relationships. We added the resistor voltage in Part B and the capacitor voltage in Part C. Part D shows the total voltage as the diagonal of a rectangle constructed from the resistor and capacitor voltages.

We'll use the sine function to find the hypotenuse of this right triangle. The hypotenuse, or side opposite the right angle, represents the total applied voltage.

$$\sin(A) = \frac{\text{opposite}}{\text{hypotenuse}} \qquad \text{(Equation 4)}$$

$$\text{hypotenuse} = \frac{\text{opposite}}{\sin(A)}$$

$$E_{TOTAL} = \frac{6.25\ V}{\sin(15.54°)} = \frac{6.25\ V}{0.268}$$

$$E_{TOTAL} = 23.32\ V$$

Angle A on our graph represents the phase angle between the total applied voltage and the circuit current. We calculated the phase angle to be 15.54° earlier. The current leads the total voltage by 15.54°. We also say the voltage lags the current by 15.54°.

We usually call this a negative angle because it goes clockwise from 0°. We measure positive angles counter-clockwise from 0°. To be completely correct, then, we will specify this voltage as 23.32 V at –15.54°.

There is another way to arrive at this negative angle. We draw the voltage across the capacitor along the 270° line, but this also is the negative reactance axis. In fact, normal practice is to label capacitive reactance as a negative value. You will normally see the minus sign followed by a lower case *j*. We won't go into the mathematical significance of using *j* here. Just think of it as a way to remind you that capacitive reactance goes along the graph's negative reactance axis. A reactance written as *j*250 Ω or +*j*250 Ω refers to an inductive reactance. In this case the plus sign indicates you should put this reactance along the positive reactance axis on your graph.

Going back to the Ohm's Law equation when we calculated the voltage, we could put a minus sign in front of the capacitive reactance. This would give a negative value for the voltage, showing that we should draw this value along the negative reactance axis on our graph. Then if we include the minus sign with the voltage when we calculate the angle, we get a negative result. Here is what those calculations look like.

$$E_C = I X_C \qquad \text{(Equation 2)}$$

$$E_C = 50 \times 10^{-3}\ \text{A} \times -j125\ \Omega$$
$$E_C = -j6.25\ \text{V}$$

$$\tan(A) = \frac{\text{opposite}}{\text{adjacent}} \qquad \text{(Equation 3)}$$

$$\tan(A) = \frac{-j6.25\ \text{V}}{22.5\ \text{V}} = -j0.278$$

You can drop the *j* here because the minus sign just indicates we measure the angle clockwise from 0° instead of counterclockwise.

$$A = \tan^{-1}(-0.278)$$
$$A = -15.54°$$

There are several ways to solve any trigonometry problem. You may want to use different functions than we used in this example. You may want to find the total voltage using the Pythagorean Theorem. If you follow the steps carefully, all these options will give the same results.

Suppose you want to calculate the circuit impedance instead of the total applied voltage. You can follow a procedure similar to the one we used to find the voltage. Circuit impedance is the combination of resistance and reactance found using a right triangle and trigonometry.

We always place the resistance along the 0° line when we draw an impedance triangle. We draw capacitive reactance along the 270° line for a series circuit. These directions correspond to the directions of the voltages across the resistor and capacitor.

Figure 3 shows the impedance graph for our problem. The phase angle goes in the negative direction again, because the reactance is capacitive. The diagonal line represents the circuit impedance, Z. Use the tangent function to calculate the phase angle and the sine function to calculate the impedance value.

$$\tan(A) = \frac{\text{opposite}}{\text{adjacent}} \qquad \text{(Equation 3)}$$

$$\tan(A) = \frac{-j125\ \Omega}{450\ \Omega} = -0.278$$

Next, use your calculator to find the inverse tangent of −0.278.

$$A = \tan^{-1}(-0.278)$$
$$A = -15.54°$$

Since we measure this angle clockwise from 0°, this is a negative angle.

$$\sin(A) = \frac{\text{opposite}}{\text{hypotenuse}} \qquad \text{(Equation 4)}$$

$$\text{hypotenuse} = \frac{\text{opposite}}{\sin(A)}$$

$$Z = \frac{-j125\ \Omega}{\sin(-15.54°)} = \frac{-j125\ \Omega}{-0.268}$$

$$Z = 466\ \Omega$$

Notice the minus signs cancel here. The impedance is positive, but at a negative angle. We must include the angle to specify the impedance completely.

$Z = 466\ \Omega$ at $-15.54°$

Let's confirm that this impedance gives the same circuit voltage we found earlier. We'll use the circuit impedance and the circuit current in Ohm's Law.

$$E_{TOTAL} = I Z \qquad \text{(Equation 5)}$$

$$E_{TOTAL} = 50\ \text{mA} \times 466\ \Omega$$

$$E_{TOTAL} = 50 \times 10^{-3}\ \text{A} \times 466\ \Omega$$

$$E_{TOTAL} = 23.3\ \text{V}$$

As you might guess, this voltage lags the circuit current by 15.54°. We can write the total voltage as:

$E_{TOTAL} = 23.3$ V at $-15.54°$.

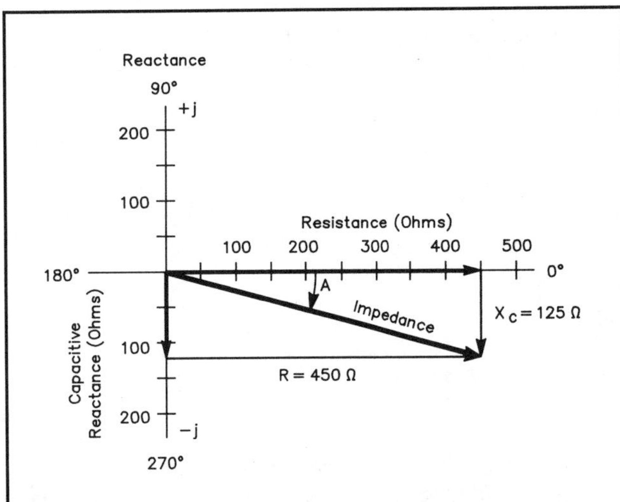

Figure 3—This graph shows resistance and capacitive reactance plotted on a graph. We use the diagonal of a rectangle to combine these values and find the circuit impedance. The text explains the calculations.

Voltage and Current in an RLC Circuit

In this section we will study circuits with resistors, inductors *and* capacitors in series. We will combine the techniques you learned for circuits with either inductors *or* capacitors and resistors in series. If you understand how to solve the earlier problems, these will seem easy. If you still have some difficulty with those problems, this section will give you some additional practice to build your confidence.

Figure 1 shows a circuit with one inductor, one capacitor and one resistor all in series. A sine-wave signal generator connects to these components. We show the inductive reactance and capacitive reactance to save you the work of calculating those values. If we gave you inductance and capacitance instead, you also would need to know the generator frequency to calculate the reactance. The values given apply only to one frequency. If we change the generator frequency, both reactances will change.

What is the generator output voltage for the Figure 1 circuit? What is the total circuit impedance? Assume we measure the total circuit current, and find it to be 450 mA.

We plot inductive reactance and capacitive reactance in opposite directions on the reactance axis. We plot the voltages across the inductor and capacitor in the same direction as their reactances. This means the voltages also go in opposite directions along the Y or reactance axis.

Suppose our circuit had several resistors instead of just one. You know we would add all the resistor values, combining them into a single resistor value. Likewise, if we have more than one inductor, we add all the inductive reactances, to get a single value. If the circuit had several capacitors, we would add their reactances, to have one total capacitive reactance value.

Let's calculate the circuit impedance first. Figure 2A shows a graph of resistance, capacitive reactance and inductive reactance. Remember that we write inductive reactance with a $+j$ to indicate it goes along the positive reactance axis. Capacitive reactance goes along the negative reactance axis, so we write it with a $-j$. Our graph shows the following values:

$R = 8660\ \Omega$

$X_L = +j25{,}000\ \Omega$

$X_C = -j20{,}000\ \Omega$

The next step is to combine the two reactance values. Adding a negative number to a positive number is the same as subtracting the negative value.

$$X_{TOTAL} = X_L + X_C \qquad \text{(Equation 1)}$$

$$X_{TOTAL} = +j25{,}000\ \Omega + (-j20{,}000\ \Omega)$$

$$X_{TOTAL} = +j25{,}000\ \Omega - j20{,}000\ \Omega$$

$$X_{TOTAL} = +j5{,}000\ \Omega$$

Figure 2B is a revised graph, showing the resistance and this combined reactance. We also drew a rectangle from these two values, and added the impedance diagonal. Now we have a right triangle, and we use our trigonometry equations to solve for the impedance, Z. We also use trigonometry to calculate the phase angle, A.

Let's use the tangent function to find the phase angle first.

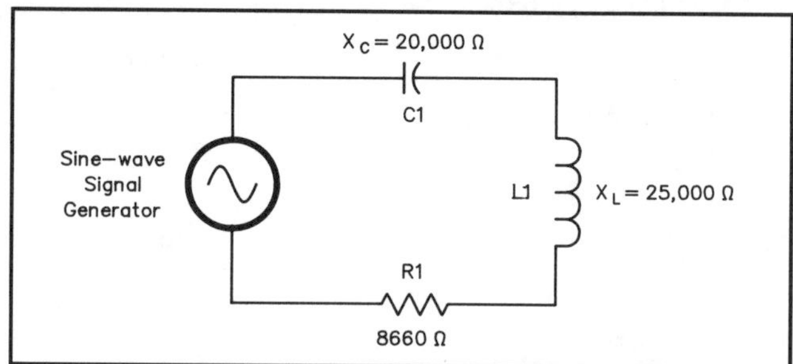

Figure 1—We'll use this circuit with one capacitor, one inductor and one resistor to learn how to calculate series circuit impedance. We use similar methods to calculate the voltage across each circuit element and the total voltage applied by the signal generator.

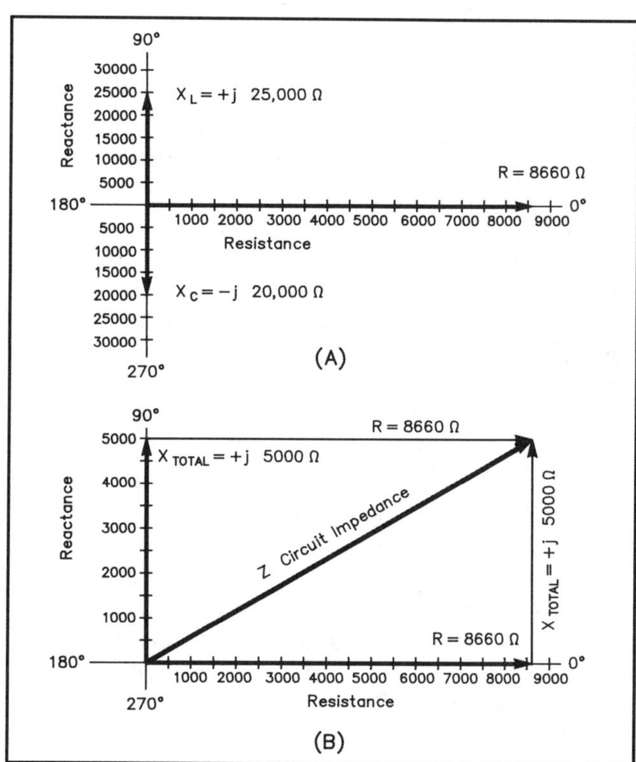

Figure 2—Part A is a graph showing the resistance, inductive reactance and capacitive reactance. Notice that inductive reactance is along the positive Y axis, or in the $+j$ direction. The capacitive reactance is along the negative Y axis, or in the $-j$ direction.

$$\tan(A) = \frac{\text{opposite}}{\text{adjacent}} \qquad \text{(Equation 2)}$$

$$\tan(A) = \frac{+j5000\ \Omega}{8660\ \Omega} = 0.577$$

Then find the inverse tangent of 0.577 to calculate the angle.

$A = \tan^{-1}(0.577) = 30.0°$

Now let's calculate the impedance value using the sine function. Solve Equation 3 for the hypotenuse, or impedance value.

$$\sin(A) = \frac{\text{opposite}}{\text{hypotenuse}} \qquad \text{(Equation 3)}$$

$$Z = \frac{X_{TOTAL}}{\sin(A)}$$

$$Z = \frac{+j5000\ \Omega}{\sin(30°)} = \frac{+j5000\ \Omega}{0.500}$$

$Z = 10{,}000\ \Omega$

When we include the phase angle, we will specify this circuit impedance as:

$Z = 10{,}000\ \Omega$ at 30°

Now let's use this impedance value in Ohm's Law with the circuit current to calculate the total applied voltage from the generator.

$E_{TOTAL} = I Z$ (Equation 4)

$E_{TOTAL} = 450\text{ mA} \times 10{,}000\ \Omega$

$E_{TOTAL} = 450 \times 10^{-3}\text{ A} \times 10{,}000\ \Omega$

$E_{TOTAL} = 4500\text{ V}$

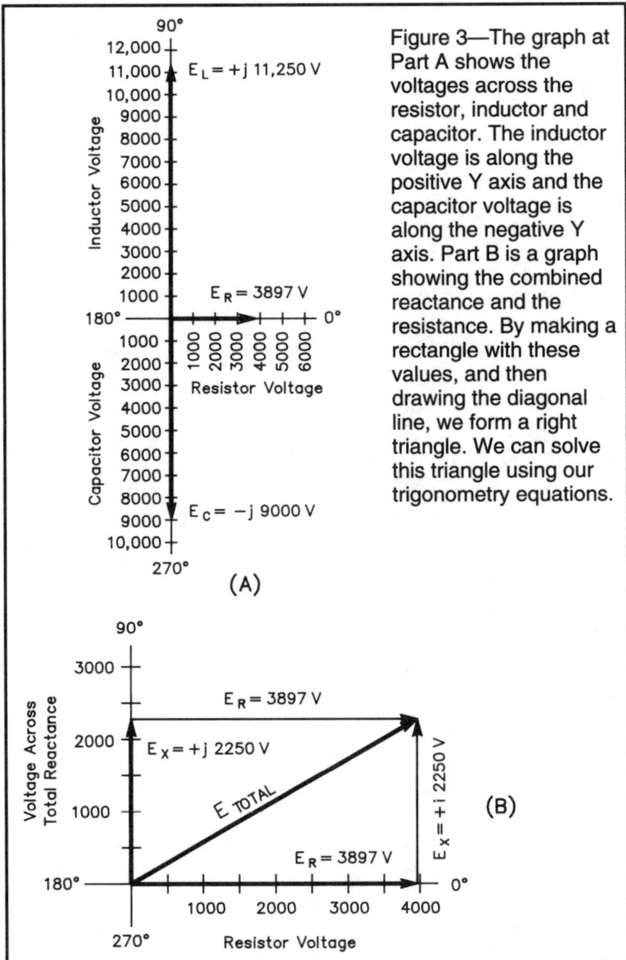

Figure 3—The graph at Part A shows the voltages across the resistor, inductor and capacitor. The inductor voltage is along the positive Y axis and the capacitor voltage is along the negative Y axis. Part B is a graph showing the combined reactance and the resistance. By making a rectangle with these values, and then drawing the diagonal line, we form a right triangle. We can solve this triangle using our trigonometry equations.

Let's calculate the voltage by finding the voltage across each component and then combining those values. Then we can compare the results found by each method. We'll use Ohm's Law to calculate each voltage.

$E_R = I R$ (Equation 5)

$E_R = 450 \times 10^{-3}\text{ A} \times 8660\ \Omega$

$E_R = 3897\text{ V}$

$E_L = I X_L$ (Equation 6)

$E_L = 450 \times 10^{-3}\text{ A} \times +j25{,}000\ \Omega$

$E_L = +j11{,}250\text{ V}$

$E_C = I X_C$ (Equation 7)

$E_C = 450 \times 10^{-3}\text{ A} \times -j20{,}000\ \Omega$

$E_C = -j9{,}000\text{ V}$

Remember the $+j$ tells us to plot this voltage along the positive Y axis and the $-j$ tells us to plot the value along the negative Y axis. Figure 3A is a graph with the three voltages plotted.

Our next step is to combine the inductor and capacitor voltages. Adding these values is the same as subtracting the 9000-V capacitor voltage from the 11,250-V inductor voltage.

$E_X = E_L + E_C$ (Equation 8)

$E_X = +j11{,}250\text{ V} + -j9000\text{ V}$

$E_X = +j11{,}250\text{ V} - j9000\text{ V} = +j2250\text{ V}$

Figure 3B shows how we combine the total reactance voltage and the resistor voltage. This result gives us the total applied voltage from the signal generator. By now these graphs, with the rectangle and right triangles must look pretty familiar. Maybe you've even worked ahead to find the voltage already. If not, let's do that now.

$$\tan(A) = \frac{\text{opposite}}{\text{adjacent}} \qquad \text{(Equation 2)}$$

$$\tan(A) = \frac{+j2250\text{ V}}{3897\text{ V}} = 0.577$$

Then find the inverse tangent of 0.577 to calculate the angle.

$A = \tan^{-1}(0.577) = 30.0°$

$$\sin(A) = \frac{\text{opposite}}{\text{hypotenuse}} \qquad \text{(Equation 3)}$$

$$E_{TOTAL} = \frac{E_X}{\sin(A)}$$

$$E_{TOTAL} = \frac{+j2250\text{ V}}{\sin(30°)} = \frac{+j2250\text{ V}}{0.500}$$

$E_{TOTAL} = 4500\text{ V}$

When we include the phase angle, we will specify this total applied voltage from the signal generator as:

$E_{TOTAL} = 4500\text{ V}$ at 30°

Surely this answer doesn't surprise you!

Have you noticed the voltage across the reactive elements (inductors and capacitors) is larger than the total voltage applied by the generator? This has to do with the way capacitors and inductors store energy. The inductor or capacitor voltage equals the applied voltage, *plus* a voltage created by the stored energy.

Voltage and Current in Parallel Circuits

All the problems we studied so far in this chapter have been with series circuits. You will frequently find parallel circuits in electronics problems, however. We can use techniques similar to the ones you learned for series circuits to solve problems involving parallel circuits. There are several important differences, though.

As always, Ohm's Law relates voltage, current and resistance, reactance or impedance. We can't find circuit impedance directly with a graph and right triangle like we did with series circuits. The process is a bit more involved. We'll skip that procedure in this book. We will show you how you can calculate circuit impedance using Ohm's Law, however.

Look at Figure 1. This circuit includes a resistor, a capacitor and an inductor. A sine-wave signal generator applies a voltage across all three components. The generator voltage goes directly to all three components. The generator output current has a choice of flowing through either the resistor, the inductor *or* the capacitor. These observations clearly indicate that this is a *parallel circuit.*

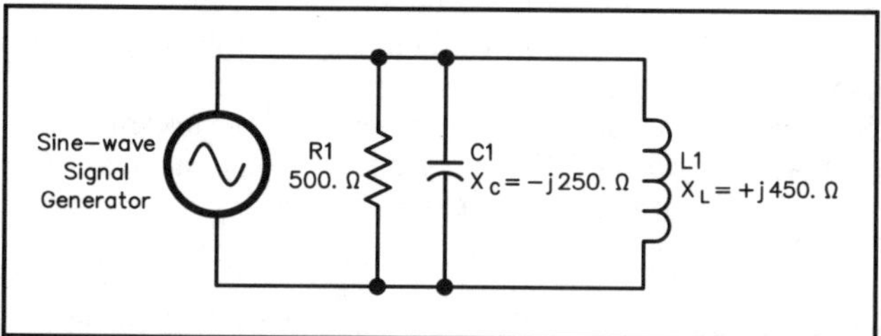

Figure 1—This parallel circuit will help us learn how to calculate circuit currents and impedance. We'll use Ohm's Law and the trigonometry equations to make our calculations. The text explains the steps involved.

Suppose we measure the signal-generator output voltage as 10.0 volts. How much current flows through each component in this circuit? Equations 1, 2 and 3 show how to write Ohm's Law for each current.

$$E_R = I_R R \qquad \text{(Equation 1)}$$

$$E_C = I_C X_C \qquad \text{(Equation 2)}$$

$$E_L = I_L X_L \qquad \text{(Equation 3)}$$

We'll solve each of these equations for current (I) and calculate the respective values.

$$E_R = I_R R \qquad \text{(Equation 1)}$$

$$I_R = \frac{E_R}{R}$$

$$I_R = \frac{10.0 \text{ V}}{500.\ \Omega}$$

$$I_R = 0.0200 \text{ A} = 20.0 \text{ mA}$$

This is the current through the resistor.

$$E_C = I_C X_C \qquad \text{(Equation 2)}$$

$$I_C = \frac{E_C}{X_C}$$

$$I_C = \frac{10.0 \text{ V}}{250.\ \Omega}$$

$$I_C = 0.0400 \text{ A} = 40.0 \text{ mA}$$

We did not include the $-j$ designator with this calculation. We will consider which direction to plot the capacitor current and the inductor current shortly.

$$E_L = I_L X_L \qquad \text{(Equation 3)}$$

$$I_L = \frac{E_L}{X_L}$$

$$I_L = \frac{10.0 \text{ V}}{450.\ \Omega}$$

$$I_L = 0.0222 \text{ A} = 22.2 \text{ mA}$$

We'll have no difficulty deciding which direction to plot the resistor current on the graph of Figure 2A. As always, the voltage and current through a resistor are in phase. The positive X axis represents a 0° phase angle, so this is where we'll plot I_R.

Where do we plot the capacitor voltage? Well, remember our friend ELI, the ICE man. Current *leads* voltage through a capacitor. The capacitor current is 90° ahead of the total voltage. We draw the capacitor current along the positive Y axis, at 90°. Notice that we called this the $+j$ axis for series problems. We still call it the $+j$ axis, but note it's the capacitor value that goes in this direction.

You probably can guess where to plot the inductor current. This current *lags* the total applied voltage, as our mnemonic reminds us. So it goes along the negative Y axis, or in the $-j$ direction. This is at 270°, 90° behind the voltage.

Figure 2B shows the three currents plotted on the graph. We combine the capacitor current and the inductor current by addition. Since the inductor value is negative, this is the same as subtracting this value from the capacitor current.

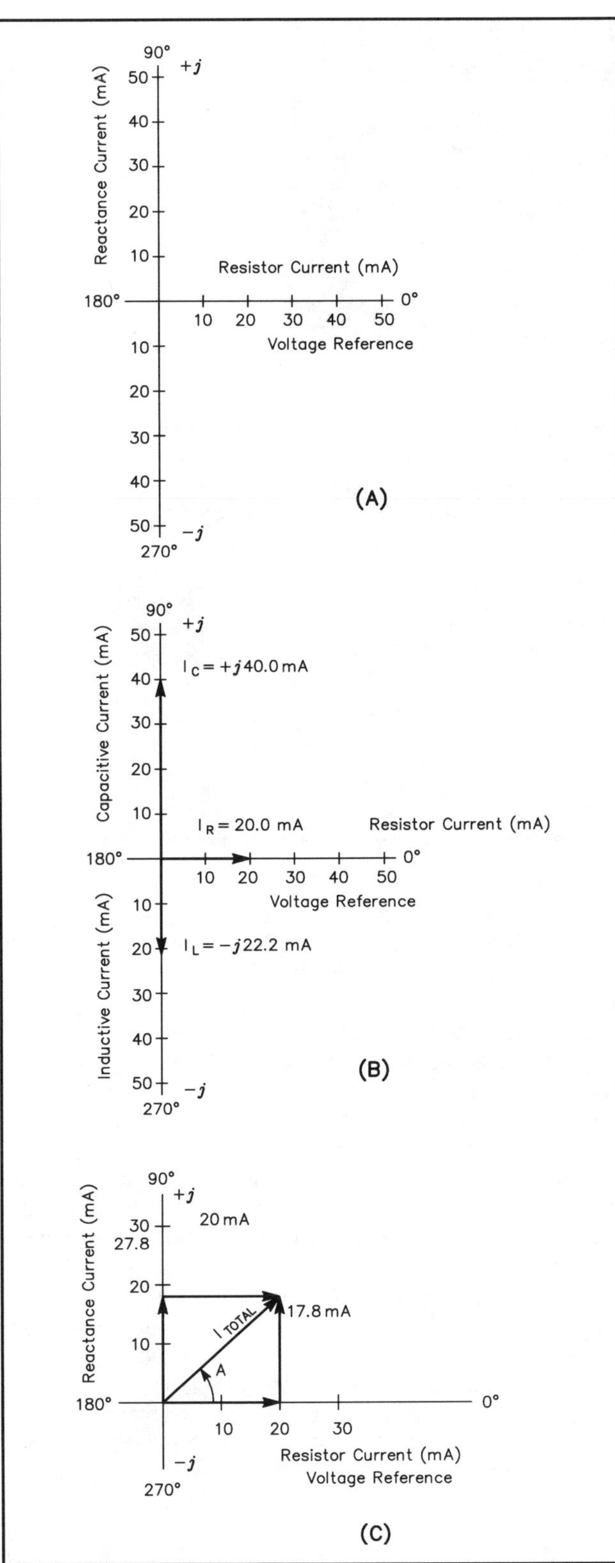

Figure 2—These graphs illustrate how to plot the current through each component. The text explains why the capacitor current is along the $+j$ axis and the inductor current is along the $-j$ axis. Part C shows the resistor current and the combined reactance current. The diagonal of this rectangle represents the total circuit current.

$$I_X = I_C + I_L \qquad \text{(Equation 4)}$$

$$I_X = +j40.0\text{ mA} + (-j22.2\text{ mA})$$

$$I_X = +j40.0\text{ mA} - j22.2\text{ mA}$$

$$I_X = +j17.8\text{ mA}$$

The capacitor current is larger, so the total reactance current is along the positive Y axis, or $+j$, at 90°. We plot this value on Figure 2C. This graph also shows the completed rectangle, with the diagonal line drawn to represent the total circuit current. Once again, we use our trigonometry equations to calculate the hypotenuse of the right triangle. This time the hypotenuse represents total circuit current, I_{TOTAL}.

$$\tan(A) = \frac{\text{opposite}}{\text{adjacent}} \qquad \text{(Equation 5)}$$

$$\tan(A) = \frac{17.8\text{ mA}}{20.0\text{ mA}} = 0.890$$

$$A = \tan^{-1}(0.890)$$

$$A = 41.7°$$

$$\sin(A) = \frac{\text{opposite}}{\text{hypotenuse}} \qquad \text{(Equation 6)}$$

$$\sin(41.7°) = \frac{17.8\text{ mA}}{I_{TOTAL}}$$

$$I_{TOTAL} = \frac{17.8\text{ mA}}{\sin(41.7°)} = \frac{17.8\text{ mA}}{0.665}$$

$$I_{TOTAL} = 26.8\text{ mA}$$

The total current is 26.8 mA at 41.7°. The total current leads the voltage by this angle.

What is the circuit impedance? The easiest way to find out is by using Ohm's Law.

$$E_{TOTAL} = I_{TOTAL} \times Z \qquad \text{(Equation 7)}$$

$$Z = \frac{E_{TOTAL}}{I_{TOTAL}}$$

$$Z = \frac{10\text{ V}}{26.8\text{ mA}} = \frac{10\text{ V}}{26.8 \times 10^{-3}\text{ A}}$$

$$Z = 373\ \Omega \text{ at } 41.7°$$

Of course, we already knew the phase angle, because we calculated it with the current.

You can use this technique to find the impedance of a parallel circuit even if you don't know the applied voltage. Just make up a voltage, and solve for the currents through each component as if you knew that was the applied voltage.

As you learn more about electronics, and begin to study more advanced texts, you will learn other methods to calculate circuit impedance. You will learn some shortcuts as you become more familiar with the mathematics of working with numbers along the resistance and reactance axes of a graph. The method shown here will help you begin to work with circuit impedances without using the advanced math, however.

CHAPTER 24 Resonant Circuits

What Happens if Inductive and Capacitive Reactances are Equal?

Inductive and capacitive reactances have opposite effects on a circuit. The voltage across an inductor is 90° out of phase with the current through the inductor. The current through a capacitor is 90° out of phase with the voltage across it. The voltage leads the current in an inductor and lags the current in a capacitor. In a series circuit, the inductor voltage and capacitor voltage are 180° out of phase. In a parallel circuit, the inductor current and capacitor current are 180° out of phase.

We plot series inductor and capacitor voltages in opposite directions along the Y axis to combine these voltages. We can write the voltage across an inductor with a *+j*. This serves to remind us the value goes along the positive Y axis. We often include a *–j* in front of the voltage across a capacitor. The *–j* tells us this value goes along the negative Y axis.

Inductor and capacitor currents for parallel circuits are 180° out of phase. We plot the capacitor current along the positive Y axis, or in the *+j* direction. Inductor current goes along the negative Y axis, or in the *–j* direction.

How does inductive reactance change when we decrease the signal frequency? Inductive reactance becomes smaller. What happens to the capacitive reactance when we decrease the signal frequency? Capacitive reactance becomes larger.

If we increase the signal frequency, opposite effects occur to inductive and capacitive reactance. Inductive reactance becomes larger and capacitive reactance becomes smaller when we increase the frequency.

Figure 1 includes a graph of inductive reactance for a coil. This line shows how inductive reactance increases as we change the signal frequency. Figure 1 also includes a graph of capacitive reactance. As we increase the signal frequency, the capacitive reactance decreases. We didn't include a frequency scale or a reactance scale because the exact values are unimportant. For any capacitor and inductor, their reactance lines will cross at only one point.

A special condition occurs when inductive and capacitive reactances are equal. This is true whether you have a series or a parallel circuit. We call this condition *resonance*.

Every combination of inductance and capacitance has one frequency at which the reactances are equal. This is the *resonant frequency* for that combination.

Resonance occurs when the reactances are equal. We can find an equation to calculate the resonant frequency of any inductor/capacitor pair. Do you remember how to calculate inductive and capacitive reactances? Equations 1 and 2 will refresh your memory.

$$X_L = 2 \pi f L \qquad \text{(Equation 1)}$$

$$X_C = \frac{1}{2 \pi f C} \qquad \text{(Equation 2)}$$

We know these values are equal at resonance, so we can set the equations equal to each other. Let's use f_r to represent resonant frequency.

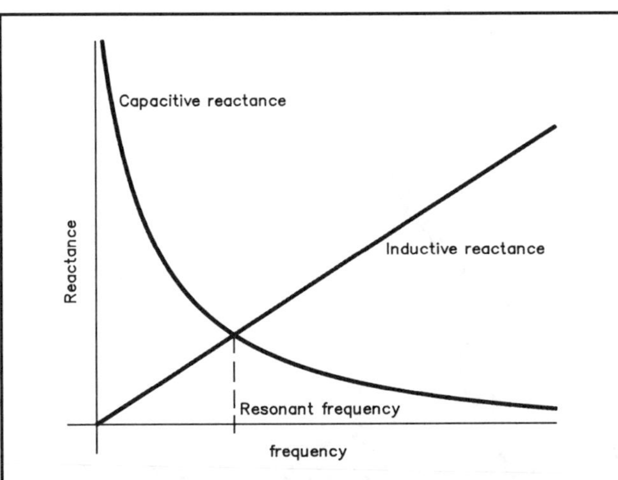

Figure 1—The straight line on this graph represents the inductive reactance for a coil. The reactance increases as the frequency increases. The curved line represents the capacitive reactance for a capacitor. This reactance decreases as the frequency increases. There are no numbers along the X and Y axes because the exact values are unimportant here. We want to show the general shapes of the lines. For a specific inductor and a specific capacitor, there is only one point where the two lines cross. This point represents the resonant frequency of the combination.

$$X_L = 2\,\pi\,f_r\,L = X_C = \frac{1}{2\,\pi\,f_r\,C} \qquad \text{(Equation 3)}$$

$$2\,\pi\,f_r\,L = \frac{1}{2\,\pi\,f_r\,C}$$

Now we'll cross multiply and combine terms to simplify the equation.

$$1 = \frac{1}{(2\,\pi\,f_r\,L)\,(2\,\pi\,f_r\,C)}$$

$$1 = \frac{1}{(2\,\pi\,)^2\,(f_r)^2\,L\,C}$$

$$f_r^2 = \frac{1}{(2\,\pi\,)^2\,L\,C} \qquad \text{(Equation 4)}$$

Finally, take the square root of both sides of the equation.

$$f_r = \frac{1}{2\,\pi\,\sqrt{LC}} \qquad \text{(Equation 5)}$$

Let's use Equation 5 to calculate a resonant frequency. Suppose you have a 20-picofarad capacitor and a 6.28-microhenry inductor. What is the resonant frequency for these two components? Remember that we must write the inductance value in henrys and the capacitance value in farads for this equation.

$$f_r = \frac{1}{2 \times 3.14 \times \sqrt{6.28 \times 10^{-6}\,\text{H} \times 20 \times 10^{-12}\,\text{F}}}$$

$$f_r = \frac{1}{6.28 \times \sqrt{125.6 \times 10^{-18}}}$$

$$f_r = \frac{1}{6.28 \times 11.2 \times 10^{-9}} = \frac{1}{7.04 \times 10^{-8}}$$

$$f_r = 14.2 \times 10^6\,\text{Hz} = 14.2\,\text{MHz}$$

You can use this equation to calculate the resonant frequency for any combination of inductor and capacitor values. You also can solve Equation 4 for values of inductance or capacitance. Given either an inductor or a capacitor, you can calculate the other component value to give a desired resonant frequency.

Resonant circuits are important in radio transmitters and receivers. A resonant circuit sets the operating frequency for an oscillator circuit. Resonant circuits also help select the desired signal and reject unwanted signals.

Figure 2 shows the resonant circuit in a simple variable-frequency oscillator, or VFO. This circuit might set the operating frequency for a simple receiver or transmitter.

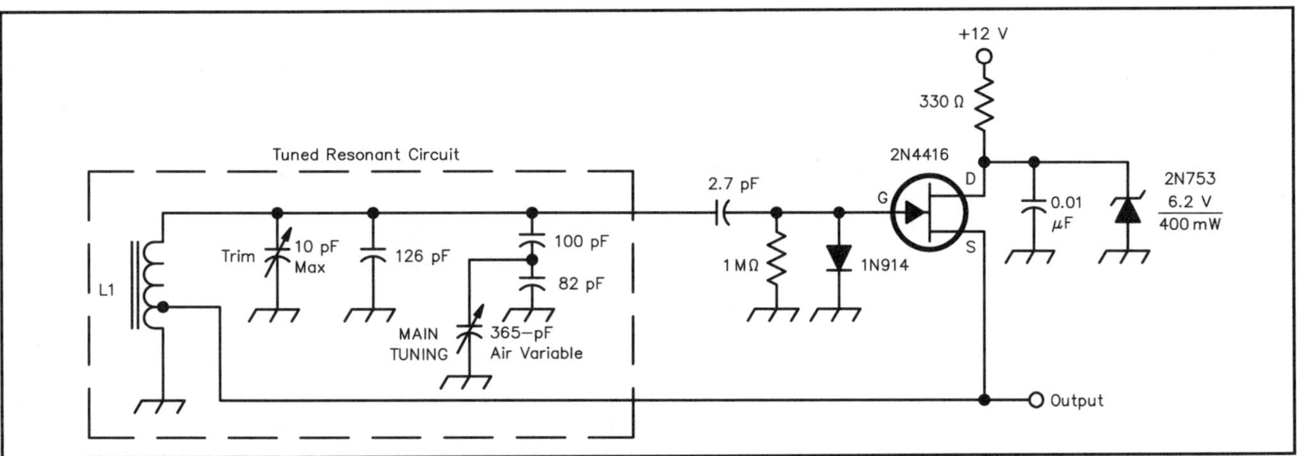

Figure 2—This circuit includes an inductor and a variable capacitor. You can change the oscillator operating frequency by adjusting the variable capacitor. Each capacitor setting produces a different resonant frequency in combination with the inductor. You can use this variable-frequency oscillator, or VFO, to set the operating frequency of a simple receiver or transmitter.

Conditions in a Series Resonant Circuit

Suppose you look around in your collection of used parts, and find a 200-pF capacitor. You also find a 2.44-μH inductor. Let's connect these two parts in series with a 100-Ω resistor and a sine-wave signal generator. Figure 1 shows this circuit. What is its resonant frequency? Equation 1 will help calculate this value.

$$f_r = \frac{1}{2 \pi \sqrt{L C}} \quad \text{(Equation 1)}$$

$$f_r = \frac{1}{2 \times 3.14 \times \sqrt{2.44 \times 10^{-6}\ \text{H} \times 200 \times 10^{-12}\ \text{F}}}$$

$$f_r = \frac{1}{6.28 \times \sqrt{488 \times 10^{-18}}}$$

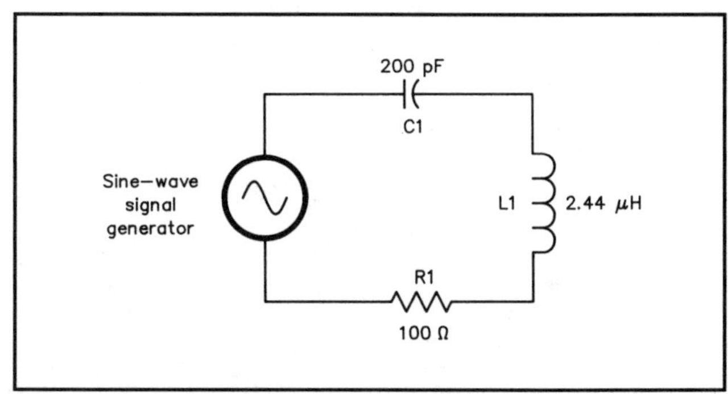

Figure 1—This diagram shows a 2.44-μH inductor, a 200-pF capacitor and a 100-Ω resistor connected in series with a signal generator. We'll use this circuit to study the effects of changing the frequency of a signal applied to a series RLC circuit.

$$f_r = \frac{1}{6.28 \times 22.1 \times 10^{-9}}$$

$$f_r = \frac{1}{1.39 \times 10^{-7}} = 7.19 \text{ MHz}$$

Set the signal generator to 7.19 MHz. What is the current through the circuit if the generator output voltage is 50 V? Remember the inductor voltage and the capacitor voltage are opposite but equal. At the resonant frequency the circuit impedance is equal to the resistor value, 100 ohms. The entire generator voltage is across the resistor at the resonant frequency. Ohm's Law tells us the circuit current.

$$E = I\,R \qquad \text{(Equation 2)}$$

$$I = \frac{E}{R} = \frac{50 \text{ V}}{100\ \Omega}$$

$I = 0.50$ A = 500 mA

Let's think about the circuit current at frequencies above and below resonance. Do you think the current will change? Will it increase or decrease?

Adjust the signal generator to produce a signal at 6.0 MHz. This is lower than the resonant frequency. You know the inductive reactance will decrease and the capacitive reactance will increase. At frequencies below the resonant frequency there is more capacitive reactance. This means the circuit impedance will increase.

Figure 2 is an impedance triangle for this circuit. The $-j$ value represents the difference between the inductive reactance and the capacitive reactance. The hypotenuse

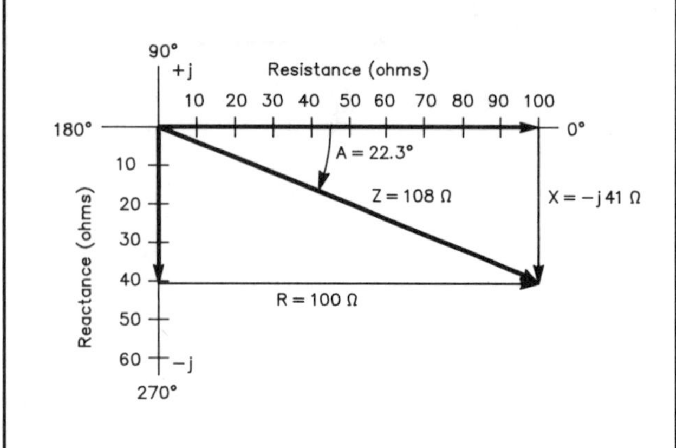

Figure 2—This diagram represents the circuit impedance when the applied signal is below the resonant frequency. Notice the capacitive reactance is larger than the inductive reactance. This results in a total reactance value that is capacitive, or along the $-j$ axis.

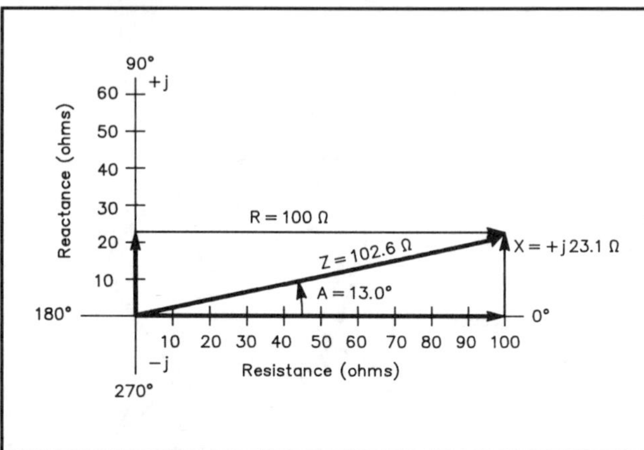

Figure 3—This diagram represents the circuit impedance when the applied signal frequency is above resonance. This time the inductive reactance is larger than the capacitive reactance. The total reactance value is inductive, or along the $+j$ axis.

of a right triangle is *always* longer than either side. Therefore, we know the circuit impedance is greater than 100 ohms.

This means the voltage across the capacitor will be higher than the voltage across the inductor. It also means the resistor voltage is less than 50, because the total generator voltage is still 50 V.

You might want to calculate the new inductive reactance and the new capacitive reactance at 6.0 MHz. This would be a good practice exercise. You also can calculate the new circuit impedance and the voltages across each part.

From this discussion, you can tell the circuit current is less at frequencies below resonance than at resonance. As you go still lower in frequency the capacitive reactance becomes larger and the inductive reactance becomes smaller. These effects produce a circuit impedance that increases as the frequency decreases. The current gets smaller as the frequency decreases.

Now let's consider the effects of increasing the generator frequency above the resonant frequency. Adjust the generator for an 8.0-MHz output signal.

This time the inductive reactance increases to a value larger than it was at resonance. The capacitive reactance decreases to a smaller value. Combining these two values gives a result along the $+j$ axis. Figure 3 shows an impedance diagram for the condition above the resonant frequency.

The triangle's hypotenuse, which represents the circuit impedance, is longer than the resistance, 100 Ω. Above the resonant frequency, the circuit impedance is larger than it is at resonance. Again, the current is smaller than at resonance.

Calculate the inductive and capacitive reactances at 7.5 MHz as another practice exercise. Then you also can calculate the new circuit current and the voltages across the components.

Figure 4 is a graph of the circuit current over a range of frequencies. Notice the shape of this graph. All series RLC circuits (those that include resistance, inductance and capacitance) have a similar current graph. The actual current value depends on the generator voltage and component values.

Circuits with less resistance have higher current at resonance. If there is no resistor, only the component resistance limits the current.

Do you remember the quality factor, or circuit Q? We calculate Q by dividing the reactance by the resistance. Use *either* the inductive reactance *or* the capacitive reactance to calculate the Q of a resonant circuit. Calculate this

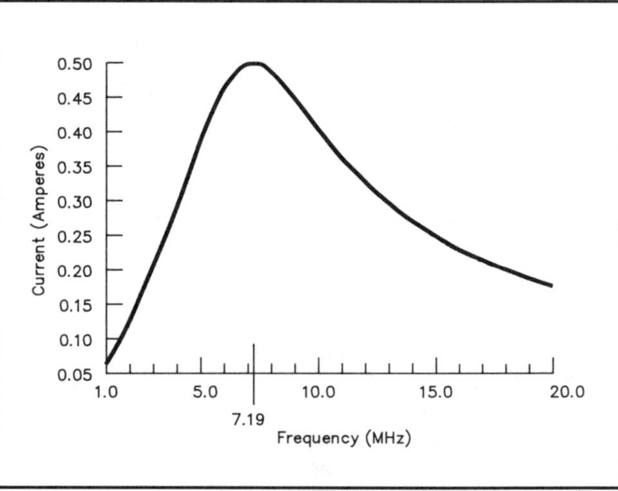

Figure 4—This graph represents the current through the Figure 1 circuit over a range of frequencies. The current reaches its highest value at the circuit resonant frequency. The circuit impedance has its smallest value at the resonant frequency.

reactance at the resonant frequency. Do not use the combined reactance, because that is zero at resonance!

Q helps determine how steep the graph's sides are. Any circuit resistance tends to widen the graph's base. Components with high inherent resistance have a wide current graph. Components with low inherent resistance have a narrow graph.

Figure 5 compares the current graphs for two circuits. One circuit has a Q of 10. The second circuit has a Q of 100. Let's assume both circuits have the same reactance and applied voltage. The circuit with a Q of 100 has ten times less resistance than the circuit with a Q of 10.

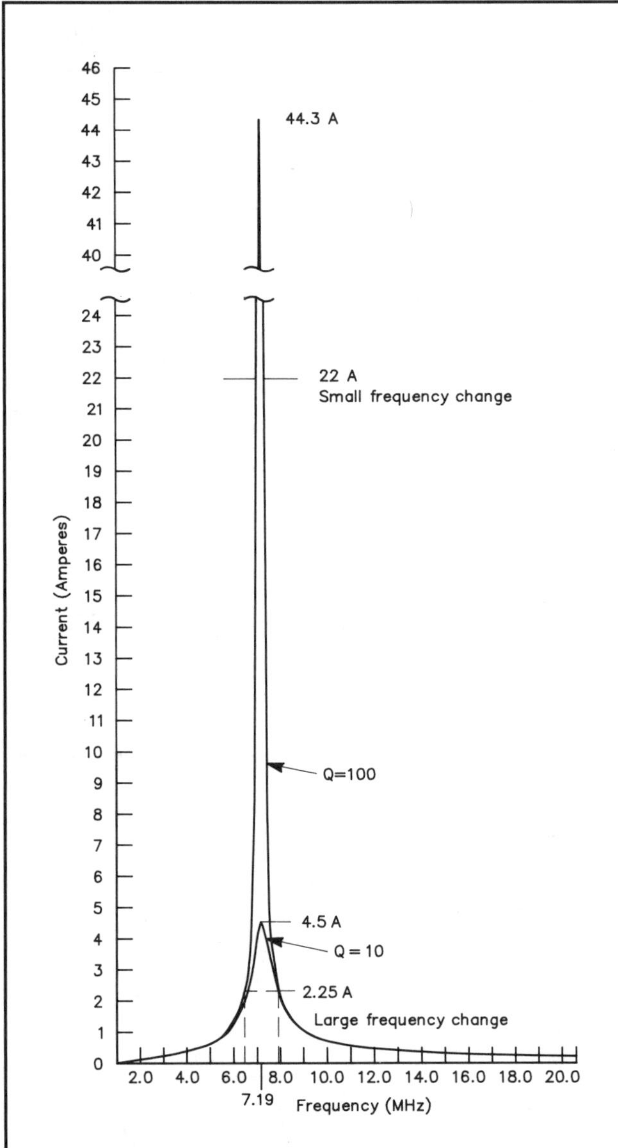

Figure 5—This graph compares the circuit current for a circuit with a Q of 10 and another circuit with a Q of 100. The high-Q circuit has a steeper graph. The current remains low until the signal frequency approaches resonance. Then the current quickly rises to its peak value and drops again at frequencies above resonance.

We can prove this by making up some numbers and calculating Q. Suppose the reactance is 100 Ω in both cases. The high-Q circuit has a resistance of 1 ohm and the low-Q circuit has a resistance of 10 ohms.

$$Q = \frac{X}{R} \qquad \text{(Equation 3)}$$

$$Q_1 = \frac{100\ \Omega}{1\ \Omega} = 100$$

$$Q_2 = \frac{100\ \Omega}{10\ \Omega} = 10$$

The high-Q circuit, with ten times less resistance, has ten times more current. You can prove this using Ohm's Law. The shape of the high-Q current curve is much sharper, however. The current is low until the signal is close to the circuit resonant frequency. Then the current shoots up to a higher peak value. It quickly drops back to a low value as you increase the frequency above resonance.

Look closely at Figure 5. Find the current that is half the peak current for each line. This will be about 22 A for the high-Q circuit and 2.25 A for the low-Q circuit. Now compare the frequency range for these values. It's hard to read the exact frequencies on this graph. You can see the high-Q circuit has a much smaller range than the low-Q circuit, however.

We call this frequency range the *bandwidth.* The signal has a value higher than some specified portion of the maximum value over this range of frequencies. In this example we specified a current of half the peak value. We also could specify a current of three fourths, one fourth or even one tenth the maximum value. Then we can measure the range of frequencies for which the signal has a level higher than the one specified.

Each of these measurements represents a bandwidth, although the frequency range will be different for each. We must always specify a signal level with a bandwidth. In our example, we measured the half-current bandwidth. Another common reference level for bandwidth measurements is half power. You will often see this written as the *–3 dB bandwidth* or just the *3 dB bandwidth.* Reducing the power by one half represents a reduction of 3 dB. (If you don't remember what a decibel (dB) is, you should refer to that section in the Math Skills unit.)

The voltage across the inductor and capacitor can be much higher than the generator voltage at resonance. You may think the voltage across the inductor or capacitor is zero at resonance. This is not true! The voltage across either part may be much higher than the generator voltage.

These large voltages develop because of energy stored in the capacitor's electric field and the inductor's magnetic field. The energy going into the electric field is coming out of the magnetic field. The energy going into the magnetic field is coming out of the electric field. The inductor and capacitor hand this energy back and forth.

Conditions in a Parallel Resonant Circuit

Figure 1 shows a parallel circuit that includes a 200-pF capacitor and a 2.44-μH inductor. The signal generator produces 50 volts at any frequency. Normally, we would include a resistor in series with the generator. This would limit the circuit current. We'll omit the resistor for this discussion because it simplifies calculations. Most practical parallel-resonant circuits include some series resistance, though.

Equation 1 helps us calculate the resonant frequency for this circuit. We used this same equation to calculate the resonant frequency of a series resonant circuit.

$$f_r = \frac{1}{2\pi\sqrt{LC}} \qquad \text{(Equation 1)}$$

$$f_r = \frac{1}{2 \times 3.14 \times \sqrt{2.44 \times 10^{-6}\ \text{H} \times 200 \times 10^{-12}\ \text{F}}}$$

$$f_r = \frac{1}{6.28 \times \sqrt{488 \times 10^{-18}}}$$

$$f_r = \frac{1}{6.28 \times 22.1 \times 10^{-9}}$$

$$f_r = \frac{1}{1.39 \times 10^{-7}} = 7.19\ \text{MHz}$$

Generator current has two paths to choose. There is some current through the inductor and some current through the capacitor. The two paths are *branch circuits*, and the current through each part is a *branch current.*

The inductor current lags the applied voltage by 90°. (Remember ELI the ICE man.) The capacitor current leads the applied voltage by 90°. This means the inductor current and capacitor current are 180° out of phase. Figure 2 is a graph showing the branch currents and the voltage reference. The inductor current is along the negative Y axis, or at *–j*. The capacitor current is along the positive Y axis, or at *+j*.

Points A and B on Figure 1 give us another way to think about the branch currents. At one instant the alternating current through the inductor seems to flow from A to B. At this same instant the current through the capacitor seems to flow from B to A. During the next half cycle of the ac waveform, these directions reverse.

We find the total circuit current taken from the generator by adding the two branch currents. (This means the inductor-current value subtracts from the capacitor-current value.)

When you change the applied-signal frequency, the inductive and capacitive reactances change. Ohm's Law tells us the branch currents also change. Figure 3 graphs the capacitor current, the inductor current and the total generator current. Remember, the capacitor current and the inductor current are in opposite directions at any instant. They are 180° out of phase.

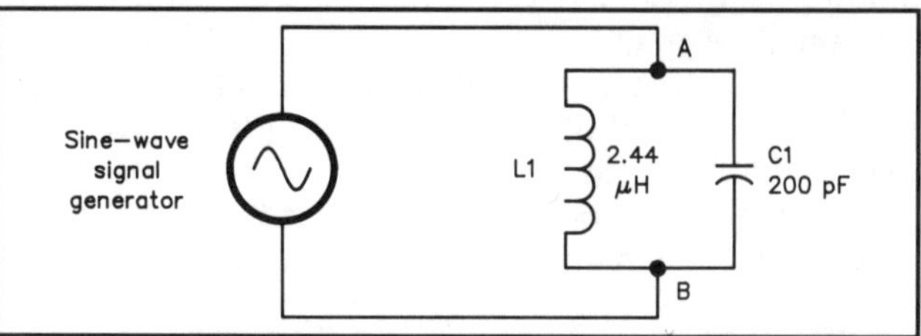

Figure 1—The inductor and capacitor in this circuit connect to the generator in parallel. We will use this circuit to study the conditions in a parallel-resonant circuit.

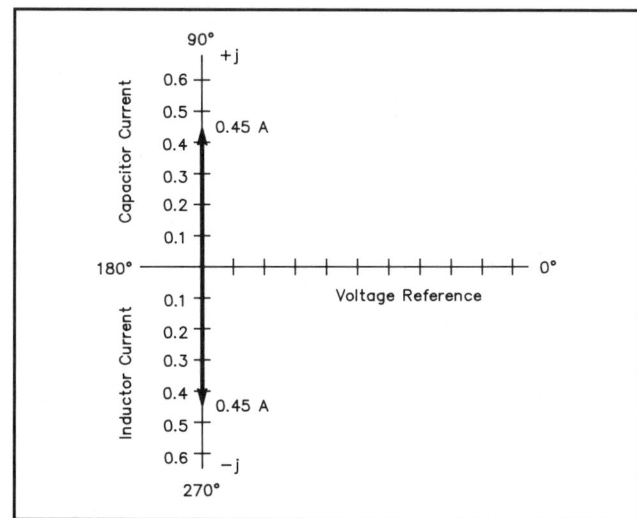

Figure 2—Inductor current lags applied voltage by 90°, so we plot it along the *–j* axis. The 0° axis represents the reference-voltage direction. Capacitor current leads applied voltage by 90°, so we plot it along the *+j* axis.

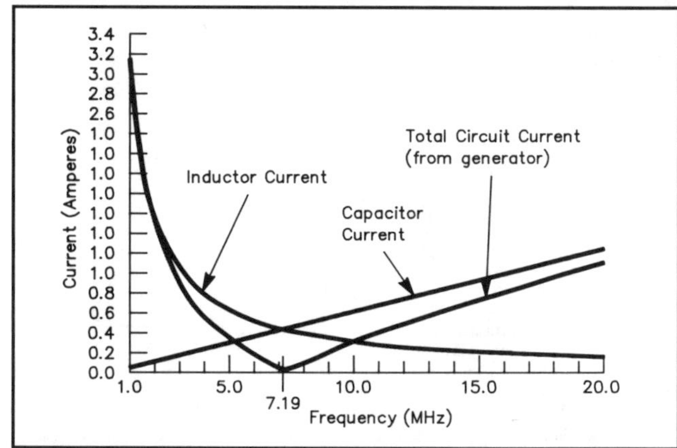

Figure 3—This graph compares the inductor current, capacitor current and total generator current. Generator current is the difference between the capacitor and inductor current values. Notice that at the circuit resonant frequency the total current becomes zero.

When you examine Figure 3, you will see the total current becomes zero at the resonant frequency. This does not mean the current through the inductor or capacitor is zero, however. It only means the current into or out of the generator is zero at this point.

Well, the generator current is zero if we use an ideal inductor and capacitor. Real inductors and capacitors always have some resistance. The generator will supply the small current needed to replace the loss from this resistance in a practical circuit.

The generator current is very small at resonance. There can be large currents through the branches, however. The branch currents seem to circulate through the inductor and the capacitor.

The inductor and capacitor hand energy back and forth between them. The inductor's magnetic field stores energy, then returns it to the circuit. The capacitor stores the returned energy in its electric field, then hands it back to the inductor.

You may want to calculate the inductive and capacitive reactances for several frequencies. Then use Ohm's Law to calculate the current through those reactances. The generator in our circuit applies 50 volts to the circuit. Combine the capacitor current and the inductor current to find the total generator current. If you calculate enough values you could plot a graph like Figure 3. Table 1 summarizes the values at several frequencies. Equations 2, 3 and 4 will help you with these calculations.

$$X_L = 2 \pi f L \qquad \text{(Equation 2)}$$

$$X_C = \frac{1}{2 \pi f C} \qquad \text{(Equation 3)}$$

$$I = \frac{E}{X} \qquad \text{(Equation 4)}$$

When you know the total circuit current and the applied voltage, Ohm's Law will help you calculate the circuit impedance. Divide the voltage by the current.

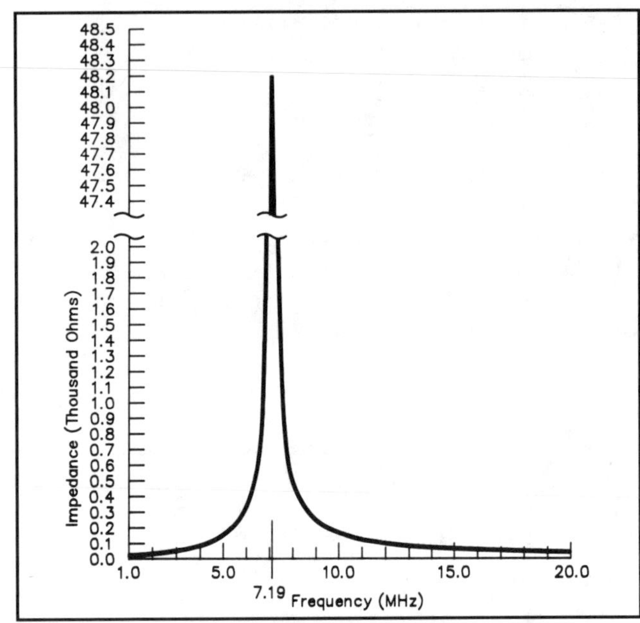

Figure 4—This graph shows how the parallel-circuit impedance changes as the frequency changes. The impedance is low above and below the resonant frequency, but becomes very large at resonance.

$$Z = \frac{E_{APPLIED}}{I_{TOTAL}} \qquad \text{(Equation 5)}$$

Figure 4 is a graph of the circuit impedance over a wide frequency range. At frequencies below the resonant frequency, the impedance is low. The capacitor has a high reactance so its current is small. The inductor has a low reactance at low frequencies, however, so most of the circuit current goes through it.

At frequencies above the resonant frequency, the impedance is also low. Now the inductor has a high impedance, and a small current. The capacitor has a low impedance, though, so most of the circuit current goes through it.

At the resonant frequency, the reactances are opposite. The inductor current and the capacitor current are equal but opposite. The circuit current is zero with ideal components. When a large applied voltage produces no current, or a very small current, the impedance is large.

Circuit Q is important with parallel resonant circuits, just as with series resonant circuits. High-Q inductors and capacitors have little loss. Low-Q components have larger losses. High-Q circuits have a narrower bandwidth than low-Q circuits.

Table 1
Calculated Circuit Values

Frequency (MHz)	*Capacitive Reactance X_C (ohms)*	*Inductive Reactance X_L (ohms)*	*Capacitor Current I_C (A)*	*Inductor Current I_L (A)*	*Generator Current I_T (A)*	*Circuit Impedance Z (ohms)*
2.0	398	31	0.126	1.613	1.487	33.6
3.0	265	46	0.189	1.087	0.898	55.7
4.0	199	61	0.251	0.820	0.569	87.9
5.0	159	77	0.314	0.649	0.335	149.3
6.0	133	92	0.376	0.543	0.167	299.4
7.0	114	107	0.439	0.467	0.028	1785.7
7.19	110.7	110.2	0.452	0.454	0.002	25000.
8.0	100	123	0.500	0.407	0.093	537.6
9.0	88	138	0.568	0.362	0.206	242.7
10.0	80	153	0.625	0.327	0.298	167.8
11.0	72	169	0.694	0.296	0.398	125.6
12.0	66	184	0.758	0.272	0.486	102.9
13.0	61	199	0.820	0.251	0.569	87.9
14.0	57	214	0.877	0.234	0.643	77.8

UNIT 4

A Few More Building Blocks

Have you been studying this book through from th beginning? If so, you know many important electronics prir ciples. You can calculate circuit conditions with a steady d signal applied. You also can calculate many circuit condition with an ac signal applied.

So far we only studied *passive* components in ou circuits. Passive components simply take an input signal an create an output signal. They require no additional voltag source. The output signals are normally smaller than the inpu signals. This means that the components lose some signa (Actually, the components change some of the signal into hea or other energy forms.)

Passive components are important. Every electronic circuit includes passive components. There are electronic functions that passive components can't perform, howeve Passive components can't amplify signals, or make ther stronger. Passive components can't form an oscillator t generate an ac signal from a dc input. There are other type of signal processing that passive components can't do, eithe These functions require another type of component—th *active component.*

Active Components require an input voltage t make them work. Usually this is a dc signal that se the operating point and gives the component th energy it needs to work properly. Active component use this input voltage to amplify signals, create a signals and perform other functions.

Active components are an important part of mar electronics circuits. Without active components ther would be no radio and TV transmitters or receivers a we know them. There would be no computers electronic calculators.

Let's turn our attention to some active compo nents. We'll learn about several types of components and ho they work. We'll see how to use those components in electro ics circuits. We also will learn about some of the functior these components perform in electronics circuits.

In This Unit You Will Learn:

- What semiconductors are.
- That diodes allow current in only one direction and how we use this principle in electronics circuits.
- About bipolar and field-effect transistors and their use in circuits.
- That integrated circuits (ICs) contain entire circuits on a tiny piece of semiconductor material.
- How vacuum tubes operate.

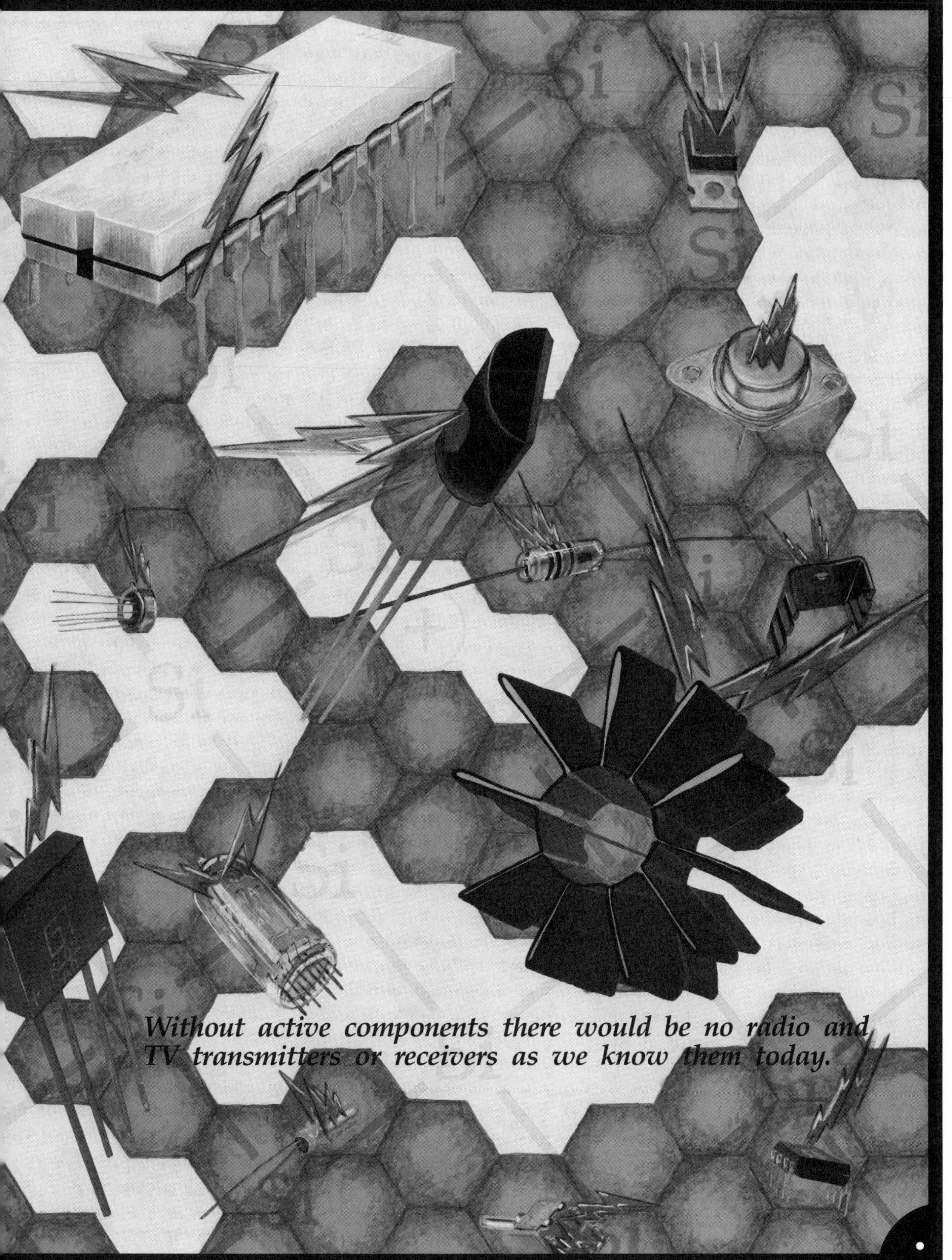

Without active components there would be no radio and TV transmitters or receivers as we know them today.

CHAPTER 25 Semiconductor Materials

Let's Start with Atoms

Many modern electronics circuits contain materials that we call *semiconductors*. Manufacturers make many different devices with semiconductor materials. It is important that you have a basic understanding of how these semiconductor materials work. That will help you learn about modern electronics.

We would have to study some physics and chemistry to gain a complete understanding of what is going on inside semiconductor materials. Don't worry, though. You can understand the basics of how the materials work even if you never studied these sciences! In fact, we have already covered most of what you need to know in another section of this book. We described a simple model of an atom in the section about insulators and conductors.

Remember that tiny particles called *atoms* make up all matter. These atoms are so small that no one can see them without the help of a very powerful microscope. Atoms are the basic building blocks of all matter. Scientists call each different type of atom an *element*. They have found more than 100 elements and are always looking for more.

Atoms form larger building blocks, which we call *molecules*. Molecules are combinations of atoms from several elements or several atoms of a single element. Water molecules, for example, contain one oxygen atom and two hydrogen atoms. Oxygen and hydrogen are elements. The oxygen that we breathe, however, consists of oxygen molecules. Two oxygen atoms make up every oxygen molecule. Figure 1 shows how atoms form molecules.

There are particles that are even smaller than atoms. *Protons*, *neutrons* and *electrons* combine to form atoms. Protons have a positive electric charge. Electrons have a negative charge. Neutrons have no charge; they are *neutral*. You probably remember seeing drawings that show a cluster of protons and neutrons in the center, surrounded by a "cloud" of electrons. Drawings like these represent the structure of atoms. (See Figure 2.) Scientists call the central cluster or ball of protons and neutrons the *nucleus* of the atom.

Every type of atom (or element) has a cer-

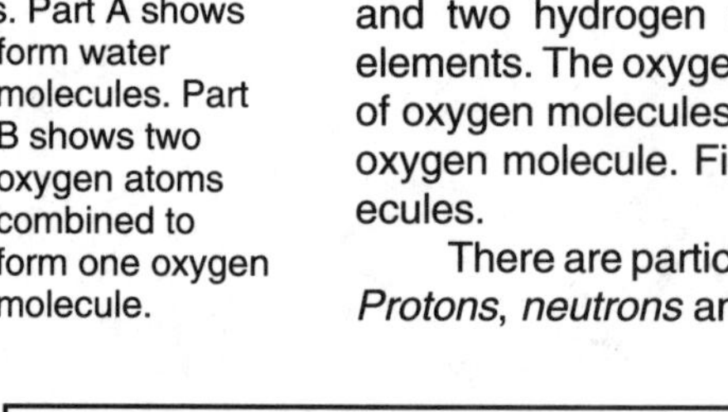

Figure 1—Atoms are the basic building blocks of all matter. Several atoms can combine to form molecules. Part A shows how hydrogen and oxygen atoms combine to form water molecules. Part B shows two oxygen atoms combined to form one oxygen molecule.

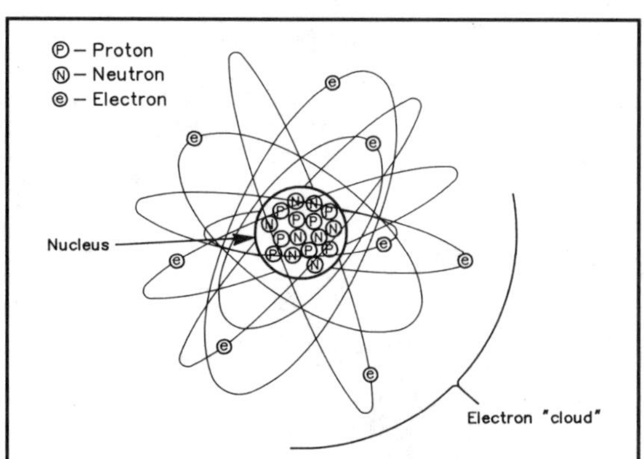

Figure 2—A simple picture of an atom includes protons and neutrons in the nucleus. The electrons surround the nucleus like a cloud. This drawing represents an oxygen atom. It has eight protons and eight neutrons in the nucleus, and eight electrons surrounding the nucleus.

tain number of protons. A hydrogen atom has one proton, a helium atom has two and a copper atom has 29, for example. Atoms are electrically neutral, so they must contain just as many electrons as they do protons. This means that a hydrogen atom contains one electron, a helium atom has two and a copper atom has 29 electrons, then.

Sometimes it is helpful to relate a new concept to something familiar. Think of the nucleus as the sun. Then imagine the electrons as planets revolving around the sun. With this kind of picture, the electrons appear to be different distances from the nucleus. Scientists use a picture like this to describe atoms. The distances from the nucleus describe *orbitals* or layers into which the electrons fit. Each orbital can hold a certain number of electrons. (Some orbitals can hold only two electrons while others can hold six, ten or fourteen.) Actually, the picture is a bit more complicated than the one shown in Figure 3. The electrons don't really fit into exact layers the way we have pictured them, either. This model is useful to help explain the inner workings of an atom, though.

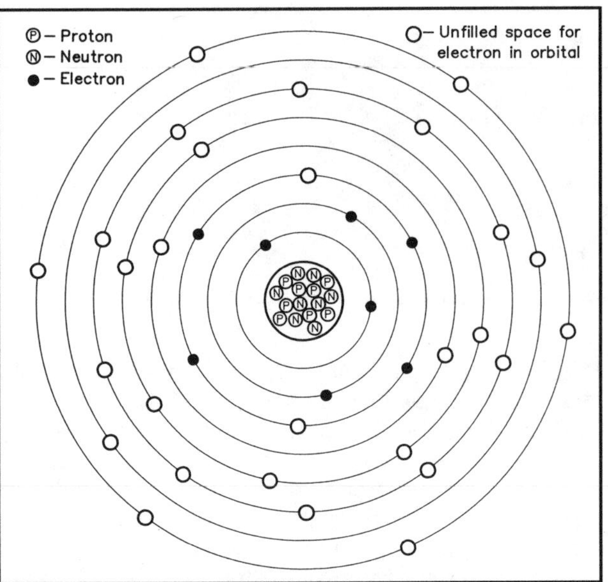

Figure 3—Our picture of an oxygen atom can be drawn to show the electrons in shells, or layers, called orbitals. These orbitals can hold certain numbers of electrons each. Electrons have a specific amount of energy to be in a particular orbital. Orbitals farther from the nucleus have higher-energy electrons in them.

The distance of the orbital from the nucleus is a measure of how much energy an electron in that orbital will have. Orbitals that are farther from the nucleus contain higher-energy electrons. This is the reason scientists often refer to such pictures as *energy-level diagrams*.

These orbitals are important. If the number of electrons in an atom just fills an orbital, that element will hold tightly to its electrons. Such atoms form good insulators. Suppose the number of electrons in an atom fills one orbital, with one or two additional electrons to go into the next orbital. The atoms will share those additional electrons or give them up to other atoms. Such atoms make good conductors. The elements used to make semiconductor materials have four electrons in the last orbital. From this discussion you can see that they are neither good conductors nor good insulators. Can you guess why we call them semiconductors?

The electrons for an element follow a specific pattern to fit into this electron-orbital structure. This pattern helps determine how the atoms combine with other atoms. There are several ways for these combinations to occur, but you won't need to understand them. It *will* be helpful, though, for you to realize that the electron structure plays an important part.

Doping the Crystals

Silicon and germanium are the materials normally used to make semiconductors. (The element silicon is not the same as the household lubricants and rubber-like sealers called silicone.) Silicon has 14 protons and 14 electrons, while germanium has 32 of each. Silicon and germanium each have an electron structure with four electrons in the outer energy layers. The silicon and germanium atoms will share these four electrons with other atoms around them.

Some atoms share their electrons so the atoms arrange themselves into a regular pattern. We say these atoms form crystals. Figure 1 shows how silicon and germanium atoms produce crystals. (Different kinds of atoms might arrange themselves into other patterns.) The crystals made by silicon or germanium atoms do not make good electrical conductors. They aren't good insulators either, however. That's why we call these materials semiconductors. Sometimes they act like conductors and sometimes they act like insulators.

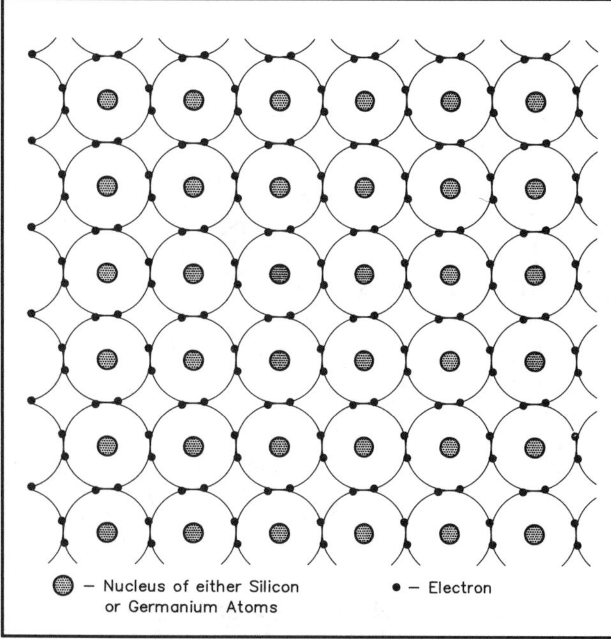

Figure 1—Silicon and germanium atoms arrange themselves into a regular pattern, called crystals. Notice that each atom in this crystal structure is sharing four electrons with other nearby atoms.

Manufacturers add other atoms to these crystals through a carefully controlled process, called doping. The atoms added in this way produce a material that is no longer pure silicon or pure germanium. We call the added atoms *impurity atoms.*

As an example, the manufacturer might add some atoms of arsenic or antimony to the silicon or germanium while making the crystals. Arsenic and antimony atoms each have five electrons to share.

Figure 2 shows how an atom with five electrons in its outer layer fits into the crystal structure. In such a case, there is an extra electron in the crystal. We refer to this as a *free electron*, and we call the semiconductor material made in this way *N-type material.* (This name comes from the extra negative charge in the crystal structure.)

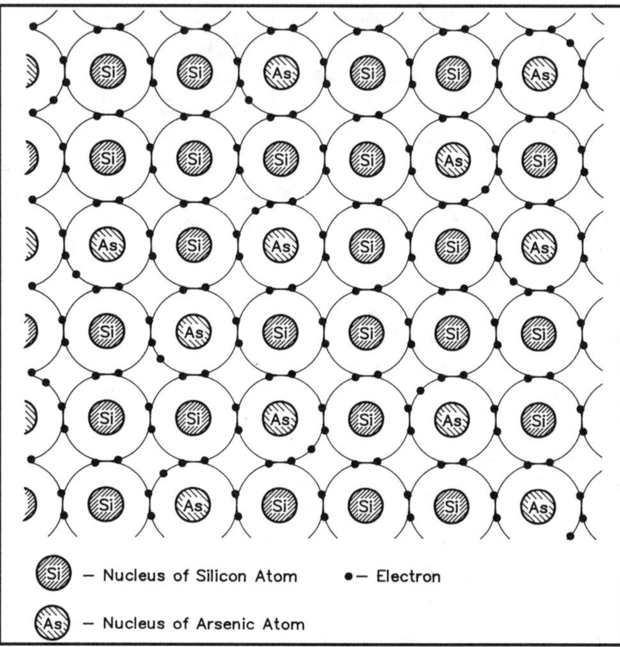

Figure 2—Manufacturers can add antimony or arsenic atoms to the silicon or germanium crystals. These *impurity atoms* add an extra electron to the crystal structure, producing N-type semiconductor material.

The impurity atoms are electrically neutral, just as the silicon or germanium atoms are. The free electrons result from the crystal structure itself. These impurity atoms create (donate) free electrons in the crystal structure. That is why we call them *donor atoms.*

Now let's suppose the manufacturer adds some gallium or indium atoms instead of arsenic or antimony. Gallium and indium atoms only have three electrons that they can share with other nearby atoms. When there are gallium or indium atoms in the crystal there is an extra space where an electron could fit into the structure.

Figure 3 shows an example of a crystal structure with an extra space where an electron could fit. We call this space for an electron a *hole.* The semiconductor material produced in this way is *P-type material.* These impurity

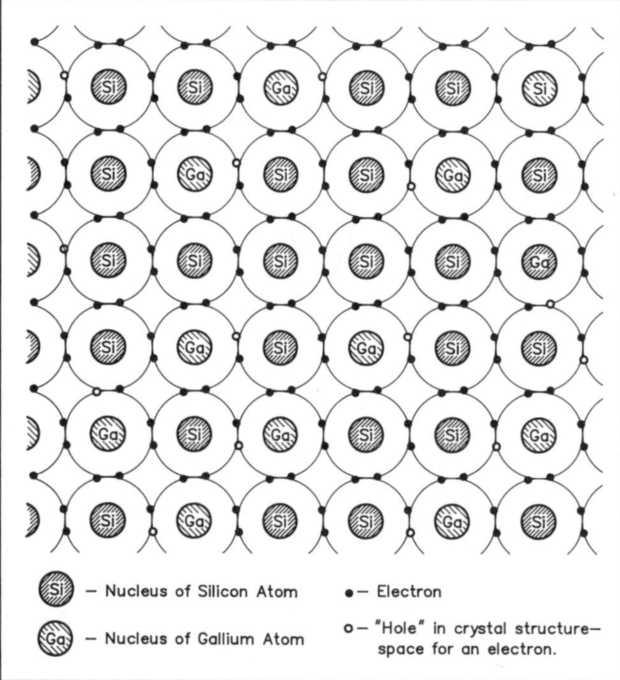

Figure 3—Manufacturers can add gallium or indium atoms to the silicon or germanium crystals. These *impurity atoms* have only three electrons to share. This leaves a *hole* or space for another electron in the crystal structure. This produces P-type semiconductor material.

atoms produce holes that will accept extra electrons in the semiconductor material. That is why we call them *acceptors*.

Again, you should realize that the impurity atoms have the same number of electrons as protons. The material is still electrically neutral. The *crystal structure* is missing an electron in P-type material. Similarly, the *crystal structure* has an extra electron in N-type material.

Suppose we apply a voltage across a crystal of N-type semiconductor. The positive side of the voltage attracts electrons. The free electrons in the structure move through the crystal toward the positive side.

Next suppose we apply a voltage to a crystal of P-type material. The negative voltage attracts the holes, and they move through the material toward the negative side.

Free electrons and holes move in opposite directions through a crystal. This is why we call the electrons and holes *charge carriers*.

Manufacturers can control the electrical properties of semiconductor materials that they make. They do this by carefully controlling the amount and type of impurities that they add to the silicon or germanium crystals.

Manufacturers build N- and P-type semiconductor materials next to each other in various combinations. They produce a wide variety of electronic devices this way.

Semiconductors are solid crystals. They are strong and not easily damaged by vibration or rough handling. We refer to electronic parts made with semiconductor materials as *solid-state devices*.

CHAPTER 26 Diodes

Diodes Allow Current in only One Direction

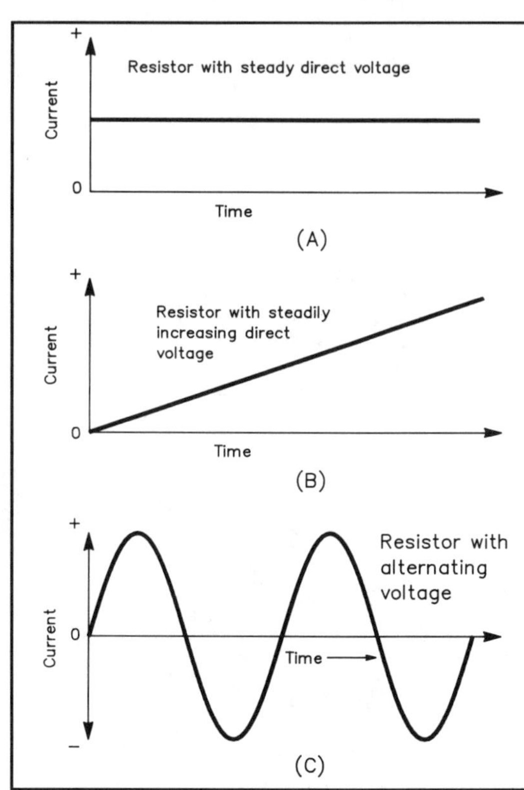

Figure 1—Part A shows the current through a resistor when we apply a steady direct voltage to the circuit. The current through the resistor caused by a steadily increasing direct voltage is shown at B. Part C shows an alternating current through the resistor if we apply an alternating voltage.

Remember that with a perfect resistor, it doesn't matter whether you pass dc or ac through it. The current through the resistor just depends on the applied voltage and the amount of resistance. Figure 1A shows the current through a resistor when we apply a steady direct voltage. Part B shows the current if we apply an increasing direct voltage. Part C shows the current through a resistor when we apply an alternating voltage.

With an ideal capacitor, however, you'll get different results. If you apply an alternating voltage to the capacitor, an alternating current will flow. When we apply a direct voltage, there will be a sudden rush of current while the capacitor plates charge. After the capacitor is fully charged, however, there will be no current. Figure 2 shows capacitor operation with both dc and ac applied.

This chapter describes devices that allow current to flow in only one direction. It doesn't matter which type of voltage you apply to these devices. We refer to such devices as *diodes*. (The word diode comes from two Greek words. Di, used as a prefix, means two. Ode means a path or way. So diode refers to a device with two parts or connections inside.) There are several different materials and construction methods used to make diodes. Diodes also serve a wide variety of purposes in modern electronic circuits. All diodes have one characteristic in common, however; they allow current to flow in only one direction.

A simple type of diode occurs when two different metals touch each other, and make a poor electrical connection. Early radio receivers used a type of simple diode. A small galena crystal in a lead holder made part of the diode. (Lead is a soft gray metal, and galena is an ore that contains lead.) A thin wire touched the surface of the galena. If you found just the right spot on the galena, radio

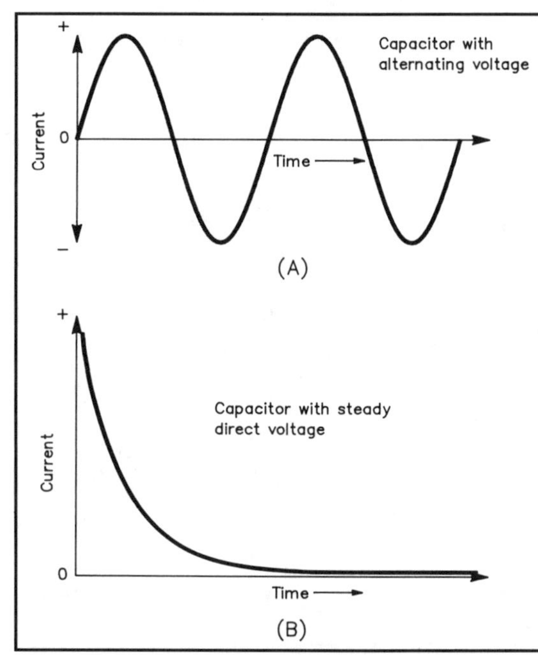

Figure 2—When we apply an alternating voltage to a capacitor there is an alternating current through it, as shown at A. When we apply a direct voltage to the capacitor, as at B, there is no current, at least after the capacitor charges.

signals would come alive in your earphones. See Figure 3. We still use diodes to detect the presence of radio signals in a receiver. A diode changes the signals into a form you can hear.

Figure 4 shows the symbol used to represent diodes on schematic diagrams. We call the end with the bar, at the arrow point, the cathode. The opposite end, at the back of the arrow point, is the anode.

How do diodes work? Let's look at a simple circuit. First, connect the anode to the positive side of a battery and the cathode to the negative side. See Figure 5. With the diode connected in this way, electrons will flow through the diode. We say the diode is *forward biased.*

Anode Cathode

Figure 4—The symbol used to represent a diode on schematic diagrams looks like an arrow point stuck into a block of wood. The end with the bar, which the arrow point appears to have hit, is the cathode. The opposite end, at the back of the arrow point, is the anode.

Now let's reverse the battery connections to the diode. Figure 6 shows the new circuit. This time the positive battery terminal connects to the cathode of the diode and the negative terminal connects to the anode. No current will flow in the circuit under these conditions.

Your next question might be, "What happens if we apply an alternating voltage?" The diode voltage polarity keeps changing. As you might expect, with a positive voltage on the anode, there is a current in the circuit. When the polarity reverses, there is a negative voltage on the anode. Now the current stops. When the next half cycle comes along, the anode connects to the positive voltage again. Then there is a current again. See Figure 7. This process continues as long as we apply the alternating voltage. Current flows through a diode in only one direction. The electrons move in a direction opposite to the way the arrow points in the diode symbol.

Diodes take an alternating voltage input and produce a direct voltage output. We call this process *rectification.* We sometimes refer to diodes as *rectifiers* because of this.

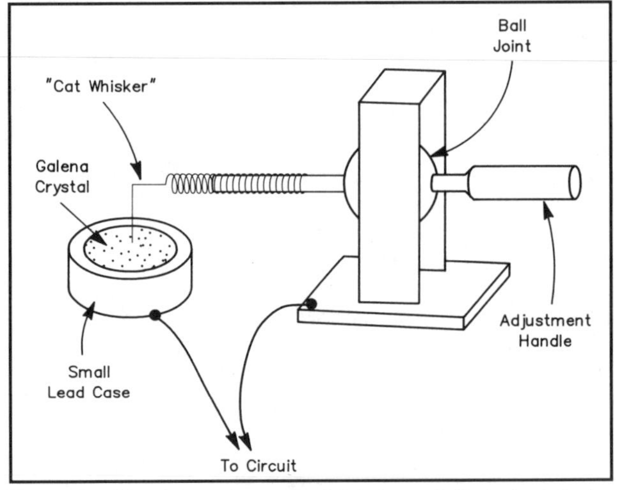

Figure 3—Simple receivers used this type of diode to detect radio signals during the early days of radio.

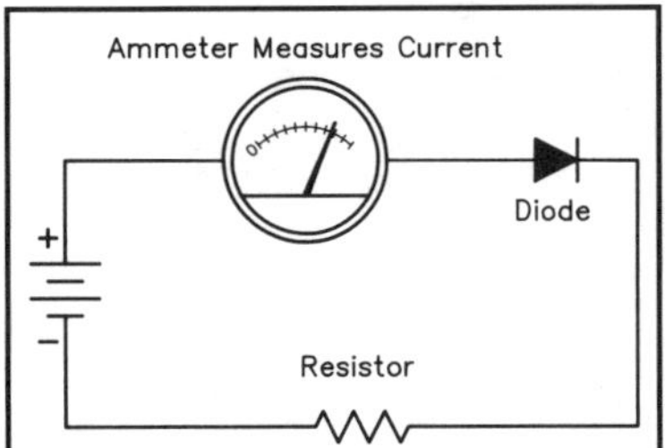

Figure 5—When you connect a diode into a circuit as shown, there will be a current through the diode (and the circuit).

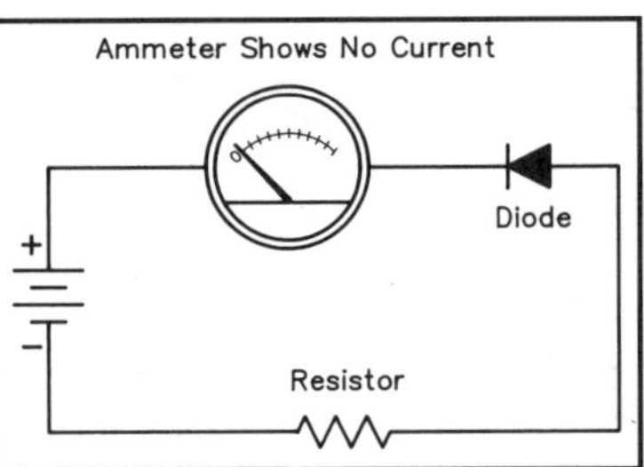

Figure 6—Here we connected the diode in the opposite direction to that shown in Figure 5. The positive battery terminal connects to the cathode and the negative terminal connects to the anode. In this case, there will be no current through the diode (or the circuit).

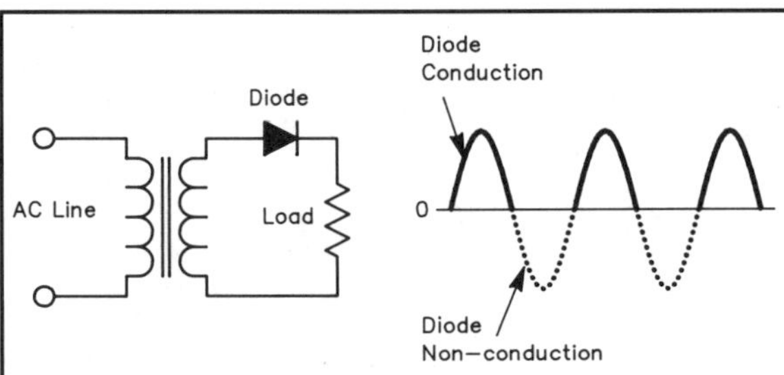

Figure 7—This drawing summarizes how a diode works when we apply an alternating voltage to the terminals. When the top of the ac supply is positive, the diode anode is positive. There is a current. When the top of the ac supply is negative, the diode anode is negative. Then there is no current. B shows the applied voltage and the circuit current over several cycles of applied voltage.

Solid-State Diodes from Semiconductor Materials

Most of the diodes in modern electronic circuits use semiconductor materials. Semiconductors normally have low operating-voltage requirements. They are rugged and reliable. In addition, they waste very little energy in the form of excess heat. These are a few advantages to using semiconductor diodes. There are still some circuits that use vacuum-tube diodes, although these tubes aren't very popular.

In the first chapter of this unit, you learned a little about manufacturing semiconductor materials. By now you are probably wondering how we can use these materials to make a diode. Well, the first step is to remember that a diode has two parts or sections. To make a solid-state diode, we need two pieces of semiconductor material. Since there are only two kinds of material (P and N type), we will need one piece of each type.

We make a diode by joining the two semiconductor-material blocks. This forms a single piece of material. The N- and P-type materials contact each other in the center, forming a *PN junction*. It is this junction between the two materials that interests us.

How do manufacturers form a PN junction? They usually start with a piece of either P- or N-type semiconductor material. Then they change a small area on the surface of this material into the other type of material. Figure 2 shows an example. There are several processes that manufacturers use to form the PN junction, but you don't have to worry about those details.

Okay. You have some idea how manufacturers put a diode together. Now you are probably wondering how it works. Well, before we even connect our diode into a circuit, something interesting happens to it. Some *free electrons* in the N-type material travel across the PN junction. There they fill some *holes* in the P-type material! This causes two other effects. First, it forms a region around the PN junction that has no holes or free electrons. We call this area the *depletion region*. This is because the electrons and holes, or semiconductor *charge carriers*, are depleted. See Figure 3.

You might ask, "Why don't all the free electrons in the semiconductor move across the junction and fill holes in the P-type material?" It's a good question. As some electrons cross the junction, the impurity atoms that they came from become positively charged. (An atom is electrically neutral. If it loses an electron, it has a positive charge. If it gains an electron it has a negative charge.) As free electrons fill some holes in the P-type material, the acceptor atoms have a negative charge. This is because they have an extra electron now. Holes from the P side can move to the N-type material also. This movement leaves a negative charge on the P material and a positive charge on the N side.

Electron and hole movement across the PN junction both affect the semiconductor material the same way. The charge movement leaves a positive charge on the N-type material and a negative charge on the P-type material. As the depletion region grows larger, the negative charge on the junction's P side increases. This repels more electrons that might try to cross the junction. The positive charge on the junction's N side also

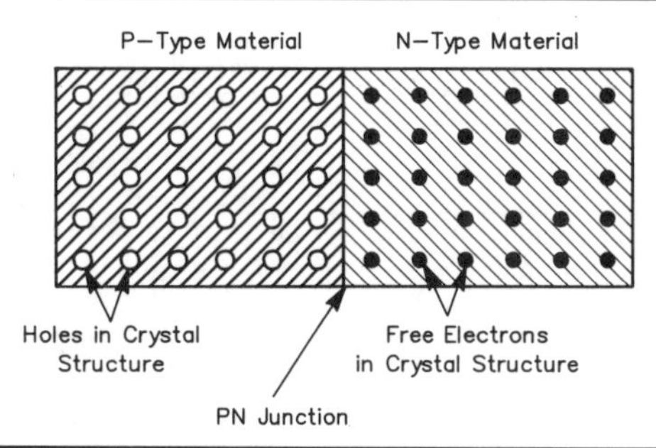

Figure 1—A block of P-type semiconductor and a block of N-type semiconductor join, to form a PN junction. We call a junction like this a solid-state diode.

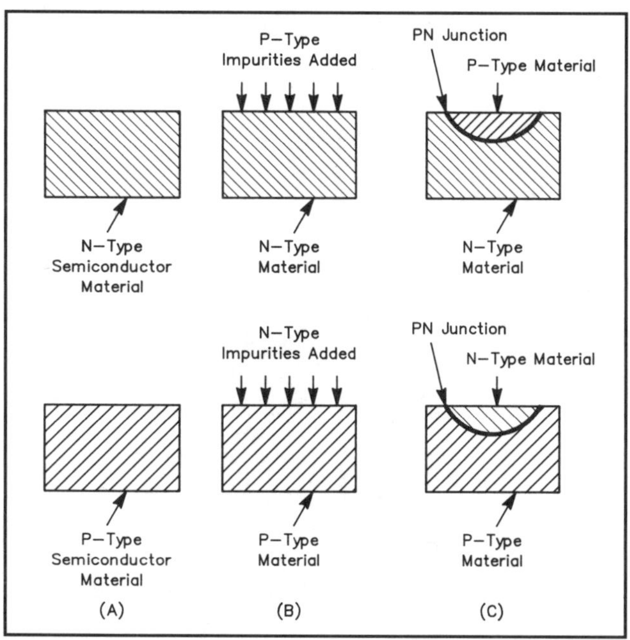

Figure 2—Manufacturers usually form a PN junction by starting with a piece of P- or N-type semiconductor material. Then they change a small layer on one surface of that material into the opposite type of material.

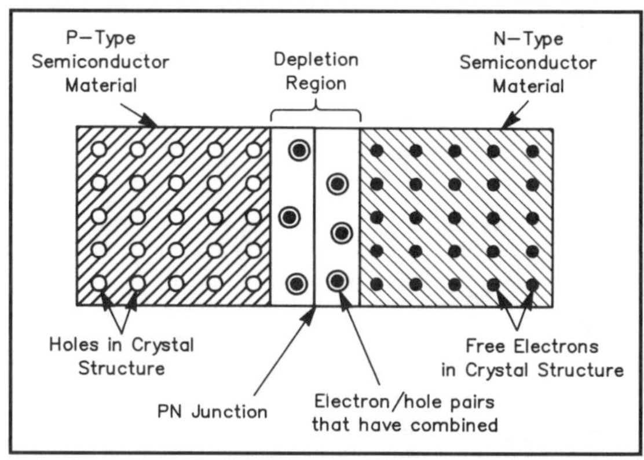

Figure 3—Free electrons and holes near the PN junction will cross the junction and combine with each other. This forms a *depletion region* around the junction inside the diode.

increases, and that attracts electrons back to the N material.

The electric charges also prevent too many holes from crossing the junction into the N-type material. Figure 4 shows the effects of these electric charges on the depletion region. The electric charges keep the depletion region from getting any larger. This electric charge is the second effect caused by electron and hole movement across the PN junction.

Now you have begun to understand the structure of a solid-state diode. So let's connect one into a circuit and see what happens. Figure 5 shows a diode and an ammeter connected in series with a battery. The positive battery terminal connects to the P-type semiconductor material. The negative terminal connects to the N-type material. You already know that like electric charges repel and opposite charges attract. So the negative side of the battery repels the extra electrons in the N-type material. Similarly, the positive side repels the holes (which act as positive charges) in the P-type semiconductor.

The battery voltage must be high enough to push (and pull) the charge carriers across the depletion region. (This takes about 0.3 volts if the semiconductor is germanium and about 0.7 volts if it is silicon.) Once the electrons and holes cross the depletion region, the voltage will attract them the rest of the way through the semiconductor. The positive voltage attracts electrons from the N-type material all the way across the junction, through the P-type material. Then the electrons travel through the circuit to the positive battery terminal.

In the same way, the negative battery terminal attracts the holes across the junction and through the N-type material. (Holes flow in the opposite direction from electrons in a circuit.) The meter in the circuit of Figure 5 indicates some current through the circuit. The diode acts like a small resistance in this example.

You may be wondering if all the free electrons can leave the semiconductor. Will the current stop then? Will that ruin the semiconductor? No, don't be afraid that all the electrons will get away! For every free electron that leaves the semiconductor on the P side, a new one comes onto the N side from the negative battery terminal. One purpose of the battery is to supply the electrons needed to maintain a current through the circuit.

The acceptor and donor impurity atoms in the crystal structure try to keep as many electrons as they started with. So when we remove the voltage, the semiconductor material will return to the same condition as before we applied the voltage. You can't remove all the free electrons or holes from a semiconductor by connecting it into a circuit.

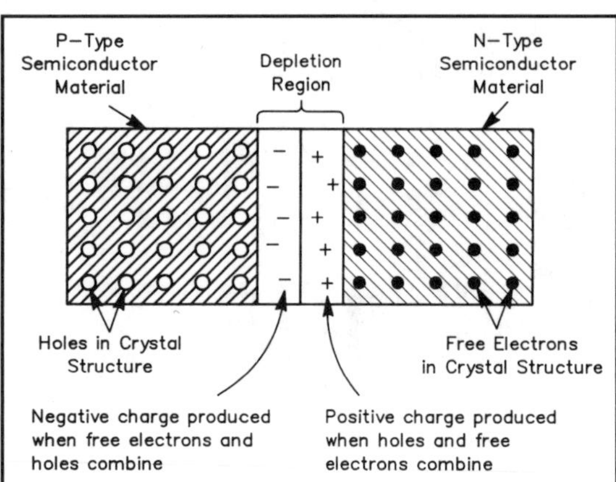

Figure 4—Holes and free electrons move across the diode junction so electrons can fill the holes. This leaves a positive and negative electric charge on opposite sides of the material, however. This electric charge repels additional charge carriers and keeps the depletion region from becoming too wide.

Now let's reverse the battery connections to our diode, as Figure 6 shows. This time the negative battery terminal connects to the P-type semiconductor. The positive terminal connects to the N-type material. Remember that like charges repel and opposite charges attract. The negative battery voltage repels any free electrons in the P-type material. The negative voltage also attracts holes in the P-type semiconductor. The positive battery voltage repels any holes in the N-type semiconductor and attracts electrons in the N-type material. The result is that the depletion region gets wider because the holes and electrons are pulled away from the junction. There is no current because electrons and holes do not make it all the way through the semiconductor. The depletion region becomes too wide for them to cross.

The diode junction is *forward biased* with positive voltage connected to P material and negative voltage connected to N material. A forward-biased diode allows current through it. With the voltage polarity reversed, we say the diode junction is *reverse biased*. A reverse-biased diode does not allow current through it.

Actually, the last sentence isn't completely true. There will often be a very small current going through a reverse-biased diode. The current is so small that we usually just ignore it, however. (There may be more than 10 milliamps of current flowing through a forward-biased diode. The reverse current in the same diode may be less than 10 microamps!) The *forward current* is at least 10,000 times greater than the *reverse*, or *leakage current*.

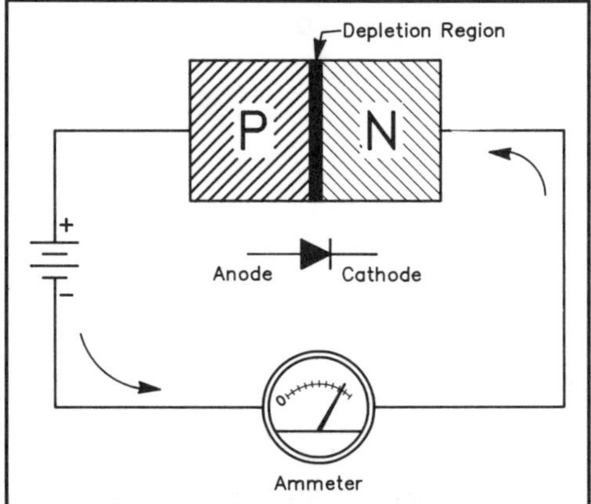

Figure 5—When connected to a voltage source as shown, a diode allows current through it. Notice the schematic symbol for a diode drawn next to the semiconductor block. With the negative voltage connected to the cathode and the positive voltage to the anode, we say the diode is *forward biased*.

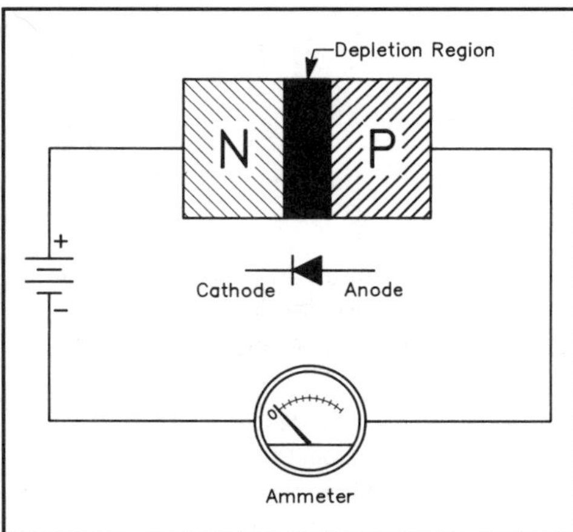

Figure 6—When we connect a diode to a voltage source as shown, there is no current in the circuit. With the anode connected to the negative voltage supply and the cathode connected to the positive supply, we say the diode is *reverse biased*.

Diode Characteristics and Ratings

When you connect a diode in a circuit so it is *forward biased*, there will be current through the circuit. (The positive battery terminal connects to the diode's anode and the cathode connects to the negative battery terminal.) If we connect the diode so it is *reverse biased*, there is no current through the circuit. (In this case we connect the diode with its anode to the negative side of the voltage source. The cathode connects to the positive side.)

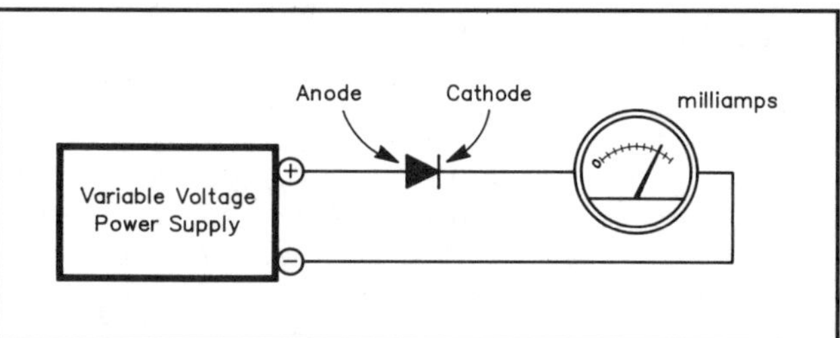

Figure 1—Connect the anode of a diode to the positive terminal of a variable voltage power supply as shown. The meter will indicate a current through the circuit.

Let's look at what happens to the current through a forward-biased diode as we vary the voltage applied to it. We'll start with a diode connected in a circuit, as Figure 1 shows. The power supply has a control to vary the output voltage. This provides a convenient way to adjust the voltage applied to the diode. With the power supply output set to 0 volts, we won't measure any current. Now we can gradually increase the voltage. A small current (less than one milliamp) will begin to flow. If we are using a germanium diode, the current will start to increase as the voltage approaches about 0.3 volts. With a silicon diode, the voltage must increase to about 0.7 volts before this current

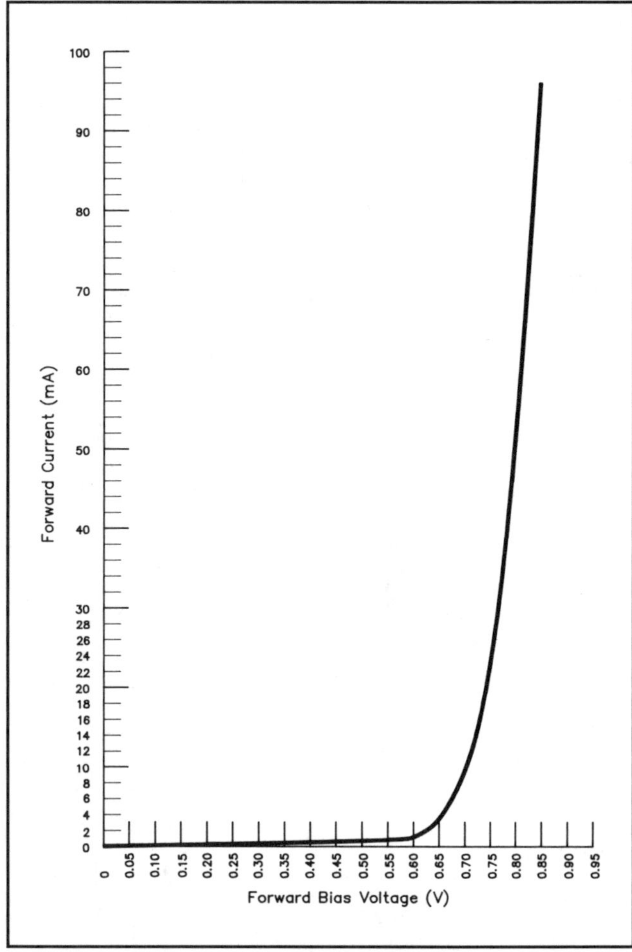

Figure 2—This is a typical silicon-diode characteristic curve, with a forward bias voltage applied to the diode. When the voltage applied to a germanium diode reaches about 0.3 volts, the forward current begins to increase. If you allow more current through the diode than the average forward current rating specifies, the diode will usually destroy itself.

increase occurs. (These voltages represent a *threshold* or changing point, where the semiconductor material begins to conduct electricity.) As you increase the supply voltage beyond this point, the current increases very rapidly. Figure 2 shows part of a typical diode *characteristic curve*.

Our diode is acting like a variable resistor. At voltages just above zero the resistance might be more than 100 ohms. As the voltage crosses the threshold value, the resistance decreases to only a few ohms. Take a look at the Figure 2 graph. The current gets very large for even a small voltage increase when you go above the threshold.

This brings us to an important diode rating. There is a practical limit to how much current can flow through a diode. Since the diode has some internal resistance, it will produce heat when there is current through it. When there is more current through the diode, it becomes hotter. If we allow the diode to become too hot, we will destroy it. Manufacturers describe a diode's ability to handle current with several different ratings. The one we are talking about here is the *average forward current*.

The average forward current rating is the largest current you should allow through the diode. (Under some conditions you might have a larger current pulse through the diode. This can only last a brief fraction of a second, however.) If the current through your diode exceeds the forward current rating, the diode is likely to destroy itself.

You must know a diode's average forward current rating before you select it for a particular circuit application. Some diodes will only handle a maximum average forward current of 100 milliamps. Some diodes can't even handle this much current. Other diodes can easily handle 10 amps or more. In general, diodes designed to handle larger currents are physically bigger. With a larger piece of semiconductor material and more surface area, the diode can get rid of more heat.

Manufacturers make some diodes with a bolt as one lead. With this lead, you can attach the diode to a large piece of metal, or *heat sink*. The heat sink helps conduct heat away from the diode, so you can allow larger amounts of current to pass through it. Figure 3 shows some typical semiconductor diodes. The photo includes a high-current device designed to mount on a heat sink.

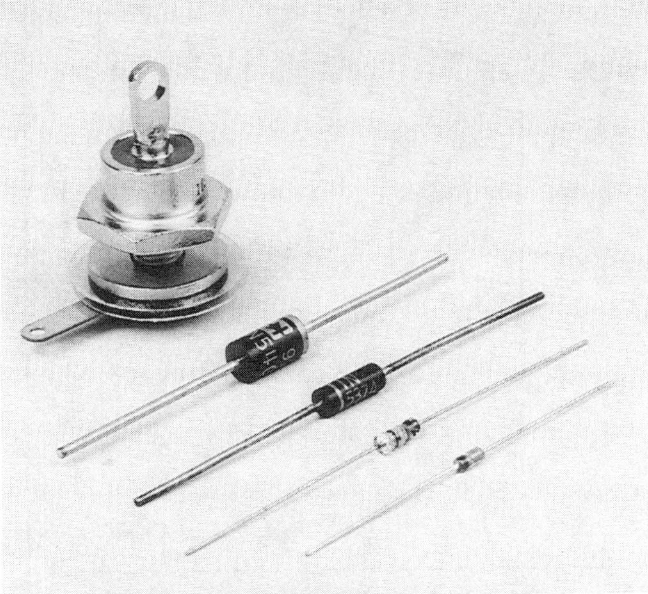

Figure 3—Solid-state diodes are available in a variety of sizes and shapes. In general, the larger the diode, the more current it can handle. Diodes made with a bolt for one lead can usually conduct at least several amperes safely.

Now let's change the direction of our diode in the test

circuit. Figure 4 shows the new circuit. This time the diode is reverse biased. We will have to use a meter that is very sensitive, so we can measure the small current. Again, we will start to increase the voltage applied to the diode. The diode current may be only a few microamperes now. As the voltage across the diode increases, there will be almost no increase in current through it.

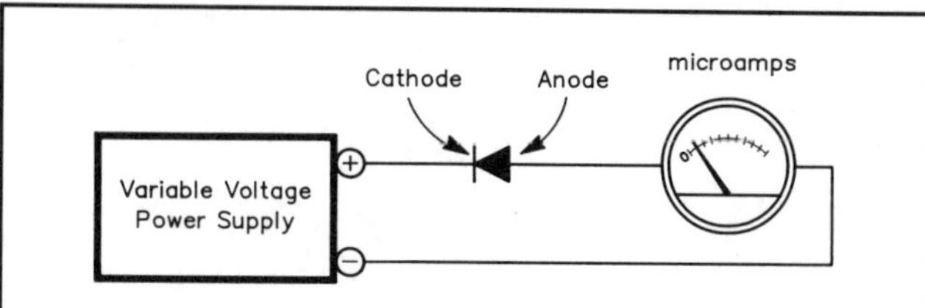

Figure 4—Connect the cathode of a diode to the positive terminal of a variable-voltage power supply, as shown. A very sensitive meter will indicate only a few microamperes of *reverse current* in the circuit.

Eventually, as we increase the voltage, we will find a point where the diode suddenly allows current to pass through it. The *reverse current* will become very large unless there is enough resistance in the circuit to limit that current. You must limit the circuit current or the diode will quickly destroy itself with the heat it produces. Figure 5 shows the characteristic curve for a reverse biased diode.

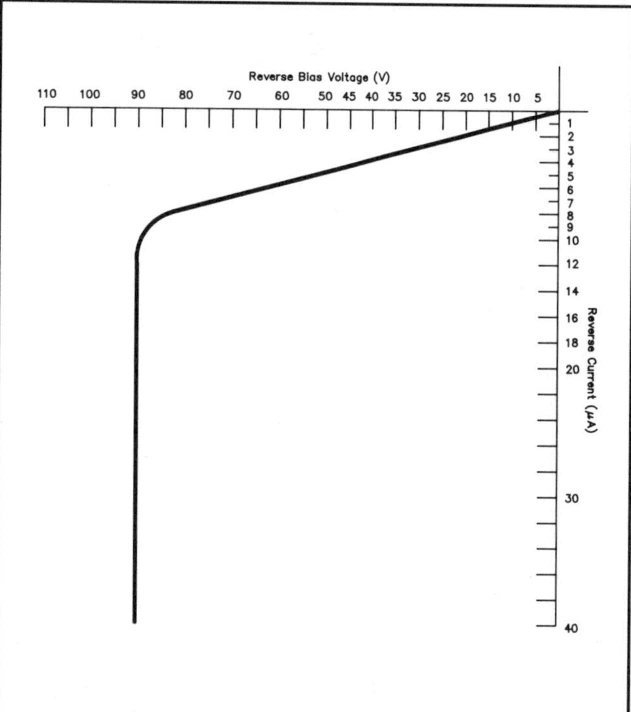

Figure 5—This graph shows the reverse-bias characteristic curve for a typical diode. The reverse bias voltage that causes the current through the diode to increase rapidly depends on the specific diode type. When this reverse conduction occurs, the diode will usually destroy itself.

This brings up another important diode rating: *peak inverse voltage (PIV)*. This is the highest reverse-bias voltage that we can safely apply to the diode. The PIV is a voltage that's a bit less than one that would cause the diode to conduct a large reverse current. A diode with the reverse-current characteristic shown in Figure 5 has a PIV of about 85 volts. When you select a diode for a circuit, be sure to pick one with a PIV rating much higher than any reverse bias voltage that your circuit may apply to it. Otherwise, there may be a large reverse current through the diode, destroying it.

In summary, if we forward bias a diode, it acts like a small resistance, or a short circuit. If we reverse bias the diode, it acts like a very large resistance, or an open circuit. So far we have been talking about the diode operation with direct voltages applied. Now we'll take a brief look at what happens when we apply an alternating voltage.

Figure 6A shows an alternating-voltage waveform. Let's begin at the point where the voltage is zero and beginning to increase in a positive direction. (This is the case at the left side of the graph.) With no voltage applied to the diode, it does not conduct. As the alternating voltage begins increasing, it forward biases the diode. The diode begins to conduct electricity, and as the voltage increases, so will the current. Figure 6B is a graph of the diode current. When the voltage reaches its positive peak and begins to decrease, the current through the diode also will decrease. As the alternating voltage goes to zero again, the current through the diode will stop. Next the voltage will start to increase in a negative direction, reverse biasing the diode. This time, though, the diode does not conduct. There is no current through the circuit. When the voltage again starts

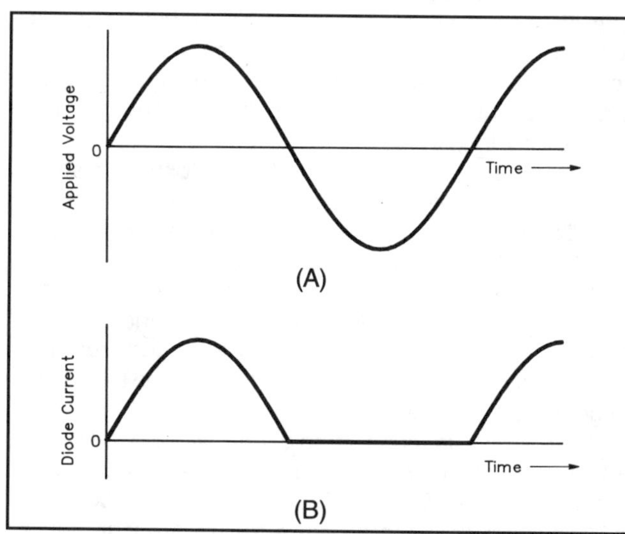

Figure 6—The graph at A shows an alternating voltage waveform, which we connect to a diode. Part B shows the current through the diode that results from this alternating voltage. The diode conducts only when its anode voltage is positive.

to become positive, the diode will conduct again. The Figure 6 graphs illustrate this whole process.

This discussion shows a very important use for diodes. By placing a diode in a circuit, we can apply an alternating voltage and the diode will turn that into a series of current pulses. These pulses all flow in one direction instead of reversing direction as with an alternating voltage. Figure 7A shows a resistor connected in series with a diode. Part B shows the waveform of an alternating voltage applied to the circuit. Part C shows the waveform of the voltage across the resistor. The resistor voltage results from the diode current pulses.

The diode in the Figure 7 circuit is forward biased only half the time. This is when it conducts current through the circuit. Each pulse of current through the diode must be twice the *average forward current* through the diode.

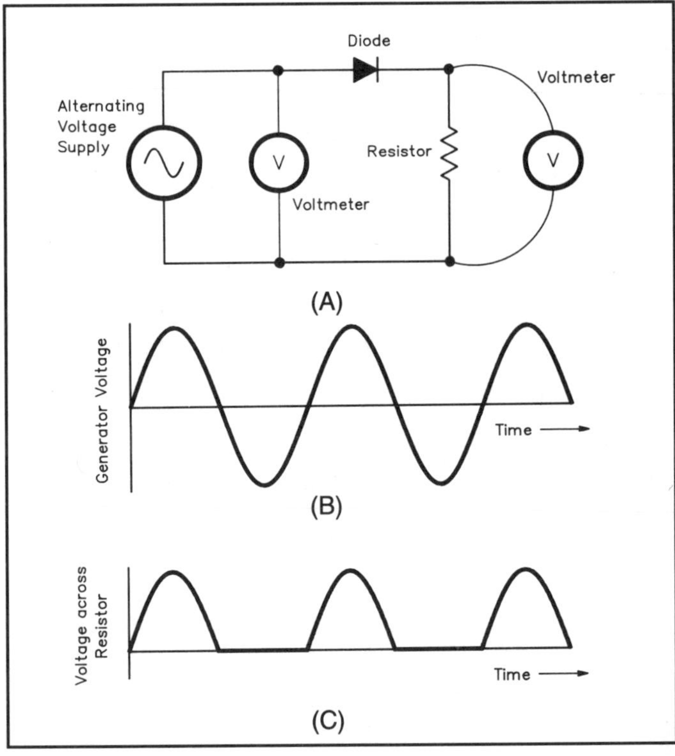

Figure 7—Part A shows how to measure the voltage across a resistor in a diode circuit. There is a voltage across the resistor only when there is current through the circuit. Part B shows the alternating voltage applied to the circuit. Part C shows the voltage measured across R as a result of this alternating voltage.

The Half-Wave Rectifier Circuit

Let's connect a single diode in series with a resistor and an alternating voltage supply. See Figure 1. The "output voltage," taken across the resistor, is a series of pulses. A single pulse occurs every time the diode is forward biased by the alternating voltage. There is one pulse for every cycle of the input waveform. Between each pulse is a time with no output voltage, because the diode is reverse biased. Figure 2 shows the input and output waveforms.

This circuit provides an output during only half of the input waveform. We call this circuit a *half-wave rectifier.* (Remember, the term *rectifier* refers to a device or circuit that takes an alternating voltage input and produces a direct voltage output.) A big advantage of a half-wave rectifier is that it is simple to build. The circuit makes inefficient use of the input voltage, however, and that is a big disadvantage.

The Figure 2 output waveform shows that the output voltage is always on the same side of the zero axis. (It can be either positive or negative with respect to ground. We can change the output-voltage polarity by reversing the diode direction.) The output voltage varies with time, and the current through the circuit changes with these variations. The current flows through the circuit in only one direction, however. Our circuit takes the alternating voltage input and changes it to a varying dc output.

The *rectified* output voltage is not a steady value. It varies from 0 to a maximum (peak) value for each pulse. The peak value of the direct output voltage is equal to the peak value of the alternating voltage supply. You can calculate the output peak voltage if you know the effective (RMS) value of the input voltage. Simply multiply the RMS value times the square root of two (1.414):

$$E_{peak} = E_{RMS} \times 1.414 \qquad \text{(Equation 1)}$$

For example, let's suppose we have an ac generator with an RMS output of 120 volts. See Figure 3. What will be the *peak* output voltage from a half-wave rectifier circuit connected to this

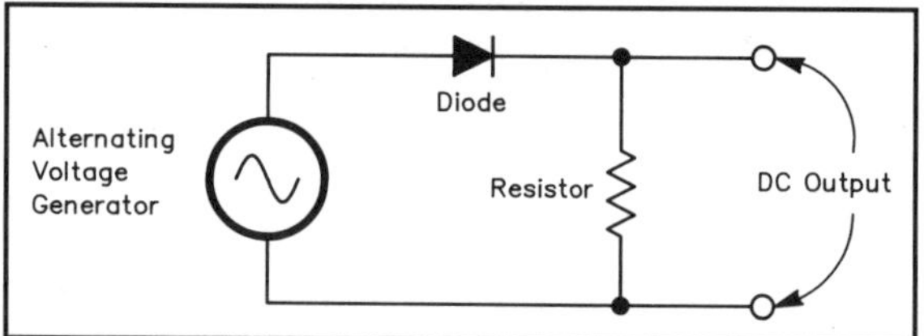

Figure 1—A half-wave rectifier is a simple circuit that changes an alternating voltage to a direct voltage. A single diode and a resistor are all you need.

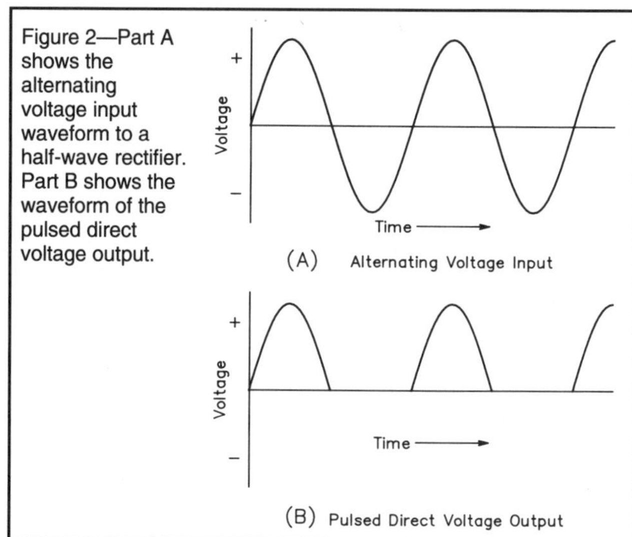

Figure 2—Part A shows the alternating voltage input waveform to a half-wave rectifier. Part B shows the waveform of the pulsed direct voltage output.

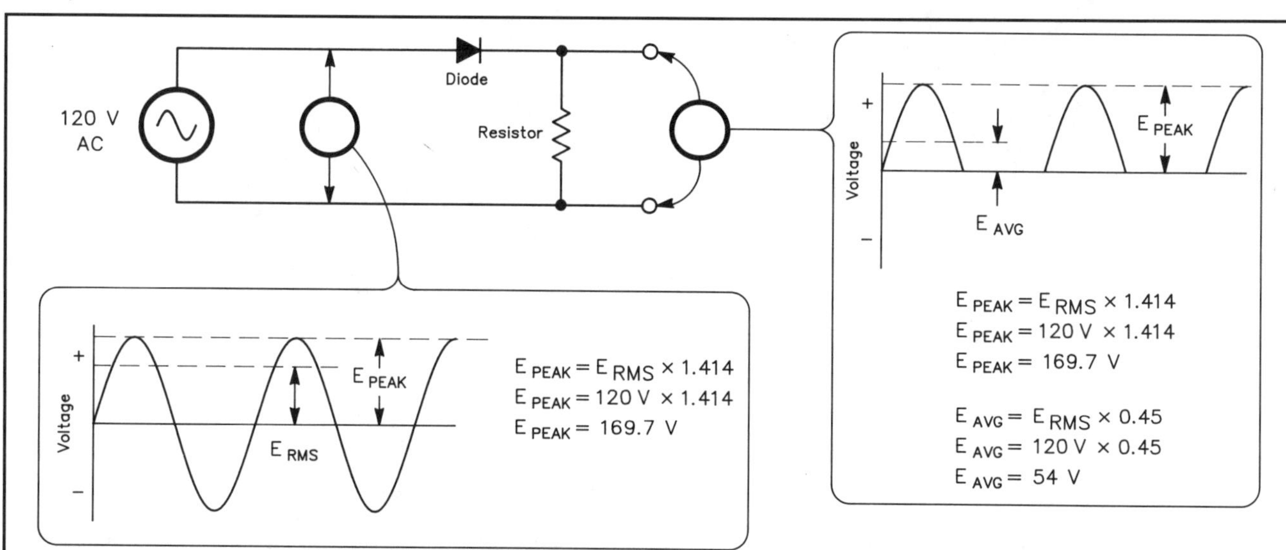

Figure 3—This circuit diagram shows a half-wave rectifier connected to an alternating voltage generator with an RMS output of 120 volts. The input and output waveforms are also shown, with their peak and average values.

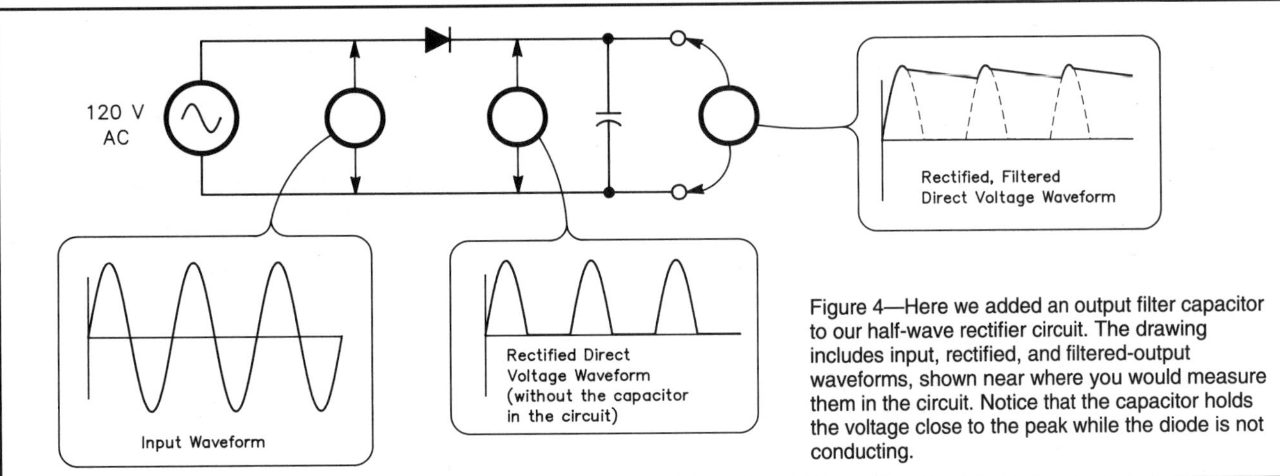

Figure 4—Here we added an output filter capacitor to our half-wave rectifier circuit. The drawing includes input, rectified, and filtered-output waveforms, shown near where you would measure them in the circuit. Notice that the capacitor holds the voltage close to the peak while the diode is not conducting.

generator? Equation 1 will help us answer this question.

$E_{peak} = E_{RMS} \times 1.414 = 120\ V \times 1.414 = 169.7\ V$

A half-wave rectifier connected to a 120-V (RMS) supply will produce a dc output with 169.7-V peak output.

We also want to know the *average* value of the direct voltage produced by a half-wave rectifier. Remember, there is a time of zero output between each pulse. Also, the output is at the peak value for only a brief instant during each cycle. On the average, then, the output voltage is much less than the peak value. In fact, you can calculate the average value by multiplying the RMS value of the input voltage by 0.45:

$E_{avg} = E_{RMS} \times 0.45$ (Equation 2)

For the example used above, we find the average direct output voltage by using Equation 2:

$E_{avg} = E_{RMS} \times 0.45 = 120\ V \times 0.45 = 54\ V$

The diode must withstand a *peak inverse voltage (PIV)* that is equal to the peak voltage value given by Equation 1. You must consider PIV when you select a diode for a certain circuit. For the circuit of Figure 3, you would choose a diode with a PIV rating higher than 170 volts. Provide a wide safety margin. Select a diode that has a PIV rating of about twice the applied peak inverse voltage. (A diode with at least a 340-V PIV rating would be a safe choice in this case.)

Many electronics devices will not work properly with the pulsed direct voltage output from a half-wave rectifier. They require a steady value of direct voltage for proper operation. How can we make use of the output from a half-wave rectifier, then? We must *filter* the output voltage before using it. A filter will help fill in the gaps in the output voltage. It also will maintain the voltage at a fairly constant level. We can make a filter simply by connecting a large-value capacitor across the output terminals, as Figure 4 shows.

When we first close the switch, the capacitor will quickly charge to a voltage equal to E_{peak}. While the diode is not conducting, the voltage across the capacitor begins to decrease. The output voltage will drop more gradually, however, than it would without the capacitor. The next time the diode conducts, the capacitor will charge up to E_{peak} again. Figure 5 shows the output voltage from the filtered half-wave rectifier.

The filter capacitor should have a large capacitance value; often 1000 microfarads or more. If the output current is small (less than a few milliamps) the output voltage will remain fairly constant. The capacitor voltage will drop while the diode is not conducting. If the circuit takes more output current, this voltage drop will be greater.

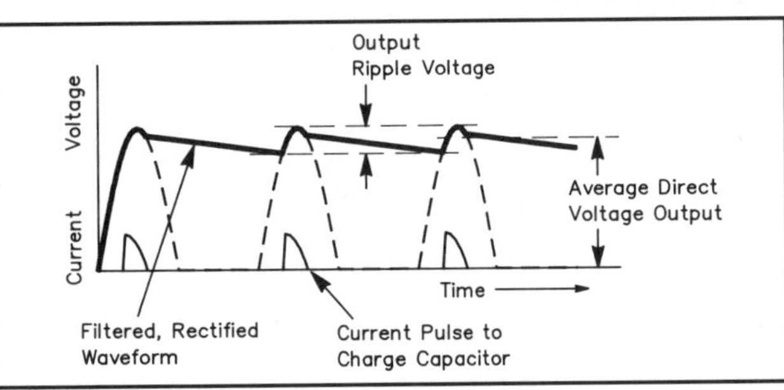

Figure 5—Each time the diode is forward biased (to conduct electricity) there is a brief pulse of current. This pulse charges the filter capacitor to the peak voltage again. During the rest of the waveform cycle, the diode is reverse biased. Then the capacitor slowly discharges. The higher the capacitance of the filter capacitor, the slower the output voltage drops during the diode off times.

Using a filter capacitor in the circuit does make one important change in the required PIV rating of the diode. Since the capacitor charges to the peak voltage, the maximum reverse voltage applied to the diode is two times the peak value. (The circuit applies the peak output voltage from the capacitor to one side of the diode. At the same time, it applies the opposite-polarity peak voltage from the generator to the other side of the diode. So, the diode is in series with these two voltages.) With a capacitive filter, the peak inverse voltage (PIV) across the diode in a half-wave rectifier is given by:

$PIV = E_{RMS} \times 2.8$ (Equation 3)

or:

$PIV = 2 \times E_{peak}$ (Equation 4)

Let's go back to our example of a 120-V generator and a peak output voltage of 169.7 volts. With a filter capacitor on the output, the peak inverse voltage applied to the diode is:

$PIV = E_{RMS} \times 2.8 = 120\ V \times 2.8 = 327.6\ V$

We should select a diode with a PIV rating of about 650 V.

The Full-Wave Rectifier Circuit

In the last section we talked about the half-wave rectifier circuit. How can we make more efficient use of the alternating input voltage than with that circuit? The *full-wave rectifier* produces a direct-voltage output pulse for each half of the input-voltage cycle. This circuit uses two diodes. We connect them so one diode conducts during the positive half cycle and the other conducts during the negative half cycle.

This increased efficiency is not without tradeoffs, however. The input-signal generator must have a terminal that is at a voltage half-way between the positive and negative voltage peaks. This terminal, called a center tap, is at ground potential. The generator alternately applies the positive and negative voltages to the two diodes. The common, or ground, connection provides a reference point between them. Figure 1 shows such a circuit.

The most common source of alternating input voltage is a transformer connected to the ac mains supply. You choose a transformer with a secondary, or output voltage to give the desired peak direct voltage from the rectifier.

A full-wave rectifier circuit requires a center-tapped transformer to supply the input voltage. The full secondary voltage must be twice the peak output voltage you want. In other words, the peak output voltage must appear on either side of the center tap. In contrast, the half-wave rectifier does not require a center-tapped transformer. In that case, the full secondary voltage only needs to be equal to the desired peak output voltage.

How does this full-wave rectifier work? Let's take another look at Figure 1. When the top of the supply is positive, diode D1 is forward biased, and conducts. The bottom half of the supply is negative with respect to ground, and D2 is reverse biased. When the polarity reverses, the bottom half of the supply is positive and the top half is negative. D2 conducts and D1 is reverse biased. Notice that the output from D2 connects to the load resistor with the same polarity as D1. So the output pulse from D2 is in the same direction as the pulse from D1. Figure 2 shows there is an output pulse for each half of the input waveform. Both pulses are in the same direction, so we have a pulsating direct-voltage output.

Figure 2 shows the output waveform from the full-wave rectifier. It has two output pulses for every complete input waveform cycle. The ripple frequency of the output is twice the input frequency. The ripple frequency for a half-wave rectifier is the same as the input frequency. The greater the ripple frequency, the easier it is to filter the output to produce a smooth direct voltage. This means it is easier to filter the output from a full-wave rectifier than a half-wave rectifier. A smaller-value filter capacitor will give the same degree of smoothing to the output waveform as with the half-wave rectifier. Figure 3A shows a full-wave rectifier circuit with a filter capacitor on the output. Figure 3B shows how the charged capacitor smooths the output ripple by holding the output voltage at a nearly constant level.

The output voltage from the full-wave rectifier is a pulsating direct voltage. This voltage varies from zero to a peak value twice during every complete input-voltage cycle. You can calculate the peak value of these pulses. You must know the RMS value of the transformer secondary voltage between the center tap and either side.

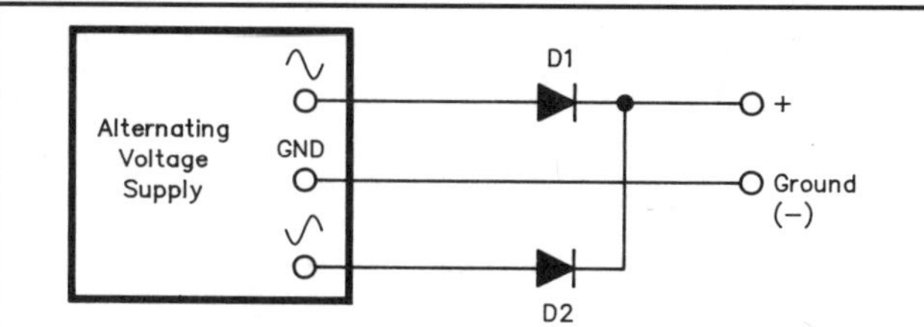

Figure 1—This diagram shows a full-wave rectifier circuit. Notice that the alternating voltage input has a center, or ground, connection. The voltage on the other two terminals changes from positive to negative with respect to the center connection.

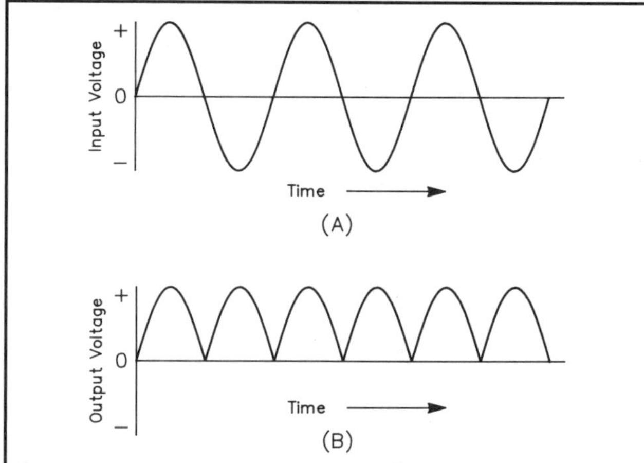

Figure 2—Part A shows the input voltage waveform and Part B shows the output voltage waveform for the full-wave rectifier. Notice the two output pulses for every input cycle. The output voltage varies between zero and a maximum or peak value. It is always on the same side of the zero line, however. The output is a pulsating direct voltage.

$$E_{peak} = E_{RMS} \times 1.414 \qquad \text{(Equation 1)}$$

We also want to know the average voltage of the output waveform. To calculate this voltage you must know the RMS value of the transformer secondary voltage between either side and the center tap.

$$E_{avg} = E_{RMS} \times 0.9 \qquad \text{(Equation 2)}$$

You may remember that the average output voltage with the half-wave rectifier was 0.45 times the RMS supply voltage. By using the entire input waveform to produce the direct voltage output, we have doubled the average output voltage! That's quite an improvement.

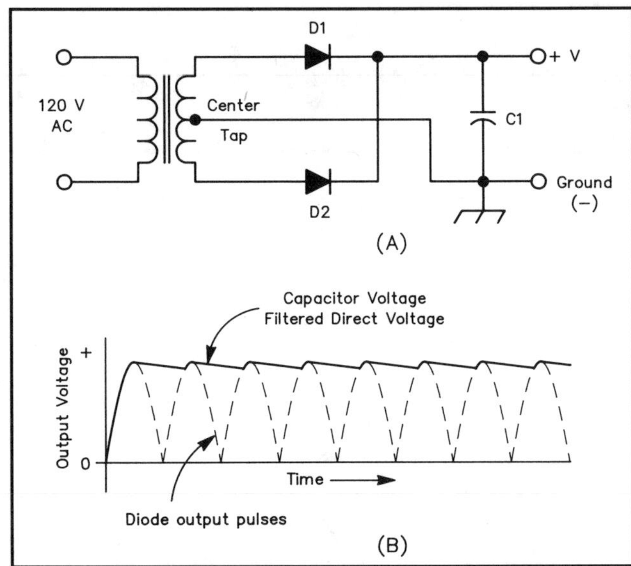

Figure 3—Part A shows a full-wave rectifier circuit with a capacitor filter on the output. Part B shows how the capacitor serves to smooth the ripple in the output voltage.

To select the proper diodes for the full-wave rectifier circuit, we must consider the peak inverse voltage across a diode. Look at Figure 4. When the top half of the transformer is positive, diode D1 conducts. The circuit applies the transformer peak positive voltage to the cathode of D2. At this instant, the bottom half of the transformer is at its peak negative value. So the circuit applies the transformer peak negative voltage to the anode of D2. D2 does not conduct. The peak inverse voltage across D2 is twice the transformer peak secondary voltage between the center tap and either side. When the transformer polarity reverses, the same conditions apply to D1.

$E_{PIV} = E_{RMS} \times 2 \times 1.414 = E_{RMS} \times 2.828$ (Equation 3)

When selecting diodes for the circuit, allow a safety factor of about two times the calculated E_{PIV}.

Let's work through an example to show how we use these equations to calculate practical information about a rectifier circuit. A center-tapped transformer supplies the input voltage to the rectifier circuit of Figure 4. The RMS output voltage of this transformer is 18 volts on either side of the center tap. First, let's calculate the peak output voltage from the circuit. Equation 1 will help with this calculation.

$E_{peak} = E_{RMS} \times 1.414$ (Equation 1)

$E_{peak} = 18\ V \times 1.414 = 25.5\ V$

Next, we will use Equation 2 to calculate the average output voltage.

$E_{avg} = E_{RMS} \times 0.9$ (Equation 2)

$E_{avg} = 18\ V \times 0.9 = 16.2\ V$

Finally, we will use Equation 3 to calculate the peak inverse voltage across the diodes.

$E_{PIV} = E_{RMS} \times 2 \times 1.414 = E_{RMS} \times 2.828$ (Equation 3)

$E_{PIV} = 18\ V \times 2.828 = 50.9\ V$

We should select a diode with a PIV rating of about 100 V to provide the suggested safety factor.

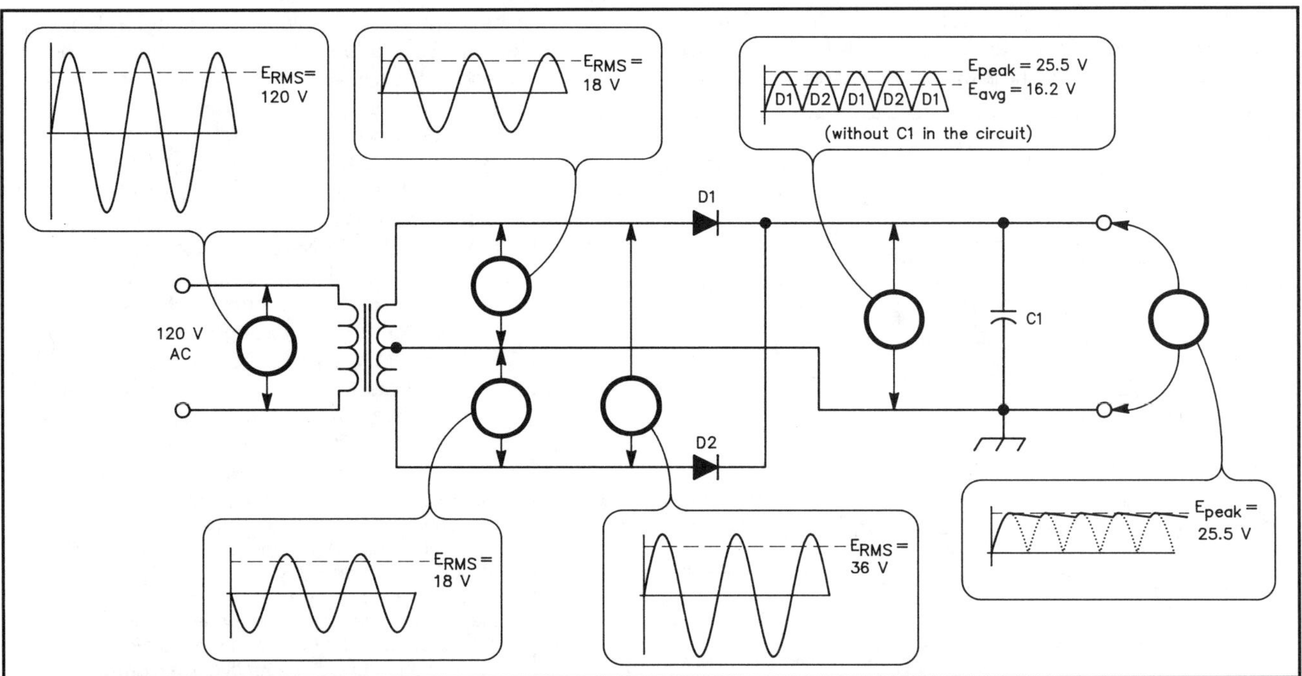

Figure 4—This is an example of a full-wave rectifier circuit. It shows the input and output voltage waveforms at various points throughout the circuit. You can calculate peak and average output voltages from the RMS value of the input voltage. A large-value capacitor across the output takes the rectified pulses and turns them into a smooth output.

The Bridge Rectifier Circuit

The *bridge rectifier* offers the best features of the full-wave rectifier circuit without the major disadvantage. It gives two output pulses for each cycle of the input voltage, like the full-wave rectifier does. The bridge rectifier is a form of full-wave rectifier. The bridge rectifier does not require a center-tapped transformer, though. The peak alternating voltage across the entire transformer secondary appears at the rectifier output. Figure 1 shows a bridge rectifier circuit with input and output voltage waveforms.

A bridge rectifier has one possible disadvantage: the circuit requires four diodes. Remember that a full-wave rectifier requires a transformer with a center-tapped secondary, though. Two extra diodes don't cost much compared to the cost of a center-tapped transformer with twice the full secondary voltage. So this is a fair trade-off. In fact, bridge rectifier circuits are so popular that manufacturers make units with the four diodes built into a single package.

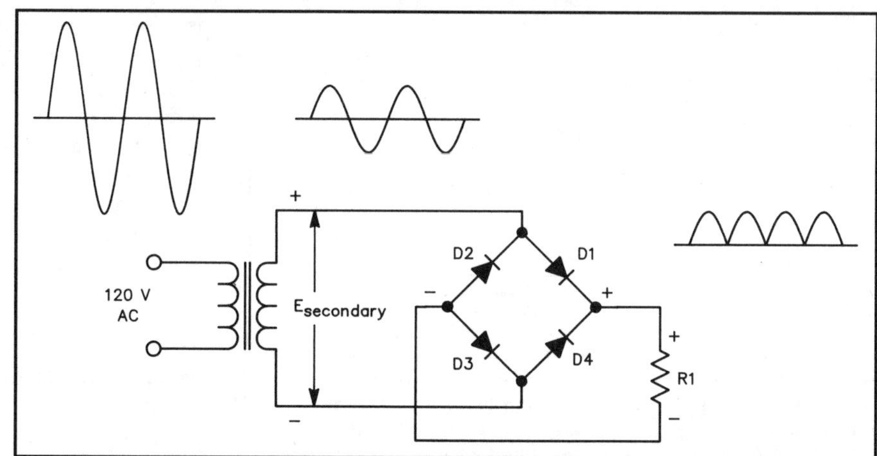

Figure 1—This circuit is a bridge rectifier. The drawing includes alternating voltage waveforms at the transformer primary and secondary. It also shows the pulsed direct-voltage output waveform from the rectifier.

Let's analyze how our bridge rectifier circuit works. The circuit may look complicated at first glance. It's easy to understand if we take it one step at a time, though.

First, we redraw the circuit from Figure 1 as shown in Figure 2A. Notice that the top transformer lead is positive and the bottom one is negative at this instant. This means that D1 is forward biased and D2 is reverse biased. We drew D2 in a dot pattern to show that it is not part of the circuit in this case. Since D1 is forward biased, the transformer voltage connects to the load, represented by R1. This voltage drops across R1, but is still more positive than the voltage on the bottom transformer lead. Therefore, D3 is forward biased and D4 is reverse biased. (We also drew D4 with a dot pattern to show that it is not part of the circuit in this case.) The electrons flow out the negative transformer lead, through D3, the load resistor, D1, and finally back to the positive transformer lead.

What happens when the voltage polarity on the transformer reverses? Take a look at Figure 2B. Now the bottom transformer lead is positive and the top one is negative. D4 is forward biased and D3 is reverse biased. The transformer voltage connects to the load resistor, R1, through D4. There is a voltage drop across the resistor. The voltage at the junction of D1 and D2 is still more positive than the top transformer lead, however. D2 is forward biased this time, and D1 is reverse biased. The circuit is again complete, and there is a current. This time, electrons flow out the top transformer lead, through D2, the load resistor, D4, and finally to the bottom transformer lead.

Figures 1 and 2 show the bridge rectifier circuit drawn in a traditional way. The diodes form

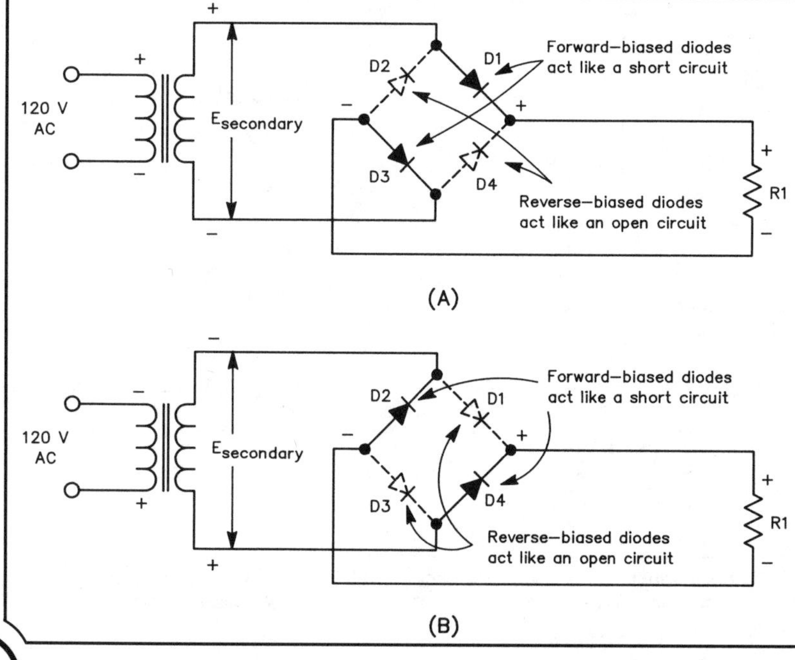

Figure 2—Part A shows the rectifier circuit when the top of the transformer is positive and the bottom is negative. Diodes D1 and D3 are forward biased and conduct. Diodes D2 and D4 are reverse biased and do not conduct. Part B shows the circuit conditions when the input polarity changes. Then the bottom of the transformer is positive and the top is negative. Both drawing parts show the reverse-biased diodes drawn with a dot pattern. This illustrates that those diodes act like an open circuit, effectively removing them from the circuit.

the sides of a diamond. You may find the circuit operation easier to understand if we redraw the circuit like Figure 3 shows. Again, D1 and D3 conduct on one half of the input waveform. D2 and D4 conduct on the opposite half cycle of the input waveform.

As you might expect, the output voltage from the bridge rectifier is a pulsating direct voltage. This voltage varies from zero to a peak value twice during every complete input-voltage cycle. You can calculate the peak value of these pulses if you know the RMS value of the transformer secondary voltage.

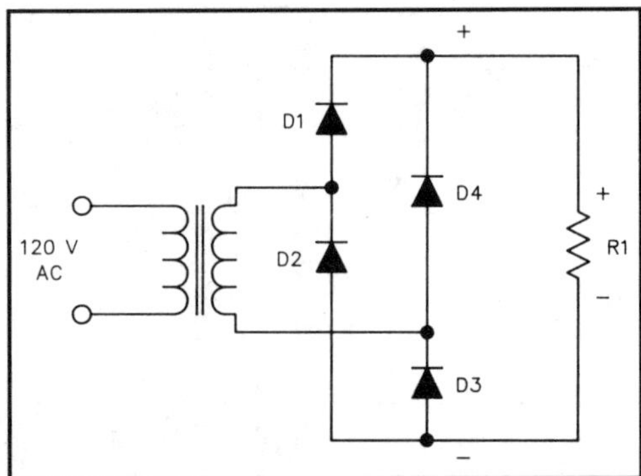

Figure 3—Drawing the diodes as shown here may make it easier for you to understand how a bridge rectifier circuit works. The circuit operation described in the text still applies.

$E_{peak} = E_{RMS} \times 1.414$ (Equation 1)

This equation applies any time you want to calculate the peak value of a sine-wave alternating voltage.

We also would like to know the average voltage of the output waveform. Calculate the average voltage using the RMS value of the transformer secondary voltage.

$E_{avg} = E_{RMS} \times 0.9$ (Equation 2)

This equation also gives the average output voltage from a full-wave rectifier circuit.

As with other rectifiers, we normally want to smooth the output pulses before applying the direct voltage to an electronic circuit. We can smooth these pulses by using a filter capacitor. Figure 4 shows such a capacitor across the rectified output from a bridge rectifier. The capacitor provides a smoothed, steady direct voltage. There is only a small amount of ripple, or variation, in the output. In general, larger values of capacitance provide an output voltage with less ripple.

To select the proper diodes for the bridge rectifier circuit, we must consider the peak inverse voltage across a diode. Look at Figure 2 again. When the top of the transformer is positive, diodes D1 and D3 conduct. We are applying the transformer peak positive voltage across D2 and D4. When the transformer polarity reverses, diodes D2 and D4 conduct. We are applying the peak secondary voltage to D1 and D3. Each diode must withstand the peak secondary output voltage.

$E_{PIV} = E_{RMS} \times 1.414$ (Equation 3)

When selecting diodes for a bridge rectifier circuit, allow a safety factor of about two times the calculated E_{PIV}.

An example will show how to use these equations to calculate practical information about a rectifier circuit. Let's suppose the input voltage to our bridge rectifier circuit comes from a transformer with an RMS output of 18 volts. First, we'll calculate the peak output voltage from the circuit. Equation 1 will help with this calculation.

$E_{peak} = E_{RMS} \times 1.414$ (Equation 1)

$E_{peak} = 18 \text{ V} \times 1.414 = 25.5 \text{ V}$

Next, we'll use Equation 2 to calculate the average output voltage.

$E_{avg} = E_{RMS} \times 0.9$ (Equation 2)

$E_{avg} = 18 \text{ V} \times 0.9 = 16.2 \text{ V}$

Finally, use Equation 3 to calculate the peak inverse voltage across the diodes.

$E_{PIV} = E_{RMS} \times 1.414$ (Equation 3)

$E_{PIV} = 18 \text{ V} \times 1.414 = 25.5 \text{ V}$

We should select a diode with a PIV rating of about 50 V to provide the suggested safety factor.

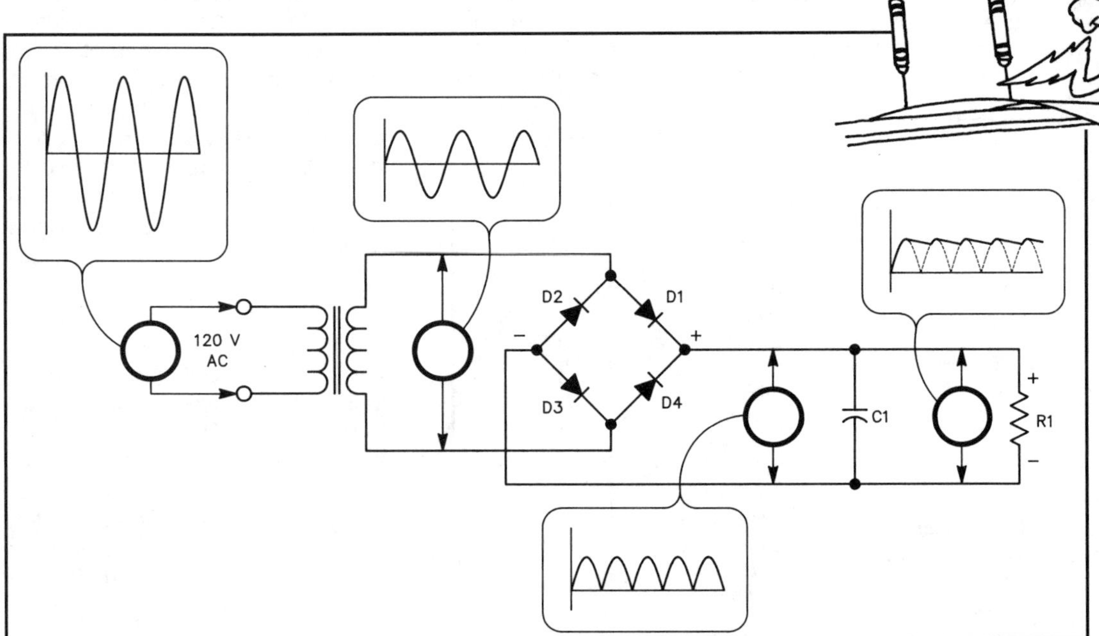

Figure 4—A bridge rectifier produces two output pulses for each complete input-waveform cycle. A large-value capacitor connected across the output serves to smooth the pulses, leaving us with a smoothed, steady direct voltage.

We Can Use Diodes as Switches

A forward-biased diode acts like a short circuit. (Remember that a forward-biased diode has a positive voltage connected to its anode and a negative voltage connected to its cathode.) Similarly, a reverse-biased diode acts like an open circuit. In that case, the diode cathode is positive and the anode is negative. This description begins to make a diode sound a bit like a switch. When the diode is forward biased, it is a closed switch. When it is reverse biased, it is an open switch.

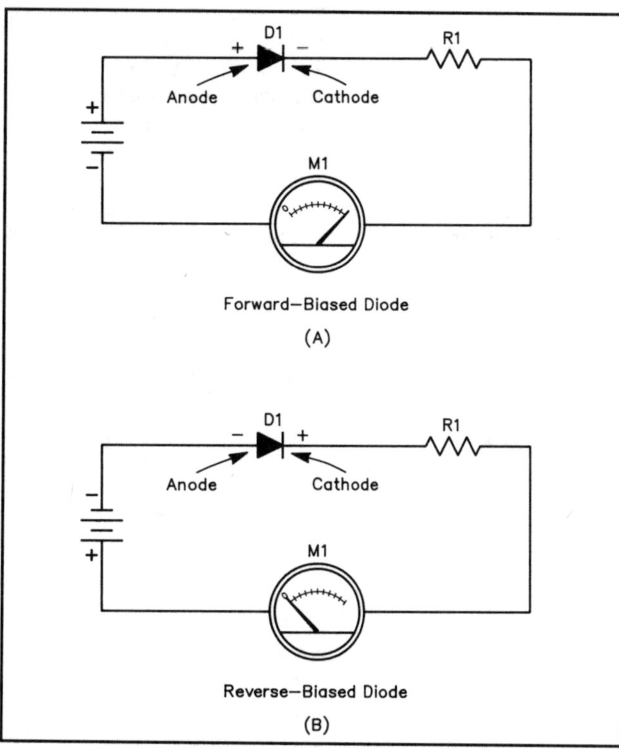

Figure 1—A forward-biased diode acts like a closed switch, allowing current through the circuit, as Part A shows. A reverse-biased diode acts like an open switch, preventing current through the circuit, as Part B shows.

In fact, it is common to use diodes as switches to turn RF currents on and off in electronic circuits. You can turn such a switch off or on by controlling the diode bias voltage. Figure 2 shows one simple example of such a *diode switch*. D2 serves as a switch to turn on a type 47 bulb with a 50-MHz RF source. When S1 is open, D2 is reverse biased. It is an open switch, and there is no current through the bulb. When S1 is in the closed position, the cathode of D2 is more negative than its anode. The diode is forward biased. It acts like a closed switch, and conducts current. Now there is a complete circuit for the RF supply. There is current through the bulb, C1 and D2 to ground. The bulb lights.

You are probably saying, "Hey, wait a minute! What good is a diode switch, if it takes a mechanical switch to turn it on and off?" That's a good question. For a simple circuit like the one shown in Figure 2, it really has no advantage. You could just as easily use a mechanical switch in place of D2 in this circuit. The switch would then control whether the bulb was lit or not. (See Figure 3.)

Let's suppose our circuit was a bit more complex. What if our electronic circuit was a radio receiver, and we wanted to select between several different filter circuits in the receiver. We might have four or five wires to connect or disconnect for each filter. We must switch each of these wires at the same time to change from one filter to the other. We could use a five-pole, double-throw switch (if we could find one). A few diodes also can do the switching. Now, one mechanical switch connects the bias voltage to all the diodes, and they will switch the circuit for us.

There is another important advantage of diode switches. You can place the diode right at the point in the circuit where you want to do the switching. Then you can apply the dc bias voltage through a mechanical switch located some distance away. The mechanical switch can be on a control panel or another convenient location. Otherwise, we would have to run long lengths of

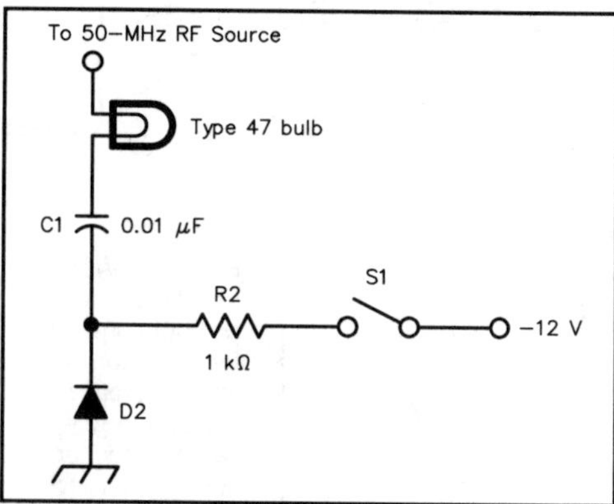

Figure 2—Here is a simple diode switch. Closing S1 applies forward-bias voltage to D2. The diode conducts, allowing current through the bulb.

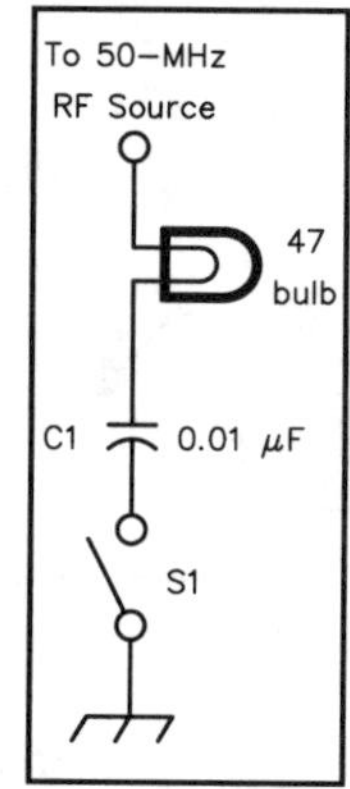

Figure 3—There may not be much advantage to using a diode switch in the circuit of Figure 2. You could use a mechanical switch instead, and the circuit looks much simpler. There are several big advantages to using diode switches in circuits that are more complex, however.

Figure 4—Part A shows a portion of an amplifier circuit, with a series diode switch. When S1 is in the position shown, D1 is forward biased. The signal flows through the diode to the amplifier. When the switch is in the opposite position, the diode is reverse biased. In this case the signal can't reach the amplifier. Resistors R1 and R2 serve to limit the diode current and isolate the signal current from the biasing circuit. Part B shows a section of an oscillator circuit. The shunt diode switch adds C2 to the circuit when it is forward biased (with +12 volts applied). The oscillator frequency decreases when the circuit capacitance increases. This diode switch provides two frequency ranges for the oscillator by changing the tuned-circuit component values. This type of diode switching does not work well with dc or at frequencies below the VHF range.

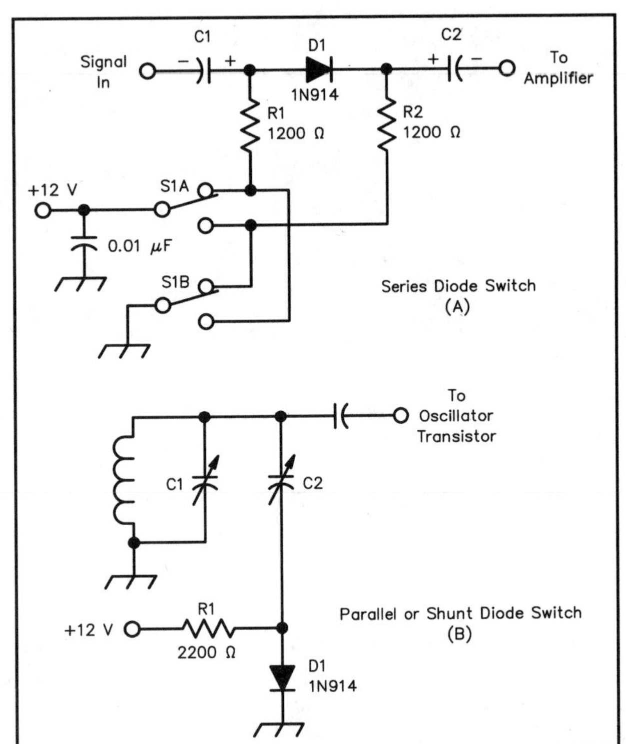

wire from the circuit to the switch location. Long leads may be okay for direct voltages. We want to keep the lead lengths as short as possible when we are switching audio or radio frequencies, however.

Perhaps the biggest advantage of a diode switch is the way it can operate automatically. Suppose part of the circuit measures a certain condition, and operates the switch based upon that condition. For example, you might have a circuit that would detect the presence of radio frequency (RF) energy from a transmitter. This circuit could operate a diode switch to connect the antenna lead to the transmitter. When the circuit finds no RF, the switch connects the antenna to the receiver.

We can connect diode switches in either of two ways. One is to wire the diode in *series* with a signal lead. The other is to wire the diode from the signal lead to ground (in *shunt*). Figure 4 shows partial circuits using these two diode-switching techniques.

When selecting a diode to use as a switch, you should select a diode with a *peak inverse voltage (PIV) rating* suitable for your circuit. You should also pay attention to the *maximum forward current ratings* of the diode. This will depend on how much current you expect your circuit to draw. The most important rating for a switching diode is its *speed*, or *switching time*. Manufacturers make some diodes specifically for use as *rectifier diodes* and others for use as *electronic switches*. You should select a diode designed for your intended application.

When a diode goes from forward to reverse bias, it takes some time before the diode cuts off. During this time, there is reverse current through the diode. In general, diodes designed for rectifier circuits have slower switching times than diodes designed as switches.

CHAPTER 27 Bipolar Transistors

A Three-Layer Sandwich

Scientists experimented with semiconductor materials in the 1930s and 1940s. Three of them made a very important discovery in 1947 at the Bell Telephone Laboratory. William Shockley, John Bardeen and Walter Brattain discovered *transistor action*. They used this term to describe the action of "transferring current across a resistor." This discovery is the basis for modern solid-state electronics. The discovery was so important that in 1956 the three men received the Nobel Prize in Physics.

Current passes through PN-junction diodes only in one direction. Connect the positive side of a battery or other voltage supply to the P-type material. The negative side connects to the N-type material. This puts a forward-bias voltage on the diode and there is a current through it. If we reverse these battery connections, the diode has a reverse-bias voltage. In that case there is almost no current.

There is a very small *leakage current* through a reverse-biased diode. If we apply a large enough voltage, we can push a large *reverse current* through the diode. We call this reverse voltage the *breakdown voltage*. Figure 1 is a graph of the relationship between the voltage and current for a typical diode.

Shockley, Bardeen and Brattain made their discovery by forming a single piece of semiconductor with two diodes back-to-back. Figure 2 shows three layers of semiconductor material. There is a thin layer of N-type material between two layers of P-type material. The diode cathodes connect in this example.

We can connect either diode to a voltage source, and this "sandwich" behaves like a diode. Figure 3 shows the diode junctions connected to a battery one at a time. The resistors limit the forward current so we don't destroy the diodes.

Now we'll try another experiment. Connect one diode to a forward-bias voltage and the other to a reverse-bias voltage. Figure 4 shows these circuit connections. You should expect a current through the forward-biased diode circuit. There should not be any current through the reverse-biased diode circuit, however.

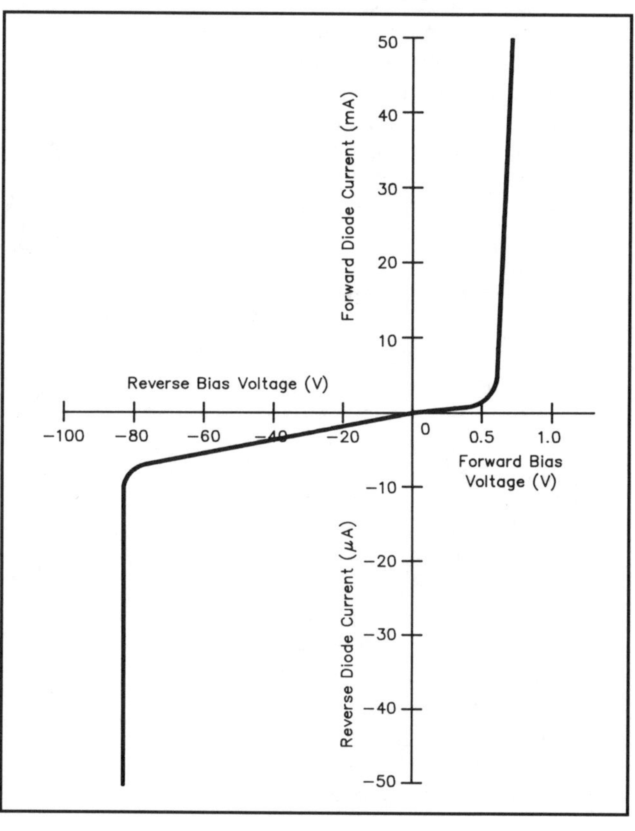

Figure 1—This graph shows how the current through a diode changes as the applied voltage changes. When the voltage applied to the anode, or P-type material is positive, the forward current rises rapidly. For a silicon diode the forward bias voltage must be more than about 0.6 volts to cause this forward current. The forward bias voltage only needs to be more than about 0.2 volts for a germanium diode. When the voltage applied to the cathode, or N-type material is positive, there is only a small leakage current. When the reverse-bias voltage exceeds the diode's breakdown voltage, however, there is a large reverse current. This will destroy many diodes.

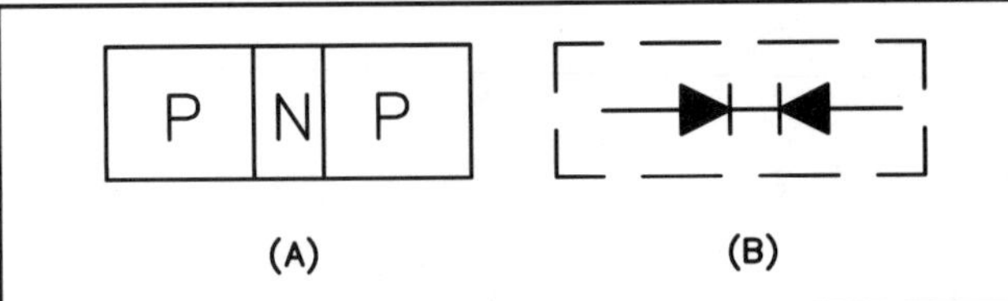

Figure 2—This drawing shows a three-layered sandwich of semiconductor materials. This sandwich forms two PN-junction diodes, as indicated by the diode schematic symbols shown at B. The dashed box indicates these diodes are coupled on one piece of semiconductor. They are not individual diodes.

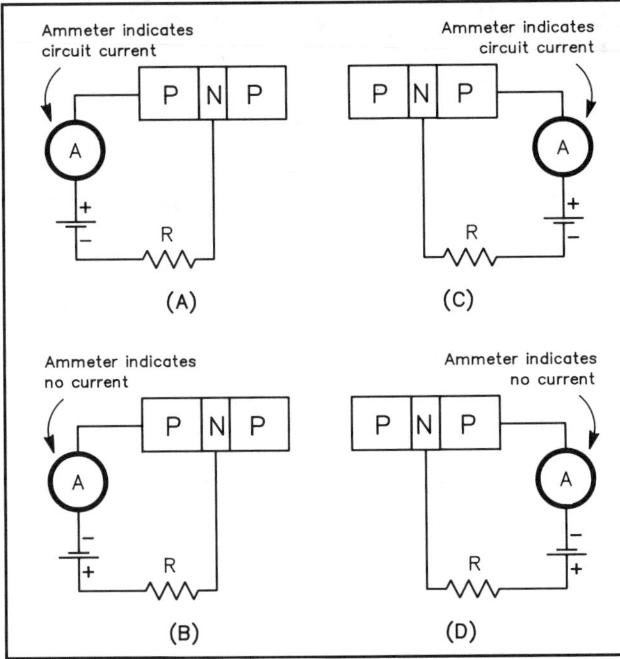

Figure 3—Each PN junction behaves like a typical diode. Parts A and B show the left junction forward and reverse biased. Parts C and D show the right junction with forward and reverse bias applied. Notice that both junctions conduct electricity when they have a forward-bias voltage. Neither junction conducts when it has a reverse-bias voltage.

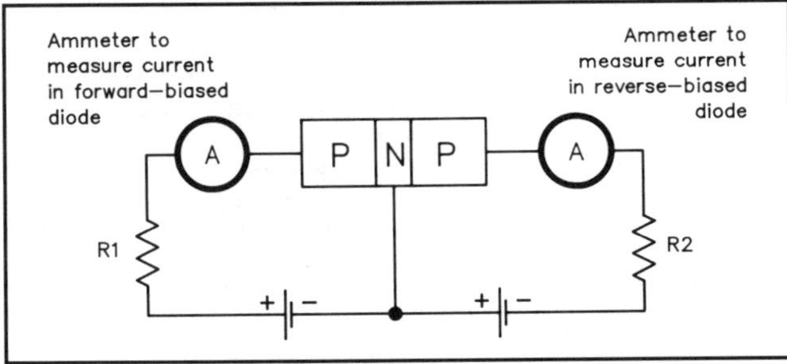

Figure 4—Connect one diode of our "sandwich" to a forward-bias voltage and the other to a reverse-bias voltage. This circuit illustrates the *transistor effect.* Current in the forward-biased diode causes current in the reverse-biased diode.

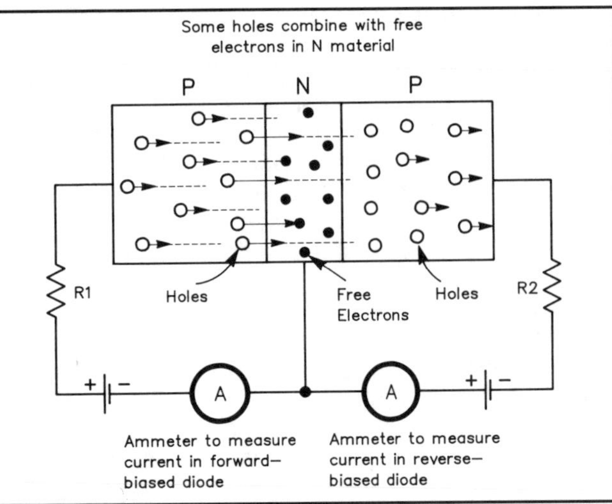

Figure 5—This illustration shows the movement of holes and electrons in our three-layered sandwich. Holes move through the thin layer of N-type material. Then the negative voltage on the other side of the second layer of P-type material pulls the holes. The third terminal collects these holes. The holes create a current through the second diode.

Something amazing happens when we conduct this experiment. With current in the forward-biased circuit there also is current in the reverse-biased one. Increasing the current in the first part of the circuit increases the current in the second part. Decreasing the current in the first part decreases the current in the second part. This is what Shockley and his coworkers called the *transistor effect.* Current in the forward-biased diode is transferred across the resistance of the reverse-biased diode.

How can this action take place? Figure 5 shows the semiconductor materials with *free electrons* in the N-type material and *holes* in the P-type material. Holes in the forward-biased P material move toward the negative voltage polarity on the N material. The voltage pulling on the holes makes them speed up as they move closer to the PN junction.

The motion of these holes causes them to move through the thin layer of N-type material without combining with free electrons there. The holes enter the second layer of P-type material. The negative voltage on that material pulls the holes all the way through the second layer of P-type material.

This device became known as a *transistor*, because it is a device that exhibits transistor action. We call the first layer of P material the *emitter.* The emitter puts current into the transistor, or emits current. The second layer of P material we call the *collector*, because it collects the holes that move through the N-type material. The section of N-type material in the middle is the *base*.

It is important to understand that you can't make a transistor by simply connecting two diodes. The diodes must be formed as a single unit. The two P layers must be in close contact with the thin N layer.

PNP and NPN Transistors

In the last section you learned that a *transistor* is a three-layered sandwich of P and N-type semiconductor material. All our examples in that section had two P-material layers and one N-material layer.

We pictured a transistor as a connection between the cathodes of two PN-junction diodes. This joint represents one transistor connection. The other two connections are to the diode anodes. Figure 1 shows diodes connected this way. Part B shows the three-layered sandwich. Part C shows the schematic-diagram symbol we use to represent such a transistor on circuit diagrams.

The Figure 1 transistor has a layer of N-type material between two layers of P-type material. This is why we call it a *PNP transistor*. You will notice that the schematic-diagram-symbol *emitter* lead has an arrow. This arrow points toward the transistor *base*. Remember that the arrow points toward the base for a PNP transistor. Just keep in mind that the arrow *Points iN Proudly*.

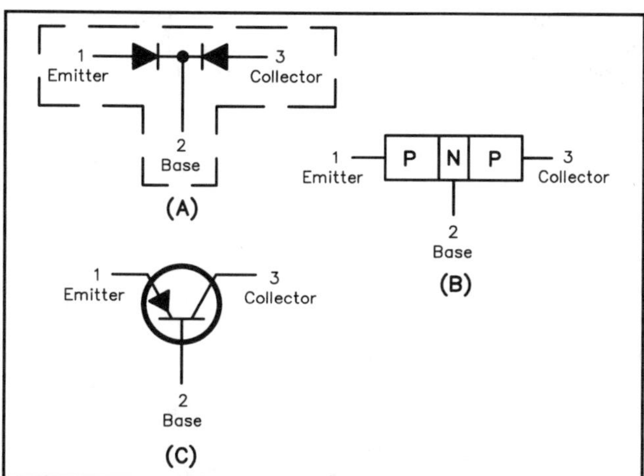

Figure 1—This drawing shows a PNP transistor as a diode pair and a semiconductor sandwich. The dashed box indicates these diodes are coupled on one piece of semiconductor. They are not individual diodes. Part C shows the schematic diagram symbol for PNP transistors. The arrow in this symbol *Points iN Proudly*.

You may be wondering if we can make a transistor from two layers of N-type material and one of P-type material. The answer is, "Yes, we can!" Figure 2 shows a

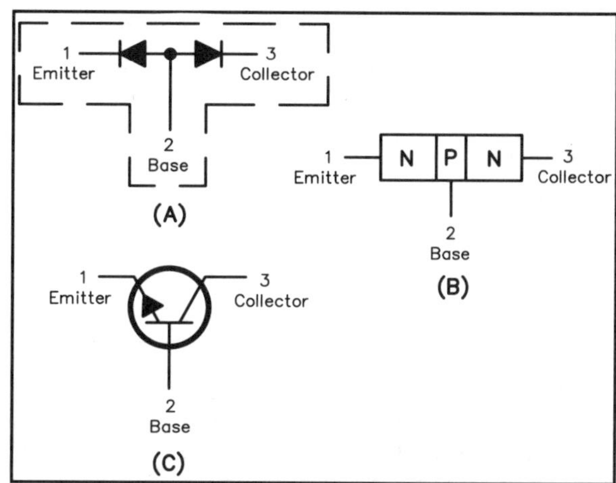

Figure 2—A diode pair connected anode to anode form an NPN transistor. The dashed box indicates these diodes are coupled on one piece of semiconductor. They are not individual diodes. Part C shows the schematic diagram symbol for NPN transistors. The arrow on the emitter does *Not Point iN* on an NPN transistor.

diode pair connected anode to anode. This forms a pair of PN junctions that have the P material in the center. We have an NPN sandwich, or *NPN transistor.* Part C shows the schematic-diagram symbol for an NPN transistor. There is an arrow on the emitter lead of this symbol. Notice that this arrow does *Not Point iN.*

The diode-symbol arrow points in the direction of *conventional current.* This is opposite to the direction of electron flow. The transistor-symbol arrows also point in the direction of conventional current. The arrows always appear on the emitter lead.

Figure 3A shows a PNP transistor connected in a test circuit. The emitter connects to the positive voltage supply. The collector connects to the negative side of a second voltage supply. The emitter/base diode has a forward-bias voltage applied. The collector/base diode has a reverse-bias voltage.

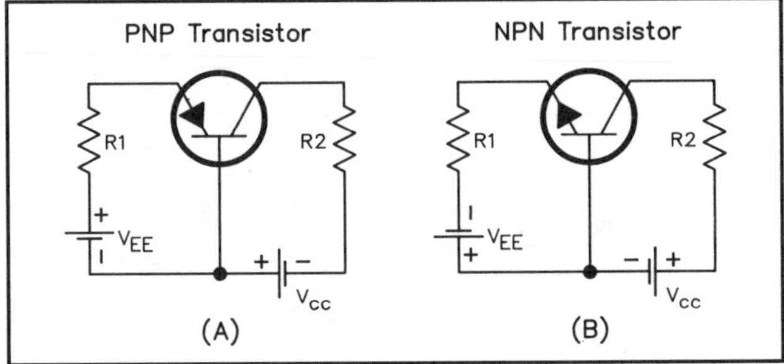

Figure 3—Part A shows a simple test circuit for a PNP transistor. V_{EE} and V_{CC} are common designators for the emitter and collector bias voltages. Resistors R1 and R2 limit the transistor current. Part B shows a similar connection for an NPN transistor.

Figure 3B is an NPN transistor test circuit. The NPN-transistor emitter connects to the negative side of the voltage supply. The collector connects to the positive side of a second voltage supply. Again, the emitter/base diode has a forward-bias voltage. The collector/base diode has a reverse-bias voltage.

Transistor action occurs in a PNP transistor when the bias voltages pull holes from the emitter material through the base region. These holes move into the collector region, where the bias voltage continues to attract the holes through the material.

Transistor action in an NPN transistor is similar. This time the emitter has a negative voltage and the base has a positive voltage. The base region attracts free electrons from the emitter. These free electrons are moving fast enough to move right through the base. Once they enter the collector material, the positive bias voltage on that region continues to attract the electrons.

When we draw a transistor as a diode pair or as a semiconductor block, the construction looks symmetrical. You might wonder if you could use either end connection as the emitter. Transistors aren't usually manufactured to be symmetrical, though. Manufacturers create better transistors by making the emitter and collector sides different.

Controlling a Current with a Smaller Current

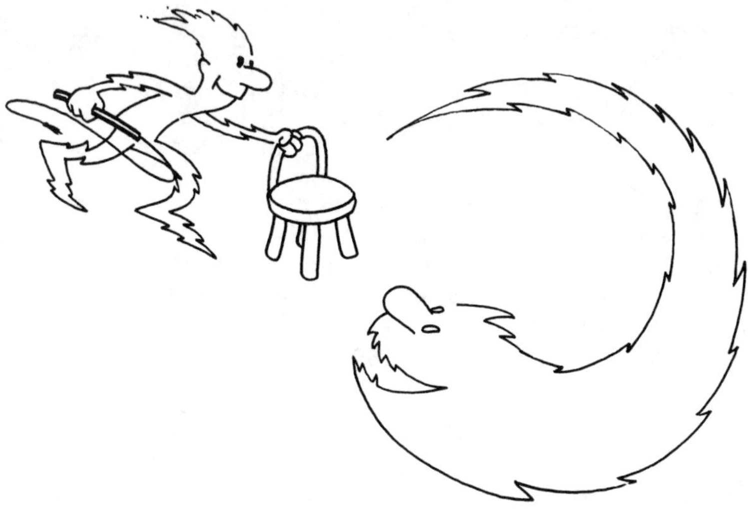

Figure 1 shows a PNP transistor circuit. The circuit includes ammeters and voltmeters to measure various circuit conditions. The transistor's emitter/base diode has a forward-bias voltage applied.

Ammeter M1 measures the current through the emitter/base junction. Ammeter M2 measures the current at the transistor base lead. Ammeter M3 measures current through the reverse-biased base/collector junction. Voltmeter M4 measures the voltage between the emitter and collector.

We will use the Figure 1 test circuit to measure circuit conditions for our transistor. By changing the bias voltages one at a time, we can learn a lot about *transistor action*. Let's conduct an experiment with this circuit and make some measurements. Assume we can vary the emitter and collector bias voltages as needed.

Begin with V_{CC} and V_{BB} both set to 0. Adjust V_{CC} until voltmeter M4 reads 5 volts. Then increase V_{BB} until ammeter M2 reads 0.2 mA. When we read ammeter M3, the current is 4.2 mA.

For the next part of our experiment, we increase V_{CC} until M4 reads 10 volts. Continue to increase this reading, 5 volts at a time, up to 40 volts. M3 displays the collector current.

It may surprise you to learn the collector current remains constant. With a collector voltage of 40, the collector current is still 4.2 mA. Figure 2 is a graph of the collector current as the collector voltage changes. Ammeter M1 reads a constant 4.4 mA during these collector-voltage changes.

The experimental results given in this example are for an ideal transistor. The collector current actually does increase slightly for real transistors. The increase is normally small, however. Let's continue our experiment with this ideal transistor.

Set V_{CC} so meter M4 reads 5 volts again. Then increase V_{BB} and watch the base-current meter, M2. Set V_{BB} so M2 reads 0.4 mA. What do you think M3 will read? Well, we doubled the base current, so if you guessed 8.4 mA, you are correct.

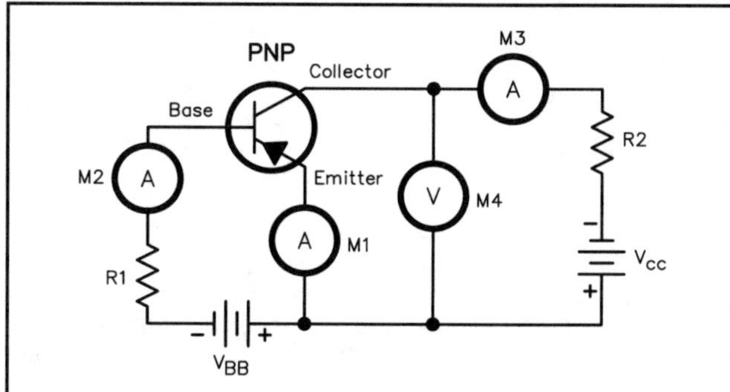

Figure 1—This test circuit will help us measure transistor circuit conditions. V_{BB} and V_{CC} allow us to vary the transistor bias voltages. The meters help us measure the current at each lead and the voltage between the collector and emitter.

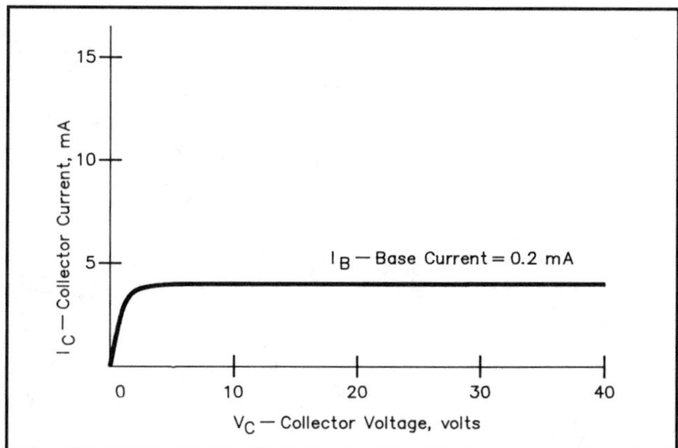

Figure 2—This graph shows measurements taken from the Figure 1 circuit. The data shows that collector current remains constant while the collector voltage varies over a wide range.

Now we are ready to increase the collector voltage again. Adjust V_{CC} for a reading of 10 volts on M4 and read the collector current on M3. Continue increasing the collector voltage 5 volts at a time and reading the collector current. This time you shouldn't be surprised to learn that the current remains at 8.4 mA. Likewise, the emitter current remains a constant 8.8 mA.

Reset V_{CC} for 5 volts on the collector. Adjust V_{BB} until the base current is 0.6 mA. Note that the collector current reads 12.6 mA. Then increase the collector voltage in 5-volt steps and read the collector and emitter currents at each step. What is the reading from the emitter-current meter, M1? If you have followed the pattern, you can estimate that the emitter current is 13.2 mA. Equation 1 shows the relationship between base current, emitter current and collector current.

$$I_E = I_B + I_C \qquad \text{(Equation 1)}$$

where:
I_E is the emitter current
I_B is the base current
I_C is the collector current.

We are measuring the transistor characteristics in this experiment. Figure 3 is a graph of collector current and collector voltage at each measured base-current value. Manufacturers make many different transistor types. The characteristic curves for each type have different specific numbers. The general shapes of the curves are similar, however.

Eventually, we find that increasing the base current further does not increase the collector current. For our sample transistor, this point might be at 1.2 mA of base current. When the collector current reaches a maximum value, we say the transistor is *saturated.* This means the collector circuit gathers all the charge carriers the emitter injects through the base. The charge carriers are holes for PNP and electrons for NPN transistors.

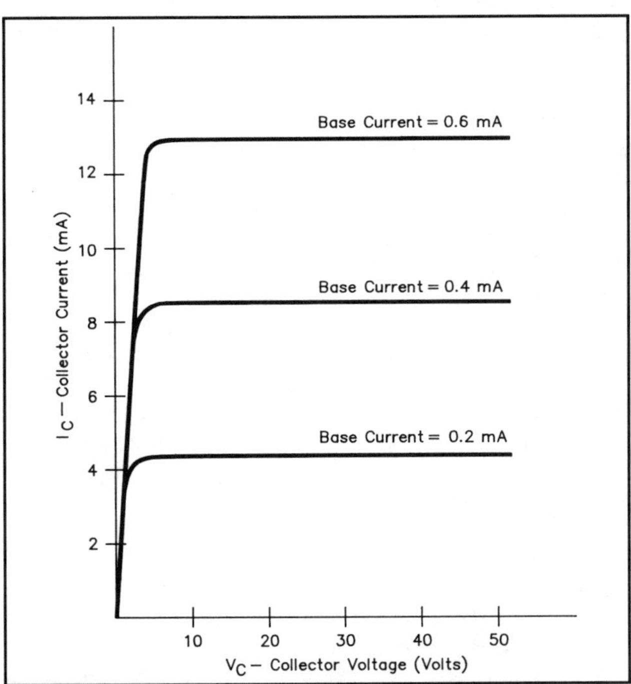

Figure 3—This graph shows the characteristic curves for our sample transistor. Characteristic curves of other transistor types have similar shapes, although the specific voltages and currents are different.

What else have we learned from our experiment? Small changes in base current produce larger changes in collector current. If we feed a signal into the transistor base, the collector current will be an enlarged version of that signal. Suppose the base signal varies between 0.2 mA and 0.6 mA. The collector current varies between 4.2 and 12.6 mA.

A small base current controls a larger collector current. The output signal is an enlarged version of the input, so the circuit *amplifies* the input signal. Because of *transistor action*, our circuit is an *amplifier.*

You will hear two terms used to describe the current gain characteristics of transistors. These gain factors are important to circuit designers and engineers. You probably won't use these factors in circuit calculations until you study more advanced electronics. A transistor's *alpha* (α) and *beta* (β) help evaluate how effective a particular transistor type will be for an amplifier circuit.

Calculate current gain α by dividing collector current by emitter current. Current gain α is always less than 1, normally in a range between 0.92 and 0.98.

$$\alpha = \frac{I_C}{I_E} \qquad \text{(Equation 2)}$$

where:
I_C is the collector current
I_E is the emitter current

Let's calculate α for the transistor in our example.

$$\alpha = \frac{I_C}{I_E} = \frac{4.2 \text{ mA}}{4.4 \text{ mA}}$$

$$\alpha = 0.9545$$

Try repeating this calculation using the collector and emitter currents that go with any of the other base currents we measured. You will find the same answer.

Calculate current gain β by dividing collector current by base current. Current gain β normally varies between 10 and 200. Current gain β is not constant for a particular transistor type. If one transistor has a β of 100, another one of the same type may have a β of 150 or 200. Current gain β even varies with operating conditions for a particular transistor. It is still an important parameter for designers and engineers.

$$\beta = \frac{I_C}{I_B} \qquad \text{(Equation 3)}$$

where I_B is the base current

Use Equation 3 to calculate β for our sample transistor.

$$\beta = \frac{I_C}{I_B} = \frac{4.2 \text{ mA}}{0.2 \text{ mA}}$$

$$\beta = 21$$

You can repeat this calculation for other collector- and base-current values from Figure 3.

Biasing the Transistors

Voltages applied to transistor diode junctions set up the circuit operating conditions. In the last section we measured transistor characteristics and plotted the values on a graph. We found that base current controls collector current. What happens when we apply a changing signal, such as an ac sine wave, to the base?

Suppose the only voltage we apply to the transistor base is an ac signal. Figure 1 is an NPN transistor circuit with no other base voltage. When the signal voltage is positive, it puts a forward-bias voltage on the base/emitter junction. This creates base current, which turns on the collector current.

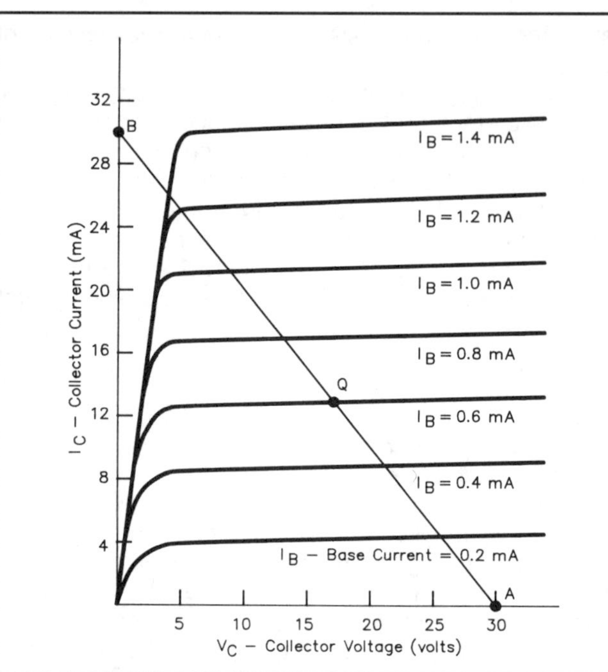

Figure 2—This graph shows a typical transistor characteristic curve. Line AB is a load line with V_{CC} = 30 volts and R2 = 1 kΩ. Q marks the operating point, or Q point.

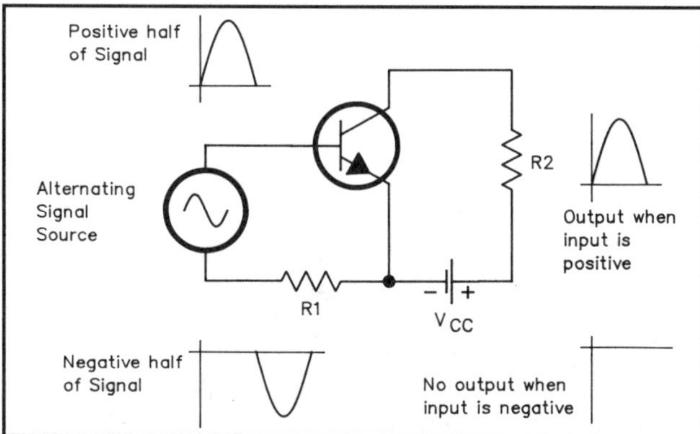

Figure 1—This NPN transistor circuit has an alternating-current signal applied to its base. When this signal is positive, the base/emitter junction has a forward-bias voltage. This junction conducts, and there is collector current. When the signal voltage is negative, the base/emitter junction has a reverse-bias voltage. With no base current there is no collector current either.

What happens when the ac signal voltage is negative? The base/emitter junction has a reverse-bias voltage. There is no base current, so the collector current also stops.

The collector-current output is a pulse for every half cycle of the input voltage. This is similar to the output from a half-wave rectifier circuit. That isn't the kind of output we usually want from a transistor circuit. The Figure 1 circuit is not producing an amplified version of the input signal.

A transistor circuit requires dc bias voltages to set the operating point on the transistor characteristic curves. Proper bias voltages set this *quiescent point* or *Q point.*

Figure 2 shows a set of characteristic curves. Q marks the desired operating point, or Q point. The dc bias voltages set up these circuit conditions with no ac signal on the input.

With a Q point established, alternating input signals vary the base/emitter junction bias voltage. If the input-signal voltage is within certain limits it won't reverse bias the base/emitter junction, however. For the characteristic curves of Figure 2, the input signal current can vary about ± 4 mA. That corresponds to a variation of about ± 13 volts, which will change the bias voltage between 33 V (20 + 13) and 7 V (20 – 13).

Figure 2 also shows a *load line* on the graph. A load line provides a way for us to picture the circuit operation. The equations and techniques for calculating the transistor bias values and circuit components are beyond the scope of this book. You can use Ohm's Law to understand the basics, however.

Look at the Figure 3 circuit and the Figure 2 characteristic curves. The collector voltage will be 30 V when there is no collector current, because there is no voltage drop across R2. That represents point A on the load line. Next consider the collector/emitter junction to be a short circuit. In that case, R2 drops 30 V. Equation 1 gives the current through R2 and the transistor.

$I = E / R$ (Equation 1)

$I = 30\text{ V} / 1000\ \Omega = 30\text{ mA}$

Point B on the load line represents a 0-V drop across the collector, and a 30 mA collector current.

Connect points A and B to form a load line. The Q point will fall somewhere along this line. For our example, we want the Q point on the $I_B = 0.6$ mA line. We use Ohm's Law again to select V_{BB} and R1 to provide 0.6 mA base current.

These simple calculations will help you understand why load lines and Q points are important. We must consider other factors to design a practical transistor circuit. Using two bias-voltage supplies is seldom practical. Variations in individual transistor characteristics and temperature stability also make a circuit like Figure 3 unreliable.

Figure 4 shows a circuit that uses a single voltage supply for both the base and collector bias voltages. The additional resistors provide temperature stability and reduce the circuit dependence on individual transistor characteristics.

Capacitor C1 is the emitter *bypass capacitor*. It routes the signal current around the emitter resistor, R3. The bypass capacitor aids circuit stability. If the signal current went through the emitter resistor, it would upset the bias conditions.

The input and output *coupling capacitors* C2 and C3 pass the alternating signal and block the dc bias voltages. We sometimes refer to these capacitors as *blocking capacitors* because they block the direct voltages.

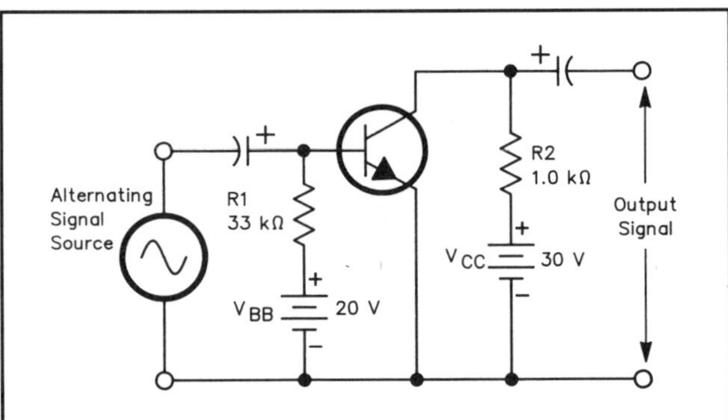

Figure 3—This schematic diagram is a revised version of Figure 1. We added a base bias supply and coupling capacitors on the input and output. The coupling capacitors block the dc bias voltages from going to the ac signal source and the output circuit.

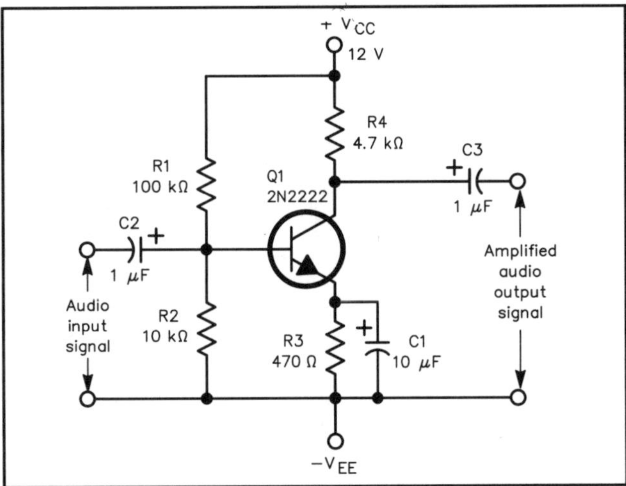

Figure 4—This audio amplifier circuit uses a common 2N2222 NPN transistor. The resistors set the proper Q point bias voltages and make the circuit nearly independent of temperature and individual transistor variations. Capacitor C1 is the emitter *bypass capacitor*. Sometimes we refer to the input and output *coupling capacitors* C2 and C3 as *blocking capacitors*.

The Common-Emitter Circuit

Bipolar transistors have three leads, the *base*, *emitter* and *collector*. For all the transistor amplifier circuits we studied so far, the input signal goes between the base and emitter. We take the output signal from the collector and emitter. The emitter lead is common to both the input and output circuits.

One transistor lead is always common to the input and output, because you need four connections and the transistor only has three. In the next two sections of this book you will learn about some other ways to connect transistors as amplifiers.

The circuit of Figure 1 shows an NPN transistor connected as a *common-emitter amplifier*. This circuit should look familiar, because we studied it earlier. Notice there is one bias-voltage supply.

Resistors R1 and R2 form a voltage divider circuit, to supply the correct base/emitter junction bias voltage. The emitter resistor, R3, helps make the circuit stable and less dependent upon the individual transistor current gain β. Collector current through resistor R4 develops the output voltage. Capacitor C1 bypasses the ac signal around the emitter resistor. Capacitors C2 and C3 couple the ac signal into and out of the amplifier. The amplified output signal is 180° out of phase with the input signal.

Figure 2 shows a PNP transistor connected as a common-emitter amplifier. Notice this circuit is nearly identical to the Figure 1 circuit. The bias-voltage supply has the opposite polarity, however.

We used a common-emitter circuit to measure the transistor characteristics in the previous sections. Manufacturers and circuit designers refer to these characteristics as *common-emitter characteristics*.

The input and output *impedances* of our transistor amplifier circuit are important. Let's begin by considering the input and output impedances of the transistor itself. Then we will learn how this relates to the circuit impedances.

The base/emitter junction is a diode with forward-bias voltage. A forward-biased diode conducts current. It has a low resistance, typically a few ohms or less. We can say, then, that a transistor has low input impedance.

The base/collector junction is a diode with reverse-bias voltage. A reverse-biased diode does not conduct easily. It presents a very high resistance, typically a million ohms (a *megohm*) or more. Bipolar transistors have very high output impedance.

Don't confuse the transistor input and output impedances with our amplifier impedances. The amplifier imped-

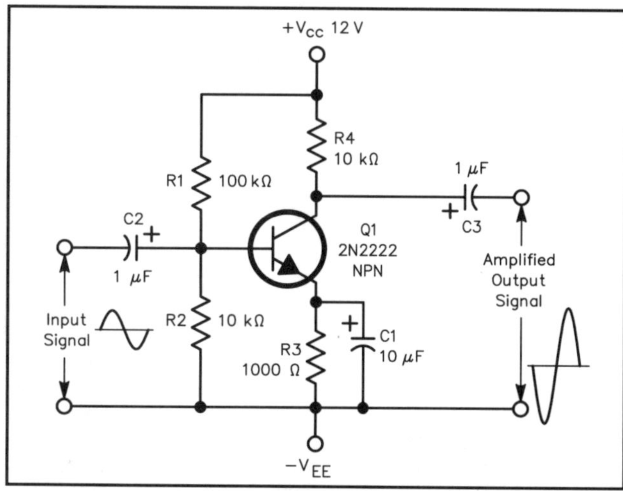

Figure 1—This diagram is a typical NPN common-emitter amplifier.

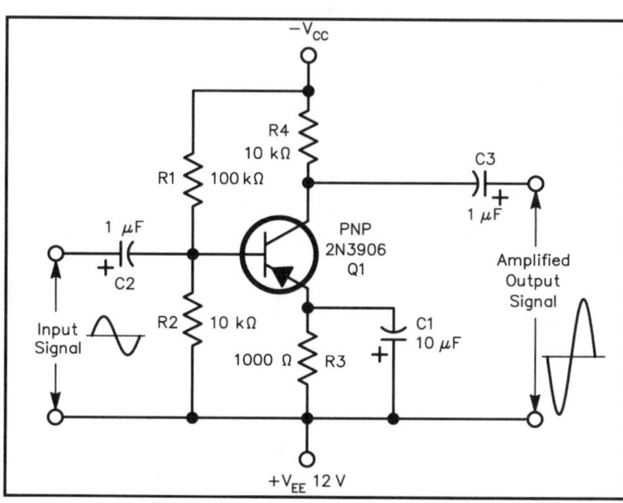

Figure 2—A typical PNP common-emitter amplifier is very similar to the NPN version.

ances depend on the bias-resistor values, if we choose those values properly.

This book won't teach you to design transistor amplifiers. That is beyond the scope of the book. You can learn more about circuit design by continuing your studies. A few rules will help you understand the circuit operation, however. These rules also will help you evaluate and modify someone else's design.

The emitter resistor, R3, will usually be between 100 Ω and 1000 Ω. R2 should be about 5 to 10 times larger than R3. R1 and R2 form a voltage divider to set the base bias voltage. R1 should be about 10 times the value of R2. The output resistor, R4, should be about 5 to 10 times the emitter resistor value, R3. Let's summarize these rules with some equations.

$$\text{R3 between } 100\ \Omega \text{ and } 10000\ \Omega \qquad \text{(Equation 1)}$$

$$\text{R2} \approx 5 \times \text{R3 to } 10 \times \text{R3} \qquad \text{(Equation 2)}$$

$$\text{R1} \approx 10 \times \text{R2} \qquad \text{(Equation 3)}$$

$$\text{R4} \approx 5 \times \text{R3 to } 10 \times \text{R3} \qquad \text{(Equation 4)}$$

where ≈ is a symbol that means "approximately equal to."

We won't go through a complete mathematical analysis here. By making a few simplifying assumptions, such an analysis would show the input impedance of our amplifier circuit nearly equal to R2. You can look at a common-emitter amplifier circuit and quickly determine the circuit input impedance. Just look at the bias resistor between the base and the ground, or common, connection.

Are you uncertain about why R2 determines the input impedance? Think of it this way. The base/emitter junction has only a few ohms of resistance. The transistor's current gain ß multiplies the emitter resistor value, however. Suppose ß is 50 for our transistor. With a 1000 Ω resistor at R3, this gives a value of 50 kΩ. In effect, R2 is in parallel with this value. You can calculate the resistance of this parallel combination. Since R3 is much larger than R2, the result will be only a little less than R2. This makes R2 a good approximation for the circuit input impedance.

You also can determine the circuit output impedance by looking at the circuit. The output impedance is approximately equal to the output resistor, R4.

To understand this output-impedance approximation, we'll consider the transistor output impedance. Remember the reverse-biased junction has a high impedance. This high impedance is in parallel with R4, which has a much smaller value. Again, the parallel combination is only a little smaller than R4's value.

Gain refers to how much an amplifier will increase the strength of an input signal. Depending on the application, you may be interested in the *current gain*, the *voltage gain* or the *power gain* of an amplifier. Equations 5, 6 and 7 show these gains depend on the circuit resistors.

$$G_i = \frac{R2}{R3} \qquad \text{(Equation 5)}$$

$$G_v = \frac{R4}{R3} \qquad \text{(Equation 6)}$$

$$G_p = G_i \times G_v = \frac{R2 \times R4}{R3^2} \qquad \text{(Equation 7)}$$

Common-emitter amplifiers have high input and output impedances. The method used to stabilize the circuit conditions and make it independent of temperature and individual transistor variations, reduces the current gain. Common-emitter amplifiers have good power gain, however. They are the most-used bipolar-transistor amplifier type.

The Common-Base Circuit

The emitter lead is common to both the input and output section for all the transistor circuits we studied so far. You may be wondering if either of the other leads can be common to both sections. Yes, any of the three leads can be common to the input and output sections.

Figure 1 shows a simple *common-base circuit*. We used a PNP transistor for this example, but we also could use an NPN transistor. Notice the emitter/base junction has a forward-bias voltage and the base/collector junction has a reverse-bias voltage. We would reverse the bias-battery polarities for an NPN transistor.

Figure 1—This diagram shows a simple common-base transistor circuit. Separate emitter/base and base/collector-junction bias supplies help you understand circuit operation.

There is nothing surprising about biasing the common-base circuit. Separate batteries supply the emitter/base and base/collector-bias voltages. R1 and R2 limit the currents. C1 and C2 couple the alternating signal voltages into and out of the circuit.

We normally want to use a single bias supply instead of the two supplies shown in Figure 1. A voltage divider works well for this circuit, as it does for the common-emitter circuit. We also want our circuit to be stable with temperature variations. The circuit should be independent of individual transistor variations. Figure 2 shows a PNP transistor with the proper biasing resistors.

Resistors R1 and R2 form a voltage divider to set the base/emitter bias voltage. The emitter resistor, R3, and the load resistor, R4, stabilize the circuit operation. The resistor combination sets the transistor Q point.

Figure 3 shows an NPN transistor connected as a common-base amplifier. The circuit is identical to the Figure 2 circuit, except the bias-voltage polarity reverses. Notice the input and output signals are in phase.

The input impedance of a common-base transistor circuit has a low value. This is because the base/emitter junction is a forward-biased diode. The impedance depends on the emitter current.

Shockley, who helped discover transistors, found that Equation 1 gives the input impedance.

$$Z_{in} = 26 / I_E \qquad \text{(Equation 1)}$$

where I_E is the emitter current in milliamps.

The input impedance of a common-base circuit is 26 ohms when the emitter current is 1 mA.

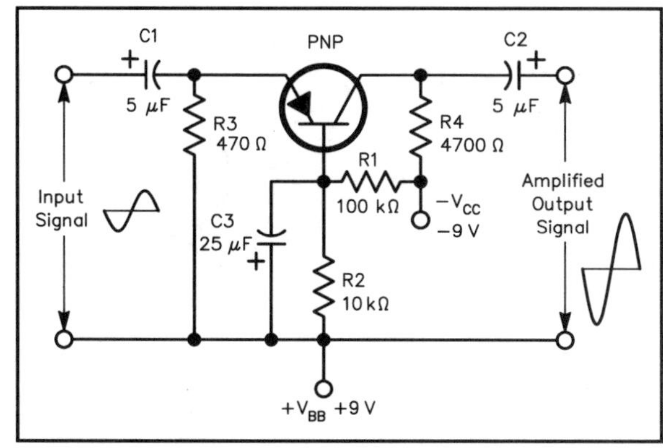
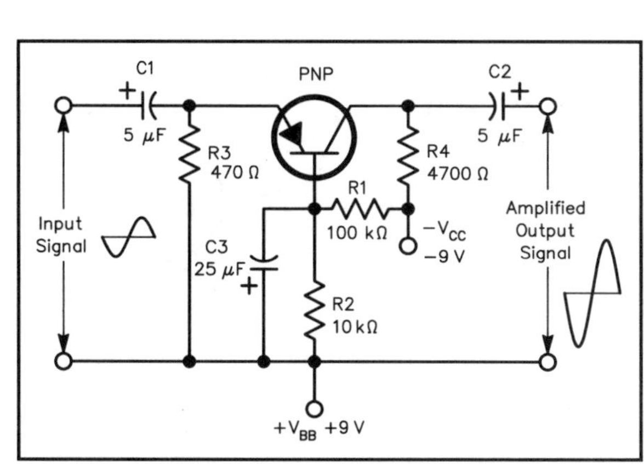

Figure 2—This PNP transistor serves in a common-base circuit. This circuit has a single bias supply. The amplifier is stable with temperature variations.

The transistor output impedance is a high value, because the base/collector junction has a reverse-bias voltage. A reverse-biased PN junction has a high impedance.

The load resistor, R4, sets the common-base *circuit* output impedance because this resistor is in parallel with the diode junction. The load resistor may vary between a few thousand ohms and 100 kΩ, depending on the desired circuit output impedance. In any case, the load resistor has a small value compared to the base/collector junction impedance. The circuit output impedance is approxi-

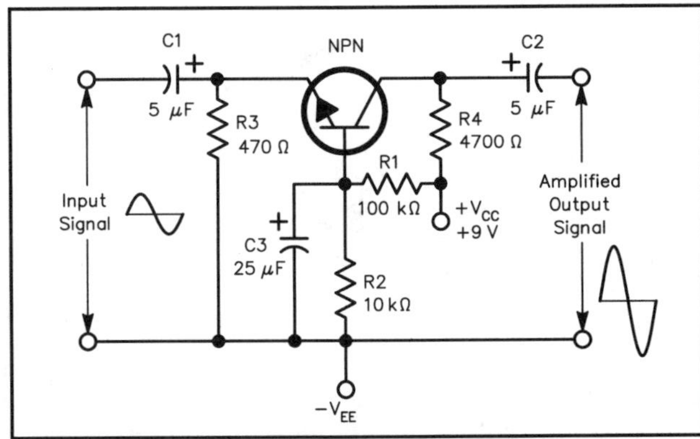

Figure 3—This common-base transistor amplifier uses an NPN transistor. The circuit is similar to the Figure 2 circuit, with the bias-voltage polarity reversed.

mately equal to R4's value.

You can calculate the amplifier current gain by dividing the output current by the input current. Output current is the same as collector current and input current is the same as emitter current. We can write Equation 2 for current gain.

$$G_i = \frac{I_o}{I_i} = \frac{I_c}{I_e} \qquad \text{(Equation 2)}$$

In an earlier section we defined a transistor's *current gain* α as collector current divided by emitter current. Equation 2 tells us that the common-base current gain equals the transistor α. Collector current is nearly equal to the emitter current. It is slightly less, though, because collector current and base current add up to the emitter current. Typical values for α range between about 0.90 to 0.99

$$G_i = \alpha \qquad \text{(Equation 3)}$$

Divide the output voltage by the input voltage to find the circuit voltage gain. Equation 4 shows another way to write the voltage gain.

$$G_v = \frac{\alpha\, R4}{Z_{in}} = \frac{\alpha\, I_e\, R4}{26} \qquad \text{(Equation 4)}$$

where

R4 is the load-resistor value

I_e is the emitter-current value

Equation 5 gives the power gain of a common-base circuit. This is the product of the current and voltage gains. Since current gain is slightly less than 1, power gain is a bit less than voltage gain.

$$G_p = G_i \times G_v \qquad \text{(Equation 5)}$$

A typical value for voltage gain is about 320. This value gives a typical power gain of about 310. The common-base circuit output power will be about 310 times greater than the input power.

A common-base amplifier has a low input impedance and a high output impedance. The circuit can be unstable, however, and may oscillate if the gain is set too high.

The Common-Collector Circuit

In this section we will study the third transistor-amplifier circuit type. These amplifiers have the collector lead common to both the input and output circuits. As you probably can guess, we call this a *common-collector circuit.*

Figure 1 shows a simple *common-collector transistor amplifier.* This example uses a PNP transistor, but we also could use an NPN transistor. The bias-battery polarities would have to reverse for a PNP circuit.

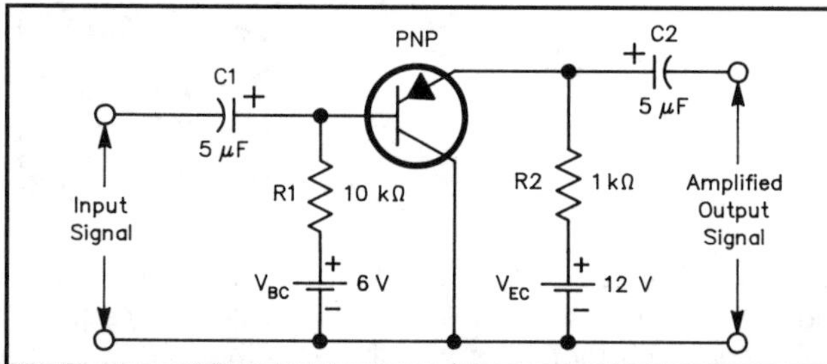

Figure 1—This diagram shows a simple common-collector transistor circuit. Separate bias supplies help you understand circuit operation. Batteries V_{BC} and V_{EC} provide the proper forward and reverse bias voltages. See the text for an explanation.

The Figure 1 circuit is not practical, because we normally want to use a single bias supply. The circuit helps us study the common-collector circuit operation, though.

Look at the bias circuit for this transistor. The base and emitter both have positive voltage relative to the collector. The voltage between the collector and emitter is more positive than the voltage between the base and collector. You might look at the circuit and decide the voltage between the transistor's collector and emitter is 12 volts. That is not the case, however. There is a voltage drop across R2. That voltage drop depends on the current through R2. (Remember Ohm's Law?) This voltage drop is important when the circuit designer sets up the bias conditions.

There is also a voltage drop across R1 in the base/collector circuit. The voltage between the base and collector is somewhat less than 6 volts. The base current is small, so we often ignore this voltage drop. The emitter/base junction does have a forward-bias voltage, because the emitter is more positive than the base. Battery V_{BC} provides a reverse-bias voltage for the base/collector junction.

This bias circuit may look a little different from the other two amplifier types. The principles are the same, though. Separate batteries supply the emitter/base and base/collector-bias voltages. R1 and R2 limit the currents. C1 and C2 couple the alternating signal voltages into and out of the amplifier.

We normally want to use a single bias supply instead of the two supplies shown in Figure 1. A voltage divider works well for the common-collector circuit. We use a similar voltage divider for the common-emitter and common-base circuits. We also want our circuit to be stable with temperature variations. The circuit should be as independent of individual transistor variations as possible. Figure 2 shows a PNP transistor with the proper biasing resistors. Notice the input and output signals are in phase.

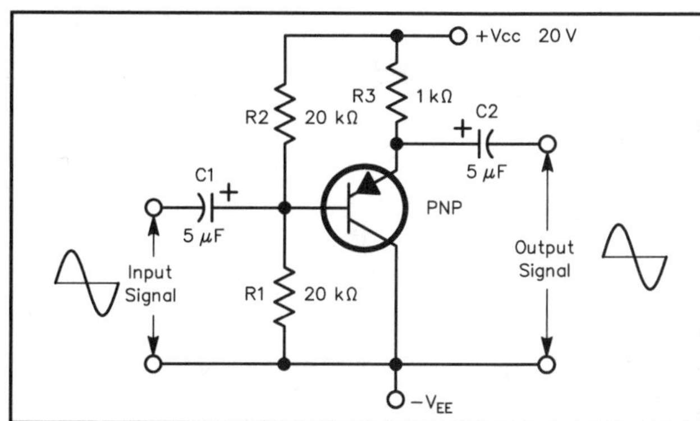

Figure 2—This PNP transistor serves in a common-collector circuit. The circuit has a single bias supply. The amplifier characteristics don't change with temperature variations.

Resistors R1 and R2 form a voltage divider to set the base voltage. The emitter resistor, R3, also serves as the load resistor. The resistor combination sets the transistor Q point.

Figure 3 shows an NPN transistor connected as a common-collector amplifier. The circuit is similar to the Figure 2 circuit. The bias-voltage polarity reverses.

You may notice another difference between the Figure 2 and Figure 3 circuits. We take the output across R3, the load resistor, in Figure 3. We show it this way to introduce you to another name for the common-collector circuit.

A mathematical analysis shows the voltage across

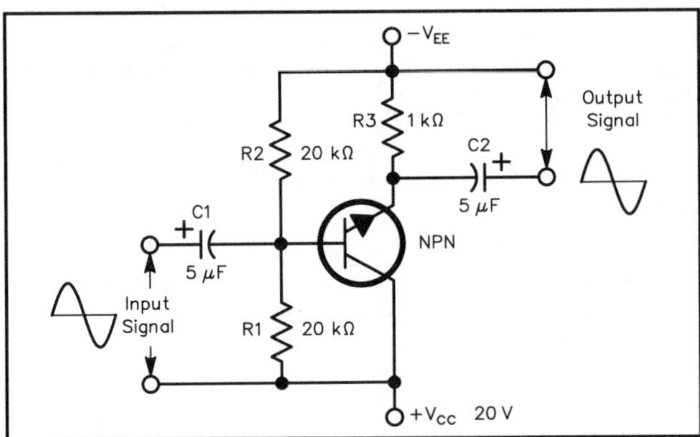

Figure 3—This common-collector transistor amplifier uses an NPN transistor. The circuit is similar to the Figure 2 circuit, with the bias-voltage polarity reversed. The output signal is taken across R3, showing why we often call the common-collector circuit an *emitter-follower amplifier.*

R3 varies directly with base-voltage variations. Suppose we feed an input signal into the amplifier. This signal causes a variation in base current, which in turn causes the emitter current to change. The emitter current is ß times the base-current. This means the emitter current changes by a factor of ß times the base-current change. The resulting emitter-current variation produces a changing voltage drop as it flows through R3.

Let's choose an input signal that varies the base voltage ± 2.5 volts. This produces a variation of almost exactly ± 2.5 volts across R3! We call the voltage across R3 the *emitter voltage*. The emitter voltage *follows* the base voltage in a common-collector circuit. For this reason, *emitter follower* is another name for the common collector.

The input impedance of a common-collector transistor circuit has a high value. This is because the collector/base junction is a reverse-biased diode. A typical transistor used in a common-collector circuit has an input impedance of 50 kΩ. The circuit input impedance is approximately equal to the parallel combination of R1, R2 and 50 kΩ. For the Figure 2 and 3 circuits, the input impedance is about 8333 Ω. This is nearly equal to the parallel combination of R1 and R2, 10 kΩ.

The transistor output impedance is a low value, because the base/emitter junction has a forward-bias voltage. A forward-biased PN junction has a low impedance.

A typical transistor has an output impedance of about 80 Ω when used in a common-collector amplifier. The circuit output impedance is approximately equal to the parallel combination of this value and the load resistor, R3. For our Figure 2 and 3 circuits the output impedance is about 74 Ω.

The voltage gain of an emitter-follower amplifier is approximately 1. This follows from our discussion about the circuit operation. The input and output voltages are nearly equal. Equation 1 shows we can find the voltage gain of an amplifier by dividing the output voltage by the input voltage.

$$G_V = \frac{V_o}{V_i} \qquad \text{(Equation 1)}$$

We also can calculate current gain as the input current divided by the output current. For our common-collector circuit, current gain is the parallel combination of R1 and R2 divided by the load resistor, R3.

$$G_i = \frac{\dfrac{1}{\dfrac{1}{R1} + \dfrac{1}{R2}}}{R3} \qquad \text{(Equation 2)}$$

For the Figure 3 circuit, current gain is 10.

Equation 3 gives an amplifier's power gain. This is the product of the current and voltage gains. Since voltage gain is 1, power gain is approximately equal to the current gain.

$$G_p = G_i \times G_V \qquad \text{(Equation 3)}$$

The Figure 3 power gain is 10.

A common-collector amplifier has a high input impedance and a low output impedance. This makes a stable amplifier circuit that is ideal as a buffer between a low-impedance load and a high-impedance source. The common-collector amplifier acts like an impedance transformer in this case.

Transistor Specifications

All electronics components have maximum voltage and current ratings. Transistors are no different. Transistors have several voltage and current ratings with which you should be familiar. There also are several other specifications that you should know about. A circuit designer must be sure not to exceed these ratings. Applying too much voltage or drawing too much current will destroy a transistor.

There is a maximum voltage rating for each pair of transistor elements. You should never apply more than 30 volts between the collector and emitter of a 2N2222 NPN general-purpose transistor, for example. You should not apply more than 60 volts across the collector/base junction of this transistor. The maximum safe voltage you can apply across the base/emitter junction of a 2N2222 transistor is 5 volts.

Other transistor types have different voltage ratings. We use the abbreviation V_{CEO} for the voltage between the emitter and collector. (The subscript letter O indicates the other terminal—the base in this case—is open, or has no voltage.) You will find transistors with a V_{CEO} rating as low as 15 volts or less, and others as high as 200 volts or more.

The abbreviation for the collector/base junction voltage is V_{CBO}. Some transistors have V_{CBO} ratings of 30 volts or less. Other transistors have V_{CBO} ratings of more than 200 volts.

We use the abbreviation V_{EBO} to mean the emitter/base junction voltage. This rating varies from 3.0 volts or less to 10 volts for some transistors. Table 1 lists these and other specifications for a few common general-purpose transistors.

The maximum collector current ($I_{c\ max}$) a transistor can handle is also important. Some transistors can only handle a collector current of 50 mA. Other transistors can safely carry 2 amperes or more of collector current.

Another important transistor specification is maximum power dissipation, P_{max}. Power dissipation means the power converted to heat inside the transistor. We calculate power by multiplying a voltage times a current. For transistor power dissipation, we multiply the collector/emitter voltage times the collector current.

$$P_{max} = V_{CE} \times I_c \qquad \text{(Equation 1)}$$

Notice that we don't calculate maximum power dissipation by multiplying the *maximum* collector/emitter voltage, V_{CEO}, times the *maximum* collector current, $I_{c\ max}$. The maximum power a transistor can dissipate is usually much less than this product.

There is a trade-off here. If your circuit will apply a voltage close to the maximum rating, you must ensure the collector current is less than maximum. If you have a circuit that requires close to the rated collector current, then you must reduce the voltage.

Let's suppose a 2N2222A transistor has 20 volts applied between its collector and emitter. How much collector current can the circuit draw without exceeding the transistor's 1.2-watt power dissipation rating? We'll solve Equation 1 for collector current. (To solve the equation,

Table 1
General Purpose Transistors

Listed numerically by device

Device	*Type*	*V_{CEO} Max. Collector-Emitter Voltage (Volts)*	*V_{CBO} Max. Collector-Base Voltage (Volts)*	*V_{EBO} Max. Emitter-Base Voltage (Volts)*	*I_c Max. Collector Current (mA)*	*P_D Max. Device Dissipation Watts*	*Current-Gain Bandwidth Product f_T* (MHz)*	*Noise Figure NF Max. (dB)*
2N918	NPN	15	30	3.0	50	0.200	600	6.0
2N2102	NPN	65	120	7.0	1000	1.0	60	6.0
2N2218	NPN	30	60	5.0	800	0.8	250	
2N2218A	NPN	40	75	6.0	800	0.8	250	
2N2219	NPN	30	60	5.0	800	3.0	250	
2N2219A	NPN	40	75	6.0	800	3.0	300	4.0
2N2222	NPN	30	60	5.0	800	1.2	250	
2N2222A	NPN	40	75	6.0	800	1.2	300	4.0
2N2905	PNP	40	60	5.0	600	0.6	200	
2N2905A	PNP	60	60	5.0	600	0.6	200	
2N2907	PNP	40	60	5.0	600	0.400	200	
2N2907A	PNP	60	60	5.0	600	0.400	200	
2N3053	NPN	40	60	5.0	700	5.0	100	
2N3053A	NPN	60	80	5.0	700	5.0	100	
2N3904	NPN	40	60	6.0	200	0.625	300	5.0
2N3906	PNP	40	40	5.0	200	1.5	250	4.0
2N4037	PNP	40	60	7.0	1000	5.0		
2N4123	NPN	30	40	5.0	200	0.35	250	6.0
2N4124	NPN	25	30	5.0	200	0.350	300	5.0
2N4125	PNP	30	30	4.0	200	0.625	200	5.0
2N4126	PNP	25	25	4.0	200	0.625	250	4.0
2N4401	NPN	40	60	6.0	600	0.625	250	
2N4403	PNP	40	40	5.0	600	0.625	200	
2N5320	NPN	75	100	7.0	2000	10.0		
2N5415	PNP	200	200	4.0	1000	10.0	15	
MM4003	PNP	250	250	4.0	500	1.0		
MPSA55	PNP	60	60	4.0	500	0.625	50	
MPS6547	NPN	25	35	3.0	50	0.625	600	

*Test conditions: I_c = 20 mAdc; V_{CE} = 20 V; f = 100 MHz

divide both sides by V_{CE}.)

$$P_{max} = V_{CE} \times I_c \qquad \text{(Equation 1)}$$

$$\frac{P_{max}}{V_{CE}} = \frac{V_{CE} \times I_c}{V_{CE}}$$

$$\frac{P_{max}}{V_{CE}} = I_c \qquad \text{(Equation 2)}$$

$$I_c = \frac{P_{max}}{V_{CE}} = \frac{1.2\ W}{20\ V} = 0.06\ A$$

$$I_c = 60\ mA$$

Table 1 shows the maximum collector current of a 2N2222A is 800 mA. Our calculation shows that the current must be much smaller for the suggested circuit. You also can calculate the maximum collector/emitter voltage for an 800-mA current. This time solve Equation 1 for V_{CE}.

$$V_{CE} = \frac{P_{max}}{I_c} \qquad \text{(Equation 3)}$$

$$V_{CE} = \frac{1.2\ W}{0.800\ A} = 1.5\ V$$

Semiconductor junctions are very sensitive to heat. Any power a circuit dissipates in the transistor heats the junction. If the transistor can't get rid of this heat fast enough, it destroys the junction. Transistors designed to handle higher power levels often require *heat sinks*. A heat sink is a piece of metal designed to conduct the excess heat away from the transistor. Some heat sinks are thin metal strips that slip over the transistor body. Others are heavy blocks of metal that attach to the cabinet or chassis sides. Figure 1 shows some common transistors and heat sinks.

You may want to build a radio-frequency amplifier as part of an Amateur Radio receiver or transmitter. You must consider the *frequency response* of the transistors you select for this circuit. Some transistors are useful to several hundred kilohertz while others operate well into the gigahertz range. You must select a transistor that will operate over the frequency range of your project. Price and other characteristics will influence your final selection.

Figure 2 shows a graph of an ideal frequency response. Resistors and capacitors in a practical circuit limit the low-frequency response. Transistor characteristics limit the high-frequency response. Ideally, the circuit should amplify all signals between the lower and upper limits the same amount. In practice, an amplifier isn't likely to have a response as flat across the top as the one we've shown in Figure 2.

We can measure the amplified signal voltage. As we vary the frequency, we find the upper and lower values at which the voltage drops to 0.707 times the top value. These frequencies define the circuit *bandwidth*. We also could measure the amplified signal power. The same upper and lower frequencies occur at the half-power points on the

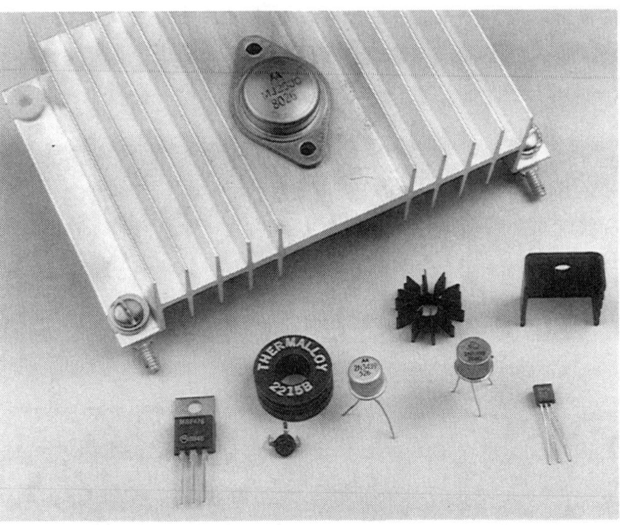

Figure 1—This photograph shows some common transistors and typical heat sinks. The heat sinks conduct excess heat away from the transistor to help cool the semiconductor junctions.

graph. When the power drops to half its original value, there is a 3-dB power decrease.

Table 1 lists the *current-gain bandwidth product*. This term specifies a transistor's useful upper-frequency limit. At this point the circuit gain drops to 1. The output signal is no stronger than the input signal, so there is no gain.

One more transistor specification is worth mentioning here. That is the transistor's noise figure, NF. *Noise figure* is a measure of how much noise the transistor adds to the amplified signal. Manufacturers specify a transistor's noise figure in decibels. They compare the transistor noise-signal power with the noise-signal power produced in a resistor as part of the test circuit.

As you continue your electronics studies you will learn about other transistor specifications. Many electronics circuits use transistors for purposes other than as amplifiers. The information you learned in this chapter will help you understand the other specifications and applications.

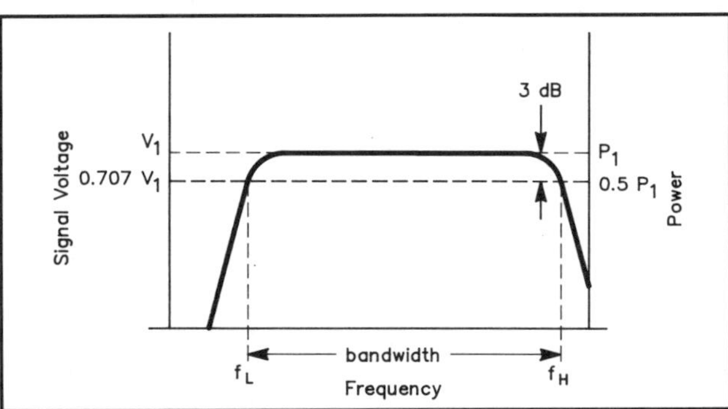

Figure 2—This graph shows an ideal transistor amplifier frequency response. We can measure the output signal voltage or power to find this response. The bandwidth is the difference between the low-frequency cutoff point and the high-frequency cut-off point.

CHAPTER 28 Field-Effect Transistors

Controlling Current by Varying an Electric Field

Bipolar junction transistors rely on an input-signal *current* to control the output current. The input-signal current variations produce a varying output current. That output current is an amplified version of the input signal.

An ideal amplifier would not draw any current from the source. Such an amplifier has a very large input impedance. A bipolar transistor amplifier circuit can have a fairly large input impedance. The practical need to draw some current always limits the actual value, though.

It is possible to make a semiconductor device that requires practically no current to operate. Such a device depends on the electric field of an applied *voltage* to control the output current. We call these devices *field-effect transistors* because they depend on an electric field for their operation. We often use the abbreviation *FET* for field-effect transistor.

Figure 1 shows a block of N-type semiconductor material. Attached to each side of the N-type material are blocks of P-type material. This drawing looks similar to the three-layered sandwich of a bipolar junction transistor. The two blocks of P-type semiconductor material have a common connection between them. This connection may be inside the transistor or it may be an external wire.

The block of N-type material provides a direct path for an electric current from one end to the other. Figure 2 shows a battery and resistor connected between the two ends. Free electrons in the semiconductor material move toward the positive battery terminal. These electrons *drain* out of the semiconductor at the positive end. We call this the *drain* connection of an FET. From the drain, electrons flow through the circuit to the other end of the block. Here

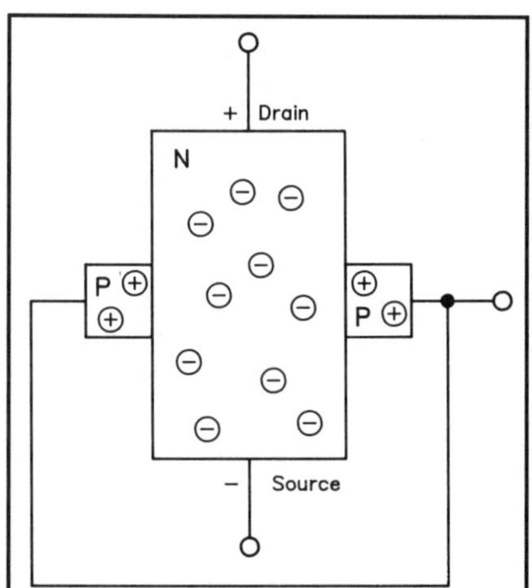

Figure 1—This diagram illustrates a *field-effect transistor* (*FET*). It has a block of N-type semiconductor material with blocks of P-type material in each side.

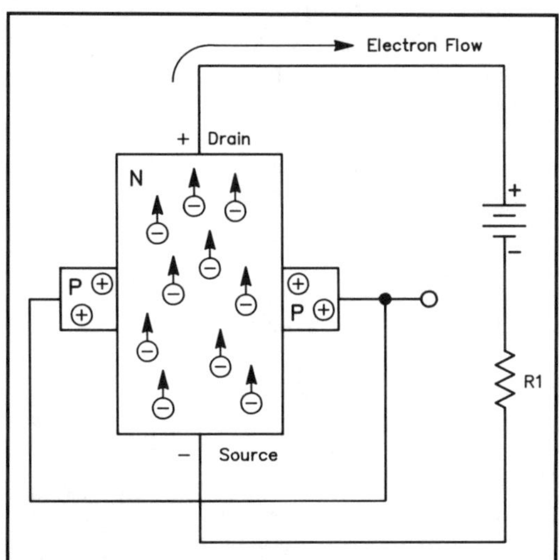

Figure 2—The N-type semiconductor material forms a direct path for an electric current through an FET. This diagram shows the electrons flowing out of the *drain* lead, through the circuit, and back into the semiconductor at the *source* lead.

they enter the N-type semiconductor. We call the end at which electrons enter the semiconductor the *source.*

In Figure 3, we connect the two P-material blocks to a negative battery terminal. The positive battery terminal connects to the source end of the N-material. The battery puts a reverse-bias voltage on the FET gate/source junction.

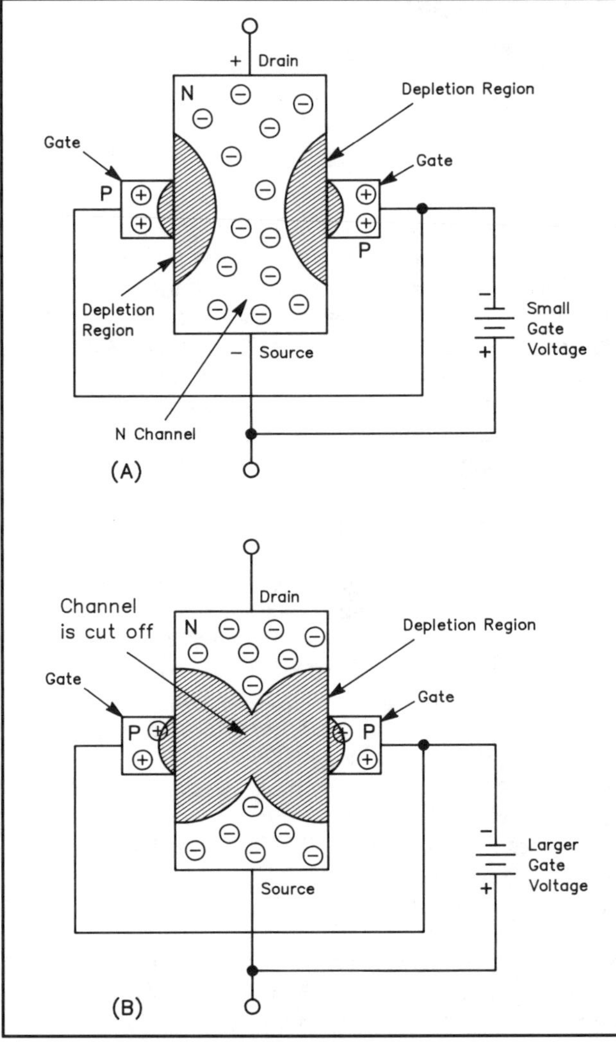

Figure 3—We connect a reverse-bias voltage across the gate/source PN junction. The depletion region begins to close the current channel through the N-type material. Part B shows that when this reverse-bias voltage becomes large enough, the depletion region closes the channel completely.

With a reverse-bias voltage on the junction, free electrons in the N-type material move away from the P-type material. Holes in the P-type material move away from the N-type material. This movement of free electrons and holes away from each other creates a wider depletion region. Figure 3A shows how the depletion region begins to close the channel for electrons to move from source to drain.

When the gate voltage is large enough, the depletion region pinches the channel completely off. Figure 3B shows this extreme condition.

Now let's see what happens when we connect a signal voltage to the FET gate. We also will connect the source and drain to a circuit, so we can use the transistor as an amplifier. Figure 4 shows the complete circuit.

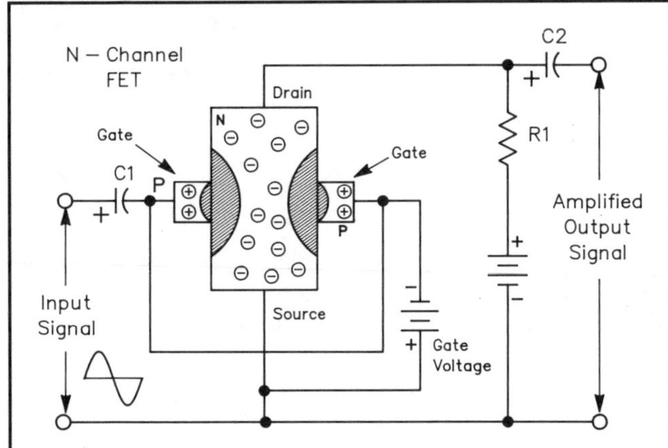

Figure 4—This circuit shows how an N-channel field-effect transistor operates. The gate-voltage battery ensures a reverse bias on the gate when the input signal is positive. The output current through R1 produces a voltage drop across the resistor. This voltage drop supplies an amplified output voltage.

When the signal voltage is negative, the gate becomes more negative. This makes the N channel smaller, reducing current in the output circuit.

Next, the input signal voltage becomes positive, making the gate less negative with respect to the source. (The gate-voltage battery ensures that the gate/source junction never has a forward-bias voltage.) The N channel becomes wider, increasing current through the output circuit.

As we learned at the beginning of this section, an FET has a large input impedance. There is almost no current across the gate/source junction because it has a reverse bias. FETs typically have an input impedance of a megohm or more.

A single block of semiconductor material controls a field-effect transistor's output impedance. This means FETs have a very low output-impedance value.

We only discussed FETs built around a block of N-type semiconductor material in this section. Perhaps you are wondering if FETs can be made with a block of P-type semiconductor material. The answer is, "Yes, there are FETs made with P-type materials." Then two blocks of N-type material form the gate. These are *P-channel FETs* instead of *N-channel FETs.* The operation is identical, except the voltage polarities reverse. A P-channel device has a positive gate voltage. The P-channel *drain* connects to the negative side of the output supply. Holes in the P-type semiconductor material carry the current, instead of free electrons.

JFET Construction

In the last section you learned about *field-effect transistors*, or *FETs*. These transistors have a PN junction, which serves as a *gate* to control the current through the channel. The transistor's PN junction gives it the name *junction field-effect transistor*, or *JFET*. In this section you will learn more about JFET construction.

Manufacturers begin with a layer of P-type semiconductor material to build an N-channel JFET. Then they form the N-material channel on the P-type base. Next, the manufacturer forms a small area of P-type material on top of the N channel. The last step is to deposit metal contacts onto the semiconductor areas to form electrodes. These electrodes provide a way for you to connect the FET into a circuit. Before sealing the transistor into a protective plastic case or other package, the manufacturer attaches wire leads to the electrodes. Figure 1A shows this construction.

Figure 1—Part A shows the construction of an N-channel junction field-effect transistor. Manufacturers form an N-type material channel into a P-type semiconductor base. Then they add a small area of P-material to form the second side of the gate. They deposit a thin metal layer onto the semiconductor segments to serve as electrodes. P-channel JFET construction is similar, but with opposite semiconductor types, as Part B shows.

Figure 1B shows the similar P-channel JFET construction. A block of N-type material holds the P-material channel. The manufacturers add the second side of the gate as a thin N-material layer on top of the channel.

Field-effect transistors made this way have symmetrical channels. There is no difference between the source and drain channel sides. You could connect either side of the channel to the circuit source connection. You also could connect either side of the channel to the drain connection.

The channel isn't always symmetrical. Variations in the manufacturing process result in differences between the source and drain terminals. These differences provide specific transistor characteristics. When you build a circuit with junction field-effect transistors, you must carefully identify the source and drain terminals. Otherwise, you may damage the transistor.

Figure 2 shows the schematic-diagram symbols that represent N-channel and P-channel junction field-effect transistors. The N-channel JFET has a P-type gate section. Notice that the arrow symbol points *away* from the P-material gate *toward* the channel in an *N*-channel FET. The P-channel JFET has an N-type gate section. The arrow points *toward* the N-material gate and *away* from the channel in a *P*-channel FET. The arrows always point *away* from the P material and *toward* the N material in semiconductor devices.

FETs come in many package styles. Larger packages generally indicate devices that handle more power.

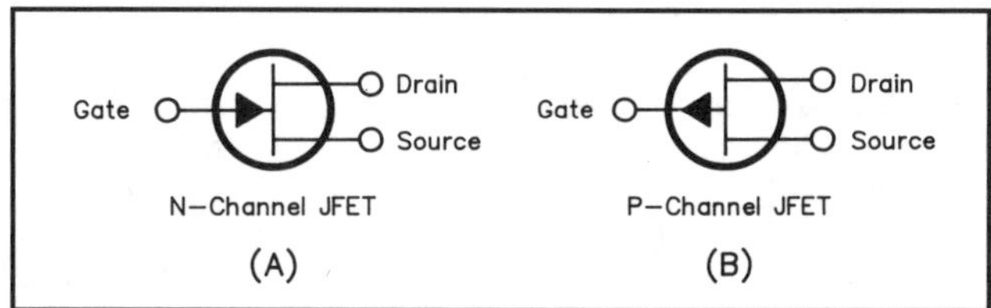

Figure 2—The symbol at A represents an N-channel junction field-effect transistor on schematic diagrams. The symbol at B represents a P-channel JFET on schematic diagrams.

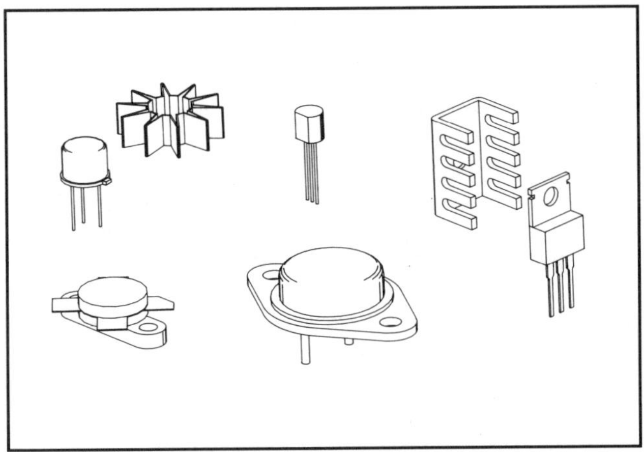

Figure 3—FET packages are similar to those used for bipolar transistors. Some applications require a heat sink to conduct excess heat away from the transistor.

Some FETs require you to use heat sinks to conduct excess heat away from the transistors. Figure 3 is a drawing of some common FET package styles.

Manufacturers use silicon and germanium as the semiconductor materials for most JFETs. These are the same semiconductor materials we learned about earlier in this Unit. They are also the materials used to make solid-state diodes and bipolar junction transistors.

Manufacturers use another semiconductor material when they design FETs for use at UHF and microwave frequencies. Transistor noise becomes an important factor at these frequencies. Gallium arsenide is a semiconductor compound. It makes FETs that produce very little transistor noise at UHF. The chemical abbreviation for gallium arsenide is *GaAs*. This is why we call field-effect transistors made with gallium arsenide, *GaAsFETs*. GaAsFETs produce noise at lower frequencies, however, and are not particularly suited for use at HF.

MOSFET Construction

The field-effect transistors we studied so far all form their gates with PN junctions. There is another common field-effect transistor construction method. This method uses a thin layer of insulating metal oxide between the channel and gate blocks. We call these transistors *insulated gate field-effect transistors*, or *IGFETs*.

The insulating layer forms a capacitor. The channel forms one capacitor plate and the gate layer forms the other plate. This construction technique solves some problems that occur with JFET gate diodes. There is always some small leakage current through the reverse-biased PN junction. The insulated gate of an IGFET does not have this problem.

The most common insulating material is a thin layer of metal oxide on the semiconductor base. This construction technique leads to our name for this type of FET: *Metal-oxide semiconductor field-effect transistor*, or *MOSFET*. MOSFETs typically have an input impedance of several megohms, which is higher than JFETs.

The insulating metal-oxide layer is very thin and has a low breakdown voltage. This makes MOSFETs sensitive to static electricity. A static charge can punch through the insulation, destroying the transistor. This is one disadvantage of MOS devices. You must be careful to discharge any static electricity from your body before you pick up a MOSFET. Handle them only by the case, not by the leads. Some MOSFETs have static-protective Zener diodes built into the device.

There are a variety of construction methods for making MOSFETs. First we'll discuss the construction and operation of a *depletion-mode* MOSFET.

We'll make an N-channel transistor by starting with a block of P-type semiconductor material. Next we'll form the channel by depositing an N-material layer on top of the base, or *substrate*. A thin metal-oxide insulating layer goes over the top of these semiconductor materials. Now we must make some holes in the insulation for the source and drain connections. Deposit a metal layer through the holes and over the center to form a gate. Add wire leads and put the transistor in a protective package. Sounds simple, doesn't it? Figure 1 shows this construction.

Figure 1B shows the schematic-diagram symbol for our N-channel depletion-mode metal-oxide-semiconductor field-effect transistor. The gate does not include a PN junction, so there is no arrow on the gate lead. There is an arrow pointing toward the channel, though, with a connection to the source. The arrow points toward the channel, indicating an N-channel device.

Figure 2 shows our MOSFET with a negative gate voltage. Notice that the negative gate voltage attracts some positive holes from the P-material substrate. These holes combine with some of the free electrons carrying current through the channel. This *depletes*, or reduces, the supply of charge carriers through the channel. This is why we call this a *depletion-mode MOSFET*.

You can imagine what happens when we apply a varying signal voltage to the gate. The varying voltage changes the current through the channel. A negative gate voltage reduces the channel width. There is less current through the channel. A positive gate voltage repels holes, increasing the channel width so more current can flow.

Figure 3 shows the construction of an *enhancement-mode MOSFET*. The most interesting aspect of the enhancement-mode construction is that there is no channel between the N-type source and drain. There is a connection between the P-type substrate and the source N-type material. For

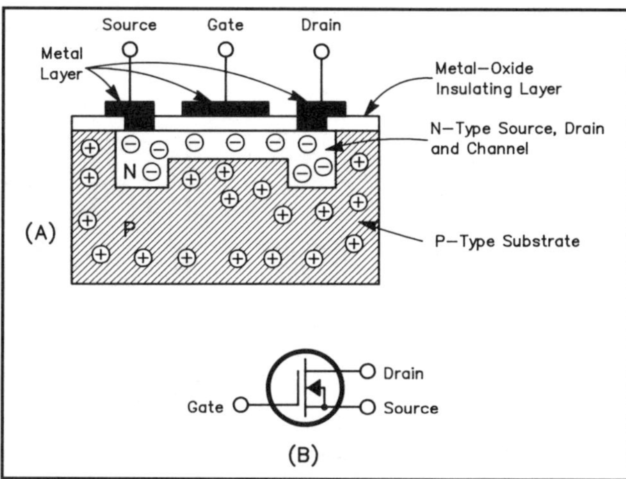

Figure 1—Part A shows the construction of an N-channel depletion-mode metal-oxide-semiconductor field-effect transistor. Part B shows the schematic-diagram symbol for this transistor. The arrow points toward the channel for N-channel devices. Notice that the arrow is not on the gate lead, however.

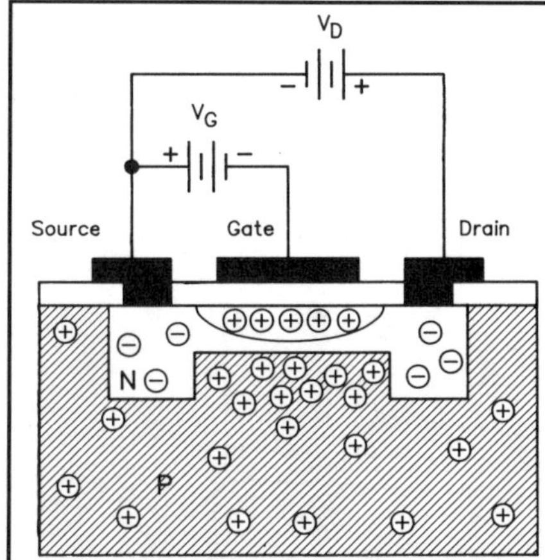

Figure 2—A negative gate voltage attracts holes from the P-type substrate material. These holes combine with free electrons in the N-type channel material, reducing the effective channel width. Less current can flow through a narrower channel. A positive gate voltage repels holes, increasing the channel width so more current can flow.

some transistors this an internal connection; for others you must add the connection as part of the external circuit. The drain and substrate form a PN junction that produces a small leakage current when the drain is positive with respect to the source. The PN junction has a reverse-bias voltage under these conditions.

Let's apply a positive gate voltage, as Figure 4 shows. The positive gate voltage attracts electrons in the P-type semiconductor material. We call these electrons *minority charge carriers* because there are more positive holes to carry charge in P-type material. The minority charge carriers collect at the P-material surface, but they cannot cross the insulating metal-oxide layer. This forms a negative channel in the P-type material between the N-type source and drain blocks. The number of minority charge carriers controls the amount of drain current. Thus, the positive gate voltage *enhances* the current-carrying ability of the transistor.

Figure 5 shows the schematic symbol for an enhancement-mode MOSFET. Notice that the center line has divisions to represent the source, drain and substrate. This is a subtle difference from the symbol for a depletion-mode MOSFET, which has a solid center line.

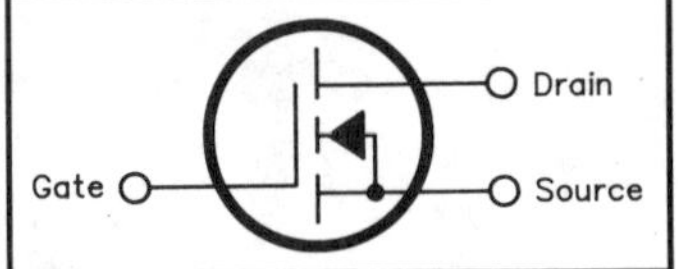

Figure 5—This is the symbol used on schematic diagrams to represent an N-channel enhancement-mode MOSFET. Notice that the center line is broken into sections to represent the drain, channel and source.

There are other construction variations for metal-oxide semiconductor field-effect transistors. We won't try to cover all the possibilities here. One variation that you should be aware of, though, is the construction of dual-gate MOSFETs. Gate 1 is the *signal gate* and gate 2 is the *control gate*. This allows a control voltage to change the amplifier characteristics.

Manufacturers can make any of these MOSFET types with either N-channel or P-channel construction. All the devices we studied in this section are N-channel types. To make P-channel devices, reverse the N and P material types and change the voltage polarities. Figure 6 shows schematic symbols for N- and P-channel depletion- and enhancement-mode MOSFETs. Figure 6 also includes symbols for dual-gate devices.

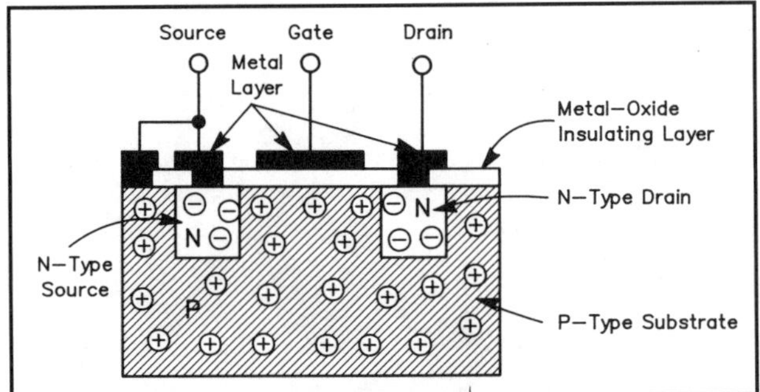

Figure 3—This drawing shows the construction of an N-channel enhancement-mode metal-oxide-semiconductor field-effect transistor.

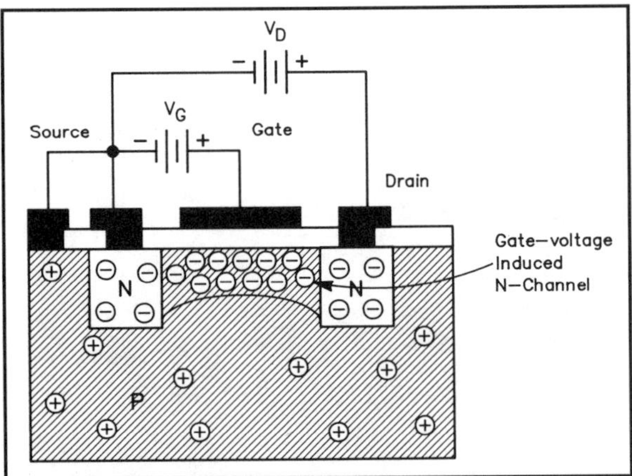

Figure 4—This diagram shows the operation of an N-channel enhancement-mode MOSFET. A positive gate voltage attracts electrons in the P-type material. We call these electrons minority charge carriers because there are fewer free electrons than holes in P-type material. The electrons form an N-channel between the source and drain.

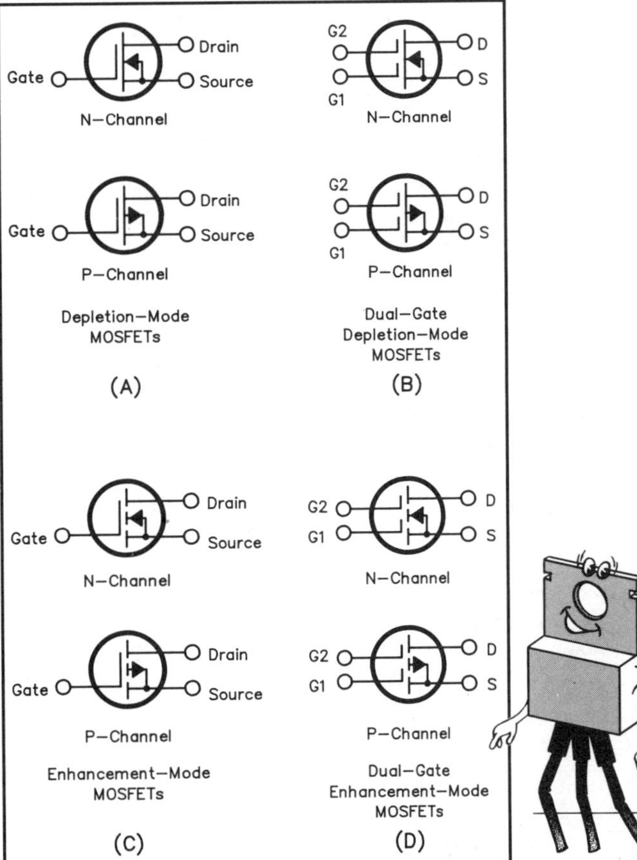

Figure 6—These symbols represent various metal-oxide semiconductor field-effect transistors. The symbols represent N- and P-channel types in both depletion- and enhancement-mode devices. We also included symbols for dual-gate MOSFETs.

CHAPTER 29 Integrated Circuits

Linear Devices

Silicon and germanium semiconductor materials form many electronics parts. These semiconductor materials form diodes, bipolar transistors, field-effect transistors and many other devices.

Manufacturers produce many transistor types. They control the amount and location of P- or N-type semiconductor materials added to the *substrate*, or base material. By carefully controlling the manufacturing process they can set the material's resistance. They use the same process to make resistors on a silicon or germanium base.

Do you remember the construction of a metal-oxide semiconductor field-effect transistor? A MOSFET's gate is a capacitor. (Figure 1 shows this construction.)

Can manufacturers place several transistors and diodes on a single substrate piece? Of course. They also can add resistors and capacitors between the transistors, as part of the semiconductor. Does this begin to sound like a complete electronics circuit on a single piece of semiconductor material?

An *integrated circuit* is a single piece of semiconductor material that replaces many parts in an electronics circuit. You will often hear *integrated circuit* abbreviated *IC*. You also will hear these devices called *chips*, because they use a small piece, or "chip," of silicon.

Manufacturers build these circuits and package them

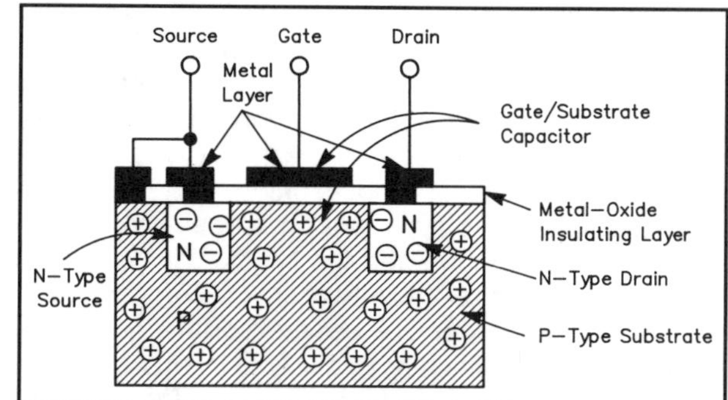

Figure 1—A metal-oxide semiconductor field-effect transistor includes a semiconductor capacitor.

Figure 2—This photo shows several IC-package styles.

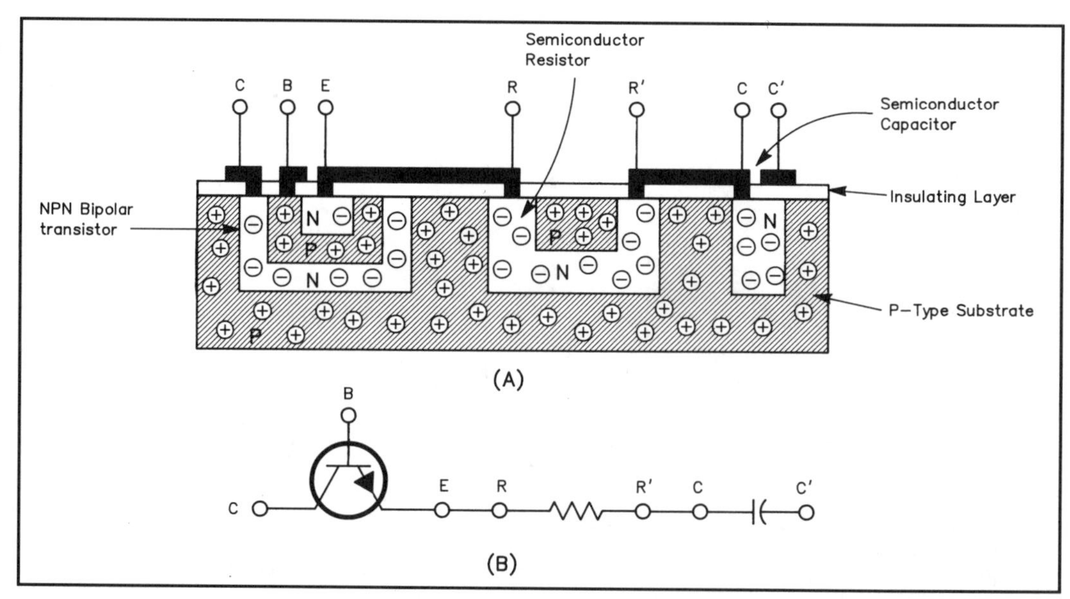

Figure 3—Part A shows a drawing of the semiconductor materials forming a simple IC. Part B shows the schematic diagram of the circuit from Part A.

in a variety of ways to make them easy to use. Figure 2 shows some common integrated circuit packages. The package must provide some way to connect the IC to the outside world. The wire leads make up a significant portion of the size of an IC. Manufacturers are always looking for new ways to package the chips to produce a smaller package.

Some ICs may have only a few transistors and other components. Many chips include hundreds or thousands of transistors, diodes, resistors and capacitors on a single block. These components consist of semiconductor-material layers formed on the substrate. Figure 3 shows this construction with a simple circuit that has a bipolar transistor, a resistor and a capacitor. Of course most ICs are much more complicated than this circuit.

Linear integrated circuits work with signals that change smoothly from one value to another. A sine wave is a good example of such a signal. You will learn about *digital integrated circuits* later in this chapter.

Linear ICs serve as amplifiers, oscillators and many sections of receivers, transmitters and other electronics circuits. There are even linear ICs that are complete receivers, requiring only a few external components such as a power supply and speaker or headphones.

Some linear ICs are general-purpose chips. These devices adapt well to many uses in various circuit types. *Operational amplifiers* are popular general-purpose devices. *Op amps* find their way into many circuits, including small-signal amplifiers and filter circuits. You will learn more about op amps in the next section.

Other linear ICs are special-purpose devices designed to perform one specific circuit function. You will find many of these special-purpose ICs in consumer-electronics devices like stereos, radios and TV sets.

Figure 4 is a diagram of the TDA7000 IC. This chip is a complete FM radio, as the block diagram shows. You only need to add a few resistors, capacitors and inductors to complete the receiver. These components provide filtering and oscillator-frequency control. A power supply to provide the necessary input voltage rounds out the external components.

You will learn about many specific ICs as you continue to study electronics. Remember that ICs are electronic building blocks. Manufacturer's data sheets and application notes are good sources of specific information about an IC. This information tells you what input-voltage range is safe for the chip. It also tells what additional circuit components you must add to complete the circuit.

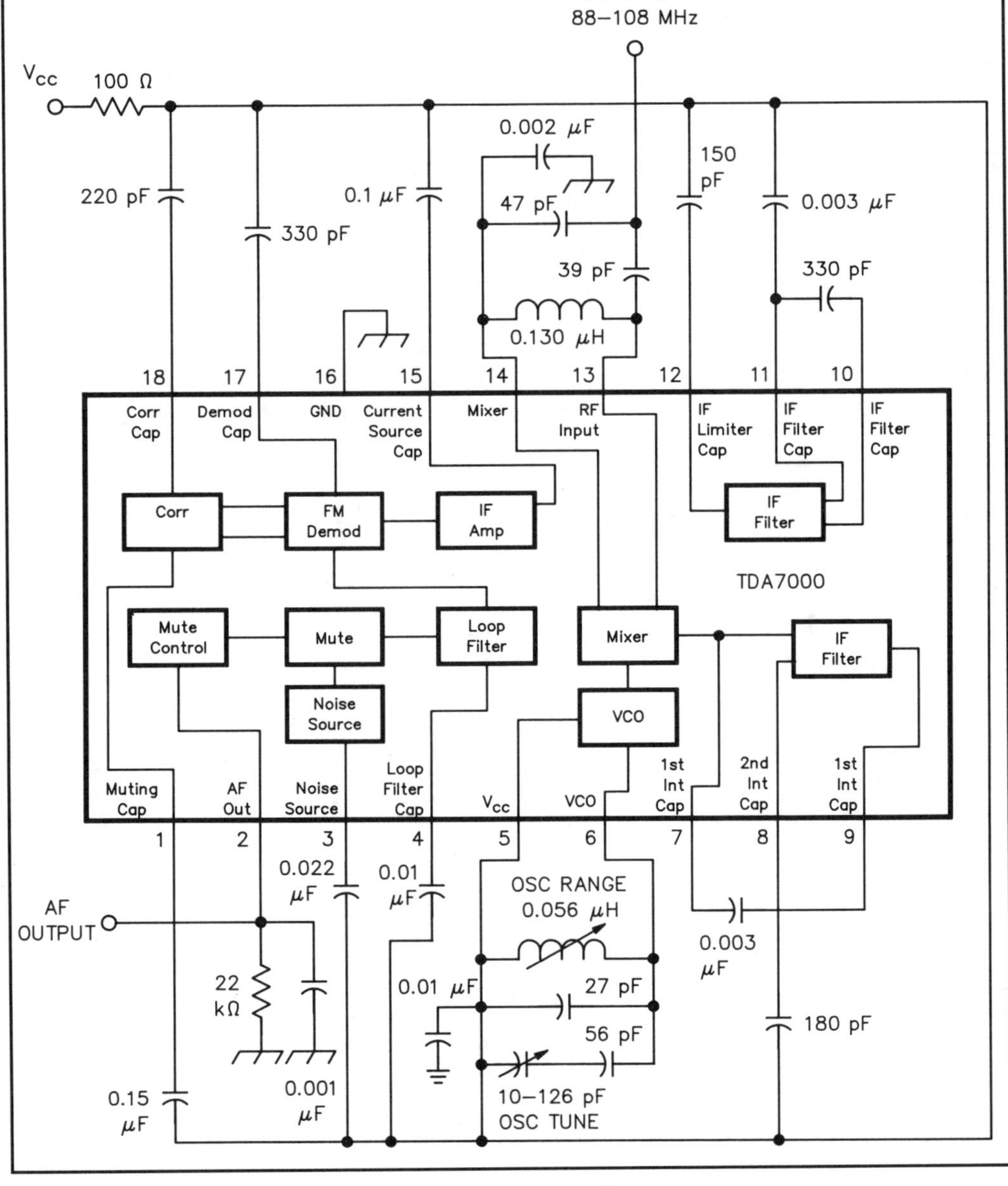

Figure 4—The TDA7000 is an FM-radio subsystem IC. You only need to add a few external parts to make a complete receiver.

Operational Amplifiers

There are many types of linear integrated circuits. One common example is the *operational amplifier.* Operational amplifiers take their name from the days of analog computers. Analog computers represented numbers with analog voltages. Larger numbers had higher voltages. Such computers used direct-current amplifiers to perform mathematical *operations,* such as addition and subtraction.

Modern IC *op amps* are high-gain dc-coupled differential amplifiers. A differential amplifier has two input connections and amplifies the *difference* between those inputs. A dc-coupled amplifier amplifies direct-current signals and alternating-current signals.

There are six ideal amplifier characteristics by which we judge operational amplifiers. An ideal op amp should have:

1) infinite voltage gain
2) infinite input impedance
3) zero output impedance
4) infinite bandwidth
5) zero noise
6) zero drift with time and temperature changes.

Modern IC op amps come close to realizing the first three conditions. They don't do quite as well with the last three, but manufacturers constantly try to improve performance in these areas.

Figure 1 shows a schematic diagram that represents the circuit inside a 741 op amp IC. Don't be concerned about understanding how this circuit operates. That's one good point about ICs. You don't have to know what goes on inside the "black box." The manufacturers' data sheets will tell you what voltages and signals to connect to each IC lead.

There are 20 transistors on this integrated-circuit chip. All these transistors are on one substrate, so their characteristics are carefully matched. You would not have much success trying to build a similar op amp from individual transistors and resistors.

You also should notice the two input connections. One has a + INPUT label, which is the *noninverting input.* The other input has a – INPUT label, which is the *inverting input.*

Figure 2 shows the schematic-diagram symbol for an op amp. Part A shows a sine-wave signal connected to the noninverting input. The amplified output signal is in phase with the input signal. Part B shows a signal connected to the inverting input. This time the amplified output signal is 180° out of phase with the input. This shows the difference between these two inputs. With the inverting input, as a signal becomes more positive, the output becomes more negative.

There are two operating-voltage inputs to the IC. Many op amps require a split voltage supply. One supply

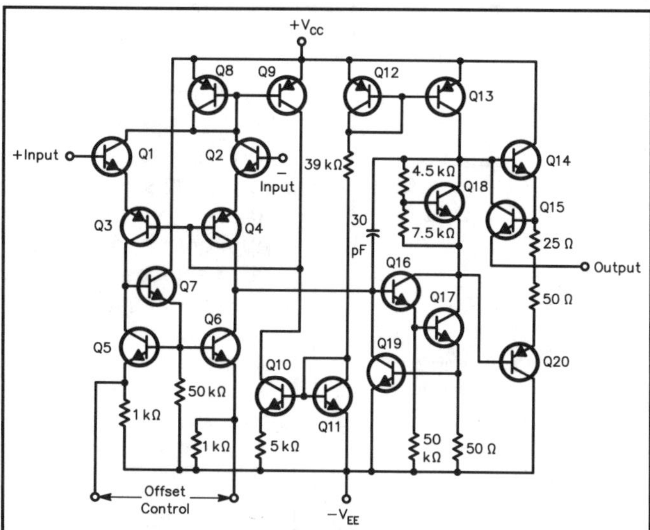

Figure 1—This diagram represents a 741 op-amp circuit. There are inverting and noninverting inputs and one output. Positive and negative voltage supply inputs allow you to use the op amp as a differential amplifier. With the offset control connections, you can adjust the amplifier so there is no output when both inputs have the same signal.

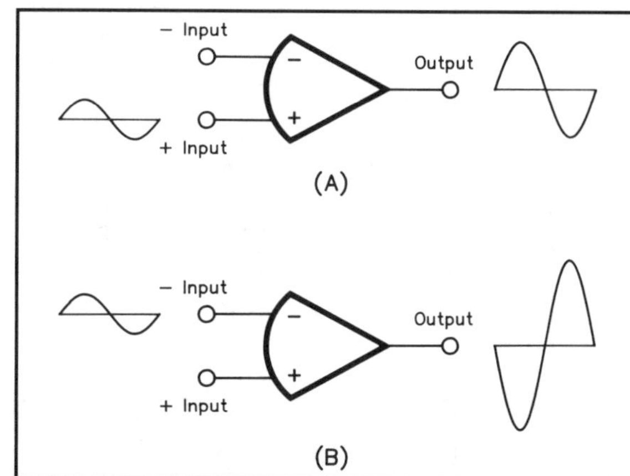

Figure 2—Part A shows a sine-wave signal applied to the op amp noninverting input. The input and output signals are in phase. When the input signal reaches its positive maximum so does the output. The output signal reaches its negative maximum at the same time as the input signal. Part B shows the input signal applied to the inverting input. In this case the input and output signals are 180° out of phase. The output signal reaches a negative maximum when the input signal reaches its positive maximum.

gives a positive voltage to ground. A second supply gives a negative voltage to ground. This is necessary to operate the differential-amplifier input circuit.

Suppose you supply one input signal to the + input and another identical signal to the – input. The difference between these signals is 0 volts. Therefore, the output signal will be 0 volts.

Sometimes the two input amplifiers aren't exactly balanced. This means there may be some output signal with identical input signals. That brings us to the last two op-amp connections. The OFFSET CONTROL leads provide an adjustment so you can set the output signal to zero when the input signals are identical.

You can make an audio amplifier using an op amp. Figure 3 shows an amplifier you can use to increase the output from a microphone. This circuit feeds a signal to a radio transmitter or other device. The negative voltage-supply lead connects to ground. You don't have to use a negative supply with this circuit because the only input signal goes to the noninverting input. The inverting input connects to the output through the gain-control resistor. This provides some negative feedback to the amplifier, stabilizing the operation.

Active filters are another popular operational amplifier application. Figure 4 shows an audio band-pass filter that you can build. This circuit has a center frequency of 750 Hz, so you could use it as an external CW filter for an Amateur Radio receiver.

There are many other op-amp applications. The ones described here will help you become familiar with this class of devices.

Figure 3—Here is the schematic diagram of an audio amplifier that you can use to increase the signal from a microphone. The input and output connectors are coaxial types, because you would use shielded, or coaxial, cable to make these connections.

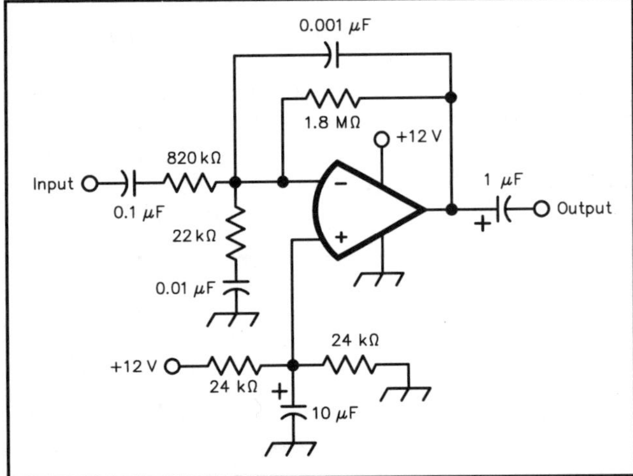

Figure 4—This active band-pass filter is suitable for use as a CW filter with an Amateur Radio receiver. It has a center frequency of 750 Hz. You can use a 741 or TL081 op amp.

Digital Devices

Digital electronics is an important branch of modern electronics. Digital circuits use signals that have specific voltage levels. Signals that fall between the specific values for the circuit round up or down to the set values. Figure 1 shows a digital representation of a sine wave. Notice the stair-step appearance to the values.

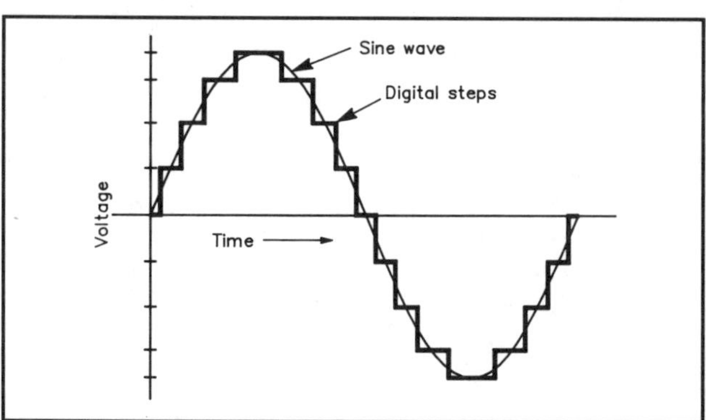

Figure 1—This "stair-step" graph is a digital representation of a sine wave. Digital signals have values at specific, defined levels. If the analog signal value falls between two steps, the digital signal rounds up or down to the closest value.

Many digital circuits use two voltage levels. We call these *binary circuits.* The *binary number system* uses only two digits, 0 and 1. We use the binary numbers, 0 and 1, to represent the two voltage levels or circuit conditions. They may represent ON and OFF circuit conditions. The binary states also may represent space and mark signals in a digital communications system such as Morse code or radioteletype.

The actual voltages depend on the circuit, but OFF is normally 0 volts. The ON condition may be around 5 volts, or it may be 12 volts or another value. Some circuits reverse these conditions. Figure 2 shows a series of digital pulses from a binary circuit.

Digital computers use digital logic circuits. These circuits represent decimal numbers using a binary coding system. The computer works with these binary numbers to perform many complex tasks.

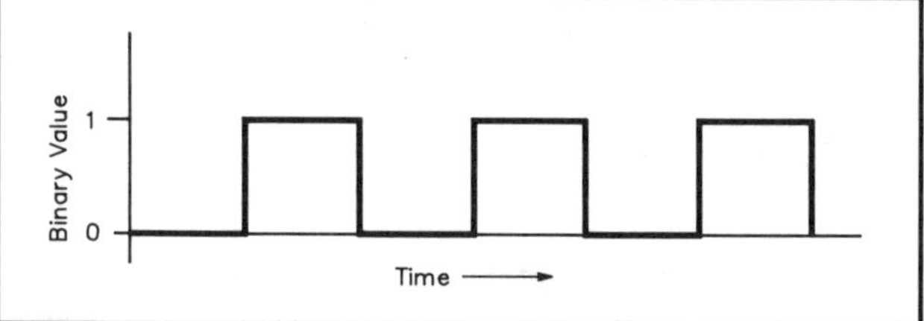

Figure 2—These pulses represent the output from a binary digital circuit. The output circuit is either ON, representing a binary 1, or it is OFF, representing a binary 0.

Digital logic circuits measure one or more inputs and produce a 1 or a 0 output. The first digital computers used vacuum tubes and mechanical relays to form logic circuits. These were large machines that required lots of power and produced lots of heat. Later, transistors replaced the vacuum tubes and relays.

Digital integrated circuits replace logic circuits built from individual transistors. A digital IC is a complete logic network. It usually needs only one or more input signals and a power source. Most digital ICs produce one or more output signals.

There are many logic functions, and there is a digital IC for each function or possible combination. There are too many ICs to discuss each one and its operation here. We will cover a few common circuits, to give you an idea of the possibilities.

One simple logic function is the *inverter.* An inverter takes a digital-signal input and creates the opposite signal output. Figure 3 shows the schematic symbol that represents an inverter. We drew the input and output waveforms with the symbol. When the input signal is a 1, the output is a 0. When the input is a 0, the output is a 1. We sometimes call this the NOT function. The small circle on the output tells you the circuit performs the NOT function, or inversion.

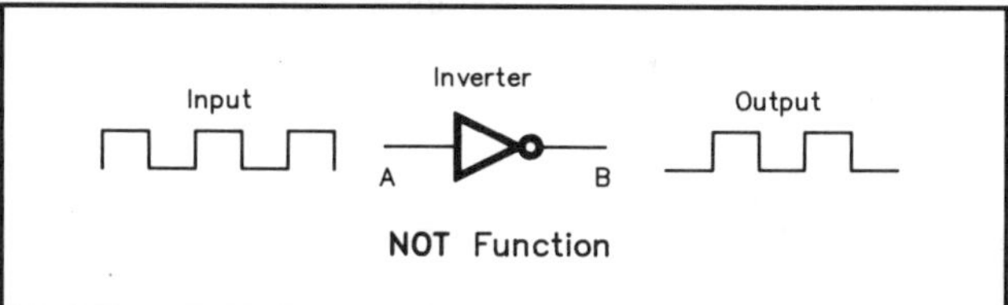

Figure 3—This symbol represents an inverter circuit on schematic diagrams. The input and output signals illustrate inverter operation.

Table 1
Inverter Truth Table (NOT Function)

Input (A)	*Output (B)*
0	1
1	0

Table 1 is an inverter *truth table.* A *truth table* lists all possible input-signal combinations. It also lists the resulting output signals for each input-signal combination. A truth table is a valuable tool to help you understand a logic circuit's operation.

A logical AND circuit has two or more inputs. An output of 1 is possible only when all the inputs are 1s. If all inputs are 0s, or some inputs are 1s and others are 0s, the output is 0. Table 2 is a truth table for the logical AND circuit.

Table 2
AND Gate Truth Table

Input (A)	*Input (B)*	*Output (C)*
0	0	0
0	1	0
1	0	0
1	1	1

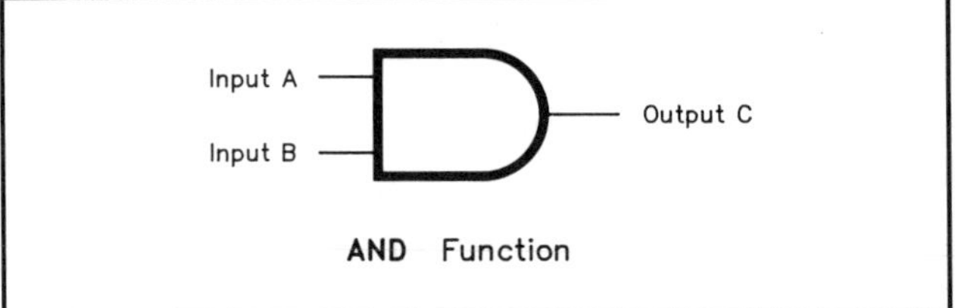

Figure 4—This symbol represents a two-input AND *gate*. Both inputs must be 1s to produce a 1 output. If either or both inputs are 0s, the output is 0.

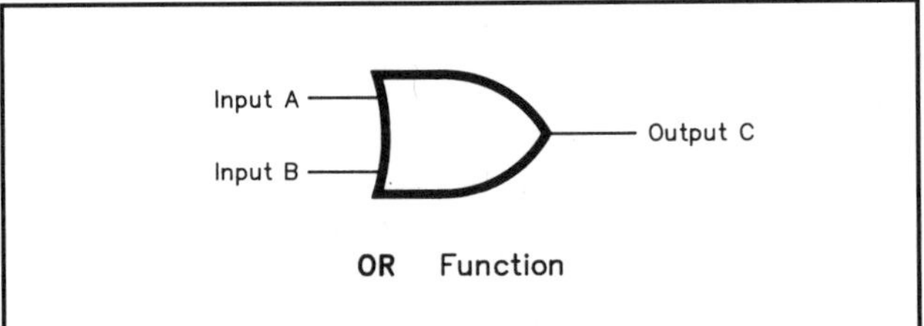

Figure 5—We use this symbol to represent a two-input OR *gate* on schematic diagrams. If either input is 1 the output will be 1.

Table 3
OR Gate Truth Table

Input (A)	*Input (B)*	*Output (C)*
0	0	0
0	1	1
1	0	1
1	1	1

Figure 4 shows the schematic-diagram symbol for a two-input AND *gate*. (*Gate* is the term used to describe digital-logic ICs.)

Another logic circuit is the OR *gate*. An OR circuit has two or more inputs and one output. If any input is a 1, the output will be 1. Figure 5 is the schematic symbol for a two-input OR *gate*. Table 3 is the truth table for this circuit.

You can use these digital IC building blocks to form nearly any logic circuit imaginable. Manufacturers combine many of these building blocks into integrated circuits. A *flip-flop* is a digital circuit that stores a single *binary digit*, or *bit* of information.

Figure 6A shows a flip-flop made from OR *gates* and inverters. B shows the schematic symbol for this flip-flop made as a single IC. Part C gives the truth table for the flip-flop. The two inputs are SET (S) and RESET (R).

When S = 1 and R = 0, the Q output is 1. When S = 0 and R = 1, the Q output is 0. Notice there is a second output, labeled $\overline{Q}$. (Read this as Q NOT, which means an inverted Q, or the opposite of Q.) The Q NOT output is always the opposite of the Q output. When both the S and R inputs are 0, the outputs don't change. The truth table also shows that if both inputs should ever be 1s, there is no way to determine the output. Remember this when you work with flip-flops.

There are several construction techniques used to make these digital ICs. Generally, you should use devices that are all made with the same construction method. One common digital IC *family* is called transistor-transistor logic, or TTL. This name comes from the bipolar transistors used to make the ICs. Part numbers in the 7400 series indicate TTL ICs. A 7404 IC has six inverters in a single package. A 7432 has four OR *gates*.

Another common family uses both P- and N-channel metal-oxide semiconductor field-effect transistors on the same substrate. We say the P- and N-channel MOSFETs are complementary. We call this construction *complementary metal-oxide semiconductor*, or *CMOS*. Part numbers in the 4000 series indicate CMOS ICs. A 4069 has six inverters, for example, and a 4081 has four AND *gates*.

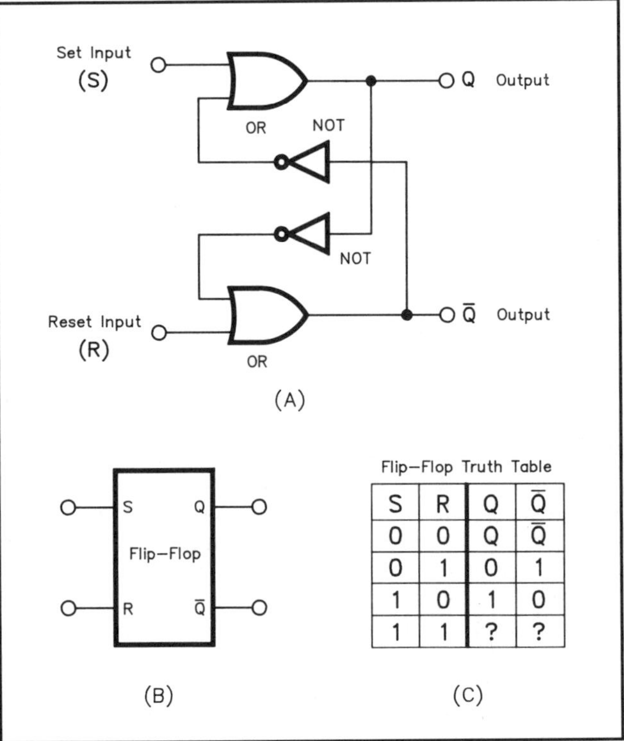

S	R	Q	$\overline{Q}$
0	0	Q	$\overline{Q}$
0	1	0	1
1	0	1	0
1	1	?	?

Figure 6—You can make a flip-flop with two OR *gates* and two inverters. Part B shows the schematic-diagram symbol that represents an IC flip-flop. Part C shows the truth table for this flip-flop.

There are many special-purpose digital integrated circuits. These are combinations of other digital devices built into a single IC. You can buy an IC to count signal pulses or to divide a series of pulses by some preset value. You can buy a chip that changes digital signals to analog signals or one to change analog signals to digital signals. There are many ICs designed for use with digital computers.

In this section we will study the timer IC. This useful building block has many applications. The most common timer IC is the 555. Figure 1 shows the schematic-diagram symbol, with pin labels.

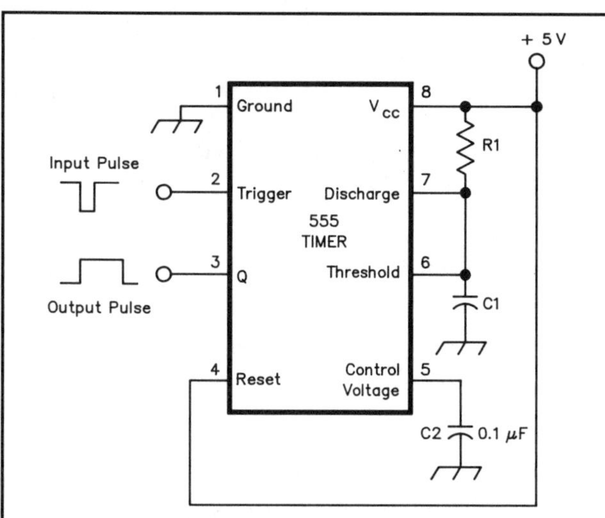

Figure 1—This 555 timer IC circuit is a monostable multivibrator, or one-shot. A negative pulse on the trigger pin produces a positive output pulse on the output, pin 3. Resistor R1 and capacitor C1 determine the output-pulse duration.

Let's study the Figure 1 circuit to learn about the timer IC. Normally pin 2, the trigger lead, is at +5 volts. We'll pull this lead to ground for an instant by applying a negative input pulse. This causes a positive output pulse on lead 3, which we call the Q output. Current from the power supply goes through R1 to charge capacitor C1. When the capacitor voltage exceeds 2/3 the supply voltage (about 3.5 volts for this example) the discharge line, lead 7, quickly discharges the capacitor to ground. Then the output returns to 0. The output remains at 0 until there is another negative trigger pulse.

The output-pulse length depends on the values used for R1 and C1. These two parts set the circuit *time constant.* When you studied capacitor circuits you learned how to calculate time constants. Multiply the resistor value, measured in ohms, times the capacitor value, measured in farads. Longer time constants produce longer output pulses.

Equation 1 shows you how to calculate the output-pulse length.

$$T = 1.1 \times R1 \times C1 \qquad \text{(Equation 1)}$$

where T is the output pulse length in seconds.
Suppose the Figure 1 circuit has a 10-kilohm resistor for R1 and a 10-microfarad capacitor for C1. How long will the output pulse last?

First change the resistance measurement to ohms and the capacitance measurement to farads.

10 kilohms = 10×10^3 ohms
10 microfarads = 10×10^{-6} farads

Now substitute these values into Equation 1.

$T = 1.1 \times 10 \times 10^3 \text{ ohms} \times 10 \times 10^{-6} \text{ farads}$
$T = 1.1 \times 100 \times 10^{-2} = 0.11 \text{ seconds}$

Each output pulse lasts 0.11 seconds, for this example.

There are two possible conditions for the Figure 1 circuit. As long as the voltage on pin 2 remains near 5 volts, the output voltage is 0. This is a *stable* circuit condition. When the pin-2 voltage goes to 0, the output voltage jumps to 5 volts. This condition only lasts a short time, however. This is an *unstable* condition.

There are many applications for this type of circuit. We call a circuit with one stable condition and one unstable condition a *one-shot* or a *monostable multivibrator. Monostable* tells us the circuit has *one* stable condition. *Multivibrator* means the circuit jumps between two output conditions. You can build a one-shot with individual components, but the 555 timer makes a convenient package.

Figure 2 shows another multivibrator circuit that has many electronics applications. It is an *astable* or *free-running multivibrator.* An astable multivibrator has two

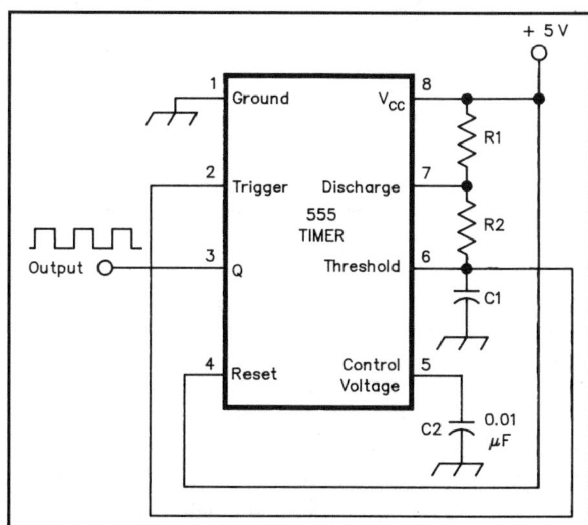

Figure 2—A 555 timer IC also can form an astable multivibrator, or free-running multivibrator. The circuit produces a string of output pulses with no input signal. Resistors R1 and R2, with capacitor C1, determine the output-pulse frequency.

unstable conditions, and it alternates between those outputs. This circuit produces a string of output pulses without any input signal. We used a 555 timer IC for this circuit also. You can make an astable multivibrator with other components, but the 555 only needs a few additional resistors and capacitors.

This circuit operation is similar to Figure 1. Capacitor C1 charges through R1 and R2 until it reaches about 2/3 the supply voltage. Then the capacitor discharges through R2 until it reaches about 1/3 the supply voltage. The values of R1, R2 and C1 determine the output pulse duration and the time between pulses. The output pulses shown in Figure 2 are on half the time and off half the time. You can set nearly any pulse times by selecting the resistor and capacitor values.

Equation 2 gives the output-pulse frequency. Resistance values must be in ohms and capacitor values must be in farads for this equation.

$$f = \frac{1.46}{[R1 + (2 \times R2)] \times C1} \qquad \text{(Equation 2)}$$

where f is the frequency in hertz.

Let's choose a 10-kilohm resistor for R1 and a 100-kilohm resistor for R2. Pick a 0.01-microfarad capacitor for C2. What frequency will the output signal have?

$$f = \frac{1.46}{[10 \times 10^3 \text{ ohms} + (2 \times 100 \times 10^3 \text{ ohms})] \times 0.01 \times 10^{-6} \text{ farads}}$$

$$f = \frac{1.46}{[10 \times 10^3 + 200 \times 10^3] \times 0.01 \times 10^{-6}}$$

$$f = \frac{1.46}{210 \times 10^3 \times 0.01 \times 10^{-6}}$$

$$f = \frac{1.46}{2.1 \times 10^{-3}} = 695 \text{ hertz}$$

The output signal from this circuit is in the range of signals you can hear. If you connect the output to a speaker, you will hear a tone.

Figure 3 shows another astable multivibrator. This is a good circuit with which to experiment. Resistors R2 and R3 replace R2 in the Figure 2 circuit. R4 serves as a volume control. C4 improves the sound by rounding the corners of the square-wave output pulses.

Try changing the values of R1, R2, R3 and C1. Listen to the effects of adjusting R3 and R4. Use Equation 2 to calculate the output frequency for this circuit. Calculate the frequency when you add the values of R2 and R3 to take the place of R2 in Equation 2. Also calculate the frequency without R3. This gives the range of output tones you can expect from the circuit.

You can use this circuit as a code-practice oscillator to learn Morse code. Replace switch S1 with a code key. You can build the circuit on a small experimenters' circuit board. The experimenters' board and all parts are available at Radio Shack stores.

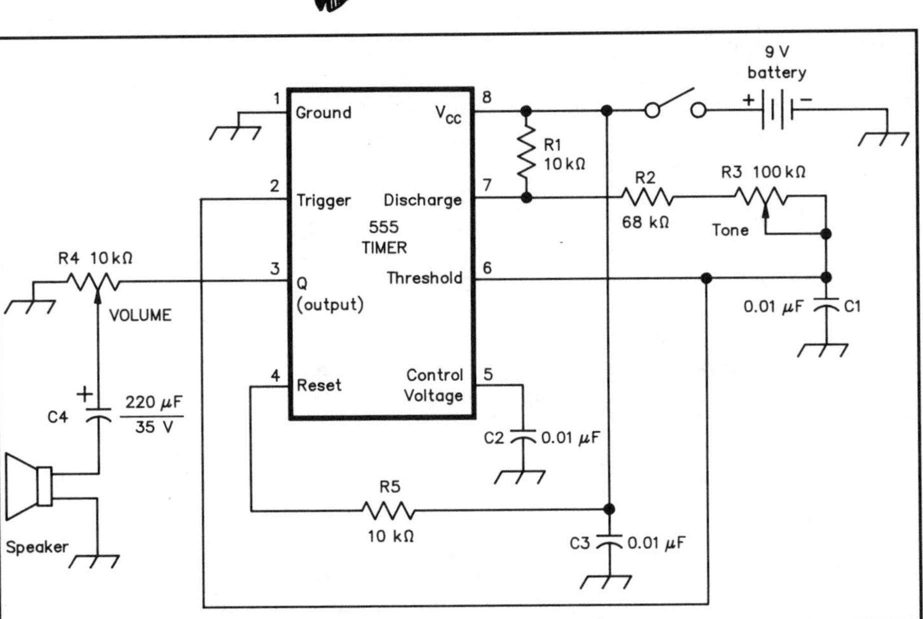

Figure 3—You can build this astable multivibrator circuit on an experimenters' circuit board. Adjust R3 and R4 while you listen to the output signal. You also can try other values for the resistors and capacitor C1. Replace the switch with a Morse code key, and you will have a code-practice oscillator.

CHAPTER 30 Vacuum Tubes

Thermionic Emission

Electric current consists of a movement of electrons. Remember that electrons bump from atom to atom in their journey along a copper wire. Each electron must gain some energy to become free of an atom before it can move along the wire. Chemical energy in a battery supplies that energy for circuits like the one shown in Figure 1.

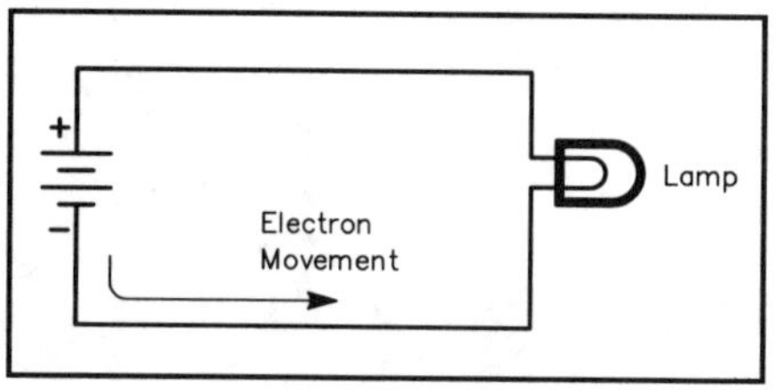

Lamp
Electron Movement

Figure 1—The battery's chemical energy helps free electrons from copper atoms in the wire. This enables the electrons to move from atom to atom along the wire.

You also can give the electrons additional energy by heating the material. When the temperature of a metal is high enough, electrons gain the energy they need to jump free. These *free electrons* can leave the solid and form a kind of cloud near the metal surface. We say the electrons *boil off* the metal. (See Figure 2.)

These electrons leave positively charged ions behind when they leave the metal. We call the electrons *negative ions.* This process of heating the metal to remove electrons is *thermionic emission.* A thermal (temperature-related) process emits the ions.

Normally, we expect these electrons to lose some of this heat energy and return to the metal. The positive ions left in the metal attract the negative electrons, pulling them back onto the metal surface.

Suppose we place another piece of metal close to the first, as Figure 3 shows. We call these elements *electrodes*, because they connect to the voltage supply in an electric circuit. We connect the first electrode to the negative battery terminal and the second one to the positive terminal. We call the first electrode the *cathode* because it

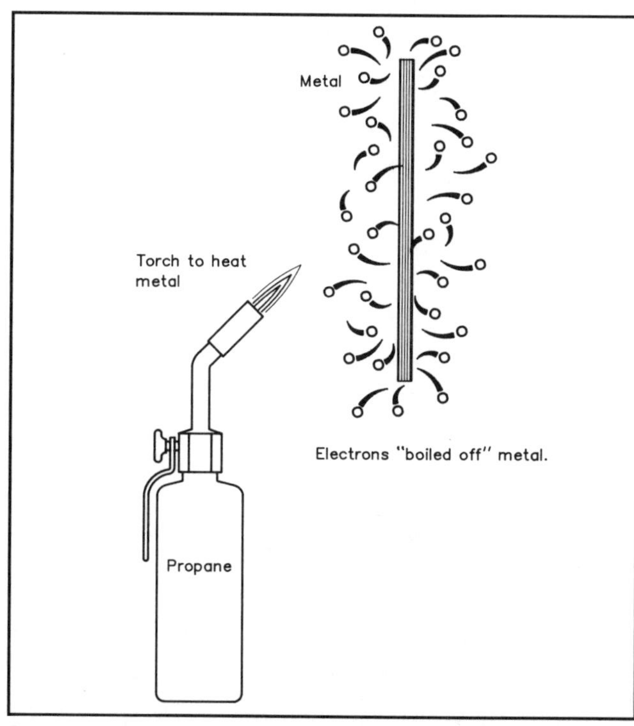

Figure 2—Heating a metal gives some electrons the energy they need to break free from the material. These electrons move a short distance into the space surrounding the metal. When an electron leaves an atom of the metal, it leaves a positive ion behind. These positive ions attract the electrons, keeping them close to the surface. These positive ions attract some of the electrons back onto the metal.

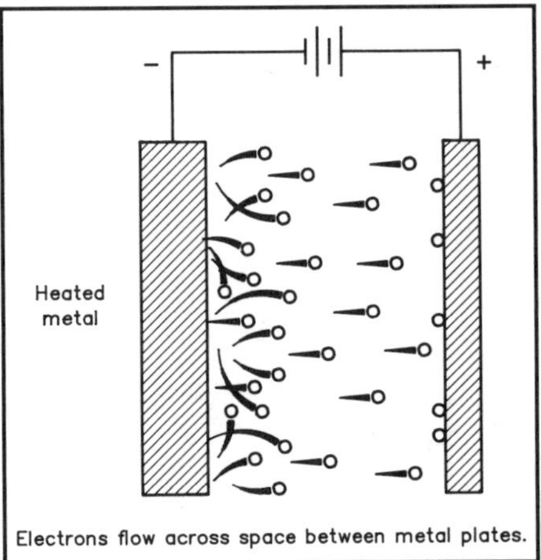

Figure 3—A second metal piece placed close to the first attracts the free electrons if we apply a positive voltage to it.

connects to the negative voltage supply. The second electrode is the *anode*, because it connects to the positive supply. Sometimes we also refer to the anode as the *plate*.

When we place the whole assembly inside a glass enclosure and remove the air inside, we have a *vacuum tube*. Removing the air ensures the free electrons can move from the cathode to the anode. Any gas molecules inside the tube would scatter the electrons and absorb energy from them. Figure 4 shows the construction of this basic vacuum tube. The cathode and anode leads go through the glass to provide connections to the external circuit.

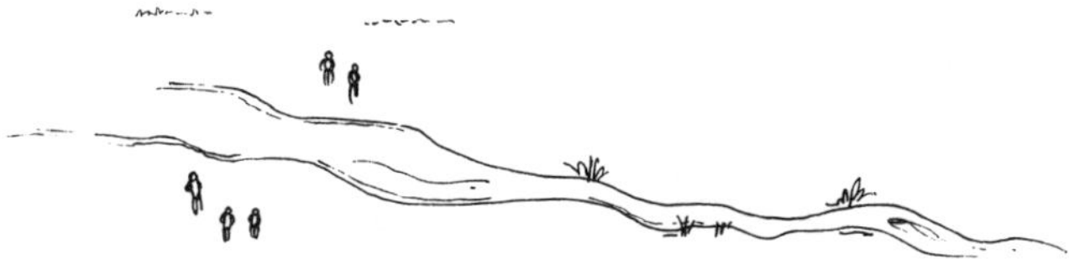

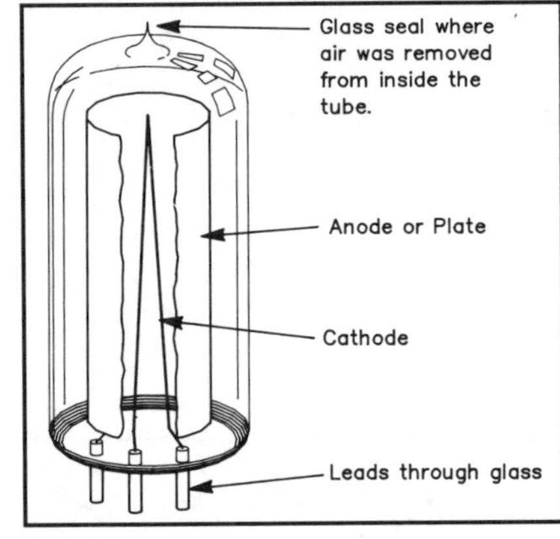

Figure 4—This diagram illustrates the construction of a simple vacuum tube. Wire leads extend through the glass envelope so you can connect the tube to a circuit. One power supply feeds a current through the cathode to heat it. A second supply puts a positive voltage on the anode.

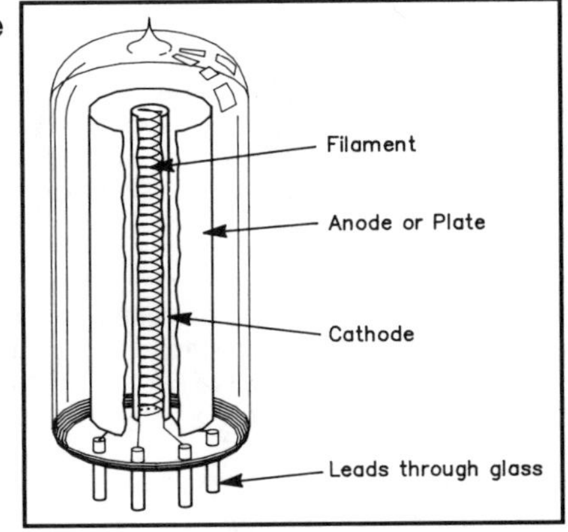

Figure 5—Some vacuum tubes use a separate filament to heat the cathode.

You may be wondering how we heat the cathode inside this vacuum tube. There are two ways we can accomplish this. Both methods involve an electric current, since current produces heat when it flows through a metal.

Passing an electric current through the cathode will heat the cathode metal. One voltage supply feeds current through the cathode and a second supply connects between the cathode and anode. Tubes using this method have a *directly heated* cathode. The diode shown in Figure 4 has a directly heated cathode.

Some vacuum tubes have a separate element to heat the cathode. A current through this *filament* heats the filament wire, which then *indirectly* heats the cathode. Figure 5 illustrates the construction of a tube with an *indirectly heated* cathode.

Diodes

A vacuum-tube diode contains two circuit elements. We sometimes call these circuit elements *electrodes*. You probably knew this tube has two electrodes because the prefix *di* means *two*.

The *cathode* gives up electrons, to produce current through the tube. The cathode connects to the negative supply lead. Current through the cathode heats the wire to "boil off" electrons in a directly heated cathode. A *filament* close to an indirectly heated cathode produces the heat needed to free electrons.

The *plate*, or *anode*, collects the electrons that come off the cathode. The plate connects to the positive supply lead. This positive voltage attracts the electrons, pulling them across the space between the two electrodes.

Figure 1 shows two symbols used to represent vacuum-tube diodes on schematic diagrams. You may wonder why we don't count three elements in the tube with an indirectly heated cathode. The filament is not part of the *circuit* in which we use the tube. The filament's only function is to heat the cathode to boil off electrons. Therefore, we don't count the filament as one of the active elements in the tube.

Figure 2 shows a diode, with filament, connected to a battery. We included an ammeter to measure current through the tube. The positive plate voltage attracts the electrons given off by the cathode. The electrons move through the vacuum inside the tube, producing a current through the tube.

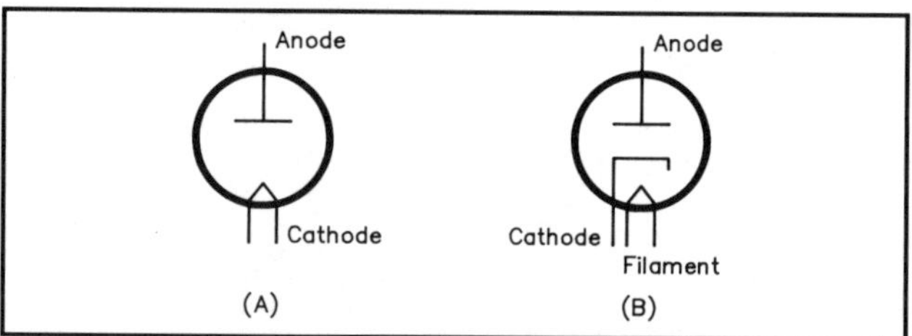

Figure 1—These symbols represent vacuum-tube diodes on schematic diagrams. The symbol at A represents a tube with a directly heated cathode. We use the symbol shown at B to represent a tube with an indirectly heated cathode.

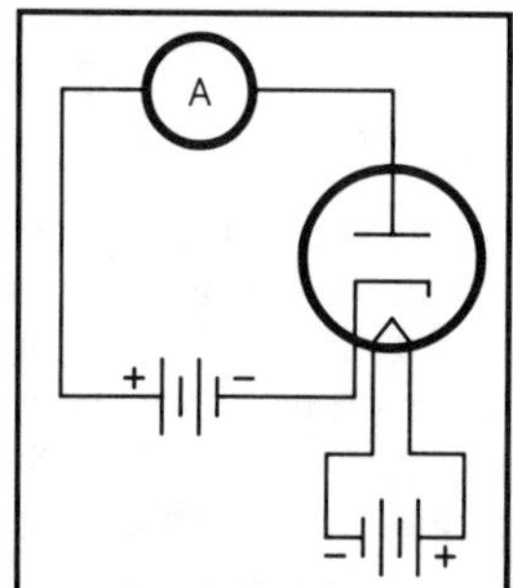

Figure 2—The positive supply lead connects to the diode plate lead. This tube has a filament to heat the cathode. A separate supply powers the filament circuit. The ammeter measures current through the tube.

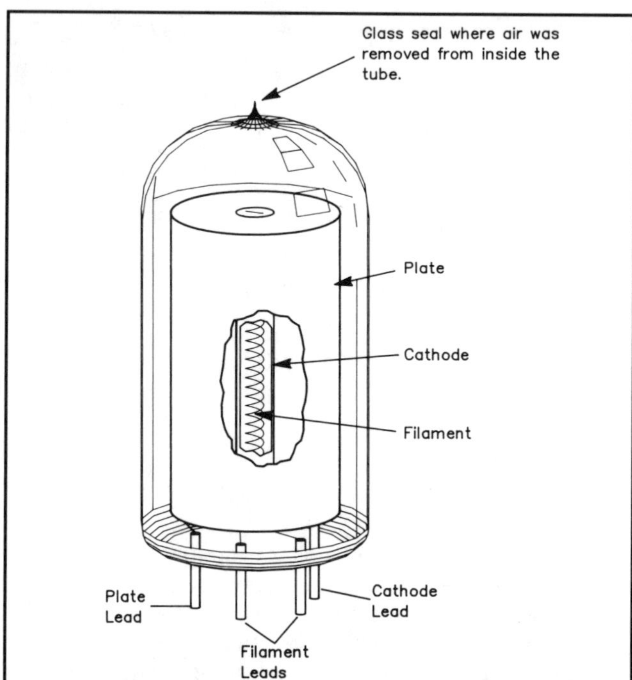

Figure 3—This cut-away view illustrates typical vacuum-tube diode construction.

Figure 3 is a cut-away view of the diode construction. The plate surrounds the cathode so electrons given off any side can travel directly to the plate. This ensures that the plate can collect all the electrons given off by the cathode.

How will the plate voltage affect the current through the tube? Let's conduct a simple experiment. Replace the battery in Figure 2 with a variable-voltage power supply. With no voltage applied to the plate, there is no current through the tube. Now increase the plate voltage to 10 volts. The ammeter indicates 25 mA through the tube.

Continue to increase the plate voltage, 10 volts at a time, and read the plate current for each new voltage

Table 1
Diode Plate Current and Voltage Characteristics

Plate Voltage (volts)	*Plate Current (milliamperes)*
0	0
10	25
20	70
30	130
40	200
50	275
60	360
70	450
80	560
90	660
100	775
110	890

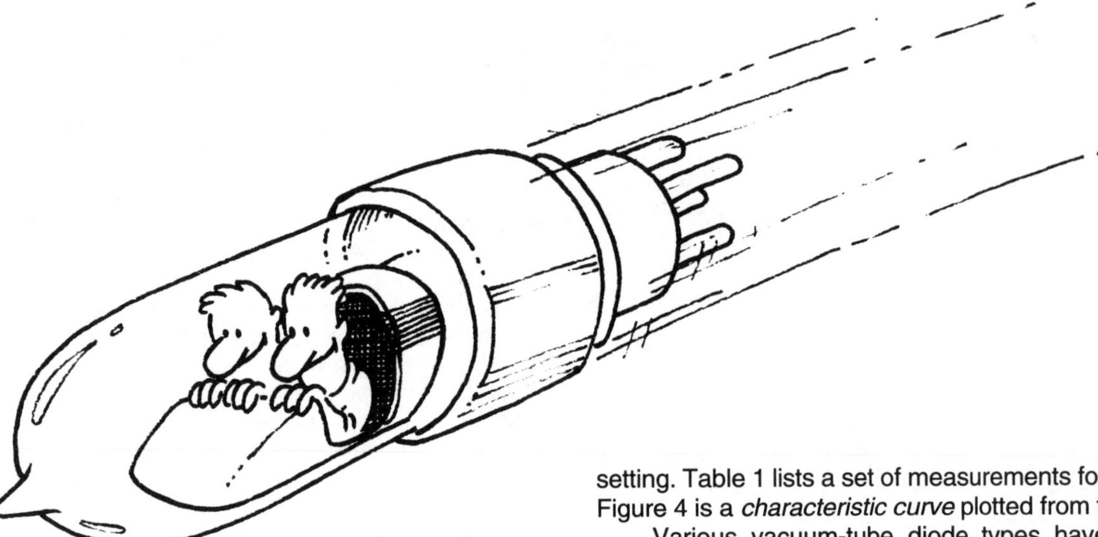

setting. Table 1 lists a set of measurements for one tube. Figure 4 is a *characteristic curve* plotted from this data.

Various vacuum-tube diode types have different characteristic curves. While the voltage and current numbers won't be the same, the general shape of the curve will be the same. Notice the current drops to zero when the plate voltage is zero.

What will happen if we connect the plate to the negative supply lead and the cathode to the positive lead? The plate will not attract electrons that come off the cathode. The negative plate will now *repel* those electrons, blocking any current through the tube. (The plate won't emit any electrons because it is not hot enough.) A diode tube with *reverse voltage* across its cathode and plate is *cut off*. That means there is no current through the tube.

A vacuum-tube diode works the same way a solid-state diode works. With an alternating voltage supply connected to the cathode and plate, the diode conducts during one-half cycle of the waveform. Figure 5 shows a diode connected to an alternating current supply. The tube output waveform is a series of pulses, which occur when the positive half-cycle connects to the plate.

Figure 4—This characteristic curve graphs the data measured for Table 1. It displays the changes in tube current as the plate voltage changes.

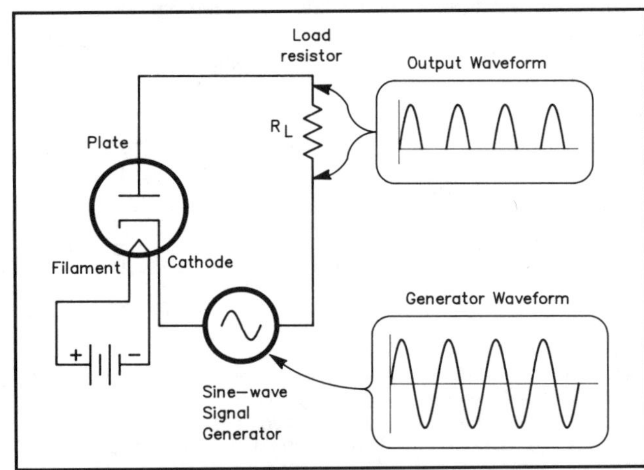

Figure 5—A vacuum-tube diode converts a sine-wave ac input to a series of dc pulses.

The Triode—A Field-Effect Device

We seldom use vacuum-tube diodes to rectify ac signals in modern electronics. They produce a lot of heat and are less reliable than solid-state diodes. These tubes illustrate the basic operation of all vacuum tubes, however. Although transistors have replaced vacuum tubes for many applications, we still use tubes for power amplifiers and other applications.

We must have some means to control the current through a tube so it can serve as an amplifier. The *triode* adds a third electrode to the cathode and anode in a diode. This third element is the *control grid.* Figure 1 shows the schematic-diagram symbols used to represent triode vacuum tubes.

The control grid is a metal screen placed between the cathode and anode, or plate. Electrons can easily pass through this screen.

We usually connect the *control grid*, or *grid*, to a negative voltage. The positive side of this supply connects to the cathode. Connected this way, the grid repels some of the electrons flowing from the cathode to the plate. Figure 2 is a circuit we might use to test the operation of a triode.

To begin our analysis of the triode, let's assume the control grid has no voltage. (Replace battery E_{grid} in Figure 2 with a piece of wire.) The cathode may be directly heated or indirectly heated. When it becomes hot, it emits electrons. Since the anode is positive when compared to the cathode, it attracts these electrons. Electrons move through the tube from cathode to plate. Does this sound identical to the vacuum-tube diode? For this experiment, we will assume the plate-current meter reads 4.5 mA.

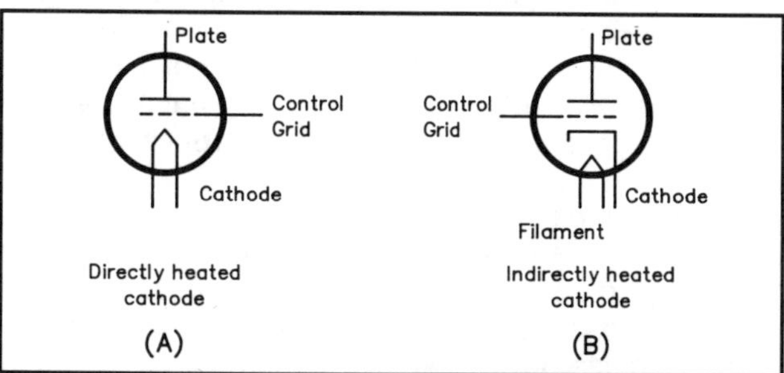

Figure 1—These symbols represent triode vacuum tubes on schematic diagrams. Some triodes have directly heated cathodes and some have indirectly heated cathodes. Notice we can draw the control-grid lead coming out of either side of the symbol.

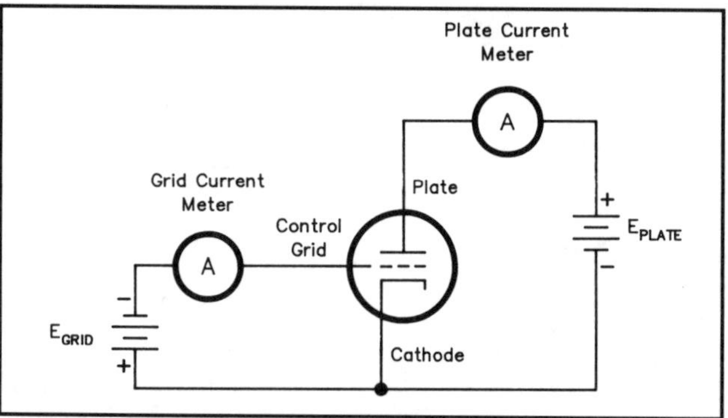

Figure 2—We can use this circuit to study the triode operation. One voltage supply connects between the cathode and plate, as it does with a diode. A second battery connects between the cathode and control grid, to make the grid negative compared to the cathode.

Now let's put the grid battery back into the Figure 2 circuit. We can adjust this voltage to any value we like for our "thought experiment," so let's set it to 3 volts. Because the negative side of this supply connects to the control grid, the voltage tries to repel electrons emitted from the cathode. This time the plate-current meter only reads 0.15 mA. Making the control grid negative reduced the plate current.

Increase the grid voltage to 10 volts. This time the grid repels all the electrons coming from the cathode, so none get through to the plate. The plate-current meter reads zero now. We say the tube is *cut off* when the control-grid voltage is high enough to block all plate current.

What do you think would happen if we made the grid positive with respect to the cathode? (In other words, what would happen if we turned battery E_{grid} around in Figure 2?) Like the plate, the grid would attract the electrons emitted from the cathode. The grid is just a wire screen, though, so most of the electrons would go right through it. Then the plate would continue to attract them, so the plate current would be even larger than when the grid voltage was zero. Some electrons would strike the grid, and this would increase the grid current also.

We call the voltages that set up the tube's operating conditions *bias voltages.* The plate gets a positive bias voltage compared to the cathode. The control grid has a negative bias voltage for normal operation.

There are two ways to change a triode's plate current. Increasing the plate voltage will increase the plate current, just as it does with a diode. (You should never apply a higher

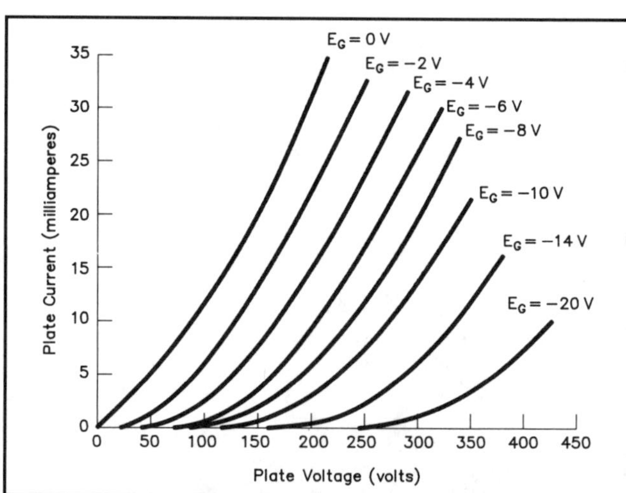

Figure 3—This graph shows a typical set of characteristic curves for a triode vacuum tube. For each grid voltage shown, we increase the plate voltage from zero.

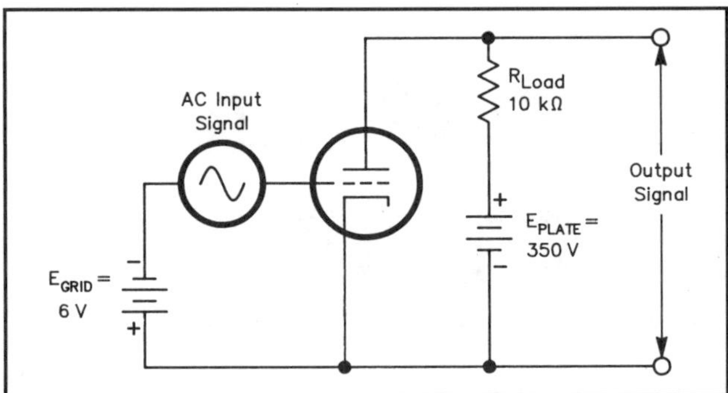

Figure 4—This simplified circuit illustrates how a triode vacuum tube amplifies the voltage of an input signal. Small variations in the input signal voltage add to and subtract from the grid bias battery, E_{grid}. The changing grid voltage produces a changing plate current, which results in a large change of the output voltage.

plate voltage than the tube manufacturer specifies. You also must use care not to allow the plate current to exceed the manufacturer's rating.) Changing the grid bias voltage also changes the plate current, as we learned in our earlier experiment.

Figure 3 is a graph that plots a typical set of triode *characteristic curves.* We could measure these curves using a circuit like the one shown in Figure 2. We would set a certain value of grid voltage, and then vary the plate voltage from zero to the maximum value. Then we would adjust the grid voltage to a new value and vary the plate voltage over the same range. After collecting the data for a range of grid voltages, we would plot the Figure 3 graph.

Each triode tube type has a particular set of characteristic curves. Different tube types will have different values of plate voltage and current, and different values of grid voltage. Even the shape of the curves may vary somewhat. Figure 3 is a representative sample of a triode characteristic curve, however.

A complete description of how a triode tube amplifies a signal is beyond the scope of this book. We'll give a brief, simplified description to help you understand why this tube is so important, however.

The circuit of Figure 4 illustrates how a triode amplifies a signal. A practical amplifier circuit requires many more parts. You wouldn't find two batteries supplying the plate and grid voltages. The grid bias battery and plate-voltage battery set the operating conditions for the tube in this example.

Figure 5 is a copy of the Figure 3 triode characteristics with the operating conditions added. The input signal voltage varies between plus 4 volts and minus 4 volts. This input signal adds to (and subtracts from) the grid battery voltage, E_{grid}. The grid bias voltage varies between –2 volts and –10 volts.

Variations in grid voltage cause changes in plate current, as Figure 5 shows. In this example, the plate current varies between 17.5 mA and 8.7 mA.

Figure 5 also shows how the output voltage changes when the plate current varies. The plate current flows through the load resistor, creating a voltage drop that you can calculate using Ohm's Law. (If the plate current is 8.7 mA, there is an 88-volt drop across R_{Load}.)

$$E = I\,R \qquad \text{(Equation 1)}$$

$$E = 8.7 \times 10^{-3}\ \text{A} \times 10 \times 10^{3}\ \Omega = 88\ \text{V}$$

This voltage drop subtracts from the battery voltage, to give the output-signal voltage. (In this case, 350 V – 88 V = 263 V.)

If the plate current increases to 17.5 mA, there is a larger voltage drop across the load resistor.

$$E = 17.5 \times 10^{-3}\ \text{A} \times 10 \times 10^{3}\ \Omega = 175\ \text{V}$$

Subtracting this value from the 350-V battery voltage, we have an output voltage of 175 volts. While the input voltage varies 8 volts between its positive and negative peaks, the output voltage varies 88 volts between peaks! The output voltage is an amplified version of the input voltage.

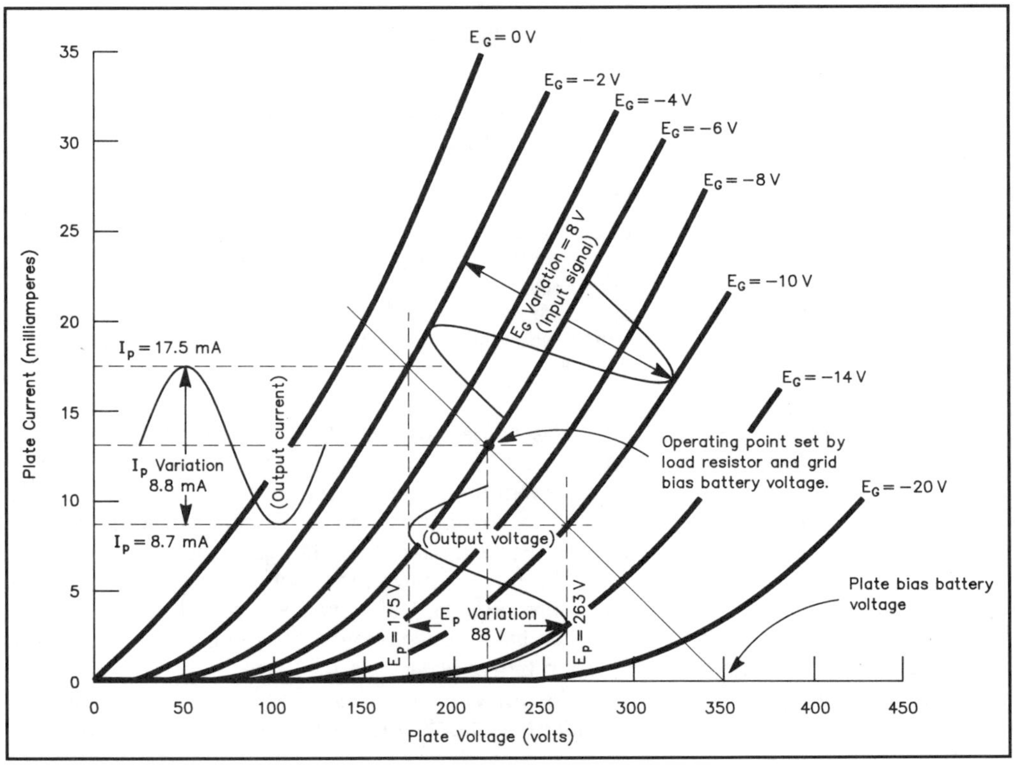

Figure 5—This graph shows the grid-voltage variations produced by the input signal. It also shows the resulting plate-current variations and output-voltage variations. This illustrates how a triode vacuum tube amplifies a signal.

The Tetrode and Pentode Have Additional Control Elements

Triode vacuum tubes make good amplifiers. One triode characteristic can lead to problems, however. The plate current is very sensitive to changes in plate voltage. Figure 1 shows a set of triode characteristic curves.

Even small plate-voltage variations can cause significant plate-current changes. This means we must have a well-regulated plate-voltage supply to produce a stable amplifier.

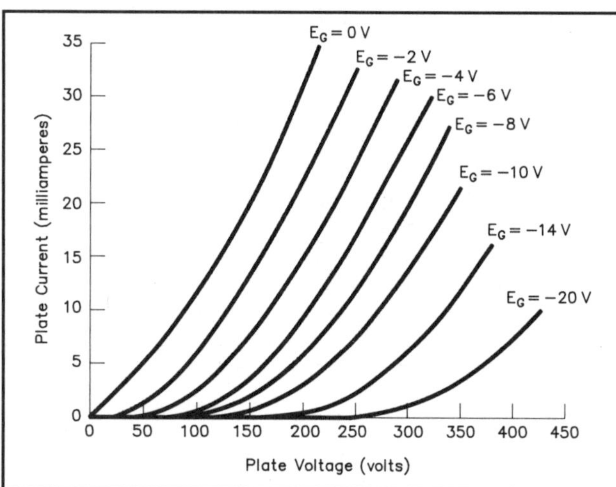

Figure 1—This graph represents the plate current and voltage characteristics of a triode vacuum tube. Notice how the plate current changes as the plate voltage changes for each grid-voltage setting.

Tube experimenters added another grid to the triode to help solve this problem. They called the new grid a *screen grid.* This tube has four elements, not counting the filament, which heats the cathode. A four-element tube became known as a *tetrode.* (*Tetra* means four in Greek.)

A tetrode has a positive voltage applied to the screen grid, as Figure 2 shows. The screen grid helps attract the electrons emitted from the cathode. As its name implies, we make the screen grid from wire mesh or screen, so electrons easily pass through it. The positive voltage applied to the plate is higher than the screen voltage. The plate attracts the electrons through the screen grid and collects them. The screen collects some of the electrons as they move through the tube. The screen current is much smaller than the plate current, however.

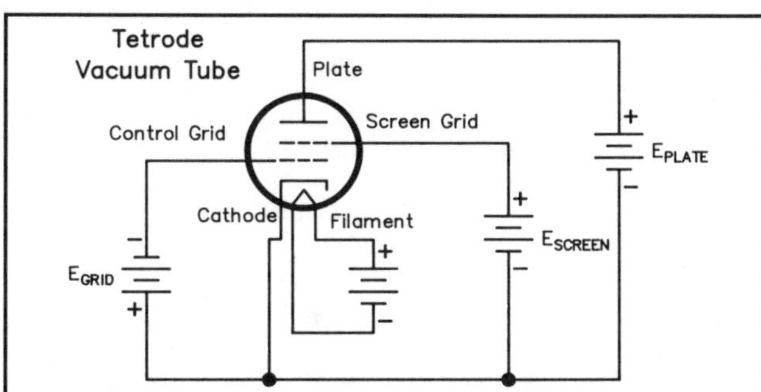

Figure 2—A tetrode vacuum tube has all the elements of a triode tube, plus an additional grid, called the screen grid. The screen grid makes the plate current almost independent of variations in plate voltage.

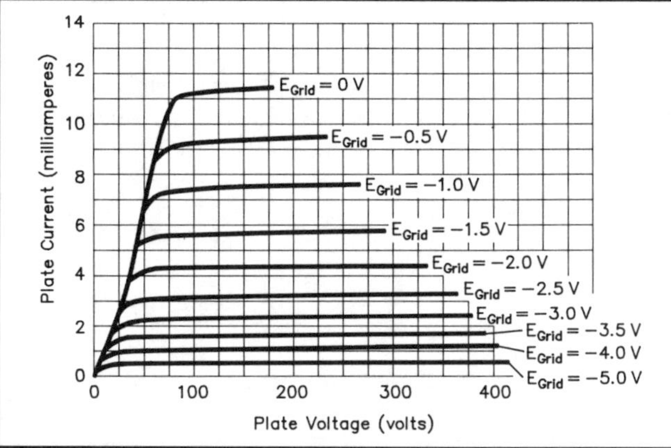

Figure 3—We seldom use tetrode vacuum tubes in modern electronics. This graph represents a set of characteristic curves for an idealized tetrode. Notice the plate current is almost independent of the plate voltage for a given control-grid voltage.

The screen grid is physically much closer to the cathode than the plate. The positive screen voltage has a larger effect on the electrons as they leave the cathode than does the plate voltage. Small variations in plate voltage produce almost no change in plate current.

The tetrode solves the biggest problem with triode amplifiers. Figure 3 shows a set of theoretical tetrode characteristic curves. Notice the plate current is almost independent of plate voltage.

We don't use tetrodes in modern electronics circuits because they create another problem. Both the screen grid and the plate have positive voltages to attract the electrons from the cathode. This attraction force pulls the electrons across the vacuum inside the tube, so electrons that hit the plate are moving very fast.

When the electrons strike the plate they knock other electrons loose from the surface. We call this process of knocking other electrons free *secondary emission.* Many of the secondary electrons fly off the plate and the positive voltage on the screen grid attracts them. These electrons travel backward through the tube, and increase the screen-grid current.

We add a fifth tube element to maintain the advantage of the tetrode, while eliminating its disadvantage. The new element fits between the screen grid and the plate. This tube is the *pentode*, because it has five electrodes. (*Penta* is Greek for five.)

This new element, called a *suppressor grid*, connects to the cathode. The connection may be inside or outside the tube. The cathode and suppressor grid have the same voltage. The suppressor grid repels secondary electrons as they travel toward the screen grid, pushing them back to the plate.

Figure 4 shows the symbol used on schematic diagrams to represent a pentode vacuum tube. Separate batteries show the bias-voltage polarities applied to a pentode.

Figure 5 shows a typical set of plate characteristics for a pentode vacuum tube. Like other tubes, each type of pentode has its own particular set of characteristic curves. The curves shown here are meant only to provide an example of the typical curve shapes.

Many high-power radio frequency amplifiers use triode or pentode tubes. The cathode, plate and control grid in a pentode operate the same way they do in a triode. A pentode amplifier works much the same as a triode amplifier.

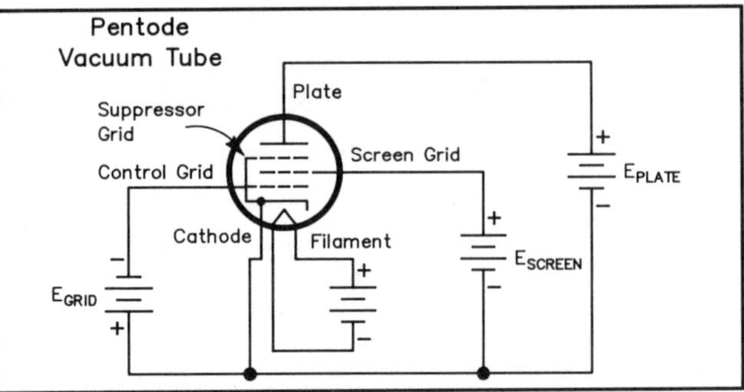

Figure 4—This drawing shows the symbol used to represent a pentode vacuum tube on schematic diagrams. The battery connections illustrate the bias polarities used in a typical amplifier.

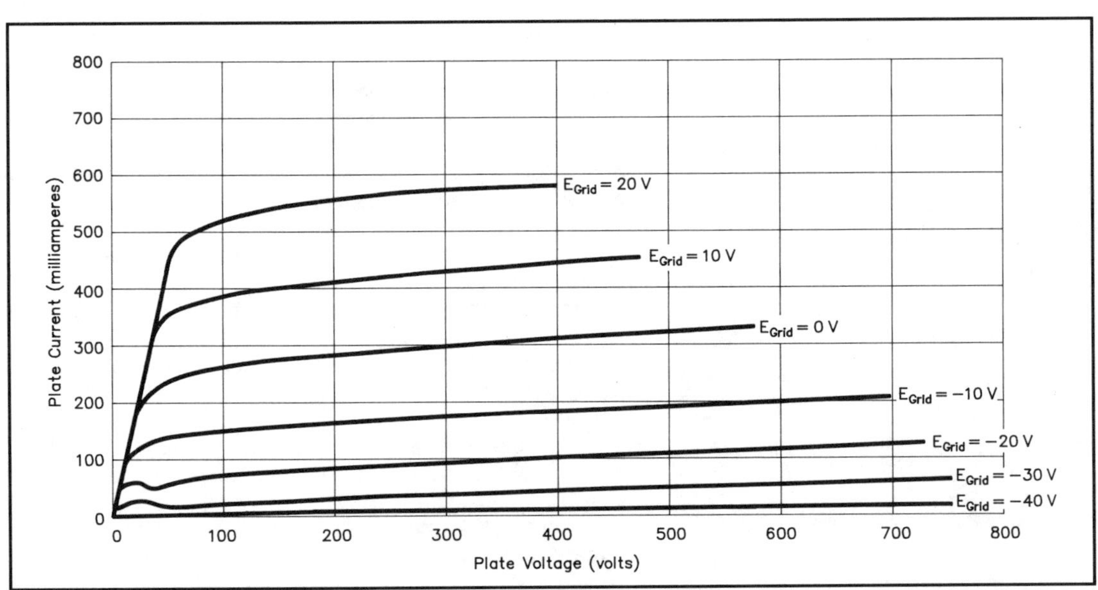

Figure 5—This graph represents a typical set of characteristic curves for a pentode vacuum tube.

The Cathode-Ray Tube (CRT) Display

Vacuum tubes have many applications in electronics circuits. Solid-state devices such as bipolar and field-effect transistors have taken the place of vacuum tubes in many applications, however. When we think about vacuum tubes today, high-power radio-frequency amplifiers will likely come to mind.

Perhaps the most widely used vacuum tube of all, though, is the *cathode-ray tube*, or *CRT*. This vacuum tube is a display device.

Several very important pieces of electronics test equipment use CRT displays. An oscilloscope uses a CRT to draw a picture of the signal waveform in a circuit. Figure 1 is a schematic diagram of a simple full-wave rectifier circuit. We connected an oscilloscope to the circuit output, and the display shows a picture of the diode output pulses.

Spectrum analyzers use CRTs to display a bar graph of the sine-wave signal frequencies that make up a waveform. Figure 2 shows the spectrum-analyzer display resulting from a radio-frequency transmitter test. We fed two audio-frequency sine-wave signals into a radio transmitter microphone input. This display shows each output-signal frequency from the transmitter. The height of each line represents the strength of that signal.

TV screens and computer monitors also use cathode-ray tube displays. These vacuum tubes can display shades of gray between black and white. They also can display color.

Figure 3 shows the construction of a basic cathode-ray tube. You should recognize the filament or heater, the cathode, the grid and the anode. These are the same basic parts used in the other vacuum tubes we studied.

The filament heats the cathode, causing it to give off electrons. The grid controls the flow of these electrons toward the anode. By increasing or decreasing the number of electrons in the beam, we can control the brightness of the display screen. The anode attracts the electrons, accelerating them so they move very fast as they approach the anode. The anode does not collect most of the electrons, however.

The anode has three sections that function as an *electron lens*. This lens focuses the electrons into a thin beam, which goes through a hole in the end of the anode. We call this assembly an *electron gun* because it shoots a thin electron beam toward the tube's front surface.

The front of the tube has a coating of a special material, called a *phosphor*. When electrons from the beam strike this phosphor, they give extra energy to electrons in the phosphor. Then these electrons radiate this energy to return to their normal position in the phosphor atoms. The energy they radiate appears as visible light, of a color specific to the phosphor used. Some phosphors give off a green glow. Others radiate white light. Still others produce the red, blue and green light needed for a color display.

After the electrons leave the electron-gun portion of the tube, they pass through two sets of *deflection plates*. One set of deflection plates moves the beam horizontally and the other set moves the beam vertically. Figure 4 shows the electron beam bent as it passes through the vertical deflection plates. Oscilloscopes and other pieces of test equipment normally use *electrostatic deflection*.

Most TV picture tubes use another method to deflect the electron beam: *magnetic deflection*. One advantage of magnetic deflection is that you can position the electromagnets around the outside of the tube neck. The magnetic field bends the electron beam, changing the position where it strikes the phosphor on the tube front. Figure 5 illustrates the electromagnetic deflection coils around the neck of a TV picture tube.

Electrostatic deflection provides better high-frequency operation, which is why oscilloscopes use this method. With no voltage applied to the deflection plates, the beam hits the center of the screen. A positive voltage on the right-hand plate and a negative voltage on the left deflects the beam to the right. The positive voltage attracts the electrons and the negative voltage repels them. The electrons don't strike the deflection plates, however, because they are moving so fast. Reversing the deflection-plate voltage polarity deflects the beam to the left. The vertical deflection plates operate the same way, moving the beam up and down from the center position.

The deflection amount depends on the applied voltage. A small voltage moves the beam a small amount and a large voltage deflects it farther.

An oscilloscope usually operates with an internally generated voltage applied to the horizontal deflection plates. The vertical deflection plates get a signal voltage from the circuit under test.

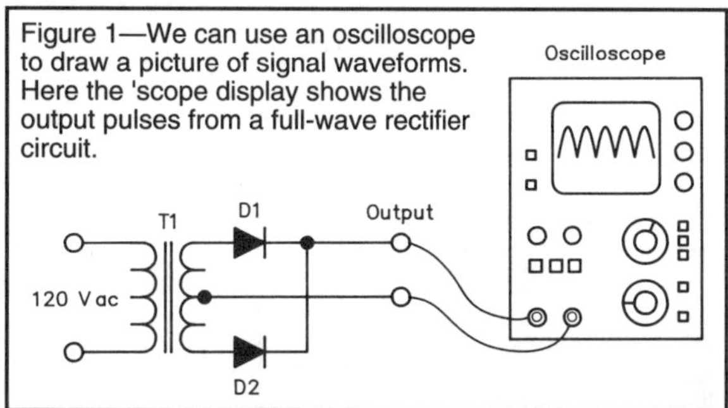

Figure 1—We can use an oscilloscope to draw a picture of signal waveforms. Here the 'scope display shows the output pulses from a full-wave rectifier circuit.

Figure 2—A spectrum analyzer shows the signal frequency and strength on a bar-graph display. This is especially useful for studying a waveform that includes several sine waves combined. Here we show the spectral display from a transmitter during a two-tone test. We feed two audio-frequency sine wave signals into the microphone input and measure the output signal frequencies and strengths.

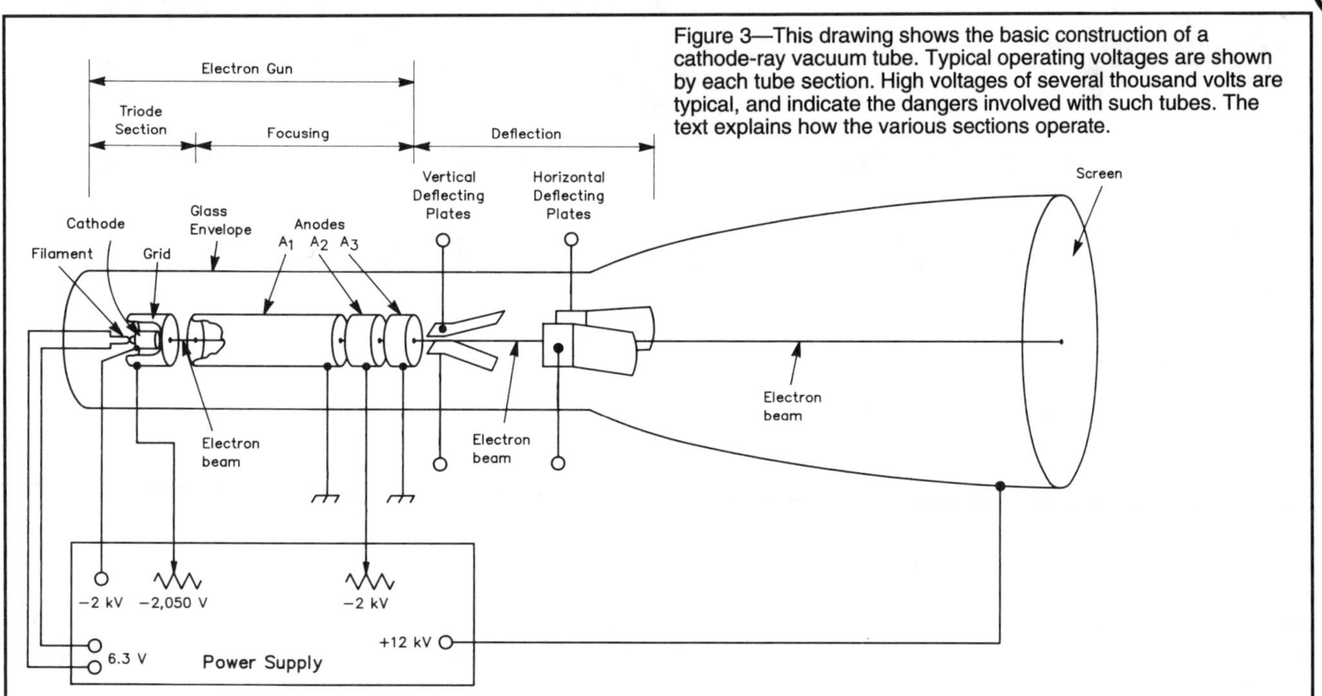

Figure 3—This drawing shows the basic construction of a cathode-ray vacuum tube. Typical operating voltages are shown by each tube section. High voltages of several thousand volts are typical, and indicate the dangers involved with such tubes. The text explains how the various sections operate.

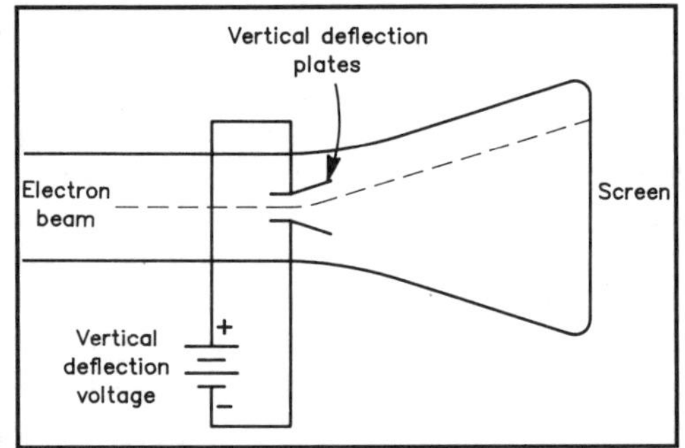

Figure 4—The electron beam bends as it passes between the deflection plates, when there is a voltage across plates. Higher voltages bend the beam farther.

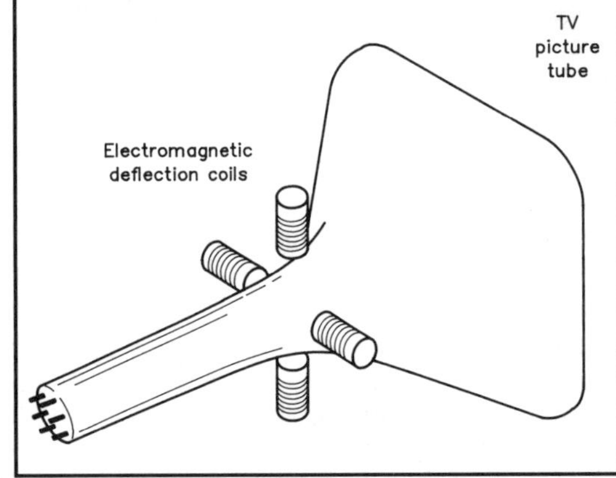

Figure 5—A TV picture tube uses magnetic deflection to move the electron beam across the tube face. These coils produce a horizontal magnetic field and a vertical magnetic field. The magnetic fields deflect the beam left to right across the screen and move the beam from top to bottom. The beam scans the entire tube surface 30 times each second.

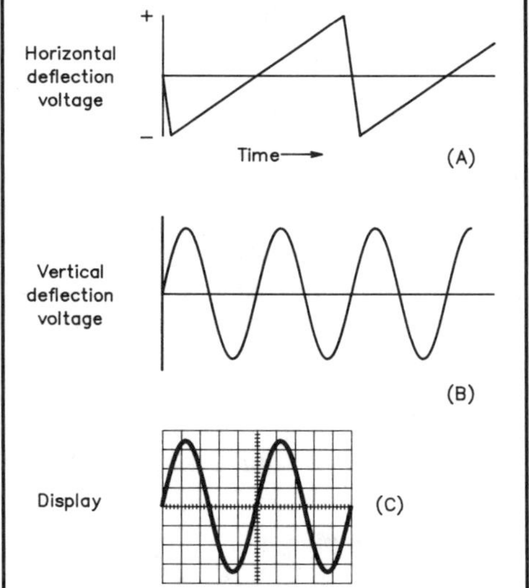

Figure 6—Part A shows the sawtooth waveform of the horizontal deflection voltage. Part B shows a sine-wave vertical input voltage. These deflection voltages combine to draw the waveform shown at C on the oscilloscope screen.

The horizontal plates have a sawtooth-waveform voltage applied. This voltage starts the beam off at the screen's left edge, then *sweeps* it horizontally across the tube. During the short time the sawtooth waveform goes from its positive peak to its negative peak, the beam jumps back to the left screen edge.

Any voltage on the vertical deflection plates moves the electron beam up and down on the screen. The beam moves up and down while it sweeps from left to right. In this way, the oscilloscope draws a picture of the input waveform. You must adjust the horizontal sweep rate to match the waveform's frequency to get a useful picture. See Figure 6.

Special circuits in a TV set move the electron beam horizontally and vertically, to scan the entire tube surface. One complete scan occurs 30 times each second. A control voltage from the received TV signal goes to the grid, changing the beam strength. This changes the brightness of the light given off from the screen phosphor, creating an image on the screen.

A Glossary of Electronics and Amateur Radio Terms

APPENDIX

A1A emission—The FCC emission designator used to describe Morse code telegraphy (CW) by on/off keying of a radio-frequency signal.

Absorption—The loss of energy from an electromagnetic wave as it travels through any material. The energy may be converted to heat or other forms. Absorption usually refers to energy lost as the wave travels through the ionosphere.

Absorption wavemeter—A device for measuring frequency or wavelength. It takes some power from the circuit under test when the meter is tuned to the same resonant frequency.

Adcock array—A radio direction finding antenna array consisting of two vertical elements fed 180° apart and capable of being rotated.

Admittance—The reciprocal of impedance, often used to aid the solution of a parallel-circuit impedance calculation.

Alpha (α)—The ratio of transistor collector current to emitter current. Alpha is between 0.92 and 0.98 for a junction transistor.

Alpha cutoff frequency—A term used to express the useful upper frequency limit of a transistor. The point at which the gain of a common-base amplifier is 0.707 times the gain at 1 kHz.

Alternating current (ac)—Electrical current that flows first in one direction in a wire and then in the other direction. The applied voltage changes polarity and causes the current to change direction. This direction reversal continues at a rate that depends on the frequency of the ac.

Alternator—A machine used to generate alternating-current electricity.

Alternator whine—A common form of conducted interference typified by an audio tone being induced onto the received or transmitted signal. The pitch of the noise varies with alternator speed.

Amateur Auxiliary—A voluntary organization, administered by ARRL Section Managers. The primary objectives are to foster amateur self-regulation and compliance with the rules.

Amateur Radio communication—Noncommercial radio communication by or among Amateur Radio stations solely with a personal aim and without pecuniary or business interest. (Pecuniary means payment of any type, whether money or other goods.)

Amateur Radio operator—A person holding a valid license to operate an Amateur Radio station. In the US this license is issued by the Federal Communications Commission.

Amateur Radio Service—A radio communication service of self-training, intercommunication, and technical investigation carried on by radio amateurs.

Amateur Radio station—A station licensed in the Amateur Radio Service, including necessary equipment at a particular location, used for Amateur Radio communication.

Ampere (A)—The basic unit of electrical current, equal to 6.24×10^{18} electrons moving past a point in one second. We abbreviate amperes as amps.

Amplifier—A device usually employing electron tubes or transistors to increase the voltage, current, or power of a signal. The amplifying device may use a small signal to control voltage and/or current from an external supply. A larger replica of the small input signal appears at the device output.

Amplitude modulation (AM)—A method of combining an information signal and an RF carrier. In double-sideband voice AM transmission, we use the voice information to vary (modulate) the amplitude of a radio-frequency signal. Shortwave broadcast stations use this type of AM, as do stations in the Standard Broadcast Band (510-1600 kHz). Amateurs seldom use double-sideband voice AM, but a variation, known as single sideband (SSB), is very popular.

Amplitude-compandored single sideband (ACSSB)—An SSB system that uses a logarithmic amplifier to compress voice signals at the transmitter and an inverse logarithmic amplifier to expand the voice signals in the receiver.

AMTOR—Amateur Teleprinting Over Radio. AMTOR provides error-correcting capabilities. See Automatic Repeat Request (ARQ) and Forward Error Correction (FEC).

AND gate—A logic circuit whose output is 1 only when both of its inputs are 1.

Anode—The terminal that connects to the positive supply lead for current to flow through a device.

Antenna—An electric circuit designed specifically to radiate the energy applied to it in the form of electromagnetic waves. An antenna is reciprocal; a wave moving past it will induce a current in the circuit also. Antennas are used to transmit and receive radio waves.

Antenna bandwidth—A range of frequencies over which an antenna will perform well. Antenna bandwidth is usually specified as a range of frequencies where the antenna SWR will be below some given value.

Antenna effect—One of two operational modes of a simple loop antenna wherein the antenna exhibits the characteristics of a small, nondirectional vertical antenna.

Antenna efficiency—The ratio of the radiation resistance to the total resistance of an antenna system.

Antenna switch—A switch used to connect one transmitter, receiver or transceiver to several different antennas.

Apogee—That point in a satellite's orbit (such as the moon) when it is farthest from the earth.

Apparent power—The product of the RMS current and voltage values in a circuit without consideration of the phase angle between them.

Ascending Pass—With respect to a particular ground station, a satellite pass during which the spacecraft is headed in a northerly direction while it is in range.

ASCII—American National Standard Code for Information Interchange. This is a seven-bit digital code used in computer and radioteleprinter applications.

Astable (free-running) multivibrator—A circuit that alternates between two unstable states. This circuit could be considered as an oscillator that produces square waves.

Asynchronous flip-flop—A circuit that changes output state depending on the data inputs, without requiring a clock signal.

Atom—A basic building block of all matter. Inside an atom, there is a positively charged, dense central core, surrounded by a "cloud" of negatively charged electrons. There are the same number of negative charges as there are positive charges, so the atom is electrically neutral.

Attenuate—To reduce in amplitude.

ATV (amateur television)—A wideband TV system that can use commercial transmission standards. ATV is only permitted on the 70-cm band and higher frequencies.

Audio frequency (AF)—The range of frequencies that the human ear can detect. Audio frequencies are usually listed as 20 Hz to 20,000 Hz.

Audio rectification—Interference to electronic devices caused by a strong RF field that is rectified and amplified in the device.

Audio-frequency shift keying (AFSK)—A method of transmitting radioteletype information by switching between two audio tones fed into an FM transmitter microphone input. This is the RTTY mode most often used on VHF and UHF.

Aurora—A disturbance of the atmosphere around the earth's poles, caused by an interaction between electrically charged particles from the sun and the earth's magnetic field. Often a display of colored lights is produced, which is visible to those who are close enough to the magnetic-polar regions. Auroras can disrupt HF radio communication and enhance VHF communication. They are classified as visible auroras and radio auroras.

Automatic gain control—An amplifier circuit designed to provide a relatively constant output amplitude over a wide range of input values.

Automatic Repeat Request (ARQ)—An AMTOR communication mode. In ARQ, also called Mode A, the two stations constantly confirm each other's transmissions. If information is lost, it is repeated until the receiving station confirms correct reception.

Autopatch—A device that allows repeater users to make telephone calls through a repeater.

Avalanche point—That point on a diode characteristic curve where the amount of reverse current increases greatly for small increases in reverse bias voltage.

Average power—The product of the RMS current and voltage values associated with a purely resistive circuit, equal to one half the peak power when the applied voltage is a sine wave.

Azimuth-equidistant projection map—A map made with its center at one geographic location and the rest of the continents projected from that point. Also called a great-circle map, this map is the most useful type for determining where to point a directional antenna to communicate with a specific location.

Back EMF—An opposing electromotive force (voltage) produced by a changing current in a coil. It can be equal to the applied EMF under some conditions.

Backscatter—A small amount of signal that is reflected from the earth's surface after traveling through the ionosphere. The reflected signals may go back into the ionosphere along several paths and be refracted to earth again. Backscatter can help provide communications into a station's skip zone.

Balanced line—A symmetrical feed line with two conductors having equal but opposite voltages. Neither conductor is at ground potential.

Balanced modulator—A circuit used in a single-sideband suppressed-carrier transmitter to combine a voice signal and the RF signal. The balanced modulator isolates the input signals from each other and the output, so that only the difference of the two input signals reaches the output.

Balun—Contraction for balanced to unbalanced. A device to couple a balanced load to an unbalanced source, or vice versa. A balun is often used to feed a balanced antenna with an unbalanced feed line.

Band-pass filter—A circuit that allows signals to go through it only if they are within a certain range of frequencies. It attenuates signals above and below this range.

Band plan—An agreement for operating within a certain portion of the radio spectrum. Band plans set aside certain frequencies for each different mode of amateur operation, such as CW, SSB, FM, repeaters and simplex.

Band spread—A receiver quality used to describe how far apart stations on different nearby frequencies will seem to be. We usually express band spread as the number of kilohertz that the frequency changes per tuning-knob rotation. Note that band spread affects frequency resolution.

Bandwidth—The frequency range (measured in hertz) over which a signal is stronger than some specified amount below the peak signal level. For example, if a certain signal is at least half as strong as the peak power level over a range of ± 3 kHz, the signal has a 3-dB bandwidth of 6 kHz.

Base loading—The technique of inserting a coil with specific reactance at the bottom of a vertical antenna in order to cancel the capacitive reactance of the antenna, producing a resonant antenna system.

Battery—A device that converts chemical energy into electrical energy. It provides excess electrons to produce a current and the voltage or EMF to push those electrons through a circuit.

Baud—The unit used to describe the transmission speed of a digital signal. For a single-channel signal, one baud is equal to one digital bit per second.

Baudot—A five-bit digital code used in teleprinter applications.

Beam antenna—A directional antenna. A beam antenna must be rotated to provide coverage in different directions.

Beamwidth—As related to directive antennas, the width (measured in degrees) of the major lobe between the two directions at which the relative power is one half (–3 dB) of the value at the peak of the lobe.

Beat-frequency oscillator (BFO)—An oscillator that provides a signal to a receiver's product detector. In the product detector, the BFO signal mixes with the incoming signal to produce an audio signal.

Beta (β)—The ratio of transistor collector current to base current. Betas show wide variations, even between individual devices of the same type.

Beta cutoff frequency—The point at which the gain of a common-emitter amplifier is 0.707 times the gain at 1 kHz.

Bipolar junction transistor—A transistor made of two PN semiconductor junctions using two layers of similar-type material (N or P) with a third layer of the opposite type between them.

Bistable multivibrator—Another name for a flip-flop circuit that has two stable output states.

Blanking—Portion of a video signal that is "blacker than black," used to be certain that the return trace is invisible.

Bleeder resistor—A large-value resistor added across the filter capacitor in a power supply. The bleeder resistor dissipates any charge left on the filter capacitor when the supply is switched off.

Block diagram—A picture using boxes to represent sections of a complicated device or process. The block diagram shows the connections between sections.

Blocking—A receiver condition in which reception of a desired weak signal is prevented because of a nearby, unwanted strong signal.

Breakdown voltage—The voltage that will cause a current in an insulator. Different insulating materials have different breakdown voltages. Breakdown voltage is also related to the thickness of the insulating material.

Butterworth filter—A filter whose passband frequency response is as flat as possible. The design is based on a Butterworth polynomial to calculate the input/output characteristics.

Bypass capacitor—A capacitor that provides an alternating-current path of comparatively low impedance around some circuit element.

Calling frequencies—Frequencies set aside for establishing contact. Once two stations are in contact, they should move their QSO to an unoccupied frequency.

Capacitive coupling (of a dip meter)—A method of transferring energy from a dip-meter oscillator to a tuned circuit by means of an electric field.

Capacitor—An electrical component composed of two or more conductive plates separated by an insulating material. A capacitor stores energy in an electrostatic field.

Capacitor-input filter—A power-supply filter having a capacitor connected directly to the rectifier output.

Capture effect—An effect especially noticed with FM and PM systems whereby the strongest signal to reach the demodulator is the one to be received. You cannot tell whether weaker signals are present.

Carbon-composition resistor—An electronic component designed to limit current in a circuit; made from ground carbon mixed with clay.

Carbon-film resistor—A resistor made by depositing a gaseous carbon deposit on a round ceramic form.

Cardioid radiation pattern—A heart-shaped antenna pattern characterized by a single, large lobe in one direction, and a deep, narrow null in the opposite direction.

Cathode—The terminal that connects to the negative supply lead for current to flow through a device.

Cathode-ray tube—An electron-beam tube in which the beam can be focused on a luminescent screen. The spot position can be varied to produce a pattern on the screen. CRTs are used in oscilloscopes and as the "picture tube" in television receivers.

Cavity—A high-Q tuned circuit that passes energy at one frequency with little or no attenuation but presents a high impedance to another nearby frequency.

Center loading—A technique for adding a series inductor at or near the center of an antenna element in order to cancel the capacitive reactance of the antenna. This technique is usually used with elements that are less than 1/4 wavelength.

Centi—The metric prefix for 10^{-2}, or divide by 100.

Ceramic capacitor—An electronic component composed of two or more conductive plates separated by a ceramic insulating material.

Characteristic impedance—The opposition to electric current that a circuit presents. Impedance includes factors other than resistance, and applies to alternating currents. Ideally, the characteristic impedance of an antenna feed line will be the same as the transmitter output impedance and the antenna input impedance.

Chassis ground—A common connection for all parts of a circuit that connect to the power supply.

Chebyshev filter—A filter whose passband and stopband frequency response has an equal-amplitude ripple, and a sharper transition to the stop band than does a Butterworth filter. The design is based on a Chebyshev polynomial to calculate the input/output characteristics.

Choke-input filter—A power-supply filter having an inductor (choke) connected directly to the rectifier output.

Circular polarization—Describes an electromagnetic wave in which the electric and magnetic fields are rotating. If the electric field vector is rotating in a clockwise sense, then it is called right-hand circular polarization. If the electric field is rotating in a counterclockwise sense, it is called left-hand circular polarization. Note that the polarization sense is determined by standing behind the antenna for a signal being transmitted, or in front of it for a signal being received.

Clipping—Occurs when the peaks of a voice waveform are cut off in a transmitter, usually because of overmodulation. Also called flattopping.

Coaxial cable—coax (pronounced ko'-aks). This is a type of feed line with one conductor inside the other. Plastic, foam or gaseous insulation surrounds the center conductor. The insulation is covered by a shielding conductor. The entire cable is usually covered with vinyl insulation.

Coaxial capacitor—A cylindrical capacitor used for noise-suppression purposes. The line to be filtered connects to the two ends of the capacitor, and a third connection is made to electrical ground. One side of the capacitor provides a dc path between the ends, while the other side of the capacitor connects to ground.

Code key—A device used as a switch to generate Morse code.

Code-practice oscillator—A device that produces an audio tone, used for learning the code.

Coefficient of coupling—A measure of the mutual inductance between two coils. The greater the coefficient of coupling, the more mutual inductance between the coils. It is a measure of the amount of energy that will be transferred from one coil to another.

Coil—A conductor wound into a series of loops.

Color code—A system where numerical values are assigned to various colors. Colored stripes are painted on the body of resistors and sometimes other components to show their value.

Communications terminal—A computer-controlled device that demodulates RTTY and CW for display by a computer or ASCII terminal. The communications terminal also accepts information from a computer or terminal and modulates a transmitted signal.

Compandoring—In an ACSSB system, the process of *com*pressing voice signals in a transmitter and ex*pand*ing them in a receiver.

Complementary metal-oxide semiconductor (CMOS)—A type of construction used to make digital integrated circuits. CMOS is composed of both N-channel and P-channel MOS devices on the same chip.

Complex number—A number that includes both a real and an imaginary part. Complex numbers provide a convenient way to represent a quantity (like impedance) that is made up of two different quantities (like resistance and reactance).

Composite video signal—A complete video signal, consisting of the actual picture information, blanking pulses and sync pulses.

Computer-Based Message System (CBMS)—A system in which a computer is used to store messages for later retrieval. Also called a RTTY mailbox.

Conductance—The reciprocal of resistance. This is the real part of a complex admittance.

Conducted noise—Electrical noise that is imparted to a radio receiver or transmitter through the power connections to the radio.

Conductor—A material that has a loose grip on its electrons, so that an electrical current can pass through it.

Connected—The condition in which two packet-radio stations are sending information to each other. Each is acknowledging when the data has been received correctly.

Constant-k filter—A filter design based on the image-parameter technique. The product of the series and shunt impedances is independent of frequency within the filter passband.

Contests—On-the-air operating events. Different contests have different objectives: contacting as many other amateurs as possible in a given amount of time, contacting amateurs in as many different countries as possible or contacting an amateur in each county in one particular state, to name only a few.

Continuous wave (CW)—A term used by amateurs as a synonym for Morse code communication. Hams usually produce Morse code signals by interrupting the continuous-wave signal from a transmitter to form the dots and dashes.

Control operator—A licensed amateur designated to be responsible for the transmissions of an Amateur Radio station.

Coordinated Universal Time (UTC)—A system of time referenced to time at the prime meridian, which passes through Greenwich, England.

Core—The material in the center of a coil. The material used for the core affects the inductance value of the coil.

Corona discharge—A condition when a static-electricity charge builds up on an antenna, usually a mobile antenna, and then discharges as the air insulation around the antenna becomes ionized and glows light blue.

Counter (divider, divide-by-n counter)—A circuit that is able to change from one state to the next each time it receives an input signal. A counter produces an output signal every time a predetermined number of input signals have been received.

Coupling capacitor—A capacitor that provides a relatively low-impedance path from one stage to the next for ac signals above some frequency, while providing a high impedance to lower-frequency ac and dc signals. Also called a blocking capacitor because it blocks dc and low-frequency ac signals.

CQ—The general call when requesting a conversation with anyone.

Critical angle—If radio waves leave an antenna at an angle greater than the critical angle for that frequency, they will pass through the ionosphere instead of returning to earth.

Critical frequency—The highest frequency at which a vertically incident radio wave will return from the ionosphere. Above the critical frequency, radio signals pass through the ionosphere instead of returning to the earth.

Cross modulation—A type of intermodulation caused by the carrier of a desired signal being modulated by an unwanted signal.

Crystal oscillator—An oscillator in which the main frequency-determining element is the mechanical resonance of a piezoelectric crystal (usually quartz).

Crystal-lattice filter—A filter that employs piezoelectric crystals (usually quartz) as the reactive elements. They are most often used in the IF stages of a receiver or transmitter.

Crystal-controlled marker generator—An oscillator circuit that uses a quartz crystal to set the frequency, and which has an output rich in harmonics that can be used to determine band edges on a receiver. An output every 100 kHz or less is normally produced.

Cubical quad antenna—An antenna built with its elements in the shape of four-sided loops.

Current—A flow of electrons in an electrical circuit.

Cutoff frequency—In a high-pass, low-pass, or band pass filter, the cutoff frequency is the frequency at which the filter output is reduced to 1/2 of the power available at the filter input.

D layer—The lowest layer of the ionosphere. The D layer contributes very little to short-wave radio propagation. It acts mainly to absorb energy from radio waves as they pass through it. This absorption has a significant effect on signals below about 7.5 MHz during daylight.

D'Arsonval meter movement—A type of meter movement in which a coil is suspended between the poles of a permanent magnet. DC flowing through the coil causes it to rotate an amount proportional to the current. A pointer attached to the coil indicates the amount of deflection on a scale.

Dash—The long sound used in Morse code. Pronounce this as "dah" when verbally sounding Morse code characters.

Decibel (dB)—The smallest change in sound level that can be detected by the human ear. Power gains and losses are also expressed in decibels. One tenth of a bel, denoting a logarithm of the ratio of two power levels. dB = 10 log (P2/P1)

Declination angle—The angle measured north or south from the celestial equator to an object in the sky.

Delta loop antenna—A variation of the cubical quad antenna with triangular elements.

Delta match—A method for impedance matching between an open-wire transmission line and a half-wave radiator that is not split at the center. The feed-line wires are fanned out to attach to the antenna wire symmetrically around the center point. The resulting connection looks somewhat like a capital Greek delta.

Depletion mode—Type of operation in a JFET or MOSFET where current is reduced by reverse bias on the gate.

Depletion region—An area around the semiconductor junction where the charge density is very small. This creates a potential barrier for current flow across the junction. In general, the region is thin when the junction is forward biased, and becomes thicker under reverse-bias conditions.

Descending pass—With respect to a particular ground station, a satellite pass during which the spacecraft is headed in a southerly direction while it is in range.

Desensitization—A reduction in receiver sensitivity caused by the receiver front end being overloaded by noise or RF from a local transmitter.

Detector—The stage in a receiver in which the modulation (voice or other information) is recovered from the RF signal.

Deviation—The peak difference between an instantaneous frequency of the modulated wave and the unmodulated-carrier frequency in an FM system.

Deviation ratio—The ratio of the maximum frequency deviation to the maximum modulating frequency in an FM system.

Dielectric—An insulating material. A dielectric is a medium in which it is possible to maintain an electric field with little or no additional energy supplied after the field has been established.

Dielectric constant—A property of insulating materials that serves as a measure of how much electric charge can be stored in the material with a given voltage.

Dielectric materials—Materials in which it is possible to maintain an electric field with little or no additional energy being supplied. Insulating materials or nonconductors.

Digipeater—A packet-radio station used to retransmit signals that are specifically addressed to be retransmitted by that station.

Digital communications—The term used to describe Amateur Radio transmissions that are designed to be received and printed automatically. The term also describes transmissions used for the direct transfer of information from one computer to another.

Digital IC—An integrated circuit whose output is either on (1) or off (0).

Dip meter—A tunable RF oscillator that supplies energy to another circuit resonant at the frequency that the oscillator is tuned to. A meter indicates when the most energy is being coupled out of the circuit by showing a dip in indicated current.

Dipole antenna—An antenna with two elements in a straight line that are fed in the center; literally, two poles. For amateur work, dipoles are usually operated at half-wave resonance.

Direct-conversion receiver—A receiver that converts an RF signal directly to an audio signal with one mixing stage.

Direct current (dc)—Electrical current that flows in one direction only.

Directive antenna—An antenna that concentrates the radiated energy to form one or more major lobes in specific directions. The receiving pattern is the same as the transmitting pattern.

Directivity—The ability of an antenna to focus transmitter power into a beam. Also its ability to enhance received signals from specific directions.

Director—A parasitic element located in front of the driven element of a beam antenna. It is intended to increase the strength of the signals radiated from the front of the antenna. Typically about 5% shorter than the driven element.

Direct sequence—A spread-spectrum communications system where a very fast binary bit stream is used to shift the phase of an RF carrier.

Direct waves—Radio waves that travel directly from a transmitting antenna to a receiving antenna. Also called "line-of-sight" communications.

Distributed capacitance—The capacitance that exists between turns of a coil.

Doping—The addition of impurities to a semiconductor material, with the intent to provide either excess electrons or positive charge carriers (holes) in the material.

Doppler shift—A change in the observed frequency of a signal, as compared with the transmitted frequency, caused by satellite movement toward or away from you.

Dot—The short sound used in Morse code. Pronounce this as "dit" when verbally sounding Morse code characters if the dot comes at the end of the character. If the dot comes at the beginning or in the middle of the character, pronounce it as "di."

Double-pole, double throw (DPDT) switch—A switch that has six contacts. The DPDT switch has two center contacts. The two center contacts can each be connected to one of two other contacts.

Doubly balanced mixer (DBM)—A mixer circuit that is balanced for both inputs, so that only the sum and the difference frequencies, but neither of the input frequencies, appear at the output. There will be no output unless both input signals are present.

Drain—The point at which the charge carriers exit an FET. Corresponds to the plate of a vacuum tube.

Drift—As related to op amps, the change of offset voltage with temperature changes, typically a few microvolts per degree Celsius.

Driven element—Any antenna element connected directly to the feed line.

Dual-trace oscilloscope—An oscilloscope with two separate vertical input circuits. This type of oscilloscope can be used to observe two waveforms at the same time.

Duct—A radio waveguide formed when a temperature inversion traps radio waves within a restricted layer of the atmosphere.

Dummy load (dummy antenna)—A resistor that provides a transmitter with a proper load. The resistor gets rid of the transmitter output power without radiating a signal. A dummy load is used when testing transmitters.

Duplexer—A device, usually employing cavities, to allow a transmitter and receiver to be connected simultaneously to one antenna. Most often, as in the case of a repeater, the transmitter and receiver operate at the same time on different frequencies.

DX—Distance, foreign countries.

DX Century Club (DXCC)—A prestigious award given to amateurs who can prove contact with amateurs in at least 100 DXCC countries.

Dynamic (edge-triggered) input—An input, such as for a flip-flop, that is sampled only when the rising (or falling) edge of a clock pulse is detected.

Dynamic range—A measure of receiver performance. The difference, in decibels, between the minimum usable signal level and the maximum signal that does not produce distortion in the audio output.

E layer—The second lowest ionospheric layer, the E layer exists only during the day, and under certain conditions may refract radio waves enough to return them to earth.

Earth ground—A circuit connection to a cold-water pipe or to a ground rod driven into the earth.

Earth-moon-earth (EME)—A method of communicating with other stations by reflecting radio signals off the moon's surface.

Earth operation—Earth-to-space-to-earth Amateur Radio communication by means of radio signals automatically retransmitted by stations in space operation.

Effective isotropic radiated power (EIRP)—A measure of the power radiated from an antenna system. EIRP takes into account transmitter output power, feed-line losses and other system losses, and antenna gain as compared to an isotropic radiator.

Effective radiated power (ERP)—The relative amount of power radiated in a specific direction from an antenna, taking system gains and losses into account.

Effective voltage—The value of a dc voltage that will heat a resistive component to the same temperature as the ac voltage that is being measured.

Electric field—A region through which an electric force will act on an electrically charged object. An electric field exists in a region of space if an electrically charged object placed in the region is subjected to an electrical force.

Electric force—A push or pull exerted through space by one electrically charged object on another.

Electrolytic capacitor—A polarized capacitor formed by using thin foil electrodes and chemical-soaked paper.

Electromagnetic radiation—Another term for electromagnetic waves, consisting of an electric field and a magnetic field that are at right angles to each other.

Electromagnetic waves—A disturbance moving through space or materials in the form of changing electric and magnetic fields.

Electromotive force (EMF)—The force or pressure that pushes a current through a circuit.

Electron—A tiny, negatively charged particle, normally found in an area surrounding the nucleus of an atom. Moving electrons make up an electrical current.

Electronic keyer—A device used to generate Morse code dots and dashes electronically. One input generates dots, the other, dashes. Character speed is usually adjusted by turning a control knob. Speeds range from 5 or 10 words per minute up to 60 or more.

Elliptical filter—A filter with equal-amplitude passband ripple and points of infinite attenuation in the stop band. The design is based on an elliptical function to calculate the input/output characteristics.

Emission—The transmitted signal from an Amateur Radio station.

Emission designator—A symbol made up of two letters and a number, used to describe a radio signal. A3E is the designator for double-sideband, full-carrier, amplitude-modulated telephony.

Emission privilege—Permission to use a particular emission type (such as Morse code or voice).

Energy—The ability to do work; the ability to exert a force to move some object.

Enhancement mode—Type of operation in a MOSFET where current is increased by forward bias on the gate.

Equinoxes—One of two spots on the earth's orbital path around the sun at which it crosses a horizontal plane extending through the center of the sun. The vernal equinox marks the beginning of spring and the autumnal equinox marks the beginning of autumn.

Exclusive OR gate—A logic circuit whose output is 1 when either of two inputs is 1 and whose output is 0 when neither input is 1 or when both inputs are 1.

F1B emission—The FCC emission designator used to describe frequency-shift keyed (FSK) digital communications.

F2B emission—The FCC emission designator used to describe audio-frequency shift keyed (AFSK) digital communications.

F3E emission—The FCC emission designator used to describe FM voice communications.

F layer—A combination of the two highest ionospheric layers, the F1 and F2 layers. The F layer refracts radio waves and returns them to earth. The height of the F layer varies greatly depending on the time of day, season of the year and amount of sunspot activity.

Facsimile (fax)—The process of scanning pictures or images and converting the information into signals that can be used to form a likeness of the copy in another location.

False or deceptive signals—Transmissions that are intended to mislead or confuse those who may receive the transmissions. For example, distress calls transmitted when there is no actual emergency are false or deceptive signals.

Farad—The basic unit of capacitance.

Faraday rotation—A rotation of the polarization plane of radio waves when the waves travel through the ionized magnetic field of the ionosphere.

Fast-scan TV (ATV)—A television system used by amateurs that employs the same video-signal standards as commercial TV.

Feed line (feeder)—The wires or cable used to connect your transmitter and receiver to an antenna. Also see Transmission line.

Ferrite loop (loopstick)—A loop antenna wound on a ferrite rod to increase the magnetic flux. Also called a loopstick antenna.

Field—The region of space through which any of the invisible forces in nature, such as gravity, electric force or magnetic forces act.

Field Day—An annual event in which amateurs set up stations in outdoor locations. Emergency power is also encouraged.

Field-effect transistor (FET)—A voltage-controlled semiconductor device. Output current can be varied by varying the input voltage. The input impedance of an FET is very high.

Field-effect transistor volt-ohm-milliammeter (FET VOM)—A multiple-range meter used to measure voltage, current and resistance. The meter circuit uses an FET amplifier to provide a high input impedance. This leads to more accurate readings than can be obtained with a VOM. This is the solid-state equivalent of a VTVM.

Field-strength meter—A simple test instrument used to show the presence of RF energy and the relative strength of the RF field.

Fills—Repeats of parts of a previous transmission—usually requested because of interference.

Filter—A circuit that will allow some signals to pass through it but will greatly reduce the strength of others. In a power supply circuit, a filter smooths the ac ripple.

Filter choke—A large-value inductor used in a power-supply filter to smooth out the ripple in the pulsating dc output from the rectifier.

Fist—The unique rhythm of an individual amateur's Morse code sending.

Fixed resistor—A resistor with a nonadjustable value of resistance.

Flip-flop (bistable multivibrator)—A circuit that has two stable output states, and which can change from one state to the other when the proper input signals are detected.

Flying-spot scanner (FSS)—A device that uses a moving light spot to scan a page. The intensity of the reflected light is sensed by a photoelectric cell, generating a signal that contains information about what is on the page.

Folded dipole—An antenna consisting of two (or more) parallel, closely spaced half-wave wires connected at their ends. One of the wires is fed at its center.

Forward bias—A voltage applied across a semiconductor junction so that it will tend to produce current.

Forward biased—A condition in which the anode of a diode is positive with respect to the cathode, so it will conduct.

Forward Error Correction (FEC)—A mode of AMTOR communication. In FEC mode, also called Mode B, each character is sent twice. The receiving station checks the mark/space ratio of the received characters. If an error is detected, the receiving station prints a space to show that an incorrect character was received.

Frequency—The number of cycles of an alternating current that occur per second.

Frequency bands—A group of frequencies where amateur communications are authorized.

Frequency coordinator—A volunteer who keeps records of repeater input and output frequencies and recommends frequencies to amateurs who wish to put new repeaters on the air.

Frequency counter—A digital-electronic device that counts the cycles of an electromagnetic wave for a certain amount of time and gives a digital readout of the frequency.

Frequency deviation—The amount the carrier frequency in an FM transmitter changes as it is modulated.

Frequency discriminator—A circuit used to recover the audio from an FM signal. The output amplitude depends on the deviation of the received signal from a center (carrier) frequency.

Frequency domain—A time-independent way to view a complex signal. The various component sine waves that make up a complex waveform are shown by frequency and amplitude on a graph or the CRT display of a spectrum analyzer.

Frequency hopping—A spread-spectrum communications system where the center frequency of a conventional carrier is altered many times a second in accordance with a pseudo-random list of channels.

Frequency modulation (FM)—The process of varying the frequency of an RF carrier in response to the instantaneous changes in a modulating signal. The signal that modulates the carrier frequency may be audio, video, digital data or some other kind of information.

Frequency privilege—Permission to use a particular group of frequencies.

Frequency resolution—The space between markings on a receiver dial. The greater the frequency resolution, the easier it is to separate signals that are close together. Note that frequency resolution affects band spread.

Frequency-shift keying (FSK)—A method of transmitting radioteletype information by switching an RF carrier between two separate frequencies. This is the RTTY mode most often used on the HF amateur bands.

Front-to-back ratio—The energy radiated from the front of a directional antenna divided by the energy radiated from the back of the antenna.

Full-break-in (QSK)—With QSK, an amateur can hear signals between code characters. This allows another amateur to break into the communication without waiting for the transmitting station to finish.

Full-wave bridge rectifier—A full-wave rectifier circuit that uses four diodes and does not require a center-tapped transformer.

Full-wave rectifier—A circuit composed of two half-wave rectifiers. The full-wave rectifier allows the full ac waveform to pass through; one half of the cycle is reversed in polarity. This circuit requires a center-tapped transformer.

Fuse—A thin strip of metal mounted in a holder. When too much current passes through the fuse, the metal strip melts and opens the circuit.

GaAsFET—Gallium-arsenide field-effect transistor. A low-noise device used in UHF and higher frequency amplifiers.

Gain (Antenna)—An increase in the effective power radiated by an antenna in a certain desired direction, or an increase in received signal strength from a certain direction. This is at the expense of power radiated in, or signal strength received from, other directions.

Galactic plane—An imaginary plane surface extending through the center of the Milky Way galaxy. This plane is used as a reference for some astronomical observations.

Gamma match—A method for matching the impedance of a feed line to a half-wave radiator that is split in the center (such as a dipole). It consists of an adjustable arm that is mounted close to the driven element and in parallel with it near the feed point. The connection looks somewhat like a capital Greek gamma.

Gate—Control terminal of an FET. Corresponds to the grid of a vacuum tube. Gate also refers to a combinational logic element with two or more inputs and one output. The output state depends upon the state of the inputs.

General-coverage receiver—A receiver used to listen to both the shortwave-broadcast frequencies and the amateur bands.

Geomagnetic disturbance—A dramatic change in the earth's magnetic field that occurs over a short period of time.

Giga—The metric prefix for 10^9, or times 1,000,000,000.

Gray line—A band around the earth that is the transition region between daylight and darkness.

Gray-line propagation—A generally north-south enhancement of propagation that occurs along the gray line, when D layer absorption is rapidly decreasing at sunset, or has not yet built up around sunrise.

Gray scale—A photographic term that defines a series of neutral densities (based on the percentage of incident light that is reflected from a surface), ranging from white to black.

Great circle—An imaginary circle around the surface of the earth formed by the intersection of the surface with a plane passing through the center of the earth.

Great-circle path—Either one of two direct paths between two points on the surface of the earth. One of the great-circle paths is the shortest distance between those two points. Great-circle paths can be visualized if you think of a globe with a rubber band stretched around it connecting the two points.

Greenwich Hour Angle (GHA)—The angle measured east or west from the prime meridian to an object in the sky.

Grid—The control element (or elements) in a vacuum tube.

Ground connection—A connection made to the earth for electrical safety.

Ground-plane antenna—A vertical antenna built with a central radiating element one-quarter-wavelength long and several radials extending horizontally from the base. The radials are slightly longer than one-quarter wave, and may droop toward the ground.

Ground waves—Radio waves that travel along the surface of the earth.

Ground-wave signals—Radio signals that are propagated along the ground rather than through the ionosphere or by some other means.

Guided propagation—Radio propagation by means of ducts.

Half-power points—Those points on the response curve of a resonant circuit where the power is one half its value at resonance.

Half section—A basic L-section building block of image-parameter filters.

Half-wavelength dipole antenna—A fundamental antenna one-half wavelength long at the desired operating frequency, and connected to the feed line at the center. This is a popular amateur antenna.

Half-wave rectifier—A circuit that allows only half of the applied ac waveform to pass through it.

Ham-bands-only receiver—A receiver designed to cover only the bands used by amateurs.

Hand key—A simple switch used to send Morse code.

Heat sink—A piece of metal used to absorb excess heat generated by a component, and dissipate the heat into the air.

Henry—The basic unit of inductance.

Hertz (Hz)—An alternating-current frequency of one cycle per second. The basic unit of frequency.

High-pass filter—A filter that allows signals above the cutoff frequency to pass through. It attenuates signals below the cutoff frequency.

Horizontally polarized wave—An electromagnetic wave with its electric lines of force parallel to the ground.

Horizontal polarization—Describes an electromagnetic wave in which the electric field is horizontal, or parallel to the earth's surface.

Horizontal synchronization pulse—Part of a TV signal used by the receiver to keep the CRT electron-beam scan in step with the camera scanning beam. This pulse is transmitted at the beginning of each horizontal scan line.

Hot-carrier diode—A type of diode in which a small metal dot is placed on a single semiconductor layer. It is superior to a point-contact diode in most respects.

Image-parameter technique—A filter-design method that uses image impedance and other fundamental network functions to approximate the desired characteristics.

Imaginary number—A value that sometimes comes up in solving a mathematical problem, equal to the square root of a negative number. Since there is no real number that can be multiplied by itself to give a negative result, this quantity is imaginary. In electronics work, the symbol *j* is used to represent $\sqrt{-1}$ since *i*, used in mathematics, is already used for current in electronics. Other imaginary numbers are represented by $j\sqrt{x}$, where x is the positive part of the number. The reactance and susceptance of complex impedances and admittances are normally given in terms of *j*.

Impedance—A term used to describe a combination of reactance and resistance in a circuit.

Impedance-matching network—A device that matches the impedance of an antenna system to the impedance of a transmitter or receiver. Also called an antenna-matching network or Transmatch.

Inclination—The angle at which a satellite crosses the equator at an ascending node. Inclination is also equal to the highest latitude reached in an orbit.

Induced EMF—A voltage produced by a change in magnetic lines of force around a conductor. When a magnetic field is formed by current in the conductor, the induced voltage always opposes changes in that current.

Inductive coupling (of a dip meter)—A method of transferring energy from a dip-meter oscillator to a tuned circuit by means of a magnetic field between two coils.

Inductor—An electrical component usually composed of a coil of wire wound on a central core. An inductor stores energy in a magnetic field.

Input frequency—A repeater's receiving frequency.

Insulator—A material that maintains a tight grip on its electrons, so that an electrical current cannot pass through it.

Integrated circuit—A device composed of many bipolar or field-effect transistors manufactured on the same chip, or wafer, of silicon.

Intercept point—As related to receiver performance, it is the input (or output) signal level at which the desired output power and the distortion product power have the same value.

Intermediate frequency (IF)—The output frequency of a mixing stage in a superheterodyne receiver. The subsequent stages in the receiver are tuned for maximum efficiency at the IF.

Intermodulation distortion (IMD)—A type of interference that results from the mixing of integer multiples of signal frequencies in a nonlinear stage or device. The resulting mixing products can interfere with desired signals on the mixed frequencies.

International Telecommunication Union (ITU)—The international body that regulates the use of the radio spectrum.

Inverted-V dipole—A half-wave dipole antenna with its center elevated and the ends drooping toward the ground. Amateurs sometimes call this antenna an "inverted V."

Inverter—A logic circuit with one input and one output. The output is 1 when the input is 0, and the output is 0 when the input is 1.

Ion—An electrically charged particle. An electron is an ion. Another example of an ion is the nucleus of an atom that is surrounded by too few or too many electrons. An atom like this has a net positive or negative charge.

Ionosphere—A region in the atmosphere about 30 to 260 miles above the earth. The ionosphere is made up of charged particles, or ions. The ionosphere bends radio waves as they travel through it, returning them to earth.

Ion trap—A magnet installed in a cathode-ray tube to prevent negative ions from burning a brown spot on the center of the CRT.

Isolator—A passive attenuator in which the loss in one direction is much greater than the loss in the other.

Isotropic radiator—An imaginary antenna in free space that radiates equally in all directions (a spherical radiation pattern). It is used as a reference to compare the gain of various real antennas.

J3E emission—The FCC emission designator used to describe single-sideband suppressed-carrier voice communications.

Joule—The unit of energy in the metric system of measure.

Junction diode—An electronic component formed by placing a layer of N-type semiconductor material next to a layer of P-type material. Diodes allow current to flow in one direction only.

Junction field-effect transistor (JFET)—A field-effect transistor created by diffusing a gate of one type of semiconductor material into a channel of the opposite type of semiconductor material.

K index—A geomagnetic-field measurement that is updated every few hours. The K index can be used to indicate HF propagation conditions.

Kepler's Laws—Three laws of planetary motion, used to mathematically describe satellite-orbit parameters.

Kilo—The metric prefix for 10^3, or times 1000.

L network—A combination of a capacitor and an inductor, one of which is connected in series with the signal lead while the other is shunted to ground.

Ladder line—Parallel-conductor feeder with insulating spacer rods every few inches.

Latch—Another name for a bistable multivibrator flip-flop circuit. The term latch reminds us that this circuit serves as a memory unit, storing a bit of information.

Libration fading—A fluttery, rapid fading of EME signals, caused by short-term motion of the moon's surface relative to an observer on earth.

Light-emitting diode—A device that uses a semiconductor junction to produce light when current flows through it.

Line of sight—The term used to describe VHF and UHF propagation in a straight line directly from one station to another.

Linear electronic voltage regulator—A type of voltage-regulator circuit that varies either the current through a fixed dropping resistor or the resistance of the dropping element itself.

Linear or plane polarization—Describes the orientation of the electric-field component of an electromagnetic wave. The electric field can be vertical or horizontal with respect to the earth's surface, resulting in either a vertically or a horizontally polarized wave.

Linear IC—An integrated circuit whose output voltage is a linear (straight line) representation of its input voltage.

Lissajous figure—An oscilloscope pattern obtained by connecting one sine wave to the vertical amplifier and another sine wave to the horizontal amplifier. The two signals must be harmonically related to produce a stable pattern.

Loading coil—An inductor that is inserted in an antenna element or transmission line for the purpose of producing a resonant system at a specific frequency.

Logic probe—A simple piece of test equipment used to indicate high or low logic states (voltage levels) in digital-electronic circuits.

Long-path communication—Communication made by pointing beam antennas in the directions indicated by the longer great-circle path between the stations. To work each other by long-path, an amateur in Hawaii would point his antenna west and an amateur in Florida would aim east.

Long-path propagation—Propagation between two points on the earth's surface that follows a path along the great circle between them, but is in a direction opposite from the shortest distance between them.

Loop antenna—An antenna configured in the shape of a loop. If the current in the loop, or in multiple parallel turns, is essentially uniform, and if the loop circumference is small compared with a wavelength, the radiation pattern is symmetrical, with maximum response in either direction of the loop plane.

Lower sideband (LSB)—The common mode of single-sideband transmission used on the 40, 80 and 160-meter amateur bands.

Low-pass filter—A filter that allows signals below the cutoff frequency to pass through and attenuates signals above the cutoff frequency.

M-derived filter—A filter, designed using image-parameter techniques, that has a sharper transition from pass band to stop band than a constant-k type. This filter has one infinite attenuation frequency, which can be placed at a desired frequency by proper design.

Magnetic field—A region through which a magnetic force will act on a magnetic object.

Magnetic force—A push or pull exerted through space by one magnetically charged object on another.

Major lobe—The shape or pattern of field strength that points in the direction of maximum radiated power from an antenna.

Major lobe of radiation—A three-dimensional area in the space around an antenna that contains the maximum radiation peak. The field strength decreases from the peak level, until a point is reached where it starts to increase again. This area is known as the major lobe.

Malicious interference—Intentional, deliberate obstruction of radio transmissions.

Marker generator—An RF signal generator that produces signals at known frequency intervals. The marker generator can be used to calibrate receiver and transmitter frequency readouts.

Matching network—A device that matches one impedance level to another. For example, it may match the impedance of an antenna system to the impedance of a transmitter or receiver. Amateurs also call such devices a Transmatch, impedance-matching network, or match box.

Matching stub—A section of transmission line used to tune an antenna element to resonance or to aid in obtaining an impedance match between the feed point and the feed line.

Maximum average forward current—the highest average forward current that can flow through a diode for a given junction temperature.

Maximum usable frequency (MUF)—The greatest frequency at which radio signals will return to a particular location from the ionosphere. The MUF may vary for radio signals sent to different destinations.

MAYDAY—From the French "m'aider" (help me), MAYDAY is used when calling for emergency assistance in voice modes.

Mechanical filter—A nonelectrical filter that uses mechanically resonant disks at the design frequency, and a pair of electromechanical transducers to change the electrical signal into a mechanical wave and back again.

Mega—The metric prefix for 10^6, or times 1,000,000.

Metal-film resistor—A resistor formed by depositing a thin layer of resistive-metal alloy on a cylindrical ceramic form.

Metal-oxide semiconductor FET (MOSFET)—A field-effect transistor that has its gate insulated from the channel material. Also called an IGFET, or insulated gate FET.

Meteor—A particle of mineral or metallic material that is in a highly elliptical orbit around the sun. As the earth's orbit crosses the orbital path of a meteor, it is attracted by the earth's gravitational field, and enters the atmosphere. A typical meteor is about the size of a grain of sand.

Meteor-scatter communication—A method of radio communication that uses the ionized trail of a meteor, which has burned up in the earth's atmosphere, to reflect radio signals back to earth.

Metric prefixes—A series of terms used in the metric system of measurement. We use metric prefixes to describe a quantity as compared to a basic unit. The metric prefixes indicate multiples of 10.

Metric system—A system of measurement developed by scientists and used in most countries of the world. This system uses a set of prefixes that are multiples of 10 to indicate quantities larger or smaller than the basic unit.

Mica capacitor—A capacitor formed by alternating layers of metal foil with thin sheets of insulating mica.

Micro—The metric prefix for 10^{-6}, or divide by 1,000,000.

Microphone—A device that converts sound waves into electrical energy.

Milli—The metric prefix for 10^{-3}, or divide by 1000.

Minimum discernible signal (MDS)—The smallest input-signal level that can just be detected by a receiver.

Minor lobe of radiation—Those areas of an antenna pattern where there is some increase in radiation, but not as much as in the major lobe. Minor lobes normally appear at the back and sides of the antenna.

Mixer—A circuit that takes two or more input signals, and produces an output that includes the sum and difference of those signal frequencies.

Mobile operation—Amateur Radio operation conducted while in motion or at temporary stops at different locations.

Modem—Short for modulator/demodulator. A modem modulates a radio signal to transmit data and demodulates a received signal to recover transmitted data.

Modulate—To vary the amplitude, frequency, or phase of a radio-frequency signal.

Modulation—The process of varying some characteristic (amplitude, frequency or phase) of an RF carrier for the purpose of conveying information.

Modulation index—The ratio of the maximum carrier frequency deviation to the frequency of the modulating signal at a given instant in an FM transmitter.

Modulator—A circuit designed to superimpose an information signal on an RF carrier wave.

Monitor mode—One type of packet-radio receiving mode. In monitor mode, everything transmitted on a packet frequency is displayed by the monitoring TNC. This occurs whether the transmissions are addressed to the monitoring station or not.

Monitor oscilloscope—A test instrument connected to an amateur transmitter and used to observe the shape of the transmitted-signal waveform.

Monostable multivibrator (one shot)—A circuit that has one stable state. It can be forced into an unstable state for a time determined by external components, but it will revert to the stable state after that time.

Moonbounce—A common name for EME communication.

Multiband antenna—An antenna that will operate well on more than one frequency band.

Multimeter—An electronic test instrument used to make basic measurements of current and voltage in a circuit. This term is used to describe all meters capable of making different types of measurements, such as the VOM, VTVM and FET VOM.

Multimode transceiver—A VHF or UHF transceiver capable of SSB, CW and FM operation.

Multipath—A fading effect caused by the transmitted signal traveling to the receiving station over more than one path.

Mutual coupling—When coils display mutual coupling, a current flowing in one coil will induce a voltage in the other. The magnetic flux of one coil passes through the windings of the other.

Mutual inductance—The ability of one coil to induce a voltage in another coil. Any two coils positioned so that the magnetic flux from one coil cuts through the turns of wire in the other are said to have mutual inductance.

N-type material—Semiconductor material that has been treated with impurities to give it an excess of electrons. We call this a "donor material."

NAND (NOT AND) gate—A logic circuit whose output is 0 only when both inputs are 1.

Negative charge—One of two types of electrical charge. The electrical charge of a single electron.

Negative-format modulation—A fax system that uses a minimum modulation level to produce white and a maximum modulation level to produce black.

Neon lamp—A cold-cathode (no heater or filament), gas-filled tube used to give a visual indication of voltage in a circuit, or of an RF field.

Nets—Groups of amateurs who meet on the air to pass traffic or communicate about a specific subject. One station (called the net control station) usually directs the net.

Network—A term used to describe several packet stations linked together to transmit data over long distances.

Neutral—Having no electrical charge, or having an equal number of positive and negative charges.

Neutralization—Feeding part of the output signal from an amplifier back to the input so it arrives out of phase with the input. This negative feedback neutralizes the effect of positive feedback caused by coupling between the input and output circuits in the amplifier.

Night effect—A special type of error in a radio direction-finding system, occurring mainly at night, when sky-wave propagation is most likely.

Node—A point where a satellite crosses the plane passing through the earth's equator. It is an ascending node if the satellite is moving from south to north, and a descending node if the satellite is moving from north to south.

Noise figure—A measurement used to characterize receiver sensitivity.

Noninverting buffer—A logic circuit with one input and one output, and whose output state is the same as the input state (Ø or 1).

Nonresonant rhombic antenna—A diamond-shaped antenna consisting of sides that are each at least one wavelength long. The feed line is connected to one end of the diamond, and there is a terminating resistance of approximately 800 ohms at the opposite end. The antenna has a unidirectional radiation pattern.

NOR (NOT OR) gate—A logic circuit whose output is 0 if either input is 1.

Nucleus—The dense central portion of an atom. The nucleus contains positively charged particles.

Offset—The difference between a repeater's input and output frequencies. On 2 meters, for example, the offset is either plus 600 kilohertz (kHz) or minus 600 kHz from the receive frequency. Also the slight difference in transmitting and receiving frequencies in a transceiver.

Offset voltage—As related to op amps, the voltage needed between the input terminals to produce Ø-V ouput. Ideally this is Ø Volts.

Ohm—The basic unit of electrical resistance, used to describe the amount of opposition to current.

Ohm's Law—A basic law of electronics. Ohm's Law gives a relationship between voltage, resistance and current ($E = IR$).

Omnidirectional—Antenna characteristic meaning it radiates equal power in all compass directions.

Open circuit—An electrical circuit that does not have a complete path, so current can't flow through the circuit.

Open-wire feed line—Parallel-conductor feeder with air as its primary insulation material.

Operational amplifier (op amp)—A linear IC that can amplify dc as well as ac. Op amps have very high input impedance, very low output impedance and very high gain.

Operator license—The portion of an Amateur Radio license that gives permission to operate an Amateur Radio station.

Open Systems Interconnection Reference Model (OSI-RM)—A seven-level computer interconnection standard developed by the International Standards Organization (ISO). OSI-RM describes the interconnection of stations in a packet-radio system.

Optical shaft encoder—A device consisting of two pairs of photoemitters and photodetectors, used to sense the rotation speed and direction of a knob or dial. Optical shaft encoders are often used with the tuning knob on a modern radio to provide a tuning signal for the microprocessor controlling the frequency synthesizer.

Optocoupler (optoisolator)—A device consisting of a photoemitter and a photodetector used to transfer a signal between circuits using widely varying operating voltages.

Orbital period—The time it takes for a complete orbit, usually measured from one equator crossing to the next. The higher the altitude, the longer the period.

OR gate—A logic circuit whose output is 1 when either input is 1.

Oscillator—A circuit built by adding positive feedback to an amplifier. It produces an alternating current signal with no input except the dc operating voltages.

Oscilloscope—A device using a cathode-ray tube to display the waveform of an electric signal with respect to time or as compared with another signal.

Output frequency—A repeater's transmitting frequency.

P-type material—Semiconductor material that has been treated with impurities to give it an electron shortage. This creates excess positive charge carriers, or "holes," so it becomes an "acceptor material."

Packet Bulletin-Board System (PBBS)—A computer system used to store packet-radio messages for later retrieval by other amateurs.

Packet radio—A system of digital communication whereby information is broken into short bursts. The bursts ("packets") also contain addressing and error-detection information.

Paper capacitor—A capacitor formed by sandwiching paper between thin foil plates, and rolling the entire unit into a cylinder.

Parabolic (dish) antenna—An antenna reflector that is a portion of a parabolic curve. Used mainly at UHF and higher frequencies to obtain high gain and narrow beamwidth when excited by one of a variety of low-gain antennas placed at the dish focus to illuminate the reflector.

Parallel circuit—An electrical circuit in which the electrons follow more than one path in going from the negative supply terminal to the positive terminal.

Parallel-conductor feed line—Feed line constructed of two wires held a constant distance apart. They may be encased in plastic or constructed with insulating spacers placed at intervals along the line.

Parallel-resonant circuit—A circuit including a capacitor, an inductor and sometimes a resistor, connected in parallel, and in which the inductive and capacitive reactances are equal at the applied-signal frequency. The circuit impedance is a maximum, and the current through the circuit is a minimum at the resonant frequency.

Parasitic element—Part of a directive antenna that derives energy from mutual coupling with the driven element. Parasitic elements are not connected directly to the feed line.

Parasitics—Undesired oscillations or other responses in an amplifier.

Path loss—The total signal loss between transmitting and receiving stations relative to the total radiated signal energy.

Peak envelope power (PEP)—The average power of the RF cycle having the greatest amplitude. (This occurs during a modulation peak.)

Peak envelope voltage (PEV)—The maximum peak voltage occurring in a complex waveform.

Peak inverse voltage—The maximum instantaneous anode-to-cathode reverse voltage that is to be applied to a diode.

Peak negative value—On a signal waveform, the maximum displacement from the zero line in the negative direction.

Peak positive value—On a signal waveform, the maximum displacement from the zero line in the positive direction.

Peak power—The product of peak voltage and peak current in a resistive circuit. (Used with sine-wave signals.)

Peak-to-peak (P-P) voltage—A measure of the voltage taken between the negative and positive peaks on a cycle.

Peak voltage—A measure of voltage on an ac waveform taken from the centerline (0 V) and the maximum positive or negative level.

Perigee—That point in the orbit of a satellite (such as the moon) when it is closest to the earth.

Period, T—The time it takes to complete one cycle of an ac waveform.

Phase—A representation of the relative time or space between two points on a waveform, or between related points on different waveforms. The time interval between one event and another in a regularly recurring cycle.

Phase angle—If one complete cycle of a waveform is divided into 360 equal parts, then the phase relationship between two points or two waves can be expressed as an angle.

Phase-locked loop (PLL)—A servo loop consisting of a phase detector, low-pass filter, dc amplifier and voltage-controlled oscillator.

Phase modulation—A method of superimposing an information signal on an RF carrier wave in which the phase of an RF carrier wave is varied in relation to the information signal strength.

Phone—Voice communications.

Photocell—A solid-state device in which the voltage and current-conducting characteristics change as the amount of light striking the cell changes.

Photoconductive effect—A result of the photoelectric effect that shows up as an increase in the electric conductivity of a material. Many semiconductor materials exhibit a significant increase in conductance when electromagnetic radiation strikes them.

Photodetector—A device that produces an amplified signal that changes with the amount of light-sensitive surface.

Photoelectric effect—An interaction between electromagnetic radiation and matter resulting in photons of radiation being absorbed and electrons being knocked loose from the atom by this energy.

Phototransistor—A bipolar transistor constructed so the base-emitter junction is exposed to incident light. When light strikes this surface, current is generated at the junction, and this current is then amplified by transistor action.

Pico—The metric prefix for 10^{-12}, or divide by 1,000,000,000,000.

Pilot tone—In an ACSSB system, a 3.1-kHz tone transmitted with the voice signal to allow a mobile receiver to lock onto the signal. The pilot tone is also used to control the inverse logarithmic amplifier gain.

PIN diode—A diode consisting of a relatively thick layer of nearly pure semiconductor material with a layer of P-type material on one side and a layer of N-type material on the other.

Pi network output-coupling circuits—A combination of two like reactances (coil or capacitor) and one of the opposite type. The single component is connected in series with the signal lead and the two others are shunted to ground, one on either side of the series element.

Plastic-film capacitor—A capacitor formed by sandwiching thin sheets of mylar or polystyrene between thin foil plates, and rolling the entire unit into a cylinder.

PN-junction—The contact area between two layers of opposite-type semiconductor material.

Point-contact diode—A diode that is made by a pressure contact between a semiconductor material and a metal point.

Polarization—A property of an electromagnetic wave that tells whether the electric field of the wave is oriented vertically or horizontally. The polarization sense can change from vertical to horizontal under some conditions, and even can be gradually rotating either in a clockwise (right-hand-circular polarization) or a counterclockwise (left-hand-circular polarization) direction.

Polar-coordinate system—A method of representing the position of a point on a plane by specifying the radial distance from an origin, and an angle measured counterclockwise from the 0° line.

Portable operation—Amateur Radio operation conducted away from the location shown on the station license.

Positive charge—One of two types of electrical charge. A positive charge is the opposite of a negative charge. Electrons have a negative charge. The nucleus of an atom has a positive charge.

Positive-format modulation—A FAX system that uses a maximum modulation level to produce black.

Potentiometer—Another name for a variable resistor. The value of a potentiometer can be varied over a range of values without removing it from a circuit.

Power—The time rate of transferring energy, or the time rate at which work is done. In an electric circuit, power is found by multiplying the voltage applied to the circuit by the current through the circuit.

Power factor—The ratio of real power to apparent power in a circuit. Also calculated as the cosine of the phase angle between current and voltage in a circuit.

Power supply—That part of an electrical circuit that provides excess electrons to flow into a circuit. The power supply also supplies the voltage or EMF to push the electrons along. Power supplies convert a power source (such as the ac mains) to a useful form. (A circuit that provides a direct-current output at some desired voltage from an ac input voltage.)

Prescaler—A divider circuit used to increase the useful range of a frequency counter.

Primary winding—The coil in a transformer that is connected to the energy source.

Procedural signal (prosign)—One or two letters sent as a single character. Amateurs use prosigns in CW (Morse Code) QSOs as a short way to indicate the operator's intention. Some examples are "K" for "Go Ahead," or "$\overline{\text{AR}}$" for "End of Message." (The bar over the letters indicates that we send the prosign as one character.)

Product detector—A detector circuit whose output is equal to the product of a beat-frequency oscillator (BFO) and the modulated RF signal applied to it.

Propagation—The means by which radio waves travel from one place to another.

Pseudonoise (PN)—A binary sequence designed to appear to be random (contain an approximately equal number of ones and zeros). Pseudonoise is generated by a digital circuit and mixed with digital information to produce a direct-sequence spread-spectrum signal.

Pulsating dc—The output from a rectifier before it is filtered. The polarity of a pulsating dc source does not change, but the amplitude of the voltage changes with time.

Pulse modulation—Modulation of an RF carrier by a series of pulses. These pulses convey the information that has been sampled from an analog signal.

Pulse-amplitude modulation (PAM)—A pulse-modulation system where the amplitude of a standard pulse is varied in relation to the information-signal amplitude at any instant.

Pulse-code modulation (PCM)—A modulation process where an information waveform is converted from analog to digital form by sampling the information signal and then encoding the sample value in some digital form (such as binary-coded decimal).

Pulse-position modulation (PPM)—A pulse-modulation system where the position (timing) of the pulses is varied from a standard value in relation to the information-signal amplitude at any instant.

Pulse-width modulation (PWM)—A pulse-modulation system where the width of a pulse is varied from a standard value in relation to the information-signal amplitude at any instant.

Q—A quality factor describing how closely a practical coil or capacitor approaches the characteristics of an ideal component.

Q signals—Three-letter symbols beginning with "Q." Q signals are used in amateur CW work to save time and for better communication.

QSL card—A postcard sent to another radio amateur to confirm a contact.

QSO—A conversation between two radio amateurs.

Quarter-wavelength vertical antenna—An antenna constructed of a quarter-wavelength-long radiating element placed perpendicular to the earth. (See Ground-plane antenna.)

Radiate—To convert electric energy into electromagnetic (radio) waves. Radio waves radiate from an antenna.

Radiated noise—Usually referring to a mobile installation, noise that is being radiated from the ignition system or electrical system of a vehicle and causing interference to the reception of radio signals.

Radiation resistance—The equivalent resistance that would dissipate the same amount of power as is radiated from an antenna. It is calculated by dividing the radiated power by the square of the RMS antenna current.

Radio frequency (RF)—The range of frequencies that can be radiated through space in the form of electromagnetic radiation. We usually consider RF to be those frequencies higher than the audio frequencies, or above 20 kilohertz.

Radio-frequency interference (RFI)—Interference to an electronic device (radio, TV, stereo) caused by RF energy from an amateur transmitter or other source.

Radio horizon—The position at which a direct wave radiated from an antenna becomes tangent to the earth's surface. Note that as the wave continues past the horizon, the wave gets higher and higher above the surface.

Radio-path horizon—The point where radio waves are returned by tropospheric bending. The radio-path horizon is 15 percent farther away than the geometric horizon.

Radioteletype (RTTY)—Radio signals sent from one teleprinter machine to another machine. Anything that one operator types on his teleprinter will be printed on the other machine. (A type of digital communications.)

Ragchew—A lengthy conversation (or QSO) between two radio amateurs.

Random-length wire antenna—An antenna having a length that is not necessarily related to the wavelength of a desired signal.

Ratio detector—A circuit used to demodulate FM signals. The output is the ratio of voltages from either side of a discriminator-transformer secondary.

Reactance—The property of an inductor or capacitor (measured in ohms) that opposes current in an ac circuit without converting power to heat.

Reactance modulator—A device capable of modulating an ac signal by varying the reactance of a circuit in response to the modulating signal. (The modulating signal may be voice, data, video or some other kind, depending on what type of information is being transmitted.) The circuit capacitance or inductance changes in response to an audio input signal.

Reactive power—The apparent power in an inductor or capacitor. The product of RMS current through a reactive component and the RMS voltage across it. Also called wattless power.

Real power—The actual power dissipated in a circuit, calculated to be the product of the apparent power times the phase angle between the voltage and current.

Receiver—A device that converts radio signals into audio signals.

Receiver incremental tuning (RIT)—A transceiver control that allows for a slight change in the receiver frequency without changing the transmitter frequency. Some manufacturers call this a Clarifier (CLAR) control.

Rectangular coordinates—A graphical system used to represent length and direction of physical quantities for the purpose of finding other unknown quantities.

Rectangular-coordinate system—A method of representing the position of a point on a plane by specifying the distance from an origin in two perpendicular directions.

Rectifier—An electronic component that allows current to pass through it in only one direction.

Reflected wave—A radio wave whose direction is changed when it bounces off some object in its path.

Reflectometer—A test instrument used to indicate standing wave ratio (SWR) by measuring the forward power (power from the transmitter) and reflected power (power returned from the antenna system).

Reflector—A parasitic antenna element that is located behind the driven element to enhance forward directivity. The reflector is usually about 5% longer than the driven element.

Refract—To bend. Electromagnetic energy is refracted when it passes through a boundary between different types of material. Light is refracted as it travels from air into water or from water into air.

Repeater—An amateur station that receives a signal and retransmits it for greater range.

Resistance—The ability to oppose an electric current.

Resistor—Any material that opposes a current in an electrical circuit. An electronic component specifically designed to oppose current.

Resonant frequency—That frequency at which a circuit including capacitors and inductors presents a purely resistive impedance. The inductive reactance in the circuit is equal to the capacitive reactance. Also the desired operating frequency of a tuned circuit. In an antenna, the resonant frequency is one where the feed-point impedance contains only resistance.

Resonant rhombic antenna—A diamond-shaped antenna consisting of sides that are each at least one wavelength long. The feed line is connected to one end of the diamond, and the opposite end is left open. The antenna has a bidirectional radiation pattern.

Reverse bias—A voltage applied across a semiconductor junction so that it will tend to prevent current flow.

Reverse biased—A condition in which the cathode of a diode is positive with respect to the anode, so it will not conduct.

RF burn—A flesh burn caused by exposure to a strong field of RF energy.

Rig—The radio amateur's term for a transmitter, receiver or transceiver.

Ripple—The amount of change between the maximum voltage and the minimum voltage in a pulsating dc waveform.

Root-mean square (RMS) voltage—Another name for effective voltage. The term refers to the method of calculating the value.

Rotary switch—A switch that connects one center contact to several individual contacts. An antenna switch is one common use for a rotary switch.

Rotor—The movable plates in a variable capacitor.

RST System—The system used by amateurs for giving signal reports. "R" stands for readability, "S" for strength and "T" for tone.

RF envelope—The shape of an RF signal as viewed on an oscilloscope.

S meter—A meter in a receiver that shows the relative strength of a received signal.

Sawtooth wave—A waveform consisting of a linear ramp and then a return to the original value. It is made up of sine waves at a fundamental frequency and all harmonics.

Scanning—The process of analyzing or synthesizing, in a predetermined manner, the light values or equivalent characteristics of elements constituting a picture area. Also the process of recreating those values to produce a picture on a CRT screen.

Schematic symbol—A drawing used to represent a circuit component on a wiring diagram.

Secondary station identifier (SSID)—A number added to a packet-radio station's call sign so that one amateur call sign can be used for several packet stations.

Secondary winding—The coil in a transformer that connects to the load.

Selective-call identifier—A four-character AMTOR station identifier.

Selective fading—A variation of radio-wave field intensity that is different over small frequency changes. It may be caused by changes in the material that the wave is traveling through or changes in transmission path, among other things.

Selectivity—A measure of the ability of a receiver to distinguish between a desired signal and an undesired one at some different frequency. Selectivity can be applied to the RF, IF and AF stages.

Self-resonant antenna—An antenna that is to be used on the frequency at which it is resonant, without the use of any added inductive or capacitive loading elements.

Semiconductor material—A material with resistivity between that of metals and insulators. Pure semiconductor materials are usually doped with impurities to control the electrical properties.

Semidiameter—The apparent radius of a generally spherical object (such as the moon) in the sky. Semidiameter is usually expressed in degrees of arc as measured between imaginary lines drawn from the observer to the center of the object and from the observer to one edge.

Sensing antenna—An omnidirectional antenna used in conjunction with an antenna that exhibits a bidirectional pattern to produce a radio direction-finding system with a cardioid pattern.

Sensitivity—A measure of the minimum input-signal level that will produce a certain audio output from a receiver.

Sequential logic—A type of circuit element that has at least one output and one or more input channels, and in which the output state depends on the previous input states. A flip-flop is one sequential-logic element.

Series circuit—An electrical circuit in which all the electrons must flow through every part of the circuit. There is only one path for the electrons to follow.

Series-resonant circuit—A circuit including a capacitor, an inductor and sometimes a resistor, connected in series, and in which the inductive and capacitive reactances are equal at the applied-signal frequency. The circuit impedance is at a minimum, and the current is a maximum at the resonant frequency.

Shack—The room where an Amateur Radio operator keeps his or her station equipment.

Short circuit—An electrical circuit where the current does not take the desired path, but finds a shortcut instead. Often the current goes directly from the negative power-supply terminal to the positive one, bypassing the rest of the circuit.

Short-path communication—Communication made by pointing beam antennas in the direction indicated by the shorter great-circle path.

Sidebands—The sum or difference frequencies generated when an RF carrier is mixed with an audio signal.

Signal generator—A test instrument that produces a stable low-level radio-frequency signal. The signal can be set to a specific frequency and used to troubleshoot RF equipment.

Signal tracer—A test instrument that shows the presence of RF or AF energy in a circuit. The signal tracer is used to trace the flow of a signal through a multistage circuit.

Silicon-controlled rectifier—A bistable semiconductor device that can be switched between the off and on states by a control voltage.

Simplex operation—A term normally used in relation to VHF and UHF operation. Simplex means you are receiving and transmitting on the same frequency.

Sine wave—A single-frequency waveform that can be expressed in terms of the mathematical sine function. A smooth curve, usually drawn to represent the variation in voltage or current over time for an ac signal.

Single sideband (SSB)—A common mode for voice operation on the amateur high-frequency bands. This is a variation of amplitude modulation.

Single-pole, double-throw (SPDT) switch—A switch that connects one center contact to either of two other contacts.

Single-pole, single-throw (SPST) switch—A switch that only connects one center contact to another contact.

Single-sideband, suppressed-carrier signal—A radio signal in which only one of the two sidebands generated by amplitude modulation is transmitted. The other sideband and the RF carrier wave are removed before the signal is transmitted.

Single-sideband, suppressed-carrier, amplitude modulation (SSB)—A technique used to transmit voice information in which the amplitude of the RF carrier is modulated by the audio input, and the carrier and one sideband are suppressed.

Skin effect—A condition in which ac flows in the outer portions of a conductor. The higher the signal frequency, the less the electric and magnetic fields penetrate the conductor, and the smaller the effective area of a given wire for carrying the electrons.

Skip—Radio waves that are bent back to earth by the ionosphere. Skip is also called sky-wave propagation.

Skip zone—A region between the farthest reach of ground-wave communications and the closest range of skip propagation.

Sky temperature—A measure of relative background noise coming from space, often expressed in kelvins. The sun is a source of much background noise, and so has a high noise temperature.

Sky waves—Radio waves that travel from an antenna upward to the ionosphere, where they either pass through the ionosphere into space or are refracted back to earth.

Sky-wave signals—Radio signals that travel through the ionosphere to reach the receiving station. Sky-wave signals will cause a variation in the measured received-signal direction, resulting in an error with a radio direction-finding system.

Slew rate—As related to op amps, a measure of how fast the output voltage can change with changing input signals.

Slope detection—A method for using an AM receiver to demodulate an FM signal. The signal is tuned to be part way down the slope of the receiver IF filter curve.

Sloper—A 1/2-wave dipole antenna that has one end elevated and one end nearer the ground.

Slow-scan television (SSTV)—A television system used by Amateurs to transmit pictures within a signal bandwidth allowed on the HF bands by the FCC. It takes approximately 8 seconds to send a single frame with SSTV.

Smith Chart—A coordinate system developed by Phillip Smith to represent complex impedances on a graph. This chart makes it easy to perform calculations involving antenna and transmission-line impedances and SWR.

Solar flux index—A measure of solar activity. The solar flux index is a measure of the radio noise on 2800 MHz.

Solar wind—Electrically charged particles emitted by the sun, and traveling through space. The wind strength depends on how severe the disturbance on the sun was. These charged particles may have a sudden impact on radio communications.

Solid-state devices—Circuit components that use semiconductor materials. Semiconductor diodes, transistors and integrated circuits are all solid-state devices.

SOS—A Morse code call for emergency assistance.

Source—The point at which the charge carriers enter an FET. Corresponds to the cathode of a vacuum tube.

Space operation—Space-to-earth and space-to-space Amateur Radio communication from a station that is beyond the major portion of the earth's atmosphere.

Space wave—A radio wave arriving at the receiving antenna made up of a direct wave and one or more reflected waves.

Spectrum analyzer—A test instrument generally used to display the power (or amplitude) distribution of a signal with respect to frequency.

Speech processor—A device used to increase the average power contained in a speech waveform. Proper use of a speech processor can greatly improve the readability of a voice signal.

Spin modulation—Periodic amplitude fade-and-peak variations resulting from the Phase III satellite's 60 r/min spin.

Splatter—The term used to describe a very wide bandwidth signal, usually caused by overmodulation of a sideband transmitter. Splatter causes interference to adjacent signals.

Sporadic-E propagation—A type of radio-wave propagation that occurs when dense patches of ionization form in the E layer of the ionosphere. These "clouds" reflect radio waves, extending the possible VHF communications range.

Spread-spectrum modulation—A signal-transmission technique where the transmitted carrier is spread out over a wide bandwidth.

Square wave—A periodic waveform that alternates between two values, and spends an equal time at each level. It is made up of sine waves at a fundamental frequency and all odd harmonics.

SSTV scan converter—A device that uses digital signal-processing techniques to change the output from a normal TV camera into an SSTV signal or to change a received SSTV signal to one that can be displayed on a normal TV.

Stability—A measure of how well a receiver or transmitter will remain on frequency without drifting.

Standing-wave ratio (SWR)—Sometimes denoted as VSWR. A measure of the impedance match between the feed line and the antenna. Also, with a Transmatch in use, a measure of the match between the feeder from the transmitter and the antenna system. The system includes the Transmatch and the line to the antenna. The ratio of maximum voltage to minimum voltage along a feed line. Also the ratio of antenna impedance to feed line impedance when the antenna is a purely resistive load.

Standing-wave-ratio (SWR) meter—A device used for measuring SWR. SWR is a relative measure of the impedance match between an antenna, feed line and transmitter.

Static (level-triggered) input—An input, such as for a flip-flop, which causes the output to change when it changes state.

Station license—The portion of an Amateur Radio license that authorizes an amateur station at a specific location. The station license also lists the call sign of that station.

Stator—The stationary plates in a variable capacitor.

Subatomic particles—The building blocks of atoms. Electrons, protons and neutrons are the most common subatomic particles.

Sudden Ionospheric Disturbance—A blackout of HF sky-wave communications that occurs after a solar flare.

Summer solstice—One of two spots on the earth's orbital path around the sun at which it reaches a point farthest from a horizontal plane extending through the center of the sun. With the north pole inclined toward the sun, it marks the beginning of summer in the northern hemisphere.

Sunspots—Dark spots on the surface of the sun. When there are few sunspots, long-distance radio propagation is poor on the higher-frequency bands.

Superheterodyne receiver—A receiver that converts RF signals to an intermediate frequency before detection.

Susceptance—The reciprocal of reactance. This is the imaginary part of a complex admittance.

Sweep circuits—The circuits in an oscilloscope that set the time base for observing waveforms on the scope.

Switch—A device used to connect or disconnect electrical contacts.

Switching regulator—A voltage-regulator circuit in which the output voltage is controlled by turning the pass element on and off at a rate of several kilohertz.

SWR meter—A device used to measure SWR. A measuring instrument that can indicate when an antenna system is working well.

Sync—Having two or more signals in step with each other, or occurring at the same time. A pulse on a TV or FAX signal that ensures the transmitted and received images start at the same point.

Synchronous flip-flop—A circuit whose output state depends on the data inputs, but that will change output state only when it detects the proper clock signal.

Synchronous motor—An ac electric motor whose rotation speed is controlled by the frequency of the electric energy used to power it.

Telecommand operation—Earth-to-space Amateur Radio communication to initiate, modify or terminate functions of a station in space operation (an Amateur Radio satellite).

Teleprinter—A machine that can convert keystrokes (typing) into electrical impulses. The teleprinter can also convert the proper electrical impulses back into text. Hams use teleprinters for radioteletype work.

Television raster—A predetermined pattern of scanning lines that provide substantially uniform coverage of an area.

Temperature coefficient—A number used to show whether a component will increase or decrease in value as it gets warm.

Temperature inversion—A condition in the atmosphere in which a region of cool air is trapped beneath warmer air.

Terminal node controller (TNC)—A TNC accepts information from a computer or terminal and converts the information into packets by including address and error-checking information. The TNC also receives packet signals and extracts transmitted information for display by a computer.

Terminator—A band around the earth that separates night from day.

Thermal noise—A random noise generated by the interchange of energy within a circuit and surroundings to maintain thermal equilibrium.

Thevenin's Theorem—Any combination of voltage sources and impedances, no matter how complex, can be replaced by a single voltage source and a single impedance that will present the same voltage and current to a load circuit.

Third-party participation (or communication)—The way an unlicensed person can participate in Amateur Radio communications. A control operator must ensure compliance with FCC rules.

Third-party traffic—Messages passed from one amateur to another on behalf of a third person.

Thyristor—Another name for a silicon-controlled rectifier.

Ticket—The radio amateur's term for an Amateur Radio license.

Time constant—The product of resistance and capacitance in a simple series or parallel RC circuit, or the inductance divided by the resistance in a simple series or parallel RL circuit. One time constant is the time required for a voltage across a capacitor or a current through an inductor to build up to 63.2% of its steady-state value, or to decay to 36.8% of the initial value. After a total of 5 time constants have elapsed, the voltage or current is considered to have reached its final value.

Time domain—A method of viewing a complex signal. The amplitude of the complex wave is displayed over changing time. The display shows only the complex waveform, and does not necessarily indicate the sine-wave signals that make up the wave.

Top loading—The addition of inductive reactance (a coil) or capacitive reactance (a capacitance hat) at the end of a driven element opposite the feed point. It is intended to increase the electrical length of the radiator.

Toroidal inductor or toroid—A coil wound on a donut-shaped ferrite or powdered-iron form.

Traffic—Messages passed from one amateur to another in a relay system; the amateur version of a telegram.

Traffic net—An on-the-air meeting of amateurs, for the purpose of relaying messages.

Transceiver—A radio transmitter and receiver combined in one unit.

Transequatorial propagation—A form of F-layer ionospheric propagation, in which signals of higher frequency than the expected MUF are propagated across the earth's magnetic equator.

Transformer—Two coils with mutual inductance used to change the voltage level of an ac power source to one more suited for a particular circuit.

Transistor-transistor logic (TTL)—Digital integrated circuits composed of bipolar transistors, possibly as discrete components, but usually part of a single IC. The power supply voltage should be 5 V.

Transmatch—See Matching network.

Transmission line—The wires or cable used to connect a transmitter or receiver to an antenna. See Feed line.

Transmit-receive (TR) switch—A device used for switching between transmit and receive operation. The TR switch allows you to connect one antenna to a receiver and a transmitter. As you operate the switch, it connects the antenna to the correct unit. A TR switch may be a mechanical switch, relay or electronic circuit.

Transmitter—A device that produces radio-frequency signals.

Transponder—A repeater aboard a satellite that retransmits, on another frequency band, any type of signals it receives. Signals within a certain receiver bandwidth are translated to a new frequency band, so many signals can share a transponder simultaneously.

Traps—Parallel LC networks inserted in an antenna element to provide multiband operation.

Triac—A bidirectional SCR, primarily used to control ac voltages.

Triangulation—A radio direction-finding technique in which compass bearings from two or more locations are taken, and lines are drawn on a map to predict the location of a radio signal source.

Triode—A vacuum tube with three active elements: cathode, plate and control grid.

Troposphere—The region in the earth's atmosphere just above the surface of the earth and below the ionosphere.

Tropospheric bending—When radio waves are bent in the troposphere, they return to earth approximately 15 percent farther away than the geometric horizon.

Tropospheric ducting—A type of radio-wave propagation whereby the VHF communications range is greatly extended. Certain weather conditions cause portions of the troposphere to act like a duct or waveguide for the radio signals.

Tropospheric enhancement—A weather-related phenomenon. Tropo can produce unusually long propagation on the VHF and UHF bands.

True or Geometric horizon—The most distant point one can see by line of sight.

Truth table—A chart showing the outputs for all possible input combinations to a digital circuit.

Tuning-fork oscillator—An oscillator in which the main frequency-determining element is the mechanical resonance of a tuning fork.

Tunnel diode—A diode with an especially thin depletion region, so that it exhibits a negative resistance characteristic.

Twin lead—Parallel-conductor feeder with wires encased in insulation. See Parallel-conductor feed line.

Two-tone test—Problems in a sideband transmitter can be detected by feeding two audio tones into the microphone input of the transmitter and observing the output on an oscilloscope.

Unbalanced line—Feed line with one conductor at ground potential, such as coaxial cable.

Unidentified communications or signals—Signals or radio communications in which the transmitting station's call sign is not transmitted.

Unijunction transistor—A three-terminal, single-junction device that exhibits negative resistance and switching characteristics unlike bipolar transistors.

Upper sideband (USB)—The common single-sideband operating mode on the 20, 17, 15, 12 and 10-meter HF amateur bands. Hams also use upper sideband on all the VHF and UHF bands.

Vacuum-tube voltmeter (VTVM)—A multiple-range meter used to measure voltage, current and resistance. The meter circuit includes a vacuum-tube amplifier to provide a high input impedance. This leads to more accurate readings than can be obtained with a VOM. VTVMs are virtually obsolete.

Varactor diode—A component whose capacitance varies as the reverse-bias voltage is changed.

Variable capacitor—A capacitor that can have its value changed within a certain range.

Variable resistor—A resistor whose value can be adjusted over a certain range.

Variable-frequency oscillator (VFO)—An oscillator used in receivers and transmitters. The frequency is set by a tuned circuit using capacitors and inductors. The frequency can be changed by adjusting the components in the tuned circuit.

Velocity factor—An expression of how fast a radio wave will travel through a material. It is usually stated as a fraction of the speed the wave would have in free space (where the wave would have its maximum velocity). Velocity factor is also sometimes specified as a percentage of the speed of the radio wave in free space.

Vertical antenna—A common amateur antenna, usually made of metal tubing. The radiating element is vertical. There are usually four or more radial conductors parallel to or on the ground.

Vertical synchronization pulse—Part of a TV signal used by the receiver to keep the CRT electron-beam scan in step with the camera scanning beam. This pulse returns the beam to the top edge of the screen at the proper time.

Vertically polarized wave—A radio wave that has its electric lines of force perpendicular to the surface of the earth.

Vertical sync pulse—Part of a TV signal used by the receiver to keep the CRT electron-beam scan in step with the camera scanning beam. This pulse returns the beam to the top edge of the screen at the proper time.

Vestigial sideband (VSB)—A signal-transmission method in which one sideband, the carrier and part of the second sideband are transmitted. The bandwidth is not as wide as for a double-sideband AM signal, but not as narrow as a single-sideband signal.

Vidicon tube—A TV-camera tube in which a charge-density pattern is formed by photoconduction, and is stored on the surface of the photoconductor that is scanned by an electron beam. This converts the light and dark areas of a picture into a varying electric signal. A type of photosensitive vacuum tube widely used in TV cameras.

Virtual height—The height that radio waves appear to be reflected from when they are returned to earth by refraction in the ionosphere.

Voice-Operated Relay (VOX)—Circuitry that activates the transmitter when the operator speaks into the microphone.

Volt (V)—The basic unit of electrical pressure or EMF.

Voltage—The EMF or pressure that causes electrons to move through an electrical circuit.

Voltage source—Any source of excess electrons. A voltage source produces a current and the force to push the electrons through an electrical circuit.

Volt-ohm-milliammeter (VOM)—A multiple-range meter used to measure voltage, current and resistance. This is the least expensive (and least accurate) type of meter.

W1AW—The headquarters station of the American Radio Relay League. This station is a memorial to the League's cofounder, Hiram Percy Maxim. The station provides daily on-the-air code practice and bulletins of interest to hams.

Watt (W)—The unit of power in the metric system. The watt describes how fast a circuit uses electrical energy.

Wattmeter—A test instrument used to measure the power output (in watts) of a transmitter.

Wavelength—Often abbreviated λ. The distance a radio wave travels in one RF cycle. The wavelength relates to frequency. Higher frequencies have shorter wavelengths.

White noise—A random noise that covers a wide frequency range across the RF spectrum. It is characterized by a hissing sound in your receiver speaker.

Winter solstice—One of two spots on the earth's orbital path around the sun at which it reaches a point farthest from a horizontal plane extending through the center of the sun. With the north pole inclined away from the sun, it marks the beginning of winter in the northern hemisphere.

Wire-wound resistor—A resistor made by winding a length of wire on an insulating form.

Yagi antenna—The most popular type of amateur directional (beam) antenna. It has one driven element and one or more additional elements.

Zener diode—A diode that is designed to be operated in the reverse-breakdown region of its characteristic curve.

Zener voltage—A reverse-bias voltage that produces a sudden change in apparent resistance across the diode junction, from a large value to a small value.

Zero beat—The condition that occurs when two signals are at exactly the same frequency. The beat frequency between the two signals is zero. When two operators in a QSO are transmitting on the same frequency the stations are zero beat.

US Customary to Metric Conversions

International System of Units (SI)-Metric Prefixes

Prefix	*Symbol*			*Multiplication Factor*
exa	E	10^{18}	=	1 000 000 000 000 000 000
peta	P	10^{15}	=	1 000 000 000 000 000
tera	T	10^{12}	=	1 000 000 000 000
giga	G	10^{9}	=	1 000 000 000
mega	M	10^{6}	=	1 000 000
kilo	k	10^{3}	=	1 000
hecto	h	10^{2}	=	100
deca	da	10^{1}	=	10
(unit)		10^{0}	=	1
deci	d	10^{-1}	=	0.1
centi	c	10^{-2}	=	0.01
milli	m	10^{-3}	=	0.001
micro	μ	10^{-6}	=	0.000001
nano	n	10^{-9}	=	0.000000001
pico	p	10^{-12}	=	0.000000000001
femto	f	10^{-15}	=	0.000000000000001
atto	a	10^{-18}	=	0.000000000000000001

Linear

1 metre (m) = 100 centimetres (cm) = 1000 millimetres (mm)

Area

1 m^2 = 1 × 10^4 cm^2 = 1 × 10^6 mm^2

Volume

1 m^3 = 1 × 10^6 cm^3 = 1 × 10^9 mm^3
1 litre (l) = 1000 cm^3 = 1 × 10^6 mmm^3

Mass

1 kilogram (kg) = 1 000 grams (g)
(Approximately the mass of 1 litre of water)

1 metric ton (or tonne) = 1 000 kg

US Customary Units

Linear Units

12 inches (in) = 1 foot (ft)
36 inches = 3 feet = 1 yard (yd)
1 rod = 5½ yards = 16½ feet
1 statute mile = 1 760 yards = 5 280 feet
1 nautical mile = 6 076.11549 feet

Area

1 ft^2 = 144 in^2
1 yd^2 = 9 ft^2 = 1 296 in^2
1 rod^2 = 30¼ yd^2
1 acre = 4840 yd^2 = 43 560 ft^2
1 acre = 160 rod^2
1 $mile^2$ = 640 acres

Volume

1 ft^3 = 1 728 in^3
1 yd^3 = 27 ft^3

Liquid Volume Measure

1 fluid ounce (fl oz) = 8 fluidrams = 1.804 in^3
1 pint (pt) = 16 fl oz
1 quart (qt) = 2 pt = 32 fl oz = 57¾ in^3
1 gallon (gal) = 4 qt = 231 in^3
1 barrel = 31½ gal

Dry Volume Measure

1 quart (qt) = 2 pints (pt) = 67.2 in^3
1 peck = 8 qt
1 bushel = 4 pecks = 2 150.42 in^3

Avoirdupois Weight

1 dram (dr) = 27.343 grains (gr) or (gr a)
1 ounce (oz) = 437.5 gr
1 pound (lb) = 16 oz = 7 000 gr
1 short ton = 2 000 lb, 1 long ton = 2 240 lb

Troy Weight

1 grain troy (gr t) = 1 grain avoirdupois
1 pennyweight (dwt) or (pwt) = 24 gr t
1 ounce troy (oz t) = 480 grains
1 lb t = 12 oz t = 5 760 grains

Apothecaries' Weight

1 grain apothecaries' (gr ap) = 1 gr t = 1 gr a
1 dram ap (dr ap) = 60 gr
1 oz ap = 1 oz t = 8 dr ap = 480 gr
1 lb ap = 1 lb t = 12 oz ap = 5 760 gr

US Customary to Metric Conversions

Multiply ⟶

Metric Unit = Conversion Factor × US Customary Unit

⟵ **Divide**

Metric Unit ÷ Conversion Factor = US Customary Unit

Metric Unit =	Conversion Factor ×	US Unit
(Length)		
mm	25.4	inch
cm	2.54	inch
cm	30.48	foot
m	0.3048	foot
m	0.9144	yard
km	1.609	mile
km	1.852	nautical mile
(Area)		
mm^2	645.16	$inch^2$
cm^2	6.4516	in^2
cm^2	929.03	ft^2
m^2	0.0929	ft^2
cm^2	8361.3	yd^2
m^2	0.83613	yd^2
m^2	4047	acre
km^2	2.59	mi^2
(Mass)	(Avoirdupois Weight)	
grams	0.0648	grains
g	28.349	oz
g	453.59	lb
kg	0.45359	lb
tonne	0.907	short ton
tonne	1.016	long ton

Metric Unit =	Conversion Factor ×	US Unit
(Volume)		
mm^3	16387.064	in^3
cm^3	16.387	in^3
m^3	0.028316	ft^3
m^3	0.764555	yd^3
ml	16.387	in^3
ml	29.57	fl oz
ml	473	pint
ml	946.333	quart
l	28.32	ft^3
l	0.9463	quart
l	3.785	gallon
l	1.101	dry quart
l	8.809	peck
l	35.238	bushel
(Mass)	(Troy Weight)	
g	31.103	oz t
g	373.248	lb t
(Mass)	(Apothecaries' Weight)	
g	3.387	dr ap
g	31.103	oz ap
g	373.248	lb ap

Schematic-Diagram Symbols

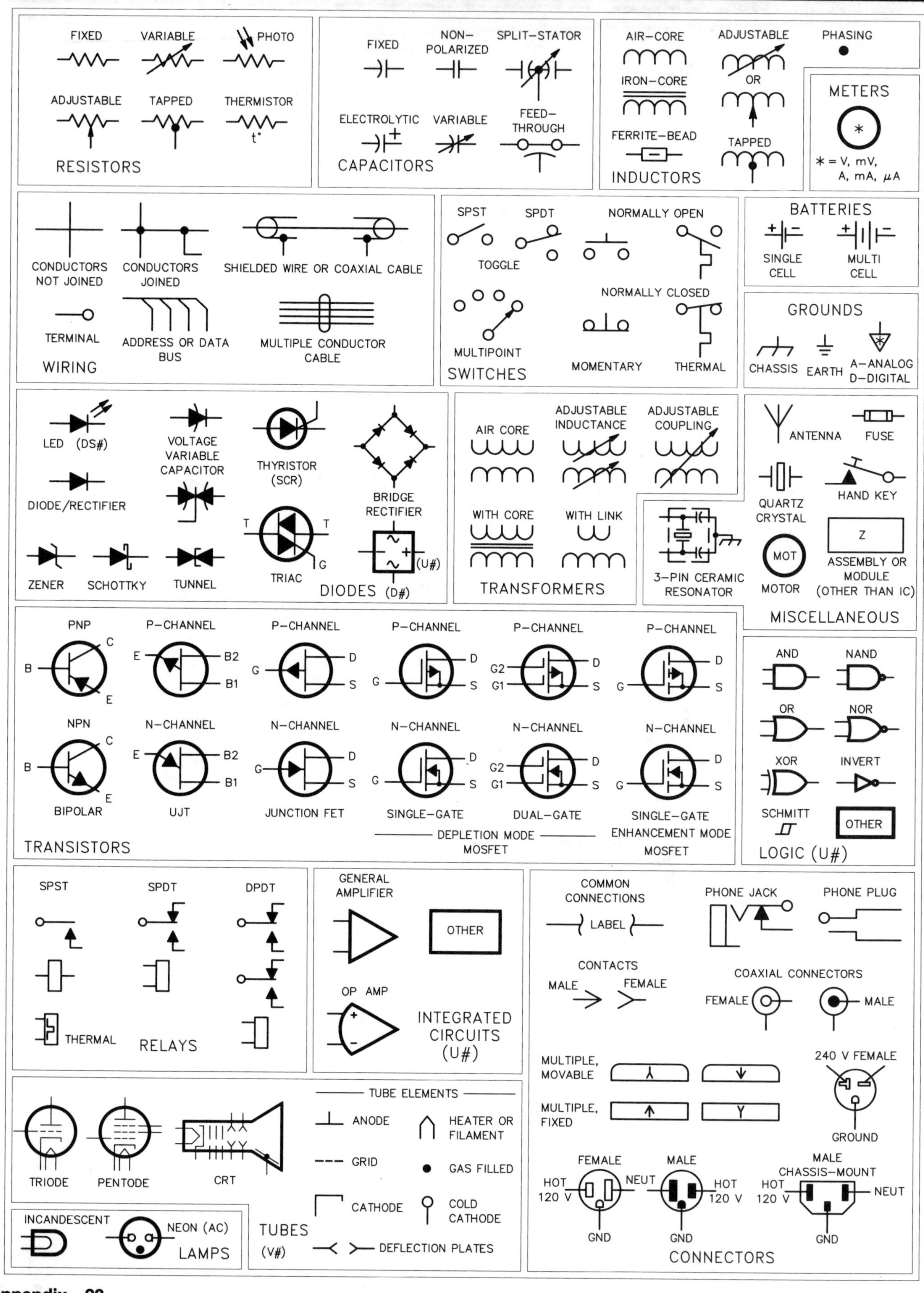

About *Amateur Radio*

There are many references to Amateur Radio throughout this book. We mentioned various types of Amateur Radio equipment in some of the examples and text discussions.

If you are a licensed Amateur Radio operator, you are familiar with the fun and excitement "hams" (as they are often called) enjoy every day! If not, you're probably wondering what this is all about.

Amateur Radio is about individuals *communicating* by means of radio signals through the air. It is about individual experimenters contributing to the advancement of the radio-communications art. Amateur Radio is about the Federal Communications Commission providing a set of rules that encourage advancement of the communication and technical aspects of radio. Amateur Radio is about a national resource: a pool of trained radio operators, technicians and electronics experts ready to lend a hand during all types of emergencies. It is also about the international goodwill that develops when people communicate across national boundaries and around the world.

These five principles define the purpose of Amateur Radio. They also give you some hint of the wide variety of interests that Amateur Radio operators bring to their hobby (or avocation as it becomes for many).

What do Amateur Radio operators actually do? Most of all, they meet other people who like to communicate. They use many methods and a wide variety of radio frequency bands to carry out this communication. Small, low-powered, hand-held radios are popular for local communications. A typical range for such radios is 50 to 150 kilometres. For longer distance communications, hams usually select other frequency bands. Hams can use a variety of frequency bands to reach just about any location on the globe. Many hams even send their signals "out of this world," communicating with astronauts in space, or using amateur-built communications satellites to relay their signals. Can you imagine actually communicating with someone by bouncing a radio signal off the **moon**? Amateur Radio operators have been doing this since the 1950s!

Amateur Radio communication takes many forms. Sure hams love to talk into a microphone. But they also thrill to the excitement of tapping out messages in Morse code, the oldest form of radio communication. Hams type messages and exchange data files over computer-to-computer networks called packet radio. Amateur Radio operators even send and receive their own television pictures.

Amateur Radio operators enjoy building all kinds of electronics devices. They often build accessories to use with their stations. Many of them build complete radio transmitters and receivers. There is nothing quite like the thrill you get when you can proudly say of your equipment, "I built it myself!"

Anyone can become an Amateur Radio operator. There are no minimum (or maximum!) age requirements. Earning a license involves passing a simple written exam covering FCC Rules, elementary electronics principles and operating practices. You can begin with a Novice license, which also includes a Morse-code exam or with a Technician license, which does not require Morse code.

For more information and answers to any questions you may have about Amateur Radio, write or call:

ARRL Educational Activities Dept
225 Main Street
Newington CT 06111
800-32-NEW HAM (800-326-3942)

Also check out our World Wide Web site:

http://www.arrl.org/

About *The American Radio Relay League*

The seed for Amateur Radio was planted in the 1890s, when Guglielmo Marconi began his experiments in wireless telegraphy. Soon he was joined by dozens, then hundreds, of others who were enthusiastic about sending and receiving messages through the air — some with a commercial interest, but others solely out of a love for this new communications medium. The United States government began licensing Amateur Radio operators in 1912.

By 1914, there were thousands of Amateur Radio operators — hams — in the United States. Hiram Percy Maxim, a leading Hartford, Connecticut, inventor and industrialist saw the need for an organization to band together this fledgling group of radio experimenters. In May 1914 he founded the American Radio Relay League (ARRL) to meet that need.

Today, with more than 170,000 members, ARRL is the largest organization of radio amateurs in the United States. The League is a not-for-profit organization that:

- promotes interest in Amateur Radio communications and experimentation
- represents US radio amateurs in legislative matters, and
- maintains fraternalism and a high standard of conduct among Amateur Radio operators.

At League headquarters in the Hartford suburb of Newington, the staff helps serve the needs of members. ARRL is also International Secretariat for the International Amateur Radio Union, which is made up of similar societies in more than 150 countries around the world.

ARRL publishes the monthly journal *QST*, as well as newsletters and many publications covering all aspects of Amateur Radio. Its headquarters station, W1AW, transmits Morse code practice sessions and bulletins of interest to radio amateurs. The League also coordinates an extensive field organization, which provides technical and other support for radio amateurs as well as communications for public-service activities. In addition, ARRL represents US amateurs with the Federal Communications Commission and other government agencies in the US and abroad.

Membership in ARRL means much more than receiving *QST* each month. In addition to the services already described, ARRL offers membership services on a personal level, such as the ARRL Volunteer Examiner Coordinator Program and a QSL bureau.

Full ARRL membership (available only to licensed radio amateurs in the US) gives you a voice in how the affairs of the organization are governed. League policy is set by a Board of Directors (one from each of 15 Divisions). Each year, half of the ARRL Board of Directors stands for election by the Full Members they represent. The day-to-day operation of ARRL HQ is managed by an Executive Vice President and a Chief Financial Officer.

No matter what aspect of Amateur Radio attracts you, ARRL membership is relevant and important. There would be no Amateur Radio as we know it today were it not for the ARRL. We would be happy to welcome you as a member! (An Amateur Radio license is not required for Associate Membership.) For more information about ARRL and answers to any questions you may have about Amateur Radio, write or call:

ARRL
225 Main Street
Newington CT 06111-1494
(860) 594-0200

Prospective new amateurs call:
800-32-NEW HAM (800-326-3942)
Email: **newham@arrl.org**
World Wide Web: **http://www.arrl.org/**

Index

F

Frequency (continued)

G

H

I

P

Q

R

S

THE AMERICAN RADIO RELAY LEAGUE, INC

225 MAIN STREET NEWINGTON, CONNECTICUT 06111 USA

A bona fide interest in Amateur Radio is the only essential requirement, but full voting membership is granted only to licensed radio amateurs of the US and possessions. Therefore, if you have a license, please be sure to indicate it below. Please print.

❑ New member ❑ Previous member ❑ Renewal ❑ Not currently licensed

Class of License Call Sign Date of Application

Name

Address

City, State, ZIP

Membership Class	❑ Regular		❑ Family ❑ Blind	❑ 65 or older
	US AND POSSESSIONS	ELSEWHERE*		US AND POSSESSIONS
❑ 1 year	$34	$47	$5	$28
❑ 2 years	65	88	10	53
❑ 3 years	92	127	15	76

*These rates include the postage surcharge which partially offsets the additional cost to mail *QST* outside the US. Write for airmail rates.

IMPORTANT: Please attach your Expiration Notice to this form if you are renewing your current membership. Payment in US funds only. Checks must be drawn on a bank within the US.

A member of the immediate family of a League member, living at the same address, may become a League member without *QST* at the special rate of $5 per year. Family membership must run concurrently with that of the member receiving *QST*. Blind amateurs may join without *QST* for $5 per year.

Persons who are age 65 or older residing in the US may upon request apply for League membership at the reduced rates shown. A one-time proof of age, in the form of a copy of a driver's license or birth certificate is required. If you are age 21 or younger, a special rate may apply. Please see reverse side.

Your membership certificate will be mailed to you in about 2 weeks from the date we receive your application. Delivery of *QST* may take slightly longer, but future issues should reach you on a regular basis. Membership is available only to individuals. Fifty percent of dues is allocated to *QST*, and the balance for membership.

DUES ARE SUBJECT TO CHANGE WITHOUT NOTICE.

If you do not wish your name and address made available for non-ARRL related mailings, please check this box ❑.

UBE98

I am donating $_____ ($1 minimum) to the Legal Research & Resource Fund.

JOIN ARRL

THE LARGEST ORGANIZATION OF RADIO AMATEURS IN THE US

- Receive Monthly *QST* Journal
- Use QSL Bureau–Clearing House for Overseas QSL cards
- Low-Cost Equipment Replacement Insurance Available for Home and Car
- Representation in Washington for All Matters Concerning Amateur Radio
- League-Member-Only Awards

Payment enclosed ❑

Charge to

❑ VISA (13 or 16 digits) ❑ MasterCard (16 digits)
❑ AMEX (15 digits) ❑ Discover (16 digits)

1 2 3 4 5 6 7 8 9 10 11 12 13 14 15 16
Card number

Card good from ________ to ________
(EXPIRATION DATE)

Signature ________

List family members on reverse side.

ARRL MEMBERS

This proof of purchase may be used as a $2.00 credit on your next ARRL purchase. Limit one coupon per new membership, renewal or publication ordered from ARRL Headquarters. No other coupon may be used with this coupon. Validate by entering your membership number from your *QST* label below:

Are you 21 or younger?
Are you the oldest licensed amateur in your household?

IF YOU CAN ANSWER "YES" TO BOTH, THESE RATES APPLY TO YOU!

Evidence of date of birth is required. Attach a copy of your birth certificate or driver's license, or have your parent or guardian complete the next line:

Applicant's date of birth ______ Signature ______ ❑ Parent ❑ Guardian

Please check the annual membership rate which applies to you:

❑ I am 13 to 21 years of age and the oldest licensed amateur in my household. $16 US & Possessions.
❑ I am 12 or younger and the oldest amateur in my household. $8.50 US & Possessions.

Name ______ Call Sign ______

Street ______

City ______ State ______ ZIP ______

Please submit this application directly to ARRL HQ with your payment. Club commissions and rebates do not apply. Family membership is not applicable, and these rates are available only in the US. You will receive all League benefits and 12 issues of *QST*. Membership is available on a yearly basis only.

Your membership certificate and first issue of *QST* will reach you shortly. Welcome to the ARRL.

Family Members:

Name ______ Call Sign ______

Name ______ Call Sign ______

Name ______ Call Sign ______

UNDERSTANDING BASIC ELECTRONICS

PROOF OF PURCHASE

FEEDBACK

Please use this form to give us your comments on this book and what you'd like to see in future editions, or e-mail us at **pubsfdbk@arrl.org** (publications feedback). If you use e-mail, please include in the body of your message: your name, call (if you are a licensed Amateur Radio operator), e-mail address and the book title, edition and printing. Also indicate whether or not you are an ARRL member.

Where did you purchase this book?

☐ From ARRL directly ☐ From an ARRL dealer

Is there a dealer who carries ARRL publications within:

☐ 5 miles ☐ 15 miles ☐ 30 miles of your location? ☐ Not sure.

License class:

☐ Novice ☐ Technician ☐ Technician Plus ☐ General ☐ Advanced ☐ Amateur Extra

---

Name ARRL member? ☐ Yes ☐ No

______________________________ Call Sign ________________

Daytime Phone ( ) ______________________ Age ________________

Address ____

City, State/Province, ZIP/Postal Code ______________________________

If licensed, how long? ______________________________

Other hobbies ______________________________

____

Occupation ______________________________

For ARRL use only	UBE
Edition	1 2 3 4 5 6 7 8 9 10 11
Printing	4 5 6 7 8 9 10 11

From ______________________________

____

____

Please affix postage. Post Office will not deliver without postage.

EDITOR, UNDERSTANDING BASIC ELECTRONICS
AMERICAN RADIO RELAY LEAGUE
225 MAIN ST
NEWINGTON CT 06111-1494

please fold and tape

Notes

Notes